AF333786

PROPERTIES AND CHEMISTRY OF BIOMOLECULAR SYSTEMS

TOPICS IN
MOLECULAR ORGANIZATION AND ENGINEERING

Volume 11

Honorary Chief Editor:

W. N. LIPSCOMB (*Harvard, U.S.A.*)

Executive Editor:

Jean MARUANI (*Paris, France*)

Editorial Board:

Henri ATLAN (*Jerusalem, Israel*)

Sir Derek BARTON (*Texas, U.S.A.*)

Christiane BONNELLE (*Paris, France*)

Paul CARO (*Meudon, France*)

Stefan CHRISTOV (*Sofia, Bulgaria*)

I. G. CSIZMADIA (*Toronto, Canada*)

P-G. DE GENNES (*Paris, France*)

J-E. DUBOIS (*Paris, France*)

Manfred EIGEN (*Göttingen, Germany*)

Kenishi FUKUI (*Kyoto, Japan*)

Gerhard HERZBERG (*Ottawa, Canada*)

Alexandre LAFORGUE (*Reims, France*)

J-M. LEHN (*Strasbourg, France*)

P-O. LÖDWIN (*Uppsala, Sweden*)

Patrick MacLEOD (*Massy, France*)

H. M. McCONNELL (*Stanford, U.S.A.*)

C. A. McDOWELL (*Vancouver, Canada*)

Roy McWEENY (*Pisa, Italy*)

Ilya PRIGOGINE (*Brussels, Belgium*)

Paul RIGNY (*Saclay, France*)

R. G. WOOLLEY (*Nottingham, U.K.*)

The titles published in this series are listed at the end of this volume.

PROPERTIES AND CHEMISTRY OF BIOMOLECULAR SYSTEMS

Proceedings of the Second Joint Greek-Italian Meeting on
Chemistry and Biological Systems and Molecular Chemical Engineering,
Cetraro, Italy, October 1992.

edited by

N. RUSSO
Universitá della Calabria, Arcavatadi Rende

J. ANASTASSOPOULOU
National Technical University, Athens

and

G. BARONE
Universitá Frederico II, Napoli

KLUWER ACADEMIC PUBLISHERS
DORDRECHT / BOSTON / LONDON

Library of Congress Cataloging-in-Publication Data

```
Chemistry and properties of biomolecular systems : proceedings of the
  second joint Greek-Italian meeting on Chemistry and Biological
  Systems and Molecular Chemical Engineering, October 1992 / edited by
  N. Russo, J. Anastassopulou, G. Barone.
       p.   cm. -- (Topics in molecular organization and engineering ;
  v. 11)
    Includes index.
    ISBN 0-7923-2666-0
    1. Biochemistry--Congresses.  2. Biomolecules--Congresses.
  3. Molecular biology--Congresses.   I. Russo, N. (Nino)
  II. Anastassopoulou, J. (Janna)  III. Barone, G. (Guido)  IV. Joint
  Greek-Italian Meeting on Chemistry and Biological Systems and
  Molecular Chemical Engineering (2nd : 1992 : Calabria, Italy)
  V. Series.
  QP501.C48   1994
  574.19'2--dc20                                           93-44247
```

ISBN 0-7923-2666-0

Published by Kluwer Academic Publishers,
P.O. Box 17, 3300 AA Dordrecht, The Netherlands.

Kluwer Academic Publishers incorporates
the publishing programmes of
D. Reidel, Martinus Nijhoff, Dr W. Junk and MTP Press.

Sold and distributed in the U.S.A. and Canada
by Kluwer Academic Publishers,
101 Philip Drive, Norwell, MA 02061, U.S.A.

In all other countries, sold and distributed
by Kluwer Academic Publishers Group,
P.O. Box 322, 3300 AH Dordrecht, The Netherlands.

''The logo on the front cover represents the generative hyperstructure of alkanes'', printed
with permission from J.E. Dubois, Institut de Topologie et de Dynamique des Systèmes,
Paris, France.

Printed on acid-free paper

All Rights Reserved
© 1994 Kluwer Academic Publishers
No part of the material protected by this copyright notice may be reproduced or
utilized in any form or by any means, electronic or mechanical,
including photocopying, recording or by any information storage and
retrieval system, without written permission from the copyright owner.

Printed in the Netherlands

CONTENTS

vi

Introduction to the Series

The Series 'Topics in Molecular Organization and Engineering' was initiated by the Symposium 'Molecules in Physics, Chemistry, and Biology', which was held in Paris in 1986. Appropriately dedicated to Professor Raymond Daudel, the symposium was both broad in its scope and penetrating in its detail. The sections of the symposium were: 1. The Concept of a Molecule; 2. Statics and Dynamics of Isolated Molecules; 3. Molecular Interactions, Aggregates and Materials; 4. Molecules in the Biological Sciences, and 5. Molecules in Neurobiology and Sociobiology. There were invited lectures, poster sessions and, at the end, a wide-ranging general discussion, appropriate to Professor Daudel's long and distinguished career in science and his interests in philosophy and the arts.

These proceedings have been arranged into eighteen chapters which make up the first four volumes of this series: Volume I, 'General Introduction to Molecular Sciences'; Volume II, 'Physical Aspects of Molecular Systems'; Volume III, 'Electronic Structure and Chemical Reactivity'; and Volume IV, 'Molecular Phenomena in Biological Sciences'. The molecular concept includes the logical basis for geometrical and electronic structures, thermodynamic and kinetic properties, states of aggregation, physical and chemical transformations, specificity of biologically important interactions, and experimental and theoretical methods for studies of these properties. The scientific subjects range therefore through the fundamentals of physics, solid-state properties, all branches of chemistry, biochemistry, and molecular biology. In some of the essays, the authors consider relationships to more philosophic or artistic matters.

In Science, every concept, question, conclusion, experimental result, method, theory or relationship is always open to reexamination. Molecules do exist! Nevertheless, there are serious questions about precise definition. Some of these questions lie at the foundations of modern physics, and some involve states of aggregation or extreme conditions such as intense radiation fields or the region of the continuum. There are some molecular properties that are definable only within limits, for example, the geometrical structure of non-rigid molecules, properties consistent with the uncertainty principle, or those limited by the neglect of quantum-field, relativistic or other effects. And there are properties which depend specifically on a state of aggregation, such as superconductivity, ferroelectric (and anti), ferromagnetic (and anti), superfluidity, excitons, polarons, etc. Thus, any molecular definition may need to be extended in a more complex situation.

Chemistry, more than any other science, creates most of its new materials. At least so far, synthesis of new molecules is not represented in this series, although the principles of chemical reactivity and the statistical mechanical aspects are included. Similarly, it is the more physico-chemical aspects of biochemistry, molecular biology and biology itself that are addressed by the examination of questions related to molecular recognition, immunological specificity, molecular pathology, photochemical effects, and molecular communication within the living organism.

Many of these questions, and others, are to be considered in the Series 'Topics in Molecular Organization and Engineering'. In the first four volumes a central core is presented, partly with some emphasis on Theoretical and Physical Chemistry. In later volumes, sets of related papers as well as single monographs are to be expected; these may arise from proceedings of symposia, invitations for papers on specific topics, initiatives from authors, or translations. Given the very rapid development of the scope of molecular sciences, both within disciplines and across disciplinary lines, it will be interesting to see how the topics of later volumes of this series expand our knowledge and ideas.

WILLIAM N. LIPSCOMB

PREFACE

During the last decade there has been an enormous increasing interest in the chemistry of Biological Systems, as well as in Molecular Chemical Engineering. Many fields of modern chemical sciences are helping to understand the elementary mechanisms of several biological processes and to discover new classes of organic and organometallic compounds with specific and high biological activity. The multidisciplinary approach leads to both great opportunities for exchange of ideas and experiences and in establishing a common language.

Following the experience of the first Joint Greek-Italian Meeting on " Chemistry and Biological Systems and Molecular Chemical Engineering" organized in Greece in 1990, the second Joint Meeting was held at Cetraro (Calabria, Italy) in October 1992, under the auspices of the Italian Society of Chemistry (S. C. I.) - Interdivisional Group for the Chemistry of Biological Systems and Processes, of the Italian National Council of Research (C.N.R.), of the Greek Chemical Society and of the National Research Center " Democritos". The meeting linked together scientists coming from different countries and disciplines.

This volume collects the major part of the plenary lectures, oral communications and selected posters presented and discussed during the Meeting.

We thank the authors for their excellent efforts and their willingness to mold their contributions to our conception of the book. We are grateful for the financial contributions from Consiglio Nazionale delle Ricerche (CNR)-Comitato per la Chimica and Comitato per le Biotecnologie, Universita' della Calabria, Universita' di Reggio Calabria (Facolta' di Medicina e Chirurgia, CZ), Dipartimento di Chimica-Universita' della Calabria, Farmitalia Carlo Erba, Deltamed, Delchimica Scientific Glassware, Analytical Control and Kontron Instruments.

Many thanks to the Organizing Committee (M. Ghedini, F. Lelj, L. Paolillo, G. Parlato, G. Sindona, D. Fessas and M. Toscano) for the preparation of the Meeting, Mr. Giovanni Marra and the staff of Grand Hotel San Michele. Mrs Franca Mele and Wendy Zupo are acknowledged for their technical assistance.

Nino Russo
Janna Anastassopoulou
Guido Barone

Arcavacata di Rende, CS, Italy
May, 1993

xi

Intrinsic and Environmental Effects on Protomeric Equilibria in the Ground and Excited Electronic States of Biological Systems

C. ADAMO and V. BARONE
Dipartimento di Chimica, Università Federico II, via Mezzocannone 4, I-80134 Napoli, Italy

1. Introduction

Proton and electron tunnelling are important features of biochemical processes[1-3] and can operate in a concerted way giving rise to enhanced efficiency of charge migration. For instance a number of intramolecular electron transfer processes, taking place in large enzymes, may be realized in a very efficient way by a combination of both kinds of tunnelling provided that some keto-enol tautomerization of the peptide bond is assumed[4,5]. As another example, tautomeric equilibria in heteroaromatic compounds, especially lactam-lactim ones, are representative of a large number of proton transfer (PT) reactions which are thought to be of importance in biological processes, such as mutagenesis[6,7]. In this case both ground and excited state PT must be taken into account. Furthermore this kind of process does not occur, of course, in the gas phase, so that environmental effects can play a significant role. While experiment provides, of course, the most definite information about any physico-chemical process, it usually detects only the overall result of a number of intervening effects. Here is exactly where theory can play its most profitable role, even at the expense of a certain degree of idealization. Only in a theoretical framework, in fact, different terms can be selectively switched on and off, to analyze general trends or to verify different hypotheses.
Here we describe a general approach to this kind of problem, with special reference to keto-enol tautomerization in formamide and lactam-lactim tautomeric equilibrium in 2-pyridone. These systems have been chosen because, together with their intrinsic interest, they represent simple models of important classes of biomolecules , proteins in the first case, and DNA bases in the second. Technical details will be reduced to a minimum, the focus being placed rather on the discussion of general features and on the proposal of suitable combinations of different quantum-mechanical tools. For instance, the combined use of small basis set ab-initio computations and semi-empirical methods, which, taken separately, have a lot of limitations, allows comprehensive studies of structural, thermodynamic and spectroscopic properties. In fact, the first class of methods often provides reliable structural and energetic data for non-covalent interactions. On the other hand, non-potential energy effects (zero point energy, entropy), and spectroscopic parameters are more conveniently obtained by specialized semi-empirical methods. In the same vein, modifications induced by environmental effects on geometrical parameters

N. Russo et al. (eds.), Properties and Chemistry of Biomolecular Systems, 1–18.
© 1994 *Kluwer Academic Publishers. Printed in the Netherlands.*

and vibrational frequencies can be described in terms of relatively simple, analytically differentiable models. On top of these calculations, energetic quantities can be refined by more sophisticated methods. Here again the combined use of different tools, namely numerical simulations and quantum continuum models, probably provides the most powerful approach.

2. Environmental effects

The most promising general approach to the problem of environmental (e.g. solvent) effects can be based, in our opinion, on a system-bath decomposition (figure 1).The system includes the part of the solute where the essential of the process to be investigated is localized plus, possibly, the few solvent molecules strongly (and specifically) interacting with it. This part is treated at the atomic (or electronic, if needed) level of resolution, and is immersed in a polarizable continuum, mimicking the macroscopic properties of the solvent. The "bath" (i.e. the remaining of the solute and the bulk solvent) can then be treated in terms of, possibly polarizable, point charges or multipoles. Further refinements are possible (e.g. use of different dielectric constants for the first few layers[8,9] and the bulk solvent, inclusion of solvent fluctuations[10], etc.), but in the present study we remain at the level of the most *naive* form of the model.

The solution process consists of inserting a solute molecule into a suitable cavity, spending energy for its creation, and *switching on* the interactions with surrounding solvent molecules. The overall change of the Gibbs free energy of solvatation, ΔG_{solv},

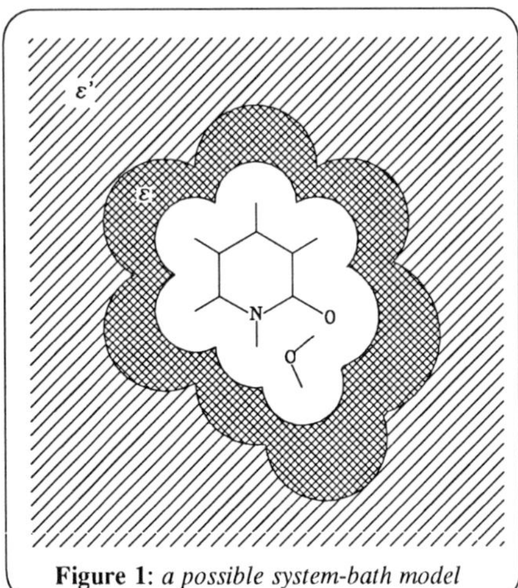

Figure 1: *a possible system-bath model*

is generally evaluated as the sum of three terms:

$$\Delta G_{solv} = \Delta G_{el} + \Delta G_{disp-rep} + \Delta G_{cav} \qquad (1)$$

There have been several attempts to account for the dominant electrostatic contribution ranging from the simple Born equation[11,12] to different reaction field models[13-19]. Relatively less attention has been devoted to the description of the remaining contributions[20-25].

Here we want to briefly report the main characteristics of two of the most successful quantum mechanical approaches to the continuum model, namely the Self Consistent Reaction Field[14,26,27] and the Polarizable Continuum Model[16,24,28]. These two approaches are usually considered as mutually exclusive, but, as we will see below, their combination can lead to an improved description of solute-solvent interactions.

2.1 THE ELECTROSTATIC TERM

In the continuum model the charge distribution ρ of the solute induces a polarization in the dielectric, so that a reaction field is generated, which acts back on the solute. The process is iterated until the charge density and the reaction field become mutually consistent.

If the cavity is modelled as a sphere and the multipolar development of solute-solvent interactions truncated at the dipole level (the Onsager model[13]), the reaction field of the solvent to the electric field generated by the solute is:

$$\Im = -g\,\mu' \qquad (2)$$

where

$$g = \frac{2\,(\varepsilon-1)}{(2\varepsilon+1)\,a_0^3} \qquad (3)$$

In the above equations ε is the dielectric constant of the solvent, a_0 the radius of the cavity, and μ' is a modified dipole moment which makes the model invariant to the choice of the origin also for charged species[27]. The Onsager theory can be introduced very easily in an SCF scheme, leading to a very effective implementation of the so called Self Consistent Reaction Field (SCRF) model[27]. In this approach, the total energy can be written as

$$E' = E^0 - \frac{1}{2}\,g\,\mu'\mu' \qquad (4)$$

where E^0 is the molecular energy without the reaction field. With an appropriate grouping of the additional terms introduced by the solvent, the total energy has the same general form as for the unperturbed solute:

$$E' = Tr\,(h'P) + \frac{1}{2}\,Tr\,[PG'\,(P)] + V' \qquad (5)$$

where $\mathbf{P}$ is the density matrix and V', $\mathbf{h'}$, $\mathbf{G'}$ differ from the standard core-repulsion (V), one-electron matrix ($\mathbf{h}$) and two-electron hypermatrix ($\mathbf{G}$) for the presence of dipole-moment integrals[27]. The main advantage of this model is the straightforward introduction of solvent effects in the SCF formalism, leading to an easy evaluation of the energy and to analytical expressions for its first and second derivatives.

When a more realistic description of the molecular cavity is required, the introduction of the classical electrostatic formalism in a quantum scheme becomes more involved[29-31]. Moreover analytical evaluation of first and second derivates is very elaborate, if not impossible[32]. In the Polarizable Continuum Model (PCM) a charge distribution arises in each point of the cavity surface from the combined effect of the electric fields generated by the polarized solute ($\Im_\rho$) and the residual charge distribution on the surface ($\Im_\sigma$):

$$\sigma(s) = -\frac{\varepsilon-1}{4\pi\varepsilon}\left[\Im_\rho(s) + \Im_\sigma(s)\right]_{n_-} \qquad (6)$$

where the subscript n_- indicates a limiting value on the concave side of the surface. The continuous charge distributions $\sigma(s)$ can be replaced by a set of discrete point charges. With this aim, the surface S of the cavity is divided into portions ΔS, small enough that

$\sigma(s)$ can be considered constant at their interiors. From a practical point of view, the cavity is represented by interlocking spheres centered on (all or part of) the atoms of the solute molecule. The surface of each sphere is, in turn, divided into a predefined number of finite elements (usually triangles), whose centers, surfaces and orientations provide all the necessary information to set up the computations[33]. The evaluation of $\sigma(s)$ at the center of each finite element can be done via eqn. (6), but an iterative procedure is necessary, because of the dependency of the electric field on $\sigma(s)$. The iterative process can be divided in the following steps:

1) an initial set of surface charge densities σ^0 (and of corresponding point charges q^0) is originated by the electric field generated by the unperturbed solute

$$\sigma^0(s) = \frac{\varepsilon - 1}{4\pi\varepsilon}\left[\Im^0_\rho(s)\right]_{n_-} \qquad (7a)$$

$$q^0 = \sigma^0(s)\,\Delta S \qquad (7b)$$

2) the charges at the center of the surface elements generate an additional contribution $\Im^0_\sigma$ to the electric field. The charge density is improved by substituting in eqn. (7a) the overall electric field

$$\sigma^1(s) = \frac{\varepsilon - 1}{4\pi\varepsilon}\left[\Im^0_\rho + \Im^0_\sigma\right]_{n_-} \qquad (8a)$$

$$q^1 = \sigma^1(s)\,\Delta S \qquad (8b)$$

This new charge distribution generates, in turn, a modified electric field $\Im^1_\sigma$, which, when introduced into eqn.(8), leads to charges q^2, and so on, up to convergence to a final set of charges q^f.

3) The electric field arising from the point charges q^f, is added to the Hamiltonian of the solute, leading to new charge distribution $\rho^1(r)$.

4) Steps 2 and 3 are iterated until self consistency is reached.

It is possible to combine the SCRF and PCM approaches to describe more efficiently solute-solvent electrostatic interactions. In fact, while the overall electrostatic interaction energy is strongly dependent on the detailed form of the cavity surface, the charge distribution and the geometrical parameters of the polarized solute are much more stable. A good estimate of the electrostatic contribution to the solvation energy can, therefore, be obtained by the PCM model using the SCRF geometry and charge distribution without further iterations. This mixed-approach combines a straightforward implementation of geometry optimization, with a detailed description of the cavity surface.

2.2 THE DISPERSION-REPULSION TERM

This contribution is usually evaluated in terms of empirical two body potentials, which are not iteratively added to the Hamiltonian of the solute[24]. The main assumptions of this approach are:
1) the use of a continuum distribution function to describe the solvent around the solute;

2) the evaluation of the energy in the form of a surface integral;
3) the use of additive atom-atom potentials.
The two body potentials are usually in the following form

$$V_{ml}\left(r_{ml}\right)= \sum_{n=6,8,10} -\frac{d_{ml}^{(n)}}{r_{ml}^{n}} + C_{ml}\,exp\left(-\gamma_{ml}\,r_{ml}\right) \qquad (9)$$

the indexes m and l running on the atoms of the solute and of the solvent. The dispersion coefficients d (n=6,8,10) and the repulsion coefficients C_{ml} and γ_{ml} are taken from literature data[34].
The evaluation of the solute-solvent interaction energy requires the knowledge of the spatial configuration of all the solvent molecules around the solute, i..e. the so-called distribution function:

$$\rho_{ml}\left(r_{ml}\right)=N_{l}\,\rho_{l}\,g_{ml}\left(r_{ml}\right) \qquad (10)$$

where N_l is the number of the atoms of type l in each molecule of the solvent, ρ_l is the macroscopic density of the solvent and $g_{ml}(r_{ml})$ is a correlation function, depending on the position of l with respect to m. In the simplest model the correlation function $g_{ml}(r_{ml})$ assumes the following values

$$g_{ml}\left(r_{ml}\right)=\begin{cases} 0 & \text{if } r_{ml} \in C_l \\ \\ 1 & \text{if } r_{ml} \notin C_l \end{cases} \qquad (11)$$

where C_l is the cavity occupied by the solute (see preceding section), and enlarged to take into account also the region of space forbidden to the nuclei of the solvent atoms. The dispersion-repulsion contribution to the solvatation free- energy of the solute M in the solvent L can be written as a sum of volume integrals:

$$G_{disp-rep} = \sum_l \sum_M \int \rho_{ml}\left(r_{ml}\right) V_{ml}\left(r_{ml}\right) dr_{ml}=$$
$$= \rho_L \sum_l N_l \sum_m \left[\sum_{n=6,8,10} -d_{ml}^{(n)} \int \frac{g_{ml}(r_{ml})}{r^n}\,dr_{ml} + \right.$$
$$\left. + C_{ml} \int g_{ml}(r_{ml})\,exp\,(-\gamma_{ml}\,r_{ml})\,dr_{ml}\right] \qquad (12)$$

where now $r_{ml}=r_l-r_m$. When the integrals in eqn. (12) are limited to the portion of space outside C_l, it is possible to transform them into surface integrals, computed on the S_l surface, which delimits the cavity C_l. As already discussed for the computation of the electrostatic term, the surface S_l of the cavity can be represented by a set of finite elements with area ΔS.

2.3 THE CAVITATION TERM

The cavitation term can be evaluated from the Scaled Particle Theory (SPT) of fluids[23], which is perfectly coherent with the continuum approach used to evaluate the other contributions. The essence of this theory is that work is required to exclude the centers of molecules from any specified region of space in a fluid. Consider a fluid consisting of spherically symmetric molecules possessing a hard core of diameter s_1 and interacting through attractive forces compatible with the molar volume V of the fluid. Imagine now to exclude the centers of solvent molecules from a spherical region of space of radius r in the volume V. This region of space would in fact be a cavity in the fluid. Suppose we denote the probability that such a cavity exists by $p_0(r, \rho)$, where ρ is the number density of the fluid. The reversible work $w(r, \rho)$ required to produce the cavity is given by

$$-\frac{w(r, \rho)}{kT} = \ln p_0(r, \rho) \qquad (13)$$

The scaled particle theory attempts to determine $p_0(r, \rho)$ by means of statistical mechanical and geometrical arguments. After suitable algebraic manipulations the final result is:

$$\frac{w(R, \rho)}{kT} = \ln (1-y) + \left(\frac{3y}{1-y}\right)R + \left[\frac{3y}{1-y} + \frac{y}{2}\left(\frac{y}{1-y}\right)^2\right]R^2 + \frac{yP}{\rho kT}R^3 \qquad (14)$$

where $y = \pi \rho a_1 /6$ is the reduced number density (a_1 being the solvent radius) and R is the ratio between solvent and cavity radii.

3. Solvent effects and time scales

In the study of dynamical processes in solution, like proton transfer or light absorption, the reorganization of the solvent around the evolving system must be taken into account[35]. The surface charge distribution σ induced by the solute can be divided in an orientational (σ_{or}) and an inductive (σ_{ind}) component. The total surface charge and the inductive contribution are formally given by the same expression (eqn.6). However the dielectric constant entering the equation is the static dielectric constant ε in the former case, and ε_∞ (the square of optical refractive index extrapolated to infinite wavelength) in the latter. The orientational contribution is then obtained as a difference. In the case of processes faster than solvent relaxation times, the following successive steps can be considered:

a) the solute in the initial state is in equilibrium with the solvent. The solute charge distribution ρ_i gives rise to a surface charge distribution σ_a

$$\sigma_a(\rho_i, \varepsilon) = \sigma_{ind}(\rho_i, \varepsilon_\infty) + \sigma_{or} \qquad (15)$$

b) the solute changes its state (e.g. jumps to an excited electronic state by absorption of a photon) and is now characterized by a different charge distribution, ρ_f. The solvent experiences the changes in ρ through its inductive part only. So the new σ can be expressed as:

$$\sigma_b = \sigma_{ind}(\rho_f, \varepsilon_\infty) + \sigma_{or} \qquad (16)$$

c) After a sufficiently long time, the solvent rearranges itself around the solute molecule in the new state:

$$\sigma_c(\rho_f, \varepsilon) = \sigma_{ind}(\rho_f, \varepsilon_\infty) + \sigma'_{or} \qquad (17)$$

Moreover we assume that the shape of the cavity, and hence its radius, remain unchanged during fast processes, like radiative transitions or proton transfers.

4. Proton Transfer Mechanism

Two kinds of Proton Transfer (PT) are, in principle, possible:
1) dissociative transfer, typical of acid-base reactions, that is the stepwise protonation and deprotonation of the substrate. An example is the PT reaction in molecules such as 4- hydroxypyridine, where the intramolecular PT it is forbidden, due to the long distance between donor and acceptor sites. Tautomerization can only take place by a solvent-mediated mechanism in which one solvent molecule accepts the proton from the donor site and another one gives a different proton to the acceptor site[36];
2) non dissociative transfer, which is important in the gas-phase, in aprotic media and in neutral aqueous solution.
In turn, this latter mechanism can be divided into three classes (see figure 2):
a) *Direct* PT between the two sites of the molecule;
b) *Dimeric* PT, in which interconversion occurs within a self-associated dimer;
c) *Assisted* PT, in which one or two amphiprotic molecules are also directly involved in the process. Experimental studies[37] have shown that this process is the most important in aqueous solution.
The mechanism of non-dissociative PT has been extensively studied at theoretical level in the gas-phase, but few attempts have been done to extend those studies to reactions in solution[38]. From another point of view, several molecules having nearby proton donor and acceptor groups have been shown to tautomerize in some excited electronic state by an Excited State Proton Transfer (ESPT) mechanism[39]. The signature of that process is the emission of a strongly Stokes red-shifted fluorescence following absorption of a UV photon. Also on this case detailed theoretical studies are scarce, especially concerning solvent effects on the kinetics of the process.

5. Case Studies.

As pointed out before, the tautomeric pairs formamide/formamidinic acid (F/FA, see figure 2), and 2-pyridone/2-hydroxypyridine (2Py/2Hy, see figure 3) provide simple models of protomeric equilibria in biomolecules. As a consequence these systems have been extensively studied both experimentally[40-44] and theoretically[38,45-47]. In the case of 2-pyridone the attention has been recently focused on the possibility of photochemical tautomerization via ESPT[41-44]. In particular, Kuzuya and co-workers[43] have found that the absorption spectrum of 2Py in cyclohexane solution is characterized by two bands, at 268 and 300 nm, corresponding to lactim and lactam form, respectively. These two bands are also present in the emission spectrum, but red-shifted to 303 and 367 nm. The red-shift of the higher wavelength band from 300 to 367 nm is similar to that occuring in other molecules and may suggest the presence of an ESPT mechanism[48].
Here we report our results for the two tautomeric reactions, paying special attention to the possibility of a solvent mediated mechanism and to the effect of bulk solvent on the

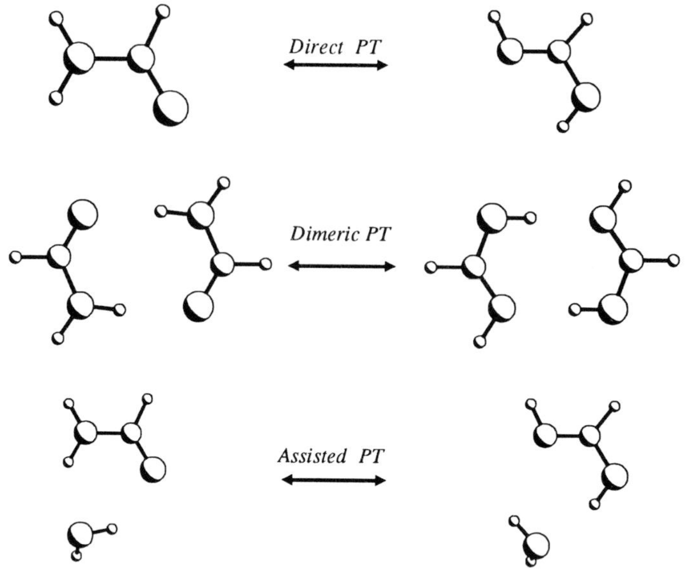

Figure 2: *various types of non dissociative PT for formamide*

energetics of the process. Moreover for 2-pyridone both ground and excited state proton transfers have been investigated.

Due to the small size of the molecule and to the necessity of investigating only the ground electronic state, formamide has been studied entirely at the ab-initio level[49,50]. On the other hand, for 2Py/2Hy we combined semi-empirical and ab-initio methods. In particular all semi-empirical computations have been carried out by the AM1 model[51], supplemented by a Configuration Interaction (CI) treatment. According to a large number of published applications, AM1 yields reliable results for ground state geometries, heats of formation, and relative stabilities of tautomeric species[52]. The quality of the AM1 approximation for excited states is more difficult to assess, since only a few papers have been published on this topic. Some attempt in this direction shows that the AM1 method allows the calculation of excited state properties with a reasonable degree of approximation[48,53]. On the other hand, the AM1 approximation overestimates hydrogen bond interactions, unduly favouring bifurcated structures with respect to linear ones. However in those cases in which bifurcated hydrogen bonds are not possible, AM1 results are of comparable quality with refined ab initio results[52]. As that concerns the calculation of energy barriers to proton transfer, test computations on malonaldehyde and literature data[54] suggest that the above quantities are significantly overestimated. General trends, which are the main concern of this study, are, anyway, correctly reproduced.

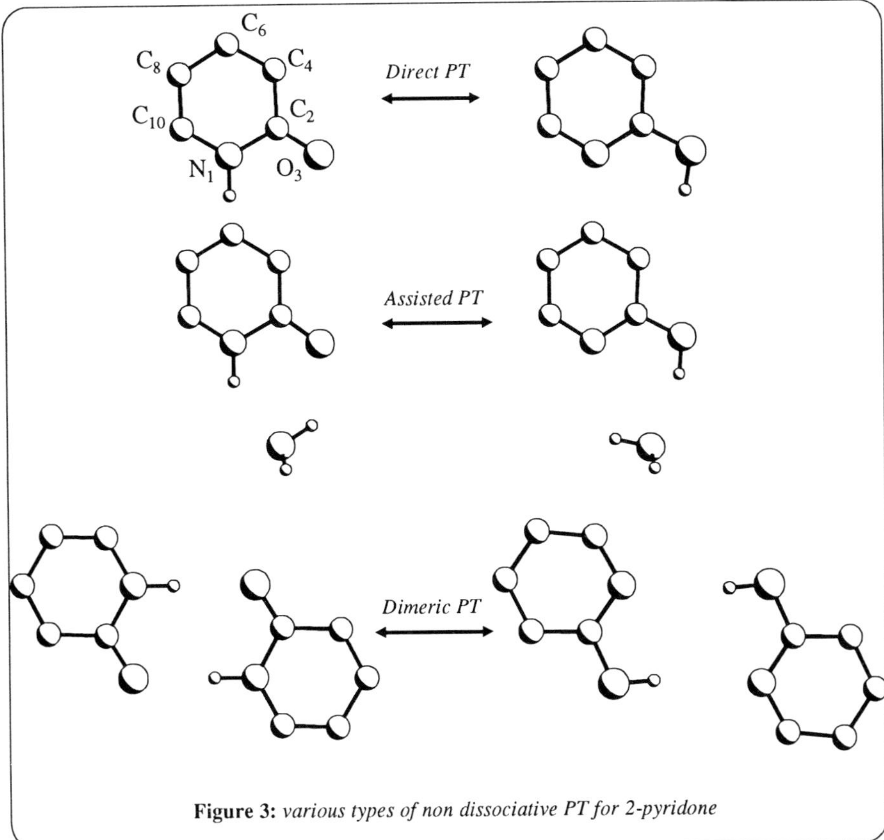

Figure 3: *various types of non dissociative PT for 2-pyridone*

5.1 FORMAMIDE

The calculated ab-initio energy difference in vacuo and in aqueous solution for the formamide tautomers are reported in table I. From these results it appears that non electrostatic, i.e. cavitation and dispersion-repulsion, terms do not modify the thermodynamics of the process, being essentially constant for the two tautomers. On the other hand, the reaction field of the solvent plays a non negligible role. In particular the endothermicity, and, to a lower extent, the activation energy of the keto-enol transformation is enhanced due to the larger dipole moment of the keto form (4.27 vs 1.31 D for the unperturtbed solute and 4.73 vs. 1.33 D in water solution).

As discussed before, tautomerization can be facilitated, for example in aqueous solution, by the transfer of a proton from a solvent molecule. As shown in table I, this model leads to a strong reduction of the activation energy, while the endothermicity is only slightly modified. The further effect of the bulk solvent on the endothermicity is the same as for direct PT within the monomer. The solvatation of the transition state (TS) resembles that of the enolic rather than ketonic form.

It is well known that formamide forms stable dimers both in the gas-phase and in the solid state. The same happens for not too low concentrations in non-polar solvents and has been postulated even for water solutions[55]. We considered, therefore, also proton transfer

Table I: endothermicities (ΔE) and activation energies ($\Delta E^\dagger$) for the tautomerization of formamide in the gas phase and in aqueous solution. Energy values (in Kcal/mol) refer to fully optimized geometries and the subscripts indicate direct (F/FA), assisted (F-H_2O/FA-$H_2$0) and dimeric (F_2/FA_2) models of proton transfer.

	MP2	HF	HF/SCRF	HF/SCRF+PCM	$\Delta G_{cav}+\Delta G_{disp\text{-}rep}$
$\Delta E(dir)$	12.5	12.9	16.5	17.8	0.14
$\Delta E^\dagger(dir)$	46.9	61.2	62.9	67.6	-0.73
$\Delta E(ass)$	10.9	11.5	13.5	16.9	-0.06
$\Delta E^\dagger(ass)$	21.9	36.3	36.5	38.8	-0.82
$\Delta E(dim)$	20.3	21.9	21.9	24.9	-0.52
$\Delta E^\dagger(dim)$	20.4	31.7	31.7	37.3	-0.74

within a symmetric dimer. The results are very similar to those obtained for the PT assisted by one water molecule (see table I). Furthermore our computations show that the mechanism of PT in the isolated dimer is concerted, the transition state being symmetric. The ketonic form is stabilized with respect to the monomer, since it is able to form a stronger hydrogen bond. The effect on the activation energy is even larger. On the other hand the effect of bulk solvent is essentially the same as that discussed for the monomer.

5.2 2-PYRIDONE

As mentioned above, the AM1 results for the ground state are quite satisfactory, especially as regards the energy differences between the two tautomeric forms (the endothermicity of the tautomerization reaction). In fact all the computed values (Table II) are comparable with those obtained by ab-initio calculations using a double zeta basis set and only slightly different with respect to more sophisticated computations[46,47].

Also the energy barrier to PT (the activation energy of the tautomerization reaction) computed for the monomer is very near to the ab-initio results. On the other hand, the barriers are overestimated by about 100% with respect to ab-initio results for the adduct with one water molecule and for the dimer.

The computations at the CI level indicate that 2Hy is the more stable form in the ground state and in the complex with one water molecule. The former result is in agreement with experiment and is reproduced by ab-initio computations using large basis sets and massive inclusion of correlation energy. In the dimer the situation is reversed, 2Py being now more stable by about 7 Kcal/mol. The addition of Zero Point Energies (ZPE) to the CI energies enforces the above trend, leading to a net stabilization of the 2Hy species and of the transition state structures with respect to the 2Py molecules. The energy barrier to PT is very high, about 50 Kcal/mol, except for the dimer in which it decreases to 34 Kcal/mol. The height of the barriers suggests that PT cannot occur in the ground electronic state. Inclusion of solvent effects (ciclohexane in the present context) does not alter this

Table II: endothermicities (ΔE) and activation energies ($\Delta E^{\dagger}$) for the tautomerization of 2-pyridone in the gas phase and in ciclohexane solution. Energy values (in Kcal/mol) refer to fully optimized geometries and the subscripts indicate direct (2Py/2Hy), assisted (2Py-H_2O/2Hy-H_2O) and dimeric (2Py$_2$/2Hy$_2$) models of proton transfer.

	AM1 ground state			Ab-initio ground state			AM1 excit.state	
	HF	CI	CI+ZPE	HF	CI	CI+ZPE	CI	CI+ZPE
ΔE(dir)	-0.5	-4.6	-5.3	1.7	3.4	2.6	15.7	17.7
$\Delta E^{\dagger}$(dir)	54.1	52.1	48.5	50.9	43.7	40.1	58.4	54.4
ΔE(ass)	2.2	-1.6	-2.3	2.9	3.1	2.4	19.6	21.3
$\Delta E^{\dagger}$(ass)	43.5	41.7	37.4	15.8	12.9	8.6	48.7	46.5
ΔE(dim)	7.1	6.9	4.9	9.6	8.4		24.6	25.5

conclusion. It is noteworthy that the relationship between dipole moment and electrostatic contribution to solvation energy is less stringent than in the case of formamide (see table III).

As expected the first excited singlet state of 2Py and 2Hy is essentially obtained by a $\pi-\pi^*$ HOMO-LUMO excitation with very small contributions of other excitations, like HOMO-(LUMO+1). In the excited state the lactam form is sensibly stabilized; for instance 2Py is more stable than 2Hy by 16 Kcal/mol and (2Py)$_2$ more stable than 2(Hy)$_2$ by 25 Kcal/mol.

Table III: dipole moments (μ in Debyes) and different contributions (Kcal/mol) to solvation free energies. Energy values are computed using AM1 optimized geometries and charges, while dipole moments are computed at the ab-initio level using a STO-3G basis set.

Parameter	Ground Electronic State			Excited Electronic State		
	Py	TS	Hy	Py	TS	Hy
μ_{gas}	3.12	2.48	1.09	1.30	2.47	1.09
μ_{sol}	3.32	2.65	1.16	1.53	2.66	1.58
ΔG_{el}	-5.11	-5.17	-3.03	-4.21	-3.66	-3.61
ΔG_{cav}	6.60	6.38	6.57	6.62	6.40	6.54
$\Delta G_{disp-rep}$	-10.13	-10.48	-10.41	-10.16	-10.42	-10.35

Energy barriers to PT remain of the same order as for the ground electronic state. These results, confirmed by some preliminary investigation at the ab-initio level[56], show that both in the ground and in the excited state direct PT is highly unlikely.

In summary, the structural origin of the strong Stoke effect observed in the spectra of pyridone should involve skeletal rather than hydrogen bridge deformations. To investigate this point we have computed the absorption and fluorescence spectra of the system 2Py/2Hy. Energy differences have been computed at the optimized geometry of the ground electronic state for absorption and at the optimized geometry of the excited state for fluorescence. The results of table IV show that AM1 results are in close agreement with experiment and, in particular, correctly reproduce all the spectral shifts.

An explication of these shifts is suggested by inspection of the characteristics of the HOMO and LUMO of 2Py and 2Hy (figure 4; see figure 3 for atom numbering). The HOMOs of the two tautomers are dominated by bonding interactions between the $2p\pi$ atomic orbitals of the pairs C_2,C_4 and C_4,C_6 (in the lactam form) or C_8,C_{10} (in the lactim form). Antibonding interactions occur within the pairs N_1,C_{10} and C_6,C_8 (2Py) or C_2,O_3 (2Hy). The HOMO of 2Hy is more stable than that of 2Py by about 0.5 eV due to a stronger N_1-C_2-C_4 bonding interaction.

Table IV: Absorption and Fluorescence spectral maxima (in nm) of the System 2Py/2Hy in the gas phase and in ciclohexane solution.

System	Absorption				Fluorescence		
	AM1	Exp., gas[a]	PCM	Exp., sol.[b]	AM1	PCM	Exp, sol.[b]
Py	311	334/335	307	300	370	368	367
Hy	281	276	282	268	286	289	303
Py-H_2O	311	328			368		
Hy-H_2O	282	282			288		
$(Py)_2$	305	325[c]	305		366	366	
$(Hy)_2$	258		258		259	261	

a) ref.44; b) ref.47; c) ref.42

The LUMO of 2Py is dominated by antibonding N_1-C_{10} and C_4-C_6 interactions, whereas C_2,C_4 and C_8,C_{10} pairs give again the only significant contributions (but now antibonding) in 2Hy. Since small bonding interactions (involving C_6-C_8 and C_2-C_4 pairs) are only operative in the lactam form, the LUMO energy of 2Py is lowered by about 0.3 eV versus that of 2Hy. So the large red-shift, about 60 nm at AM1 level, between the absorption and the emission wavelength of 2Py is strongly related to the very different structure of the HOMO and LUMO orbitals. This difference is reflected also in the molecular geometry. In the 2Py molecule, the changes produced by electronic excitation involve all the bond

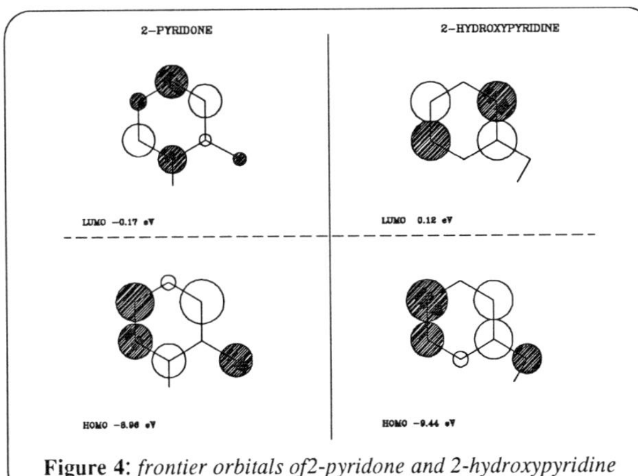

Figure 4: *frontier orbitals of 2-pyridone and 2-hydroxypyridine*

distances in the aromatic ring, the most important being the increase of C_8-C_{10} (from 1.37 to 1.45 Å) and of C_4-C_6 (from 1.36 to 1.42 Å) bond lengths and the decrease of the C_6-C_8 distance (from 1.43 to 1.37 Å). For the 2Hy molecule the changes only involve the C_2-C_4 (from 1.42 to 1.48 Å) and C_8-C_{10} (from 1.40 to 1.46 Å) bonds. This simple scheme is further supported by considerations based on the computed bond orders. The most important changes produced in the 2Py molecule by electronic excitation are the strong decrease in the bond orders of the C_8-C_{10} and C_4-C_6 double bonds, (1.65/1.12 and 1.72/1.15, respectively), and the corresponding increase of the C_6-C_8 single bond (from 1.16 to 1.54). Comparatively smaller modifications occur in the 2Hy molecule, the largest change (from 1.31 to 0.93) involving the C_2-C_4 bond order.

6. Concluding remarks

In this paper we have reported the essential results of two case studies, selected to point out a number of crucial aspects connected to the description of spectroscopic and/or reactive processes in solution. Last generation solvent models, taking into account solute polarization and quantum effects are discussed in some detail. Only these models allow, in fact, the description of feedback effects, which could be of paramount relevance especially for processes involving strong electron density redistribution in polar solvents. It has been next shown that both the kinetics and thermodynamics of protomeric equilibria can be significantly modified by the direct involvement of a single solvent molecule. The further modulation induced by bulk solvent is comparatively smaller, but, sometimes, not negligible. From another point of view, structural interpretations of spectroscopic results requires great care, since different modifications (here skeletal vs. hydrogen bridges) can sometimes lead to similar results (here red-shift of fluorescence). Once again the combined use of experimental and theoretical tools offers the best way out from this *impasse*. From a more general point of view, the most significant outcome of this and related studies is that a semiquantitative description of spectroscopic and reactive processes of biomolecules in solution is becoming more and more feasible thanks to the development of very efficient quantum-mechanical tools and the combined use of complementary methods. This offers the opportunity to start more systematic studies aimed at the recognition of general trends and thumb rules for those mechanisms in large classes of related systems.

References

1. A.G. Sykes "Electron transfer in biological systems", *Chem. Brit.*,**24**, 551-554, (1988)
2. Y. Cha, C.J. Murray, J.P. Klinmanz "Hydrogen tunneling in enzime reactions", *Science*, **243**, 1325-1328 (1989).
3. M.L. Paddock, S.H. Rongey, G. Feher, M.Y. Okamura "Pathway of proton transfer in bacterial reaction centers: replacement of glutamic acid 212 in the L subunit by glutamine inhibits quinone turnover", *Proc. Natl. Acad. Sci. USA*, **86**, 6602-6605, (1989).
4. G. Del Re, A. Peluso, C. Minichino "Hydrogen bridges and electron transfer in biomolecules. Study of a possible mechanism on a model charge-recombination system", *Can. J. Chem.*, **63**, 1850-1855, (1985).
5. G. Del Re, C. Adamo "Oximine form of the peptide bond as a transient modification in enzyme redox reactions", *J. Phys. Chem.*,**95**, 423-4238, (1991).
6. J.S. Kwiatkowski, T.J. Zielinski, R. Rein. "Quantum mechanics prediction of tautomeric equilibria", *Advan. Quantum Chem.*, **18**, 85-102, 1986.
7. M.D. Topal, J.R. Fresco "Complementary base pairing and the origin of substitution mutations", *Nature* , **263** , 285-288,(1976); W.G. Cooper "Theory of mutations. II. Physical model of genetic stability", *Intern. J. Quantum Chem.*, **14**, 71-90, 1978.
8. A. Warshel " Calculations of chemical processes in solutions", *J. Am. Chem. Soc.*, **12**,   1640-1652, (1979).
9. J.R. Bontka, P.N. Pintauro "Prediction of ion solvation free energies in a polarizable dielectric continuum", *J. Phys.Chem.*, **96**, 7778-7783, (1992).
10. R.Bianco, Tesi di dottorato, Pisa (Italy) 1992.
11. J.D. Jackson <u>Classical Electrodynamics</u>, Wiley, New York, 1975.
12. A.A. Rashin, B. Honig "Reevaluation of the Born model of hydration", *J. Phys. Chem.*, **89**, 5588-5593 , (1985).
13. L. Onsager "Electric moments of molecules in liquids", *J. Am. Chem. Soc.*, **58**, 1468-1490, (1936).
14. O. Tapia "Some remarks on the SCRF theory of solvent effects and the calculation of proton potentials", *Theoret. Chim. Acta*, **47**, 157-169, (1978).
15. P. Claverie, J.P. Daudey, J. Langlet, B. Pullman, D. Piazzola, M.J. Huron "Studies of solvent effects. 1. Discrete, continuum, and discrete-continuum models and their comparison for some simple cases: NH_4^+,CH_3OH, and substituted NH_4^{+}", *J. Phys. Chem.*, **82**, 405-418, (1978).
16. S. Miertus, E. Scrocco, J. Tomasi "Electrostatic interaction of a solute with a conti nuum. A direct utilization of ab initio molecular potentials for the prevision of solvent effects", *Chem. Phys.*, **55**, 117-129, 1981.
17. B.T. Thole, P.T. Van Duijnen "On the quantum mechanical treatment of solvent effects", *Theoret. Chim. Acta*, **55**, 307-315, (1980).
18. C.J. Cramer, D.H. Truhlar " General parametrized SCF model for free energies of solvatation in aqueous solutions", *J. Am. Chem. Soc.*,**113**,   8305-83011, (1991).
19. M. Negre, M. Orozco, F.J. Luque " A new strategy for the representation of enviroment effects in semi-empirical calculation based on a Dewar's Hamiltonian", *Chem. Phys. Lett.*, **196**, 27-31, (1992).
20. V. Frecer, S. Miertus, M. Majekova "Modelling of dispersion and repulsion interactions in liquids" , *J. Mol. strct. (THEOCHEM)*, **227**, 157-173, (1991).

21. J.M. Huron, P. Claverie "Calculation of the interaction energy of one molecule with its whole surrounding. I. Method and application to pure nonpolar compounds", *J. Chem. Phys.*, **76**, 2123-2133, (1972).
22. R.A. Pierotti "Solutions of gases in liquids", *J. Phys. Chem.*, **67**, 1840-1844, (1963); R.A. Pierotti "Aqueous solutions of non polar gases", *J. Phys. Chem.*, **69**, 281-287, (1965).
23. R. A. Pierotti "A scaled particle theory of aqueous and nonaqueous solutions", *Chem. Rev.*, **76**, 717-726, (1976).
24. F. Floris, J. Tomasi "Evaluation of the dispersion contribution to the solvation energy. a simple computational model in the continuum approximation", *J. Comp. Chem.*, **10**, 616-627, (1989).
25. D. Rinaldi, B.J. Costa Cabral, J.L. Rivail "Influence of dispersion forces on the electronic structure of a solvated molecule", *Chem. Phys. Lett.*, **125**, 495-499, (1986).
26. O. Tapia, O. Goscinski "SCRF theory of solvent effects", *Mol. Phys.*, **29**, 1653-1664, (1975)
27. M. W. Wong, K.B. Wiberg, M. Frisch "Hartree-Fock second derivatives and electric field properties in a solvent reaction field. Theory and application" , *J. Chem. Phys.*, **95**, 8991-8998, (1991).
28. J. Tomasi "Description and interpretation of molecular phenomena in solution, using effective Hamiltonian operators related to continuous solvent distributions", *Int. J. Quantum Chem. QBS*, **18**, 73-90, (1991).
29. H. Hoshi, M. Sakurai, Y. Inoue, R. Chujo "Medium effects on the molecular electronic structure. I. The formulation of a theory for the estimation of a molecular electronic structure surronded by an anisotropic medium" , *J. Chem. Phys.*, **87**, 1107-1115, (1987)
30. J.K. Kirkwood "Theory of solutions of molecules containing widely separated charges with special application to zwitterions", *J. Chem. Phys.*, **2**, 351-361, 1934.
31. D. Rinaldi, J.L. Rivail "Polarizabilites moleculaires et effet dielectrique de milieu a l'etat liquide", *Theor. Chim. Acta*, **32**, 57-65, (1973).
32. R. Bonaccorsi, R. Cammi, J. Tomasi "On the ab-initio geometry optimization of molecular solutes", *J. Comp. Chem.*, **12**, 301-307, (1991).
33. J.L. Pascual-Ahuir, E. Silla, J. Tomasi, R. Bonaccorsi "Electrostatic interaction of a solute with a continuum. Improved description of the cavity and of the surface cavity bound charge distribution", *J. Comput. Chem.* ,**8**, 778-787 (1987); J.L. Pascual-Ahuir, E. Silla "GEPOL: an improved desciption of molecular surfaces. I. Building the spherical surface set", *J. Comput. Chem.* **11**, 1047-1060, (1990); J.L. Pascual-Ahuir, I. Tunon, E. Silla "GEPOL: an improved desciption of molecular surfaces. II. Computing the molecular area and volume", *J. Comput. Chem.* **12**, 1077-1088, (1991).
34. See for example U. Sen "Study of electrolytic solution process using the scaled-particle theory. 2. The standard entropy of solvation", *J. Am. Chem. Soc.*, **102**, 2181-2188 1980.
35. S. Miertus, J. Tomasi "Approximate evaluations of the electrostatic free energy and internal energy changes in solution processes", *Chem. Phys.*, **65**, 239-245, (1982).
36. P. Beak, J.B. Covington, J.M. White "Quantitative model of solvent effects on hydroxypyridine-pyridone and mercaptopyridine-thiopyridone equilibria: correlation with reaction field and hydrogen-bonding effects", *J. Org. Chem.*, **45**, 1347-1353, (1980).

37. O. Bensaude, M. Chevrier, J.E. Dubois "Lactim-lactam tautomeric interconversion mechanism in water/polar aprotic solvent systems. 2. Hydration of 2-hydroxypyridines. Evidence for a bifunctional water-catalyzed proton transfer", *J. Am. Chem. Soc.*, 101, 2423-2429, (1979) .

38. M.W. Wong, K.B. Wiberg, M.J. Frisch "Solvent effects. 3. Tautomeric equilibria of formanide and 2-pyridone in the gas phase and solution. An ab-initio SCRF study", *J. Am. Chem. Soc.*,114, 1645-1652 , (1992) ; P. Cieplak, P. Bash, U.C. Singh, P.A. Kollman, "A theoretical study of tautomerism in the gas-phase and aqueous solution *J.Am. Chem. Soc.*, 1987, **109**, 6283.

39. See for example: P.F. Barbara,P.K. Walsh ,L.E. Brus "Picosecond kinetic and vibrationally resolved spectroscopic studies of intramolecular excited state hydrogen atom transfer" *J. Phys. Chem.*, 93,29-34, (1989).; T.Nishiya, S. Yamauchi, N. Hirota, M. Baba, I. Hanazaki "Fluorescence studies of the intramolecularly hydrogen-bonded molecules o-hydroxyacetophenone and salicylamide and related molecules", *J. Phys. Chem.*, 90 ,5730-5735, (1986); N.P. Ernsting "Dual fluorescence and excited state intramolecular proton transfer in jet-cooled 2,5-bis (2-benzoxazolyl) hydroquinone", *J. Phys. Chem.*, 89, 4932-4937, (1985); G. Smulevich, P. Foggi "Fluorescence excitation and emission spectra of 1,5 dihydroxyanthraquinone-d2 in 2 hexane at 10K"", *J.Chem. Phys.*, 87, 5657-5663, (1987).

40. J.C. Evans "Infrared spectrum and thermodynamic functions of formamide", *J. Chem. Phys.* , 22, 1228-1234, (1954); O.F. Nielsen, P.A. Lund, E. Praestgaard "Hydrogen bonding in liquid formamide. A low frequency Raman study", *J. Chem. Phys.*, 77, 3878-3883 ,(1982); A.J. Doig, D.H. Williams "Binding energy of an amide-amide hydrogen bond in aqueous and polar solvents", *J. Am. Chem. Soc.*, 114, 338-343, (1992).

41. P. Beak, F.S. Fry, J. Lee, F. Steele "Equilibration studies. Protomeric equilibria of 2- and 4- hydroxypyridines, 2- and 4-hydroxypyrimidines, 2- and 4- mercaptopyridines, and structurally related compounds in the gas phase", *J. Am. Chem. Soc.*, 98, 171-179, (1976); M.J. Nowak, L. Lapinski, J. Fulara, A. Les ,L. Adamowicz " Matrix isolation IR spectroscopy of tautomeric systems and its theoretical interpretation", *J. Phys. Chem.*, 96,1562-1569 (1992); A. Fujimoto, K. Inuzuka, R. Shiba "Electronic properties and $\pi{-}\pi^{*}$ absorption spectrum of 2-pyridone", *Bull. Chem. Soc. Jpn.*, 54 , 2801-2807, (1981);R. Tembreuil, C.H. Sin, H.M. Pang, D.M. Lubmam "Resonance enhanced multiphoton ionization spectroscopy for detection of azabenzenes in supersonic beam mass spectrometry", *Anal. Chem.*, 57, 2911-2917, (1985); A. Held ,D.W. Pratt, *J. Am. Chem. Soc.* "The 2-pyridone dimer, a model cis peptide. Gas-phase structure from high resolution laser spectroscopy", 112, 8629-8630, (1990).

42. A. Held, B.B. Champagne, D.W. Pratt "Inertial axis reorentation in the S_1-S_0 electronic transition of 2-pyridone. A rotational Duschinsky effect. Structural and dynamical consequences."*J. Chem. Phys.*, 95, 8732-8743, (1991).

43. M. Kuzuya, A. Noguchi, T. Okuda "Fluorescence spectroscopic study on tautomeric equilibria of 2(1h)-piridones, *J. Chem Soc. Perkin Trans II*, 1423-1427, (1985).
M.R. Nimlos, D.F. Kelley, E.R. Bernstein "Spectroscopy, structure and proton

44. dynamics of 2-hydroxypiridine and its clusters with water and ammonia", *J. Phys. Chem.*, 93 ,643-651, (1989).

45. Y. Sugawara, Y. Hamada, A.Y. Hirakawa, M. Tsuboi, S. Kato, K. Morokuma, *Chem.Phys.* " Ab initio MO calculation of force constants and dipole derivatives for formamide", 50, 105-111, (1980);T.J. Zielinski, R.A. Poirier, M.R. Peterson, I.G.

Csizmadia " A Water-Mediated Tautomerism Mechanism in Formamide and Amidine. An Ab Initio Study", *J. Comp. Chem.*, **4**, 419-427, (1983); J.P. Krug, P.L.A. Popelier, R.F.W. Bader, *J. Phys. Chem.*, 1992, **96**, 7604.

46. H.B. Schlegel, P. Gund, E.M. Fluder, "Tautomerization of Formamide, 2-Pyridone, and 4-Pyridone: an ab Initio Study", *J. Am. Chem. Soc.*, **104**, 5347-5351, (1982); M.J. Scanlan, I.H. Hillier " On the mechanism of proton transfer in the 2-hydroxypyridine-2-pyridone tautomeric equilibrium", *Chem. Phys. Lett.* ,**107** , 330-332, (1984); Z. Slanina, A. Les , L. Adamowicz "Dimerization in the pyridone/hydroxypyridine tautomeric systems: relative stabilities of the dimers in the 2-pyridone/2hydroxypiridine and in the 4-pyridone/4hydroxypiridine systems" ,*J. Mol. Struct. (THEOCHEM)*, **257** ,491-498, (1992); M. Moreno, W.H. Miller "On the tautomerization reaction 2-pyridone/2-hydroxypiridine: an ab initio study", *Chem. Phys. Lett.*, **171**, 475-479, (1990).

47. M.J. Field , I.H. Hillier "Non-dissociative proton transfer in 2-pyridone-hydroxypiridine. An ab initio molecular orbital study", *J. Chem. Soc. Perkin Trans. II*, 617-622, (1987).

48. A. Peluso, C. Adamo, G. Del Re " A theoretical analysis of excited state proton transfer in 3-hydroxyflavone. Promoting effect of a low frequancy bending mode", *J. Math Chem.*, **10**, 249-274, (1992), and refs. therein

49. M.J.Frisch,G.W.Trucks,M.Head-Gordon,P.M.W.Gill,M.W.Wong,J.B.Foresman, B.G.Johnson, H.B.Schlegel, M.A.Robb, E.S.Replogle, R.Gomperts, J.L.Andres, K. Raghavachari, J.S.Binkley,C.Gonzalez,R.L.Martin,D.J.Fox,D.J.DeFrees,J.Baker, J.J.P. Stewart, J.A.Pople GAUSSIAN92, Gaussian Inc., Pittsburgh, P.A. (1992).

50. P.Amodeo,V.Barone "A New General Form of Molecular Force Fields. Application to Intra and Interresidue Interactions in Peptides", *J.Am.Chem.Soc.*, **114**, 9085-9093 (1992).

51. M.J.S. Dewar, E.G. Zoebisch, E.F. Zoebisch, E.F. Healy, J.J.P. Stewart "AM1: a new general purpose quantum mechanical molecular model",*J.Am.Chem.Soc.*, **107**,3902-3909, (1985); J.J.P.Stewart MOPAC_6: Quantum Chemistry Program Exchange, program n.455.

52. W.M.F. Fabian "Tautomeric equilibria of heterocyclic molecules. A test of the semiempirical AM1 and MNDO-PM3 methods", *J. Comput. Chem.*, **12**, 17-39, (1991).

53. J. Troe, K.M. Weitzel "MNDO calculation of stilbene potential energy properties relevant for the photoisomerization dynamics", *J. Chem. Phys.*, **88**, 7030-7039, (1988); M.J.S. Dewar,C.Doubleday "A MINDO/3 Study of the Norrish Type II Reaction of Butanol",*J.Am.Chem.Soc.*,**100**, 4935-4941, (1978) ; H.E. Zimmermann, A. M. Weber "Photochemical rearrangements of molecules quenchers on a chain. Mechanism and exploratory organic photochemistry", *J.Am.Chem.Soc.*, **111**,995-1007, (1989); C. Rulliere, A. Declemy , P. Kottis , L. Ducasse "Effect of ground and excited singlet state geometry on the level ordering of 1,4 diphenylbutadiene: a theoretical study using MNDO", *Chem.Phys.Lett.*, **117**,583-589 (1982); P. Ertl "MNDO/CI study of photoisomerization about polar double bonds", *Int. J. Quantum Chem.*,**38** , 231-238, (1990).

54. E.L. Coitino, K. Irving, J. Rama, A. Iglesias, M. Paulino, O.N. Ventura "Theoretical studies of hydrogen-bonded complexes using semiempirical methods",*J. Mol. Struct. (THEOCHEM)*, **210**, 405-412, (1990).

55. P.L.Cristinziano, F. Lelj, P. Amodeo, G. Barone, V. Barone "Stability and structure

of formamide and urea dimers in aqueous solution", *J. Chem. Soc. Faraday Trans. I*, **85**, 621-632, (1989),and refs. therein
56. C.Adamo, V.Barone, C. Minichino, unpublished results.

Ternary Complexes of Pt(II) and Pt(IV) with Aminoacids and Nucleobases

V. ALETRAS, A. IAKOVIDIS and N. HADJILIADIS
Department of Chemistry, University of Ioannina, Laboratory of Inorganic and General Chemistry, Ioannina 45-110, Greece

1. Introduction

In order to study the DNA-Protein crosslinks formed by cis- and trans- $Pt(NH_3)_2Cl_2$ we used model molecules which are the nuclear bases* 1-Methylcytosine and 9-Methylguanine and the aminoacids glycine, alanine, L-2-aminobutyric acid, L-norvaline and L-valine, and we prepared the ternary compounds with the formulae cis- and trans-$[(NH_3)_2Pt(Nb)(am-ac)]^+$. We intended to study especially the hydrophobic ligand-ligand interactions occurring in these complexes, so we selected for am-ac these substituted, at the a-carbon, glycines by an aliphatic side chain increasing in a linear way from ala to nval, except the case of val which was used only for comparison reasons. These interactions were detected in aqueous solution by ^{1}H-NMR spectra and were compared with the crystal structures of the compounds cis-$[(NH_3)_2Pt(gly)(1-MeC)]^+$ [1] and trans-$[(CH_3NH_2)_2Pt(gly)(1-MeC)]^+$ [2].

2. Results and discussion

The cis-ternary complexes were prepared by two routes, the first starting from the aminoacid chelates (Route 1).

$$\text{cis -}[(NH_3)_2Pt(amac)](NO_3)+Nb \xrightarrow[\text{2-3 days}]{50°} \text{cis-}[(NH_3)_2Pt(am-ac)(Nb)](NO_3) \qquad \text{(Route 1)}$$

or from the binary nucleobase complexes, which were used also for the trans-compounds (Route 2).

$$\text{cis -,trans-}[(NH_3)_2Pt(Nb)Cl](NO_3) \xrightarrow[\text{-AgCl}]{+AgNO_3}$$

---

* Abbreviations: Nb=nucleobase, am-ac=aminoacid, 1-MeC=1-Methylcytosine, 9-MeG=9-Methylguanine, gly=glycine, ala=L-alanine, 2-aba=L-2-aminobytyric acid, val=L-valine, nval=L-norvaline.

N. Russo et al. (eds.), Properties and Chemistry of Biomolecular Systems, 19–22.

© *1994 Kluwer Academic Publishers. Printed in the Netherlands.*

$$\rightarrow \underline{cis}, \text{-}, \underline{trans}\text{ -}[(NH_3)_2Pt(Nb)]\,(OH_2)]\,(NO_3)_2 \xrightarrow[\text{pH=5.0, 35 °C, 2-3 days}]{\text{excess amacH}}$$

$$\rightarrow \underline{cis}\text{ -},\underline{trans}\text{-}[(NH_3)_2Pt(Nb)(am\text{-}ac)](NO_3) + NO_3^- + H_3O^+ \qquad\qquad \text{(Route 2)}$$

The amino-acids are N-coordinated, as indicated by the IR and Raman spectra of the ternary complexes, and the nucleobases are N(7) and N(3) coordinated for 9-MeG and 1-MeC, respectively, as indicated by the ^{1}H-NMR spectra. In the ternary systems we observed in the ^{1}H-NMR spectra a retention of N(7) and N(3) coordination of the nucleobases, compared with the binary complexes. The difference of the chemical shifts of the terminal-CH$_3$ group of the aminoacids, between anionic forms and their metal complexes, named $\Delta\delta + \delta_{am\text{-}ac} - \delta_{complex}$, is indicative of the presence of a hydrophobic ligand-ligand interaction, if it is positive. This is presented in Figure 1, as a function of the aminoacid involved.

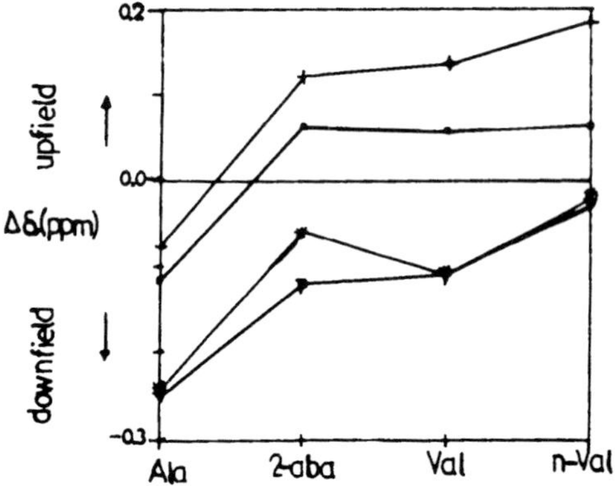

Figure 1. Variation of the chemical shift of terminal -CH$_3$ groups of the aminoacids in ternary systems, from the free aminoacid anions. am-ac/9-MeG : <u>cis</u> (+) and <u>trans</u> (*) compounds am-ac/1-MeC : <u>cis</u> (•) and <u>trans</u> (∇) compounds.

We can clearly see that such an interaction which occurs for the <u>cis</u>-compounds, and for the <u>trans</u>-compounds is very slight. The interaction also becomes larger in the case of the bulkier 9-MeG and increases with the size of the aliphatic aminoacid's side chain. The behavior of val is similar to that of 2-aba though it has the same number of C-atoms as nval. In the next Figure (Fig. 2) we have the chemical shifts of the various H-atoms of nval in the ternary complexes of 1-MeC and 9-MeG, as a function of the difference $\Delta\delta = \delta_{amacH} - \delta_{complex}$, indicating that the nearer we are to the coordination site, the larger is the hydrophobic effect. Again the strongest interaction occurs for the <u>cis</u>-compounds and for 9-MeG compound [3].

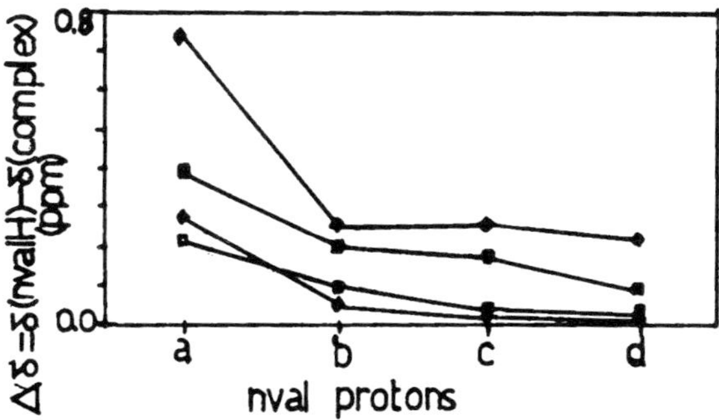

Figure 2. Variation of the chemical shifts of nval protons in the ternary systems from the free nval zwitterion, 1-MeC compounds: cis (Á), trans (◇) 9-MeG compounds: cis (◆), trans (■)

Table 1 shows the results of the calculation of the percentage of the h, t, g, rotamer populations around the Ca-Cb bond. For the aminoacids 2-aba and nval we have the h and t+g rotamers (these aminoacids have two a-H) and for val (having one a-H) the t and h+g rotamers populations. These data were obtained by the ^{1}H-NMR spectra. We observe that the population of h, t decreases, going from the aminoacid chelates to the ternary systems, almost reaching the values for the free aminoacid zwitterions. The ^{1}H-NMR spectra of the cis-compounds also revealed the presence of small amounts of the trans-complexes (2.8-13.5%) increasing with temperature. The crystal structures of the aminoacid chelates [4] show that the ligand side chain points to the direction of the metal, while in the structures of the compounds cis-[(NH$_3$)$_2$Pt(gly)(1-MeC)](NO$_3$) and trans-[(NH$_3$)$_2$Pt(gly)(1-MeC)](NO$_3$) we observe that the -COO- group of gly as well as the a-hydrogens point out of the pyrimidine plane. Finally, we oxidized cis-[(NH$_3$)$_2$Pt(gly)(1-MeC)](NO$_3$) with S$_2$O$_8$$^{2-}$ producing the Pt^{+4} species:

2 cis-[(NH$_3$)$_2$Pt(1-MeC)(N-gly)](NO$_3$)+2S$_2$O$_8$$^{2-}$ + 2H$_2$O $\longrightarrow$

cis-[(NH$_3$)$_2$Pt(1-MeC)(N,O-gly)](H$_2$O]$_2$$^{6+}$ + 4SO$_4$$^{2-}$ + 2NO$_3$$^-$

Schema

The ^{1}H-NMR spectra indicate that in the final product gly is chelated and for 1-MeC after rearrangement we obtain it's -NH_2 bound form (see scheme).

3. Conclusion

The hydrophobic interactions, and especially the DNA-Pt-Protein interactions are still under investigation. Only ^{1}H-NMR spectra indicate their existence in aqueous solution, while they seem to be absent in the solid phase, at least for the structures available until now. We have extended our study with other model molecules like nucleosides, mono- and oligo-nucleotides and peptides, in order to obtain more detailed informations about this type of interaction.

Table 1. Vicinal proton coupling constants and rotamer distribution in ternary 1-MeC compounds.

Compound	J_{AB+BC}	J_{BC}	h(%)	t(%)	t+g(%)	h+g(%)
2-abaH	11.700		37.0		63.0	
cis-[(NH$_3$)$_2$Pt(2-aba)]+	10.109		51.3		48.7	
cis-[(NH$_3$)$_2$Pt(2-aba) (1-MeC)]+	12.200		32.1		67.9	
trans-[(NH$_3$)$_2$Pt(2-aba) (1MeC)]+	12.400		30.0		70.0	
nvalH	12.200		32.0		68.0	
cis-[(NH$_3$)$_2$Pt(nval)]+	10.023		52.1		47.9	
cis-[(NH$_3$)$_2$Pt(nval) (1-MeC)]+	12.400		30.3		69.7	
trans-[(NH$_3$)$_2$Pt(val) (1MeC)]+	13.000		25.0		75.0	
valH		4.400		18.0		82.0
cis-[(NH$_3$)$_2$Pt(val)]+		3.370		8.9		91.1
cis-[(NH$_3$)$_2$Pt(val) (1-MeC)]+		4.600		20.2		79.8
trans-[(NH$_3$)$_2$Pt(val) (1MeC)]+		4.100		16.0		84.0

References

1. A. Iakovidis , N. Hadjiliadis, J.F. Britten, I.S. Butler, F. Schwarz, and B. Lippert, *Inorg. Chim. Acta* **184**, 209-220, (1991).
2. F.J. Pesch, H. Preut, and B. Lippert, *Inorg. Chim. Acta* **169**, 195-200, (1990).
3. V. Aletras, N. Hadjiliadis, and B. Lippert, *Polyhedron* **11**, 1359-1367, (1992).
4. A. Iakovidis, N. Hadjiliadis, H. Schollhorn, V. Thewalt and G. Trotscher, *Inorg. Chim. Acta* **164**, 221-229, (1989).

Free Radicals in Biological Systems

J. ANASTASSOPOULOU
*National Technical University of Athens, Radiation Chemistry-Biospectroscopy,
Zografou Campus, Zografou 157 80, Athens, Greece.*

1. Introduction

Free radicals play an important role in biological reactions. By definition, free radicals are species which contain an odd number of electrons. They may be positively or negatively charged or neutral. Free radicals can be generated in living organisms by oxidation-reduction reactions catalyzed by enzymes such as catalase, peroxidase, etc [1,2]. Also exposure of biological systems to chemical carcinogens or ionizing radiations can induce the formation of free radicals [3-5]. When a cell is irradiated many chemical reactions are induced, leading to a variety of biological changes. In the hierarchy of targets for reproductive cell death DNA should be placed at the top, due to its vast informational content. This sensitivity results from its structure. DNA is a complex molecule consisting of different chemical groups and has a large surface. The most important lesions in DNA are single and double strand breaks, single strand breaks with base damage on one strand, and base damage on two opposite strands, which may be followed by a major distortion of the DNA molecule and its molecular morphology. Since the information of each cell is stored in the bases, a base damage is a potential lesion for the disruption of cells.

2. Radiolysis of guanosine-5'-monophosphate acid

Many experiments have been carried out on radiolysis of pyrimidines [6-8] in order to shed light on the nature of the damage induced to DNA. In contrast to this, the radiation chemistry of aqueous solutions of purines is only poorly understood because these systems are more complicated.In order to study the radiolysis of purines, we irradiated aqueous solutions of 5'-guanosine-monophosphate in the acid form (5'-GMPH$_2$). The FT-IR spectra of non-irradiated 5'-GMPH$_2$ in the region 1800-1200 cm^{-1} are shown in Fig. 1. The carbonyl and amide absorptions together with those of the double bonds, C=C and C=N, as well as the water bending absorptions are observed in this region [9]. As it can be seen in Fig. 1b, the intensity of the band at 1691.6 cm^{-1} is decreased after irradiation. This band is related to the carbonyl and amide absorption νC=O and to slightly mixed carbon-carbon stretching frequencies (νC=C) of the aromatic ring [10]. When aqueous solutions are irradiated, the water decomposes according to the overall equation:

N. Russo et al. (eds.), Properties and Chemistry of Biomolecular Systems, 23–30.
© 1994 *Kluwer Academic Publishers. Printed in the Netherlands.*

$$H_2O \longrightarrow OH, H, e_{aq}^-, H_2, H_2O_2, H_3O^+ \tag{1}$$

From these species only the hydrated electrons can react by addition to the C=O bond of the guanine molecule according to the following reaction [11].

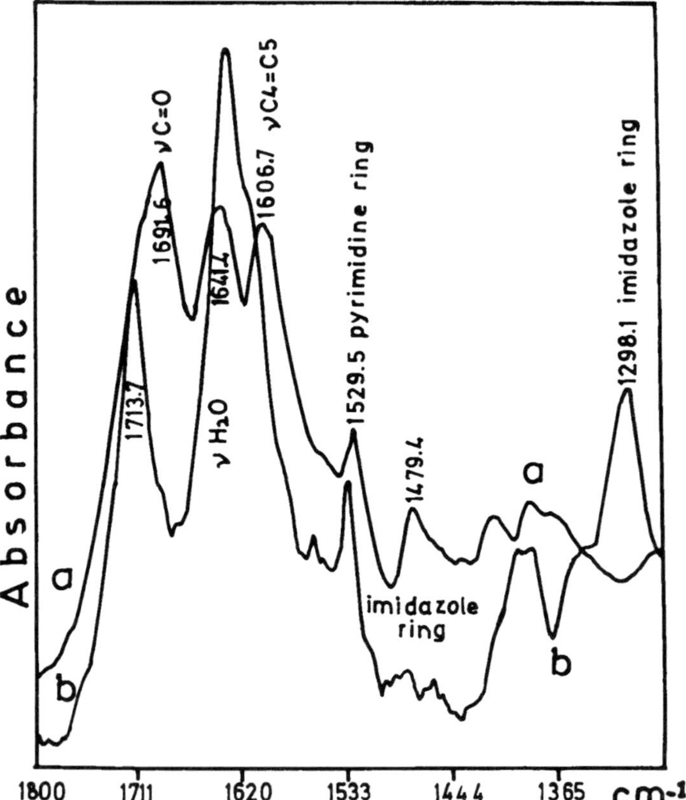

$$\tag{2}$$

where, RP= Ribose-Phosphate

Figure 1. FT-IR spectra of 5'-GMPH$_2$ in the region 1800-1267 cm^{-1}. The non-irradiated and irradiated solutions were evaporated to dryness and the solid materials were used to form KCl pellets for IR. The spectra are superimposed (ordinate, arbitary units). (a) non-irradiated sample and (b) after a total dose of 1 kGry.

This is observed in the FT-IR spectra of 5'-GMPH$_2$ by the decrease of the C=O band intensity. A shift of the carbonyl band at 1691.6 to 1713.7 cm^{-1} has also been observed, which may be explained by the presence of free C=O groups. This means that there is a disruption of the general structure of the 5'-GMPH$_2$ molecule upon irradiation. The intensity of the band at 1606.7 cm^{-1}, which is related to vC4=C5 bond is also decreased after irradiation, due most likely to the addition of OH radicals to the double carbon-carbon bonds [12,13]:

$$(3)$$

It was proposed that Guo-C5-OH radical (a) has reducing properties, while the Guo-C4-OH radical (b) has oxidizing properties [14]. Considerable changes have been also observed in the region 1350-1500 cm^{-1}, where the imidazole ring is known to absorb [15]. It seems that when the solutions are irradiated the attack of OH radicals, that are produced, occurs on the C8 position according to the reaction:

$$(4)$$

The produced Guo-C8-OH radicals are unstable and may open the imidazole ring leading to depurination as following [16]:

$$(5)$$

These changes have not been observed in the spectra when the solutions were irradiated in the presence of Mg^{2+} ions, although the OH radicals reacted with the double bonds C8=N7 of the guanine, this radical attack did not lead to depurination [17]. This result is in agreement with that which has been found in radiolysis of non deaerated aqueous solutions of deoxyguanosine-5'-monophosphate [18]. In this case the amount of the produced dGuo-C8-OH radicals reacted with oxygen in a diffusion control reaction (k=5x10^9 M^{-1}s^{-1}) and gave stable peroxyl radicals [19]. In addition, it was also found that the main oxidation product of the Fenton reaction with 2'-deoxyguanosine which involved perferryl ions or ferroxo complexes as the reactive species 8-hydroxy-2'-

deoxoguanosine was produced [20]. Furthermore, from FAB positive ions mass spectra it was found that base and sugar-phosphate moieties were released. This can be explained with the assumption that OH radicals and H-atoms reacted by H-abstraction reactions from C1' and C5' positions:

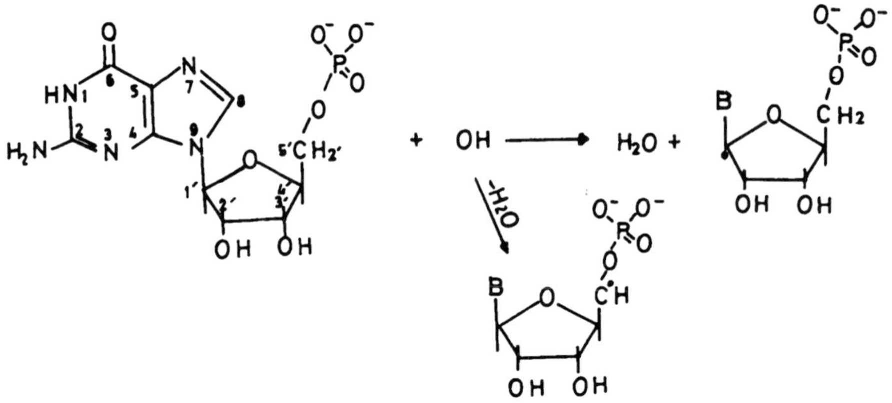

$$(6)$$

Where B=Base (guanine).

3. Radiosensitization

The fact that the presence of oxygen increases the lethal effects upon irradiation is accepted throughout radiation biology [21]. Since oxygen is a free radical, the "oxygen effect" is undoubtedly free radical in nature [22]. It is well known that the oxygen effect involves at least in part the DNA in the nucleus of the cells. A very plausible mechanism for the oxygen effect is the "radical fixation hypothesis", in which the free radicals produced in a critical biomolecule, e.g. DNA, reacted rapidly with oxygen to form stable peroxyradicals.

$$DNA^{\cdot} + O_2 \longrightarrow DNAO_2$$

$$(7)$$

In the absence of oxygen the free radicals produced by direct or indirect action of radiation in the vital molecule may undergo back reaction to the original state:

$$DNA^{\cdot} + RSH \longrightarrow DNA + RS^{\cdot}$$

$$(8)$$

$$2RS^{\cdot} \longrightarrow RS{-}SR$$

$$(9)$$

Radiobiologists have also accepted that thiols protect the living systems by the hydrogen-atom donation model [22]:

$$-\overset{|}{\underset{|}{C}}{}^{\cdot} + RSH \longrightarrow -\overset{|}{\underset{|}{C}}H + RS^{\cdot}$$

$$(10)$$

$$2 RS^{\cdot} \longrightarrow RS\text{-}SR \tag{9}$$

In an attempt to overcome the problems of radioresistance of hypoxic cells, the research work was focussed in the development of drugs that can mimic oxygen. These chemical compounds must be active only against hypoxic cells and should not increase radiation response in well oxyganated normal tissues.

It has been known that nitroxyl stable free radicals reacted as radiosensitizers [23,24]. In order to investigate the mechanism of radiosensitization we irradiated aqueous solutions of the system safranine T/thioglycolic acid (HSCH$_2$COOH) in the presence and in the absence of different amounts of 2,2,6,6-tetramethyl-piperidine-1-oxyl (TEMPO). This was due to the fact that the radiolysis of safranine T and TEMPO were well established [24,25-27]. Fig. 2 shows the effect of thioglycolic acid concentration on radiolysis of safranine T. It is observed from the changes of optical absorption that by increasing the concentration of thioglycolic acid, increases the protection of safranine T molecules from radiolysis. For a concentration of thioglycolic acid up to 5mM the protection of safranine T molecules was almost 100%. From these results it seems that the OH radicals that are produced from radiolysis of water (eq. 1) reacted with thioglycolic acid molecules according to the reaction:

$$HSCH_2COOH + OH \longrightarrow \overset{\cdot}{S}CH_2COOH + H_2O \tag{11}$$

$$2\ \overset{\cdot}{S}CH_2COOH \longrightarrow \begin{array}{c} S\ CH_2COOH \\ | \\ S\ CH_2COOH \end{array} \tag{12}$$

It is clear that reaction (11) competes with reaction (13),

$$Safranine\ T + OH \longrightarrow Products \tag{13}$$

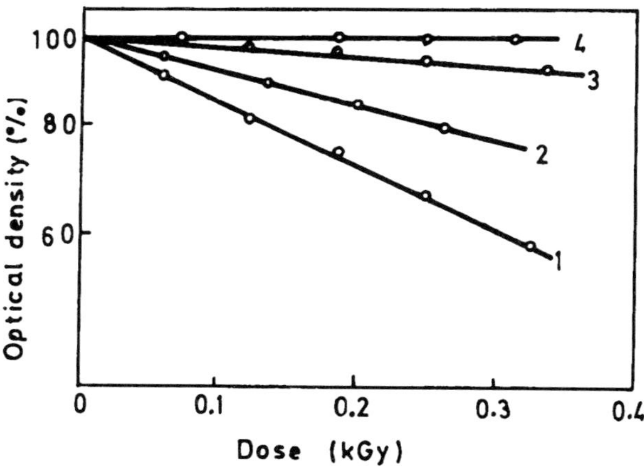

Figure 2. Concentration changes of safranine T 3x10^{-5} M irradiated in the presence of thioglycolic acid as a function of radiation dose. The curves correspond in concentration of thioglycolic acid: o, 5x10^{-5}, 1x10^{-3} and 5x10^{-3} M.

which, otherwise would lead to the disappearance of safranine T molecules.
After these results we irradiated aqueous solutions of [SafranineT]=2.7×10^{-5} M and [Thioglycolic acid]=5×10^{-3} M in the presence as well as in the absence of TEMPO. The effect of the presence of TEMPO in the radiolysis of this system is shown in Fig. 3. It is obvious that TEMPO sensitizes the safranine T molecules against radiation.

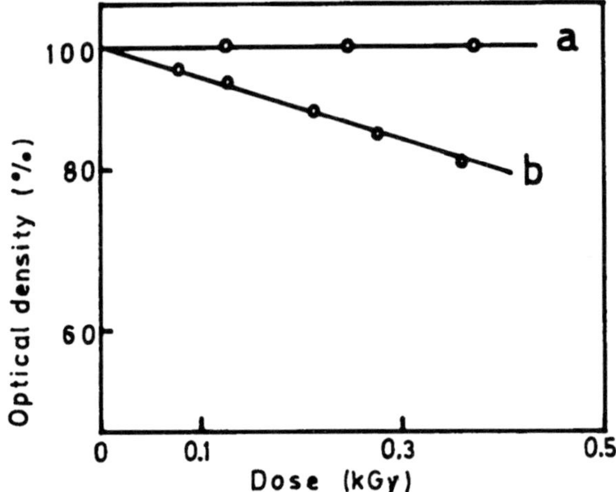

Figure 3. Radiosensitization of the system safranineT/ thioglycolic acid with TEMPO. 1. safranine T 2.7×10^{-5} M, thioglycolic acid 5×10^{-3} M. 2. safranine T 2.7×10^{-5} M, Thioglycolic acid 5×10^{-3} M and TEMPO 2×10^{-3} M.

Since TEMPO does not react with safranine T, these results are consistent with the assumption that TEMPO reacted with thioglycolic acid _via_ the following way:

$$\text{(piperidine-N-O radical)} + HSCH_2COOH \longrightarrow \text{(piperidine-N-OH)} + \dot{S}CH_2COOH \tag{14}$$

$$2\ \dot{S}CH_2COOH \longrightarrow \begin{array}{c} S\ CH_2COOH \\ | \\ S\ CH_2COOH \end{array} \tag{12}$$

Under these conditions the reaction (11), which protects the safranine T molecules from radiolysis, does not take place and the OH reacted mostly with safranine T according to the reaction (13). TEMPO appears to act as a radiosensitizer by competing with intracellular thiols _via_ the reaction (14), which otherwise, might repair some of the damage done according to the equation (10). It has been found that nitroxyl free radicals (R-NO) can also sensitize DNA by binding either through nitrogen at C6-OH position or through oxygen in the case of C5-OH adduct (Scheme 1) according to the "radical fixation hypothesis" [16,28].

Scheme 1.

There is some evidence [28,29] that nitroxyl radicals can react with free radical sites in DNA <u>in vivo</u>, to produce lethal mutation before these sites can be repaired (eq. 8). The Guo-C8-OH radicals are reduced species and can react with nitroxyls or oxygen. For the oxydized species Guo-C4-OH radicals the "oxygen effect" does not exist and they can react with thiols. Many metal ions and metal complexes show electron affinity properties due to the positive charge of the metal (M^{+n}) and they become very attractive as possible sensitizers. There are several examples in the literature which include complexes of Ag(I), Cu(II), Co(II), Pt(II) and Mg(II).It was observed <u>in vitro</u> hat Pt(II) reacted as radiosensitizer in the presence as well as in the absence of oxygen [30]. Also other transition metals as Co(III) and Fe(III) have sensitized B. megaterium spores under hypoxic and oxic conditions [31]. Recently in a study of magnesium complexes it was observed that Mg(II) ions can sensitize <u>in vitro</u> 5'-GMPNa$_2$ only in the absence of oxygen and under these conditions it doubles the radiation effect [17,30,32]. Magnesium ions show sensitizing properties approaching more closely to the definition of an ideal radiosensitizer for clinical applications.

References

1. H. Beinert, in <u>"Biological Applications of Electron Spin Resonance"</u>, eds., H.M.Swartz, J.R. Bolton and D.C. Borg, Chapter 8, Wiley, New York, (1972).
2. I. Isenberg, *Physiol. Rev.* **44**, 487, (1964).
3. A. Vithayathil, J. Ternberg and B. Commoner, *Nature* (London) **207**, 1246, (1965).
4. J. Rowlands and C. Gross, *Nature* (london) **231**, 1256, (1967) .
5. H. Swartz, *Radiat. Res.* **24**, 579, (1965).
6. G. Scholes and M. Simic, *Biochim. Biophys. Acta* **166**, 255, (1968).
7. M.N. Schuchmann and C. von Sonntag, *J. Chem. Soc.*, Perkinn Trans II, 1525, (1983).
8. J. Cadet, A. Balland and M. Barger, *Int. J. Radiat. Biol.* **39**, 119, (1981).
9. T. Theophanides, Infrared and Raman Spectroscopy of Biological Molecules, D. Reidel Publishing Co, Dordrecht, The Netherlands, pp. 185 and 205, (1979).
10. M. Tsuboi, S. Tokahoshi and I. Hodara, <u>Physicochemical Properties of Nucleic Acids</u>, Vol. **2**, ed., Dunhense, Academic Press, New York, p. 91, (1973).
11. J. Anastassopoulou, N.Th. Rakintzis and T. Theophanides, *Magnesium Res.* **3**, 15, (1989).
12. P. O'Neill, *Radiat. Res.* **96**, 198, (1983).
13. P. O'Neill and P.W. Chapman, *Int. J. Radiat. Biol.* **47**, 71, (1985).
14. E.M. Fielden and P. O'Neill, <u>Early Effects of Radiation on DNA</u>, Berlin, Springer-Verlag, (1991).

15. C. Sandorfy and T. Theophanides, <u>Spectroscopy of Biological Molecules</u>, D. Reidel Publishing Co, Dordrecht, The Netherlands, pp. 291 and 137, (1974).

16. C. von Sonntag, *The Chemical Basis of Radiation Biology*, Taylor and Francis, London, (1987).

17. J. Anastassopoulou, *Magnesium Res.* **5**, 97, (1992).

18. L.P. Candeias and S. Steeken, in "<u>The Early Effects of Radiation on DNA</u>", Eds., E.M. Fielden and P. O'Neill, Springer-Verlag, Berlin, New York, p.265, (1990).

19. S. Steeken, *Chem. Rev.* **89**, 50, (1989).

20. J. Cadet, M. Berger, J.F. Mouret, F. Odin, M. Polverelli, J.L. Ravanat, in "<u>The Early Effects of Radiation on DNA</u>", Eds., E.M.Fielden and P.O'Neill, Springer-Verlag, Berlin, New York, p.403, (1990).

21. L.H. Gray, A.d. Conger, M. Ebert, S. Hornsey and O.C.A. Scott, *Br. J. Radiol.* **26**, 638, (1953).

22. G.E. Adams, *Adv. Radiat. Chem.* **3**, 125, (1972).

23. P.T. Emmerson and R.L. Willson, *J. Phys. Chem* **72**, 3669, (1968).

24. J. Anastassopoulou, J.D. Chandrinos and N.Th. Rakintzis, *Radiat. Phys. Chem.* **17**, 55, (1981).

25. N.Th. Rakintzis and E. Papakonstantinou, *Z. Phys. Chem. Neue Folge* **44**, 257, (1965).

26. D.G. Marketos and N.Th. Rakintzis, *Z. Phys. Chem. Neue Folge* **44**, 269, (1965).

27. D.G. Marketos, *Z. Phys. Chem. Neue Folge* **65**, 306, (1969).

28. J.E. Biaglow, M.E. Clark and E.R. Epp, *Radiat. Res.* **95**, 437, (1983).

29. P.T. Emmerson and P. Howard-Flanders, *Nature* (London) **204**, 1005, (1964).

30. J. Anastassopoulou and J. Brekoulakis, *Anticancer Res.* **10**, 983, (1990).

31. B.A. Teicher, J.L. Jacobs, K.N.S. Cathcart, M.J. Abrams, J.F. Vollano and D.H. Picker, *Radiat. Res.* **109**, 34, (1987).

32. J. Anastassopoulou, in "<u>Metal Ions in Biology and Medicine</u>", Eds., Ph. Collery, L. Poirier, M. Manfait, J.-C. Etienne, John Libbey, Eurotext, London, Paris, p.525, (1990).

The Role of Oxygen in Radiolysis of Aqueous Solutions of Magnesium Guanosine-5'-Monophosphate Complexes

J. ANASTASSOPOULOU AND J. BREKOULAKIS
National Technical University of Athens, Radiation Chemistry and Biospectroscopy, Zografou Campus, Zografou 157 80, Athens, Greece.

1. Introduction

Since the appearance of life, oxygen has been utilized by aerobic organisms. It is the final electron acceptor,almost in all metabolic processes, resulting in superoxide ion (O_2^-). It is further reduced to hydrogen peroxide by enzyme superoxide dismutase (SOD) and finally to hydroxyl radical [1,2].

$$O_2 \xrightarrow{\ e\ } O_2^{\overline{\cdot}} \xrightarrow[2H^+]{\ e\ } H_2O_2 \xrightarrow{\ e\ } OH$$

$$HO^{\cdot}$$

If the cellular defensive enzyme system against free radicals is disturbed, oxygen activation causes direct cytotoxicity. The role of oxygen is more important in the radiation therapy of cancer. It is believed that in some tumors a portion of the cells becomes hypoxic because oxygen diffuses away from the blood capillaries in the tumors and is consumed by metabolizing cells. As a result of hypoxia these cells are radioresistant. The sensitizing properties of oxygen may be attributed to the ability of this element to react with a radiation-induced free radical and produce a more stable hydroperoxyl radical [3]. Clinical methods that have been tried to overcome hypoxia problems include chemical sensitizers which act against hypoxic cells. Many metal ions and metal complexes show electron affinity properties due to the positive charge of the metal, and they become very attractive as possible sensitizers. We have investigated Mg^{2+} ions, because it is known that magnesium stabilizes the B-DNA structure [4].

2. Materials and Methods

Guanosine-5'-monophosphate disodium salt (5'-GMPNa$_2$) was purchased from Sigma Chemical Co and was used without further purification. Magnesium perchlorate was obtained from Baker Co, in pure form. The aqueous solutions of concentration 1×10^{-2} M were prepared in triply distilled water, in neutral pH. They were irradiated with a ^{60}Co gamma-ray source (Atomic Energy of Canada Ltd), in the presence and in the

31

N. Russo et al. (eds.), Properties and Chemistry of Biomolecular Systems, 31–36.
© *1994 Kluwer Academic Publishers. Printed in the Netherlands.*

absence of Mg^{2+} ions. The dose of irradiation was 1 to 5 kGy, in order to avoid destruction of more than 30% of the complex. The FT-IR spectra were recorded with a Digilab FTS-15 C/D
and with a BOMEM MICHELSON 100 spectrometer, with resolution 4 cm^{-1}, as KCl or KBr pellets. The non irradiated, as well as the irradiated solutions, were evaporated to dryness and the solid materials were used to prepare pellets for IR spectra. The mass spectra were recorded on an MS 902 mass spectrometer (Kratos Ltd, Manchester, UK). The Fast Atom Bobardment (FAB) gun (Kratos Ltd.) was operated with a 5-7 kV argon beam.

3. Results and discussion

The absorption spectra of aerated aqueous solutions of Mg-5'-GMPNa$_2$ complexes for different irradiation doses showed simple first order kinetics.

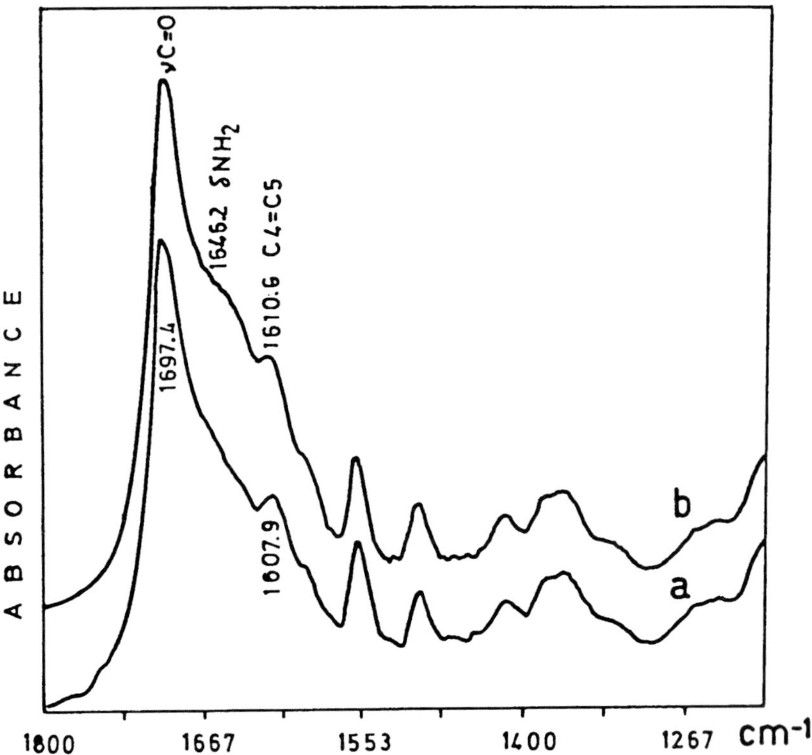

Figure 1. FT-IR spectra of the Mg(ClO$_4$)$_2$-5'GMPNa$_2$ complex in the region of 1800-1200 cm^{-1} in KCl pellets, (a) before irradiation and (b) after irradiation with total dose of 5kGy. The spectra are superimposed (ordinate, arbitrary units).

The decomposition limiting yield G(-Mg-5'-GMPNa$_2$) was found to be equal to 0.4 (G value is the number of disappeared molecules per 100 eV of absorbed energy in the solution), by measuring the decrease in the absorbance at λmax=254 nm. The G value was also found equal to 0.4 when aerated aqueous solutions of 5'-GMPNa$_2$ were irradiated in the absence of magnesium ions [5].This shows that the completion of magnesium ions is prevented by the oxygen. The FT-IR spectra of the mononucleotide and its magnesium metal adduct were recorded in the solid form before and after irradiation under anoxic conditions. The FT-IR spectra are shown in Fig. 1, in the region 1800-1200 cm^{-1}.

The appearence of the bands at 1646.2 cm^{-1} that are assigned to δNH_2+ νC_2-N_2 vibrations [7] corresponds to the release of 5'-GMPNa$_2$ from the complex. In this region there are the carbonyl and amine absorptions together with the double bond absorptions (C=C, C=N) [6]. In Fig. 2 are shown the intensity changes in the region 900-600 cm^{-1}. It is known that this region contains characteristic ring modes of the purine base which arise from coupling with vibrations of the sugar moiety.

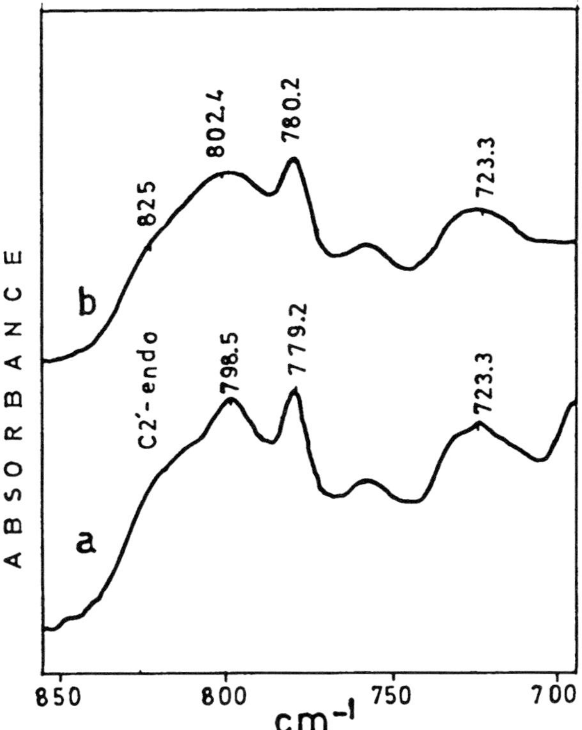

Figure 2. FT-IR spectra of the Mg(ClO$_4$)$_2$-5'GMPNa$_2$ complex, in KCl pellets in the region 900-600 cm^{-1}: (a) non-irradiated, (b) after a total dose of 5kGy. The spectra are superimposed (ordinate, arbitrary units).

From the increase, after irradiation, of the intensity at 825 cm^{-1} band that is a "marker band" for <u>C2'-endo/anti</u> conformation [7] is suggested that magnesium is not bound to the N7 atom of guanine [5]. In Fig. 3 are shown the FAB positive ions mass spectra of the complex, 5'-GMPNa$_2$-Mg(ClO$_4$)$_2$, before (a) and after irradiation (b). The fragments of the matrix glycerol are indicated by G. The weak peak of relative intensity 10% of fragment 333 is due to the [2B+O$_2$-H]$^+$. This peak indicates that the produced OH radicals from raiolysis of water,

$$H_2O \longrightarrow H, OH, e_{aq}^- \tag{1}$$

$$H + O_2 \longrightarrow HO_2 \tag{2}$$

$$e_{aq}^- + O_2 \longrightarrow O_2^- \tag{3}$$

reacted by abstraction of H atoms from the C1' site of the sugar moiety and base is released.

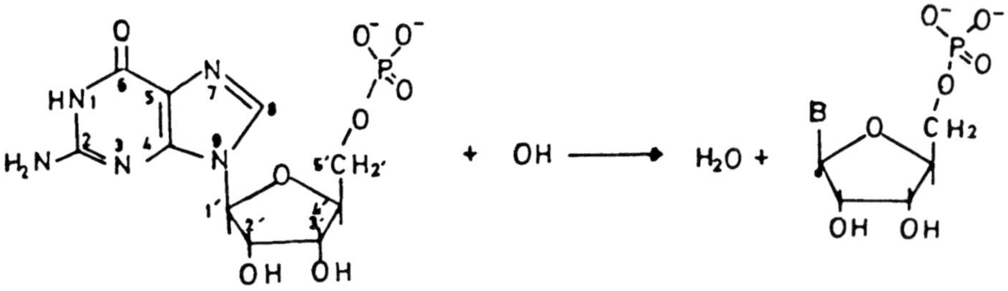

$$\tag{4}$$

Where B=Base (guanine)

The carbon-centered radicals (B) usually react with oxygen at rates that are almost diffusion-controlled (k=2x10^{-9} dm^3 mol^{-1} s^{-1}) [8,9]. In this case the produced base radicals reacted with the O$_2$ of the solution to form peroxyl radical:

$$2B^\cdot + O_2 \longrightarrow B\text{-}O\text{-}O\text{-}B \tag{5}$$

The next very weak peak of a small fragment at 425 may be the [M+OH] moiety, indicating the addition of OH radicals to the double bonds C=C of the 5'-GMPNa$_2$ molecules:

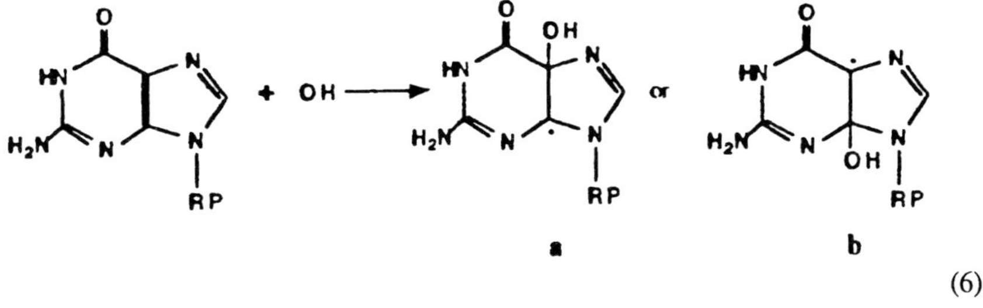

$$(6)$$

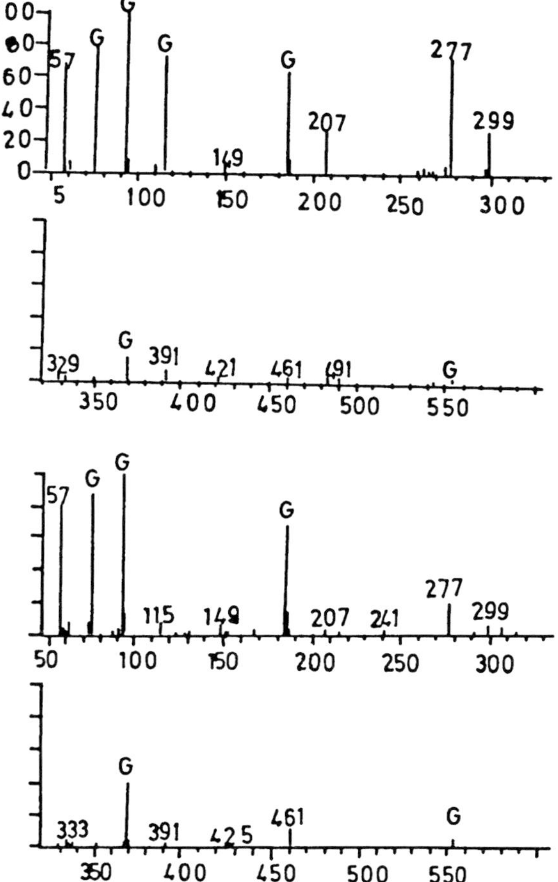

Figure 3. FAB positive ion mass spectra of the complex $Mg(ClO_4)_2$-5'-GMPNa$_2$ (a) non irradiated and (b) after a total dose 5 kGy.

In comparison of these results with what was found from radiolysis of complexes in the absence of oxygen [5], it is clearly shown that Mg^{2+} ions are not bound to the base after irradiation. This leads to the conclusion that the presence of oxygen in solution seems to prevent complexation of magnesium ions with the nucleotide, due to a compe tition substitution reaction. This observation is also in agreement with the G value 0.4 that was found when solutions were irradiated in the absence, as well as in the presence of magnesium ions. The sugar ring region at 800 cm^{-1} under oxygen also suggests that magnesium is not bound to N7 of guanine from the intesity increase of 825 cm^{-1}, band due to C2'-endo/anti conformation [7].

References

1. I. Fridovich, *Acc. Chem. Res.* **5**, 321, (1972).
2. W. Bors, M. Savan, E. Lengfelders, R. Spottler and C. Michel, *Curr. Top. Radiat. Res. Q.* **9**, 247, (1974).
3. G.E. Adams, Radiation Protection and Sensitization, Eds., H. Moroson and M.L. Quintiliani, London, Taylor and Francis, pp 1-14, (1970).
4. T. Theophanides, *Int. J. Quantum Chem.* **26**, 933, (1984).
5. J. Anastassopoulou, *Magnesium Res.* **3**, 1, (1989).
6. H.A. Tajmir-Riahi and T. Theophanides, *In. Chim. Acta* **80**, 223-230, (1983).
7. T. Theophanides, Spectroscopy of Inorganic Bioactivators, Theory and Applications-Chemistry, Physics, Biology and Medicine, Kluwer Academic Publishers, Dordrecht, 265-272, (1988).
8. G.E. Adams and R.L. Willson, *Transactions of the Faraday Society* **65**, 2981-87, (1969).
9. C. von Sonntag and H.-P. Schuchmann, *Int. J. Radiat. Biol.* 49, 1-34, (1986).

Interaction of Alkaline and Alkaline-Earth Metal Ions with Nucleosides

J. BARIYANGA[1], J. ANASTASSOPOULOU and T. THEOPHANIDES
National Technical University of Athens, Chemical Engineering Dept., Radiation Chemistry-Biospectroscopy, Zografou Campus, Zografou 15780, Athens, Greece.
[1]National University of Rwanda, Department of Chemistry, Central Africa.

1. Introduction

It is well known that metal ions play an important role in biological systems, and more particularly in the human body. No one can deny the importance of Na^+ and K^+ ions as neurotransmitters and Mg^{2+} and Ca^{2+} in the bones and cells. The metal cations are found in the body to be complexed with proteins, nucleic acids, enzymes and other biological molecules. In this short review we would like to determine which are the usual sites of interaction of these cations and how they perturb the conformations of the free nucleoside, cytidine and its derivatives when they interact with metal ions. The nucleosides of course do not have the phosphate group with which these hard acids would preferentially interact. There are numerous studies [1-3] of complexes formed in solution with alkaline and alkaline-earth cations and the nucleosides, but not as isolated compounds in the solid state and as fully characterized coordination complexes. There is an IR and NMR study of the complex, $CaCl_2.Cyt.H_2O$ [4] and its crystal structure was determined by X-ray diffraction later [5]. We have studied some metal-nucleoside complexes by NMR spectroscopy and Mass spectrometry. The nucleosides which are the building blocks of DNA are made from one of the four bases, Adenine, Guanine (purines) or Thymine (Uracil) and Cytosine (pyrimidines) with ribose or deoxyribose shown in Fig. 1. The main conformations that are present in nucleosides are [6]:
1) rotation of the base around the glycosidic bond C1'-N9 <u>syn</u>, <u>anti</u> conformations:
<u>syn</u> : when x=0+90 or x>270° and x>90°
<u>anti</u> : when x=180+90° and
2) the conformations C2'-endo and C3'-endo (Fig. 2).
In Fig. 2 the C2' or C3' and C5' carbons are at the same part of the plane passing through the atoms C1'-O1'-C4', respectively. The metal cations, Na^+, K^+ , Mg^{2+}, Ca^{2+}, Ba^{2+} are the elements of the first and second group of the Periodic Table and they are hard acids, i.e., they interact preferentially with hard bases, such as oxygen bases.

2. Metal complexes.

As we have mentioned above, metal ions are present in our body and may interact with nucleic acids. In Table 1 are given the concentrations of metal ions in plasma, the total

N. Russo et al. (eds.), Properties and Chemistry of Biomolecular Systems, 37–47.
© 1994 *Kluwer Academic Publishers. Printed in the Netherlands.*

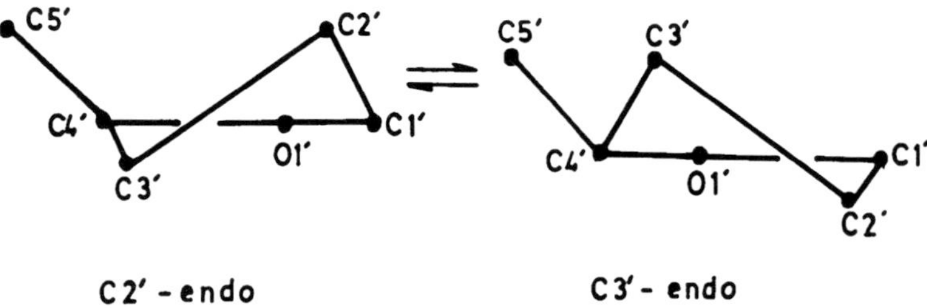

Figure 1. Structures, Atom numbering and nomenclature of the nucleic acid bases and sugars.

Figure 2. The C2'-endo (B-DNA) and C3'-endo (A-DNA) conformations of sugar.

Table 1. Metal ion concentrations in the human body, daily amounts, and sources where to find them [6-8].

Cations	Concentration in plasma (mmole/1)	Daily needs (g)	Total quantity (g/10kg)	Sources
Na^+	142	5	100	Table salt
K^+	4	3	140	Plants
Ca^{2+}	3	0.8	1100	Milk
Mg^{2+}	1	0.3	35	Cereals
Fe^{2+}	0.018	0.01	4	Liver

amount in the body, the daily requirements of an adult and where they come from [6-8]. It is shown from Table 1 that Na^+ and K^+ ions are the most abundant in the blood, whereas Ca^{2+} is the most abundant in the body.

It is well known [9] that the transition metal ions are preferentially bound to the base sites, in particular the nitrogen sites and perhaps the phosphates. From X-ray measurements these base sites are the N3 of the pyrimidine base and the N7 of the purine base (guanine). The crystal structure of $Ca(Cyt)_2 Cl_2 . 2H_2O$ has been solved by X-ray crystallography and the coordination of the Ca^{2+} ion was found to be with N3 and O2 [5]. Studies in solution [10] on the stability constants of the alkaline-earth metals (Mg^{2+}, Ca^{2+}, Ba^{2+}) with guanosine show that the metals are bound to N7 (from stability constant calculations).

3. NMR spectra

Furthermore, the sites N1H and NH_2 have also been proposed from NMR studies to react with alkaline-earth cations [11]. More NMR studies [12] of the salts, LiCl, NaCl, KCl, $MgCl_2$, $CaCl_2$ and $BaCl_2$ indicate interaction with guanosine, but did not give sites of interaction. It was also found [13,14] that Li^+ was linked to N7, while the ion Cl^- formed hydrogen-bonding which with N1H and NH_2 of guanosine and the order of complexation which was proposed [15] gives cytidine the strongest complexing molecule to an alkaline -earth metal ion: cytidine>> guanosine>adenosine>uridine [14]. Barium and calcium complexes with cytidine have been synthesized and characterized [16]. The NMR spectra of the complexes (Fig. 3) showed significant downfield shifts of H5, H6 and NH_2 protons of cytidine (Fig. 4). In Table 2 are given the NMR data.

The N3 site of cytidine is assigned as the site of coordination with the above metal ions. The structures of the complexes have been studied by Fast Atom Bombardment (FAB) spectrometry. Detailed interpretation of the spectra suggests that these hard cations are linked to cytidine.

An X-ray crystal determination of the complexes could be necessary to confirm these findings. NMR studies on Me-1-Cyd, and its complexes with Mg^{2+} Ba^{2+} and Ca^{2+} are shown in Fig. 5a. The chemical shifts are given in Table 3. The corresponding compounds with Cyd and its complexes are shown in Fig. 5b and Table 4.

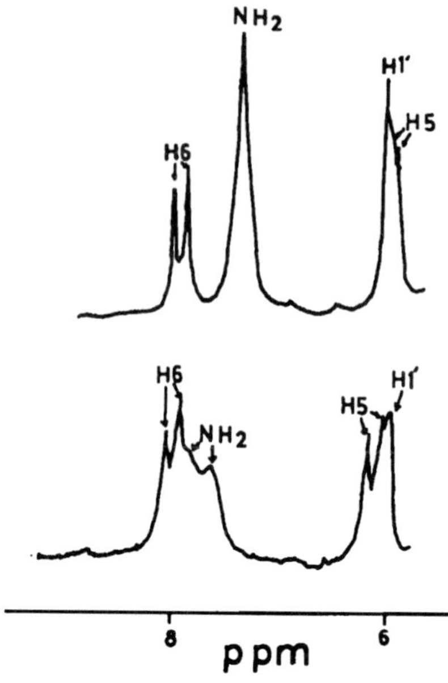

Figure 3. Chemical shifts of (a) the cytidine protons and (b) $Ba(Cyd)Cl_2.2H_2O$.

Table 2. Chemical shifts of the protons H5, H6 and NH_2 of cytidine and its complexes in DMSO in ppm.

Compound	H5	H6	NH_2
Cyd	5.76	7.88	7.27
$Ba(Cyd)Cl_2.2H_2O$	5.98	7.97	7.95
			7.56
$[Ca(Cyd)_2 (H_2O)_4]Cl_2.4H_2O$	5.94	7.93	7.57
			7.28
			7.79
$[Mg(Cyd)_2 (H_2O)_4]Cl_2.2H_2O$	5.98	7.97	7.30

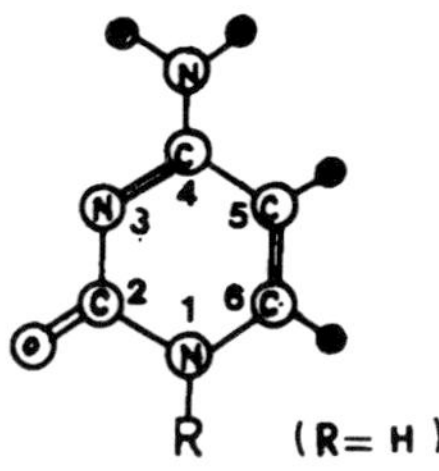

Figure 4. Cytidine molecule.

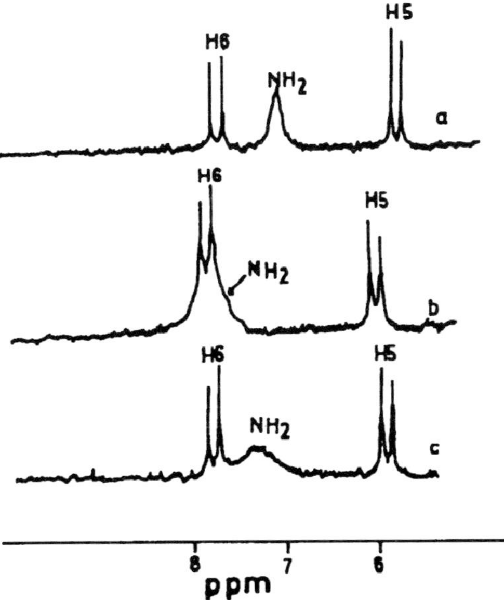

Figure 5a. Chemical shifts of protons H6, NH_2 and H5 of (a) Me-1 Cyt, (b) Ba(Me-1 Cyt)Cl_2 . H_2O.1/2 alcohol and (c) [Mg(Me-1Cyt)(H_2O)$_4$]$_2$ Cl_2 . $4H_2O$.

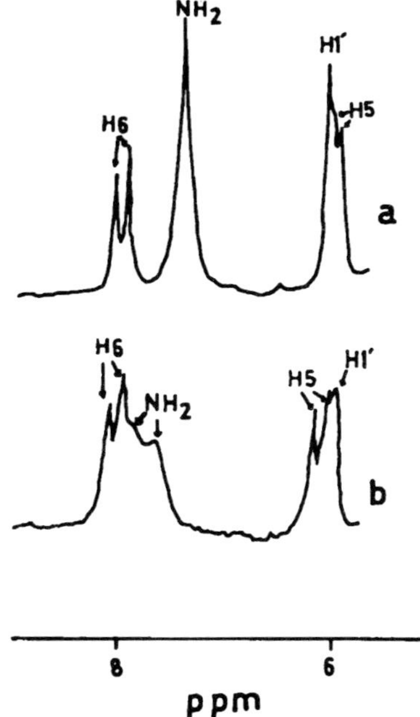

Figure 5b. Chemical shifts of the protons of: (a) cytidine (Cyd) and (b) Ba(Cyd)Cl_2.$2H_2O$.

The protons of NH_2 of the complexes are shifted downfield with respect to methyl-1 cytidine. This shift could be explained by hydrogen bonding $NH^+...Cl^-$ and coordination of the metal ion to the pyrimidine ring. The H5 protons are also shifted downfield in all the complexes. In these complexes the Ca^{2+} ion may be linked to both N3 and O_2 of cytidine (Fig. 6) [15,19].

In comparison, the complex [Pt(Me-1 Cyt)(Thymine-H)]ClO_4, shows the Pt^{2+} to be coordinated to N3. In this complex the chemical shift of NH_2 and H5 is 1.5 ppm and 0.1 ppm, respectively [17] compared to the free ligand. The two NH_2 protons are not equivalent and are separated by 0.25 ppm. In conclusion, the sites of coordination of the ions Mg^{2+}, Ba^{2+} and Ca^{2+} are suggested to be N3 and O_2 for Me-1Cyt.

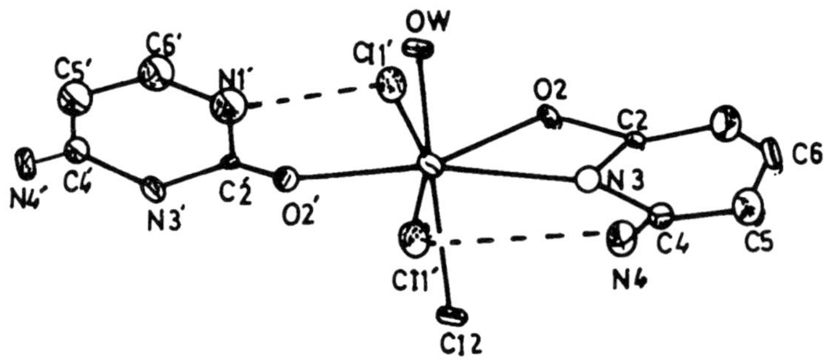

Figure 6. Crystalline structure of Ca(Cytosine)Cl_2 . $2H_2O$.

Table 3. Chemical shifts of the protons of methyl-1 cytosine and its complexes in DMSO (ppm), a=alcohol.

Compounds	H5	H6	NH$_2$
Me-1 Cyt	5.73	7.66	7.03
Ba(Me-1 Cyt)Cl$_2$. H$_2$O . 1/2a	6.00	7.81	7.75
[Ca(Me-1 Cyt) (H$_2$O)$_4$]Cl$_2$. 1/2a	5.86	7.72	7.28
[Mg(Me-1 Cyt)$_2$ (H$_2$O)$_2$]Br$_2$. 2H$_2$O	5.80	7.70	7.10
[Mg(Me-1 Cyt) (H$_2$O)$_4$]Cl$_2$. 4H$_2$O	5.86	7.71	7.25

Table 4. Chemical shifts of the protons of methyl-1 cytidine and its complexes in DMSO (ppm), a=alcohol

Compounds	H5	H6	NH_2
Cyd	5.76	7.88	7.03
			7.95
$Ba(Cyd)Cl_2 . 2H_2O$	5.98	7.97	7.56
			7.57
$[Ca(Cyd)_2 (H_2O)_4]Cl_2 . 4H_2O$	5.94	7.93	7.28
			7.79
$[Mg(Cyd)_2 (H_2O)_4]Cl_2 . 2H_2O$	5.98	7.97	7.30

4. Mass spectra

The above results have been reached also by Mass Spectrometry. The mass spectra of nucleosides have been studied previously [18,19]. The fragmentation which was proposed (Fig. 7) predicts first the loss of H2CO (m=30) and breaking of the bond C3'-C2' which gives a fragment of m=89. There is a subsequent reorganization which gives the species B+30 which again loses an H_2CO (m=30) and gives the base B.

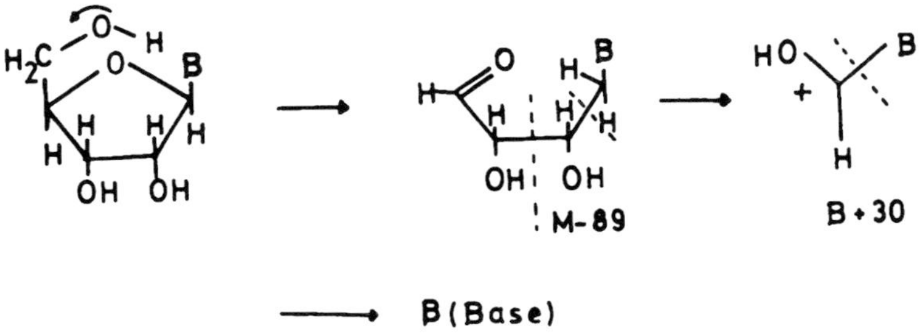

Figure 7. Fragmentation of nucleosides [19].

In negative mode the important masses in the spectrum are $[B]^-$ and $[M-H]^-$. In positive mode there are the peaks $[B+H]^+$ or $[B+H_2]^+$ and $[M+H]^+$. The mass spectra of cytosine and cytidine are known [19,21]. For cytidine in negative mode the important peaks are $[M-H]^-$ (m=242) and $[B]^-$ (m=110). The complexes of Mg^{2+}, Ba^{2+} and Ca^{2+} with Me-1 Cyt and Cyd have been obtained and interpreted. The mass spectra of the last two, i.e. $Ba(Cyd)Cl_2.2H_2O$ and $[Ca(Cyd)_2 (H_2O)_4]Cl_2$ $4H_2O$ are shown in Figs. 8 and 9, respectively. The data are given in Tables 5 and 6.

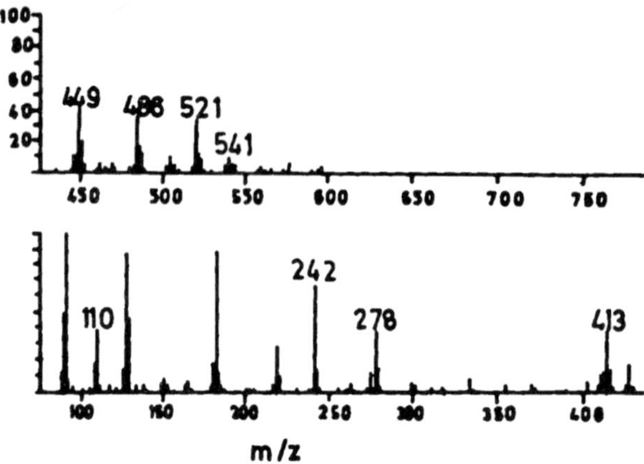

Figure 8. Mass spectra of Ba(Cyd)Cl$_2$ .2H$_2$O in negative mode.

Table 5. Masses (m/e) and relative intensities of fragments of the complex Ba(Cyd)Cl$_2$.2H$_2$O.

m/e	intensities %	species
110	61.5	[B]$^-$
242	100	[Cyd-H]$^-$
278	61.5	[Cyd Cl]$^-$
413	55.4	[Ba(Cyd)Cl-3H]$^-$
449	6.8	[Ba(Cyd)Cl 2H$_2$O-3H]$^-$
485	15.5	[(Cyd)$_2$-H]$^-$
505	1.8	[Ba(Cyd)Cl Gly-3H]$^-$
521	4.9	[(Cyd)$_2$ 2H$_2$O-H]$^-$
541	1.5	[Ba(Cyd)Cl 2H$_2$O Gly-3H]$^-$

m/e=mass, B=Base, Cyd-H deprotonated cytidine,
Gly=glycerol.

The mass spectra of Mg^{2+} with the above ligands, were not reproducible. It is concluded from these data that the Ba^{2+} ion is bound directly to cytidine, due to the presence of fragments [Ba(Cyd)Cl-3H]$^-$ (m=413) [Ba(Cyd)Cl.2H$_2$O-3H]$^-$ (m=449), [Ba(Cyd)Cl(Gly)-3H]$^-$ (m=505) and [Ba(Cyd)Cl 2H$_2$OGly-3H]$^-$ (m=541). All these fragments contain Cl, which means that there is a direct Ba-Cl bond [16] under these conditions. The presence of Cl$^-$ in the fragments has been confirmed also from the isotopic ratio 35/37 ^{35}Cl/^{37}Cl [22]. In the case of the calcium complex (Fig. 9, Table 6) the presence of fragments [Ca(Cyd) 2H$_2$O Gly-3H]$^-$, Ca(Cyd) 3H$_2$O Gly-3H]$^-$, and [Ca(Cyd)$_2$Cl-3H]$^-$ is indicated that the calcium ion is bound to cytidine. From the above data, it is concluded that the alkaline- earth elements are bound to the nucleoside cytidine. A proposed structure for the Ca^{2+} ions is shown in Fig. 10, where the metal ion is linked

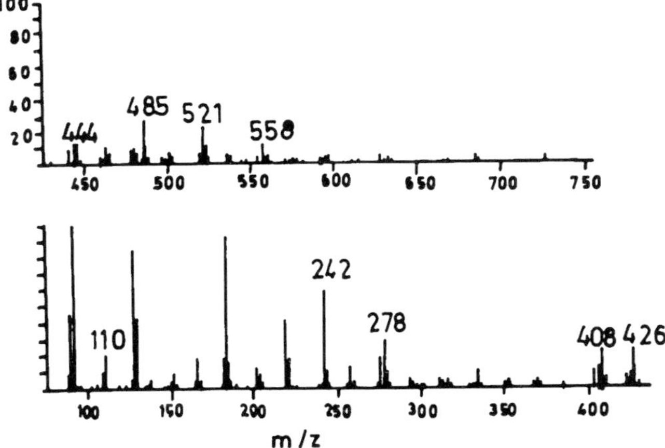

Figure 9. Mass spectra of $[Ca(Cyd)_2(H_2O)_4]Cl_2$ in negative mode.

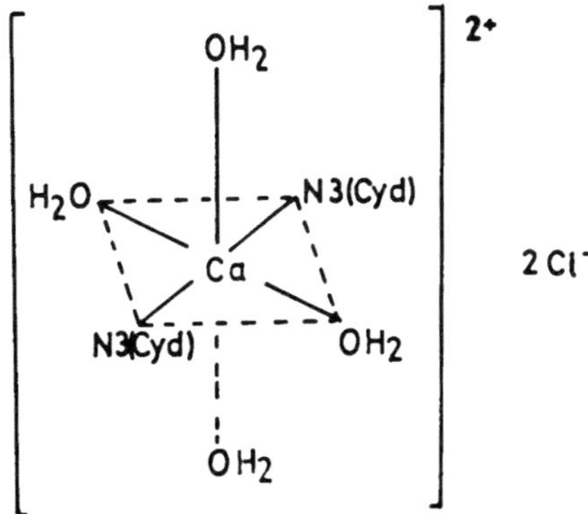

Figure 10. Proposed structure of the complex $[Ca(Cyd)_2(H_2O)_4]Cl_2$.

to two cytidines through the N3 site and four water molecules, which was suggested from the crystaline structure of $Ca(Cyt)Cl_2.2H_2O$, where the X-ray structure has been solved. However, the present complex, $[Ca(Cyd)_2(H_2O)_4]Cl_2$ with two Cyd per Ca^{2+} may be different and it is suggested also from IR data [16] that the bonding is only through N3 [16], since the carbonyl frequency has not been altered upon complexation of the nucleoside compared to complex $Ca(Cyt) Cl_2.H_2O$, where the carbonyl frequency is changed and the X-ray structure is known [5].

Table 6. Masses (m/e) and relative intensities of fragments of the complex
$[Ca(Cyd)_2(H_2O)_4]Cl_2$.

m/e	intensities %	species
110	33	$[B]^-$
242	100	$[Cyd-H]^-$
278	50	$[Cyd\ Cl]^-$
408	33	$[Ca(Cyd)\ 2H_2O\ Gly-3H]^-$
426	33	$[Ca(Cyd)\ 3H_2O\ Gly-3H]^-$
444	2.5	$[Ca(Cyd)Cl\ 2H_2O\ Gly-2H]^-$
485	5	$[(Cyd)_2-H]^-$
521	4	$[(Cyd)_2\ Cl]^-$
558	2	$[Ca(Cyd)_2-3H]^-$

References

1. H.A. Tajmir-Riahi and T. Theophanides, *Inorg. Chim. Acta* **80**, 233, (1983).
2. R. Sridharan and C.R. Krishna Moorthy, *J. Coord. Chem.* **12**, 231, (1983).
3. P.R. Reddy and V.B.M. Rao, *Polyhedron* **4**, 1603, (1985).
4. S. Shirotake, *Chem. Pharm. Bull.* **28**, 956, (1980).
5. K. Ogawa, M. Kumihashi and K.I. Tomita, *Acta Cryst. B* **36**, 1793, (1980).
6. W. Saenger, "Principles of nucleic acid Structure", Springer- Verlag, Berlin, pp. 202 (1984).
7. T. Theophanides, *Int. J.Quant. Chem.* **26**, 933, (1984).
8. T. Theophanides, *Can. J. Spectrosc.* **26**, 165, (1981).
9. T. Theophanides and J. Anastassopoulou, in "Metal-Based Antitumor-Drugs", Ed., M.F. Gielen, Freund Publishing House LTD, London, p.151, (1988).
10. H. Lonnberg and P. Vihanto, *Inorg. Chim. Acta.* **56**, 157, (1982).
11. S. Shimokawa, H. Fukui, J. Sohma and K. Hotta, *J.Amer. Chem. Soc.* **95**, 1777, (1973).
12. T. Yokono, S. Shimokawa and J. Sohma, *J. Am. Chem. Soc.* **97**, 3827, (1975).
13. L.G. Marzilli, B. De Castro, J.P. Caradonna, R.C. Stewart and C.P. van Vuuren, *J. Am. Chem. Soc.* **102**, 916, (1980).
14. L.G. Marzilli, R.C. Stewart, C.P. van Vuuren, B. De Castro and J.P. Caradonna, *J. Am. Chem. Soc.* **100**, 3967, (1978).
15. A.C. Plaush and R.R. Shapr, *J. Am. Chem. Soc.* **98**, 7973, (1976).
16. J. Bariyanga and T. Theophanides, Greek-Italian Meeting, Cetraro, 5-9 October, (1992), Italy.
17. R. Faggiani, B. Lippert, C.J.L. Lock and R. Pfab, *Inorg. Chem.* **20**, 2381, (1981).
18. K. Biemann and J.A. McCloskey, *J. Am. Chem. Soc.* **84**, 2005, (1962).
19. F.W. Crow, K.B. Tomer, M.L. Gross, J.A. McCloskey and D.E. Bergstrom, *Anal. Biochem.* **139**, 243, (1984).

20. D.L. Smith, K.H. Schram and J.A. McCloskey, *Biomed. Mass Spectr*. **10**, 269, (1983).
21. J.G. Liehr, D.L. Minden, S.T. Hattox and J.A. McCloskey, *Biomed. Mass Spectr*. **1**, 281, (1974).
22. C.K. Mann, T.J. Vickers and W.M. Gulick, "Instrumental Analysis", Harper & Ros, Publishers, New York, p. 614, (1974).

Thermal Behaviour of Three Ribonucleases

G. BARONE, P. DEL VECCHIO, D. FESSAS, C. GIANCOLA, G. GRAZIANO and
A. RICCIO
*Department of Chemistry, University "Federico II" via Mezzocannone, 4 - 80134 -
Naples, Italy*

1. Introduction

The Ribonucleases are a well known family of proteins, whose main function is the
hydrolysis of the RNAs. Their ubiquitary presence allows to re-utilize the nucleotides
from the mRNA and tRNA, after the development of their respective role during the
protein synthesis. In this manner both enormous sparing of genetic material and high
efficiency of polypeptide synthesis are accomplished.
The most studied and known, among these enzymes, is the Ribonuclease A (RNase A)
usually obtained from bovine pancreas. Its primary sequence [1] and its three-
dimensional structure (determined by X-rays diffractometry on the crystal [2-4] and
refined in solution by NMR [5]) are well known in detail from some time. In Figure 1 is
reported a scheme of the primary sequence, that emphasizes the framework of the
secondary structure and the disulphide bridge network. The Anfinsen experiments [6,7]
on the reversible reconstitution of the tertiary structure (after denaturation and reduction
of disulphide bridges) are a historical crucial demonstration that the physiological
conformation of the proteins corresponds to a minimum of the Gibbs energy function.
It is necessary to accomplish particular experiments to obtain *in vitro* some unspecific
and inactive structures with the scrambled disulphide bridges. *In vivo* however the
reconstitution of the bridges is catalyzed by an enzyme, the Protein Disulphide
Isomerase (PDI) [8,9]. It is possible to restore the biological activity of the Ribonuclease
A, after the hydrolytic selective cleavage, produced by Subtilisin, of the "S-peptide"
(the first 20 N-terminal residues) and stoichiometric recombination of it with the residual
"S-protein". The obtained complex, called RNAase S, is thermally less stable [10-12]
than RNAase A, but has practically the same activity [13].
From some years a homodimeric ribonuclease has been studied formed by two identical
chains, covalently bonded by two extra disulphide bridges at positions 31 and 32 [14-
21]. This protein (RNAase BS) is obtained from the bull seminal vesicles or directly
from the seminal plasma. In the extracellular environment the RNAase BS seems to
display antispermatogenic [22-24], immunodepressive [25-28] and antitumoral action
[29-31]. The primary sequence of half of the molecule is shown in Figure 2. Its
homology with the monomeric RNAase A is found on the maintainance of 80% of the
amino acid residues (101 on 124) in the RNAase BS.

49

N. Russo et al. (eds.), Properties and Chemistry of Biomolecular Systems, 49–65.
© *1994 Kluwer Academic Publishers. Printed in the Netherlands.*

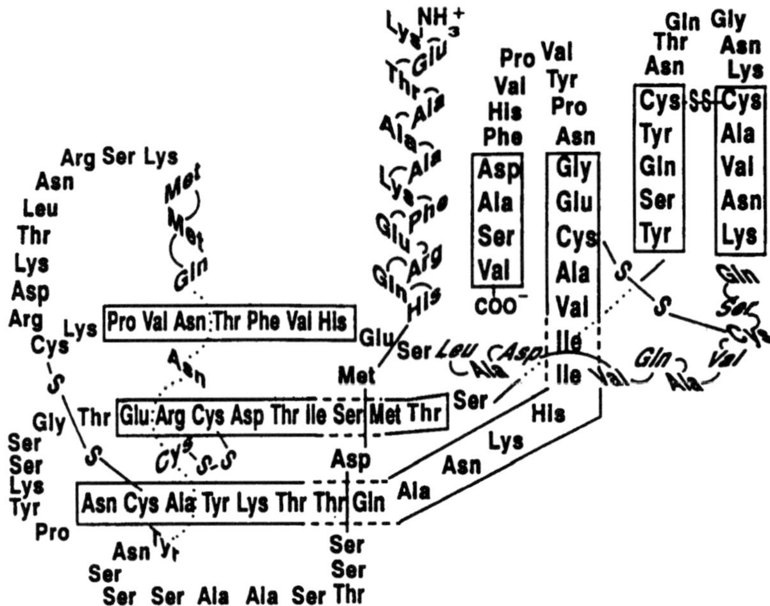

Figure 1. Amino acid sequence of pancreatic bovine Ribonuclease A. The secondary structures are emphasized by means of rectangles (β-strands), in case bent, or stylized spirals in italics (α-helices) [1].

The three-dimensional organization of each of the two domains is, on the overall, identical to that of the pancreatic protein, without any appreciable distorsion promoted by the presence of the second chain. The dimeric structure of the RNAase BS seems indispensable for the expression of the immunorepressive and antitumoral activity, probably because it is recognized by membrane receptors [32]. Monomeric derivatives of RNAase BS are inactive, except for what concerns the biocatalysis; on the contrary, artificial dimers of RNAase A acquire immunodepressive and antitumoral activity [33,34]. The X-ray diffraction studies give strong evidence that, in the crystal, parts of the two S-peptides (the 1-16 N-terminal segments) are exchanged between the two protomers. The residues 17-20 act as hinges for the rotation of each segment 1-16, so that in this way they juxtapose themselves to the channel, that in RNAase A is occupied by the S-peptide of its own subunit [18-20]. This is what probably occurs in the artificial dimer of RNAase A, obtained by lyophilization from 50-50% mixture of acetic acid and water [35-37]. The proofs of this exceptional conformational situation have been indirectly obtained by the brilliant demonstration that 50% active dimer can be obtained from a stoichiometric mixture of two RNAase A monomers, one inactivated by carboxymethylation of His 119 and the other of His 12 [36,37]. For both RNAase BS and dimeric RNAase A this characteristic tail-exchange has the consequence that three residues involved in the active site lie on the S-peptide of one subunit (Lys 7, Glu 11

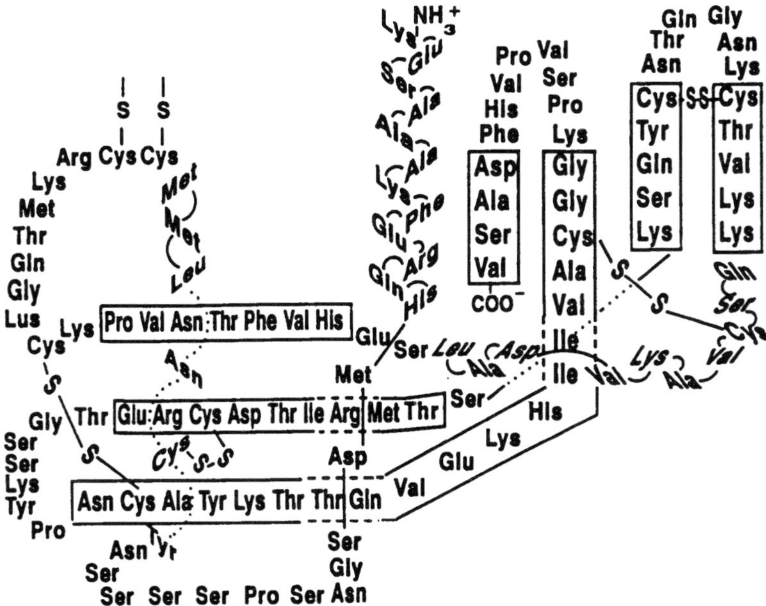

Figure 2. Amino acid sequence of one of the two identical subunits of bovine seminal Ribonuclease BS. The secondary structure is emphasized as in Figure 1.

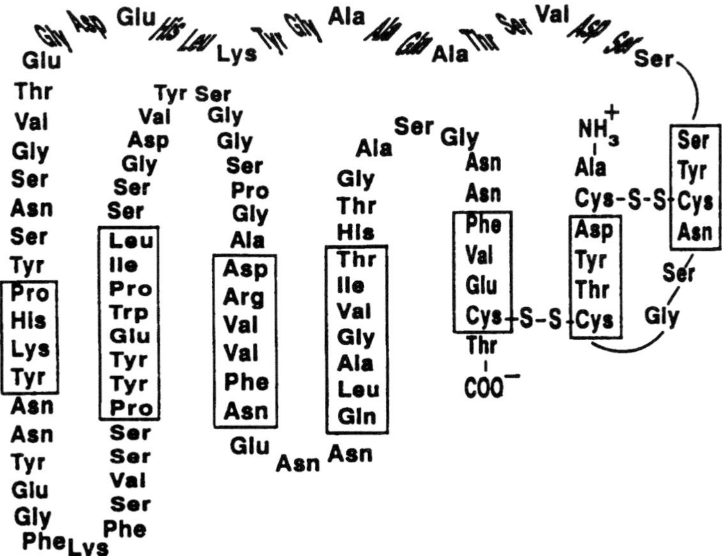

Figure 3. Amino acid sequence of fungine Ribonuclease T1. The secondary structure is emphasized as in Figure 1.

and His 12) and three other on the core of the other subunit (Lys 41, Thr 45 and His 119). A more recent work [38] has shown that RNAase BS is probably present in solution as a 2:1 mixture of one form with the tail exchanged and the other with the S-peptide tilted on its own subunit. It must be remembered that in aqueous solutions the RNAase A dimer is definitely unstable.

The RNAase BS is biosynthetized in an amidated form at Asn 67, as for RNAase A. A post-traductional event transforms Asn 67 in an aspartate residue [36,39-41]. The RNAase A can undergo this reaction *in vitro* [42,43]. This reaction seems favoured in these cases by the next Glu 68. Generally the exposition of Asn is a pre-requisite for the deamidation, that for many proteins can be a source of dramatic physiological damages [44-45-49].

Table 1. Amino acid composition of RNAase A, BS and T1 and total number of alkylic hydrogens in RNAase A and T1.

Residue	RNAase A	RNAase BS	RNAase T1	$n_{CH}(A)$	$n_{CH}(T1)$
Glycine	3	6	12	6	24
Alanine	12	8	7	48	28
Valine	9	10	8	72	64
Leucine	2	2	3	20	30
Isoleucine	3	3	2	30	20
Proline	4	5	4	28	28
Asparagine	9	6	9	27	27
Glutamine	7	7	2	35	10
Tryptophan	-	-	1	-	8
Tyrosine	6	4	9	42	63
Phenylalanine	3	3	4	24	32
Cysteine	8	10	4	24	12
Methionine	4	5	-	32	-
Serine	15	15	15	45	45
Threonine	10	9	6	50	30
Lysine	10	14	2	90	18
Arginine	4	4	1	28	7
Histidine	4	4	3	20	15
Aspartate	6	5	6	18	18
Glutamate	5	4	6	25	30

Recently some attention has been paid to a fungine Ribonuclease (RNAase T1), not homologous to the preceding two [50-57]. It has been extracted from cultures of the microorganism *Aspergillus Oryzae* [50] and more recently prepared by gene expression in *Escherichia Coli* [58]. The natural protein is a mixture of two isobaric isoforms: the RNAase T1 Lys-25 (or K-25) and the RNAase T1 Gln-25 (or Q-25). The primary sequence of Lys-25 isoform is given in Figure 3, where the secondary structure elements are also sketched. Among the 104 residue only two Lys, one Arg and three His are mantained, against a total of 18 and 22 for RNAase A and BS respectively (Table 1). However in the catalytic site are still involved two histidine residues (His 40 and His 92) together with an acidic residue (Glu 58) and probably Asn 99. This site is specific for the recognition of guanine and in part of adenine, while the RNAase A site is specific for uracil and cytidine. Only two disulphide bridges are present in RNAase T1, linking Cys 2 - Cys 10 and Cys 6 - Cys 103 respectively. However, the stability of the core of the protein is strengthened, as for other RNAases, by a thick network of H-bonds concerning the long α-helix, the short strands in the bent β-sheets and others involving bends, loops, less regular conformations and the same secondary structures [57,59,60]. In the course of a program that we are developing on the thermal stability of enzymes and other biological macromolecules, we like to present and discuss in this paper a series of results concerning the three Ribonucleases considered before. Two are the aims: to test the potentiality of the differential scanning microcalorimetry, and to show the very different denaturation mechanisms in action in a correlated family of proteins. The last aspect is in turn challenging, because it requires the development of different models, capable to interpret and to rationalize the experimental data, at this level of knowledge, in expectation of a general theory for the folding and unfolding.

2. Experimental

2.1. MICROCALORIMETRY

Differential scanning calorimetry offers some advantages with respect to other physico-chemical techniques. First of all it gives a direct measurement (per each scan) of the thermodynamic quantity $\Delta_{trans}H$ (in the present case the transition is a denaturation and will be indicated in the following with the subscript d). There is no need to introduce any model or use temperature derivates (van 't Hoff isocores). Besides, dsc allows to determine a lot of other thermodynamic and analytical parameters: the temperature of the maximum thermal effect, namely T_d, the variation of the molar heat capacity $\Delta_d C_p = C_{p,D} - C_{p,N}$ the reversibility and the analytical degree of progress of the transformation at increasing the temperature, $\theta(T)$. From the last determination it is possible to obtain (in the case of a one-step process) an analytical or a numerical expression of the apparent equilibrium constant as a function of the temperature:

$$K'(T) = \theta(T)/[1 - \theta(T)] \tag{1}$$

It is easy to control if the assumption is correct, calculating the molar denaturation enthalpy from the van 't Hoff equation:

$$[d\ln K'(T)/dT]_p = \Delta_d H^\circ_{v.H.} / RT^2 \tag{2}$$

Often a simple expression, calculated at $\theta = 0.5$ and $T = T_d$ is used [61]:

$$\Delta_d H^\circ_{v.H.} = 4RT_d^2 \, \Delta_d C_p(T_d)/\Delta_d H^\circ_{exp} \tag{3}$$

If $\Delta_dH^\circ_{v.H.}(T_d)$ coincides (within $\pm$ 10%) with the $\Delta_dH^\circ_{exp}$ directly measured by the integration of calorimetric peak area, it is possible to assume that the unfolding is a one - step process, at least as working hypothesis [61-63]. As a consequence, if the denaturation is reversible, it is possible to obtain standard thermodynamic quantities, as a function of the temperature, from classical relationships even out of the limits of the denaturation temperature range. First of all can be obtained the standard Gibbs energy per mole:

$$\Delta_dG^\circ = -RT\,\ln K'$$
(4)

the apex ' indicating that this quantity is approximated for the lack of activity coefficients. Note that Δ_dG° is null only for $T = T_d$, while Δ_dG is identically zero in the overall equilibrium range of denaturation. The value of $\Delta_dC_p(T_d)$ or, if determined, its fuctional dependence on temperature, $\Delta_dC_p(T)$, allows to calculate the denaturation standard enthalpies and entropies at every temperature:

$$\Delta_dH^\circ(T) = \Delta_dH^\circ(T_d) - \int_T^{T_d} \Delta_dC_p^\circ(T)\,dT$$
(5)

$$\Delta_dS^\circ(T) = \Delta_dS^\circ(T_d) - \int_T^{T_d} [\Delta_dC_p(T)/T]\,dT$$
(6)

and finally:

$$\Delta_dG^\circ(T) = \Delta_dH^\circ(T) - T\,\Delta_dS^\circ(T)$$
(7)

The quantity $\Delta_dG^\circ(T)$ gives a measure of the thermal stability of the protein as a function of the temperature. It is $\Delta_dG^\circ = 0$ at T_d and, at decreasing temperature, it passes through a maximum for $T_s \ll T_d$ (usually lower than the room temperature) where $\Delta_dS^\circ = 0$. In principle is another denaturation temperature forecasted, the "cold denaturation" temperature $T'_d \ll T_s$ at which $\Delta_dG^\circ = 0$ [63,64]. Only in few cases, for practical reasons, it was possible to determine experimentally the value of T_d' [65-67]. The Eqs.(1-3) give necessary conditions for the applicability of one-step model. These and other tests will be discussed in the following paper on this book [68]. If these tests give negative responses or if the denaturation is irreversible, the Eqs. (5) and (6) cannot be used. The dsc results, however, give an useful basis to identify possible realistic models for the unfolding, which will be discussed in part in the following, and in part, in the second paper [68]. A practical advantage of the dsc, as of other calorimetric methods, is that it can work even with media not optically transparent at whatever wavelength range (that is in the presence of gels, precipitates, insoluble matrices and fibers, membranes, whole cells, tissues etc.). The actual disadvantages of microcalorimetry are the unspecificity of the signal and the relative substance consume with respect to other techniques. The temperature scanning cannot be carried on at high rates, otherwise the signals are distorted, due to the relative high time constants of the instruments (≈ 100 s). On the other hand, a long time spent by the denatured sample at high temperatures can cause some damages due to secondary, undesired reactions involving exposed groups [45,49]. This is revealed coming back at room temperature and performing a new run.

2.2. DSC INSTRUMENTS

The limited availability of natural macromolecules requires calorimetric instruments with small cells and high performances. On the other hand, the necessity to operate with very low concentrations (to avoid aggregation and solubility problems) makes it practically impossible to use the sample cells so small (25-75 μl) as those for the dsc measurements on solids. These contrasting requirements lead to use ≈ 1 ml cells and 10^{-4}-10^{-5} M protein solutions and to design in a specific manner new geometries to surround, as completely as possible, the cells with a very high number of junctions (1000-10000) packed in a compact thermopile.
All modern instruments are based on the Calvet heat-flow principle and use twin cells to minimize the effects of undesired internal temperature gradients. The principle lies on the easiness with which the heat flows out of the limited place, where the thermal phenomenon occurs, to a heat sink (a large high capacity metal block in contact with a thermostat). If the heat flows fastly enough out of the measurement cell (or in them), so that the temperature change is sufficiently small ($\Delta T \approx 10^{-3}$ °C), the system can be considered isothermal, at least for what concerns the characterization of the determined thermodynamic parameters. At the same time, this ΔT is still sufficient to give an electromotive force at the terminals of the thermopile (inserted between the cell walls and the heat sink), that can be amplified by the modern electronics and measured with a very low error ($\approx 0.1\%$). The temperature program must give a linear increase (or decrease) of the temperature, with the time. This requires very good electronic control.
Only few kind of commercially dsc instruments have these characteristics, among them the DASM series of the Academy of Science of USSR, the Microcal (Amherst,MASS-USA) and Setaram (Lyon, France) apparatus. Among the home-made instruments it must be quoted the calorimeters built-up by Ackermann and coworkers at University of Regensburg (BRD) [69], by Sturtevant and coworkers at University of Yale (CONN-USA) [70,71] and by Freire and coworkers at University of Baltimore (MD-USA) [72,73]. General reviews on the dsc applied to biological systems and instrumentation appear time by time in the current literature and cannot here be quoted completely [74-78]. For all the experiments reported in the following, we made use of two Setaram instruments: micro DSC and micro DSC-II. Scanning rate used: 0.5 deg min^{-1} , slow enough to avoid signal distortion, but sufficiently fast to avoid long permanence of proteins at high temperature.

2.3. SUBSTANCES

The bovine pancreatic RNAase A was a SIGMA, high purity product, used without any special treatments. The RNAase T1, Lys 25 isoform, genetically engineered, was a kind gift of Dr. C. Nick Pace of the Texas A & M University. The purity of the isoform was tested by Prof. P.Pucci and Dr.M. Ruoppolo of the Department of Organic and Biological Chemistry of our University. The RNAase BS, from bull seminal plasma, was a kind gift of Profs. G.D'Alessio and R.Piccoli of the Department of Organic and Biological Chemistry of our University and was used without further treatment. The concentrations of the proteins (about 10^{-4}M) were determined spectrofotometrically, using ε_{278} = 9800, 18500, 12750 l cm^{-1} mol^{-1} for RNAase A [79], RNAase T1 [80,81] and RNAase BS [82] respectively. The buffers used in the increasing pH ranges were 0.1M acetate, 0.1M 2-[morpholino]ethansulphonate (MES), 0.1M N-[hydroxy-ethyl]piperazine-N'-[2-hydroxypropanesulphonate] (HEPPSO) and 0.1M 3-N-[morpholino]- propanesulphonate (MOPS), all as equimolecular mixtures of acids and sodium salts. Deionized water, twice distilled and filtered on Millipore membranes was used regularly for the preparation of all the solutions. The protein solutions were finally equilibrated by dialysis against repeatedly renewed buffer solutions. The pH was measured by means of a Schott pH-meter.

3. Results and Discussion

The thermal behaviour of RNAase A and T1 was studied in a wide range of pH and that of RNAase BS at two pH values to verify if the one-step transition model is a good representation of the unfolding process of these proteins. The dsc measurements for RNAase A were accomplished in the pH range 2.0÷8.0 (see Table 2a). It has been possible to demonstrate without doubt [81,83] that the denaturation process is well described by the one-step transition model (i.e. no intermediates are present between native and denatured forms) and is reversible. Exceptions occur at pH 2.0 where the ratio $\Delta_d H^\circ_{exp} / \Delta_d H^\circ_{v.H.}$ is less than unity, probably due to a partial swelling of the compact globular structure and at pH 8.0 where the process is irreversible due to side reactions in alkaline conditions at high temperatures. The values of thermodynamic parameters are in sufficient agreement with literature data which refer to a more narrow pH range [84-88]. The worst agreement is with the old data of Sturtevant and coworkers [89]. The behaviour of RNAase T1, investigated in the pH range 3.7÷8.0, seems more complicated because the ratio $\Delta_d H^\circ_{exp} / \Delta_d H^\circ_{v.H.}$ is always less than unity, as expected for a one-step process [81,83]. RNAase T1 has the highest thermal stability at pH 5.0, near its isoelectric point (see Table 2b). Moreover at pH 3.7 the dsc curves show a reproducible and irreversible exothermic phenomenon, due to the aggregation of denatured protein molecules. This peak is centered at about 75 °C [81]. Our dsc results are in good agreement with those recently reported both for Lys-25 and Gln-25 isoforms [80,90,91]. All these data do not support a simple one-step transition model.

For RNAase BS the dsc curves have a non symmetric shape and indeed the ratio $\Delta_d H^\circ_{exp} / \Delta_d H^\circ_{v.H.}$ results greater than unity at two investigated pHs. This and other indications [68,82] clearly exclude a one-step mechanism, and strongly suggest a two-step transition. As can be seen from the values reported in Table 2c the experimental transition enthalpy for RNAase BS is less than double of that for RNAase A at pH 5.0, probably because the two extra-disulphide bridges which link the subunits, allow a minor degree af freedom to the polypeptide chain in the unfolded state of RNAase BS. Further it must be noted that the denaturation process is irreversible at pH 7.0 and that the thermodynamic parameters strongly differ from those at pH 5.0. The reason is not yet understood.

Table 2. **a**) Thermodynamic parameters of thermal denaturation of RNAase A; **b**) Thermodynamic parameters of thermal denaturation of RNAase T1; **c**) Thermodynamics parameters of thermal denaturation of RNAase BS.

pH	T_d ℃	$\Delta_d H°$ kJ/mol	$\Delta_d C_p$ kJ/mol·K	R*
a)				
2.0	35.2	274	4.8	0.85
3.7	56.8	435	6.4	1.03
5.0	61.3	465	7.0	1.02
5.5	61.8	480	6.2	1.00
6.0	62.2	490	6.5	0.99
7.0	62.8	500	5.7	1.01
8.0	63.0	505	5.5	1.00
b)				
3.7	59.0	403	4.1	0.77
4.5	61.3	440	7.0	0.83
5.0	61.9	452	5.9	0.84
5.7	59.3	436	5.2	0.84
6.3	57.3	402	4.0	0.75
7.0	55.0	387	6.6	0.81
8.0	50.2	375	6.7	0.75
c)				
5.0	61.2	863	10.7	1.65
7.0	66.3	674	6.1	1.45

*R = $\Delta_d H°_{exp} / \Delta_d H°_{v.H.}$

3.1. THE "CRYSTAL MOLECULE" MODEL

The three ribonucleases here discussed, rich of disulphide bridges and intramolecular hydrogen bonds, are very good examples of the family of small globular proteins. This class of macromolecules assumes a very compact structure in solution, the so called native form, which is essential for their biological function. A variety of experimental investigations has evidentiated that the protein core resembles a solid rather than a liquid (i.e. the failure of the "liquid drop" model). Just the fact that a large number of proteins gives crystal structures is a clear indication of their solid-like nature. Further proteins show values of packing density and adiabatic compressibility in the range determined for organic solids [92-94]. Also it has been pointed out that a solid-like interior appears to have a selective advantage [95]. However, before these experimental evidences, Liquori advanced the hypotesis that a single protein molecule can be regarded as a "crystal molecule", whose cooperative unit is represented by the whole folded polypeptide chain [96,97]. This fascinating suggestion has been confirmed by dsc investigations, because the endothermic denaturation peak can be interpreted as the "fusion" of a crystal. In this manner the denaturation process is assimilated to a first order phase transition between two states of different symmetry. Furthermore the smallness of the cooperative unit in proteins justifies the calorimetric peak, broader than that of large real crystal.

The conclusion that the protein core is more accurately described as a solid is the starting point for trying to characterize the energetics of protein folding-unfolding transitions. Indeed it is possible to obtain detailed knowledge of hydrophobic interactions and weakly polar interactions from the thermodynamics of dissolution in water of solid uncharged model molecules of peptides [98-100]. On this line, Murphy and Gill developed a model that allows to evaluate the unfolding parameters of small globular proteins, assuming that the denaturation can be described as a transfer process of amino acid residues from the solid core to the water solvent and that the group additivity can be applied. The analogy between the two processes is strengthened by the circumstance that for these model molecules (i.e. solid cyclic dipeptides of the simple amino acids called also diketopiperazines) the crystals are very rich of *inter*molecular hydrogen bonds and van der Waals contacts between side chains as the interior of proteins. Using the group additivity principle the thermodynamic parameters characterizing the dissolution at infinite dilution of diketopiperazines can be decomposed in a sum of polar and apolar contributions, referred to the number of CONH groups and CH equivalent groups, respectively. So, the net heat capacity change associated to the unfolding process (assumed constant with respect to temperature), results composed of a positive term, $\Delta_d C_{p,ap.}$, due to water exposition of apolar CH groups (which is greater than it is usually believed), and a negative term $\Delta_d C_{p,pol.}$, due to water exposition of polar CONH groups. In order to simulate the thermodynamics of unfolding, other knowledges, are necessary, that allow to decompose the denaturation enthalpy and entropy of proteins in polar and apolar contributions per amino acid residue. The existence of convergence temperature for $\Delta_d H°$ and $\Delta_d S°$ per residue, reported by Privalov [61], and its rationalization in terms of group additivity scheme, allows to realize the requested task. In this manner Murphy and Gill were able to elaborate a series of relations for the calculation of thermodynamic potentials $\Delta_d H°$, $\Delta_d S°$ and $\Delta_d G°$ as function of temperature. Clearly information is necessary on the buried fraction of amino acid residues in the native state. The fraction of buried polar groups is assumed constant (f_{pol}=0.73) and the fraction of buried apolar groups is calculated according to the relationship f_{ap}=0.574+0.000702*N_{res}, following the indications of Chothia [92]. Thus, combining solid model peptide studies, the existence of convergence temperature for denaturation enthalpy and entropy per residue and "average structural properties" of small globular proteins, it is possible to predict their thermodynamic stability. However, this model does not consider variables such as pH or presence of denaturants or stabilizing agents.

For RNAase A we have determined $\Delta_d C_p$ and the $\Delta_d G°$ as function of temperature according to the Murphy and Gill model. The curve is reported in Figure 4 together with that obtained on the basis of our dsc data at pH 5.0 through Eqs.5-7. It is worthnoting that a very delicate point is the choice of a reliable experimental value of $\Delta_d C_p$, because the net heat capacity change markedly affects the temperature dependence of all the thermodynamic potentials. The adopted criterion is to calculate an average value of $\Delta_d C_p$ from the dependence of $\Delta_d H^\circ_{exp}$ on T_d. In this way the mean value of $\Delta_d C_p$ becomes independent on pH. The agreement between the Gibbs energy calculated from our experimental data and that calculated following the Murphy and Gill model is very satisfactory. Certainly some accidental compensations can facilitate the success of the approach, because it seems very difficult that a complex system such as a protein and a complex process as unfolding can be adequately represented by the simple sum of two contributions from solution process of small crystalline substances. It must be outlined that in the reversible denaturation the disulphide bridges are not disrupted, so their number does not influence the overall energetics of the process, that mainly consists in the swelling of the globular structure and wetting of interior groups. Finally, the important point is that from the recognition of the solid-like nature of small globular proteins it is possible to construct a model for the determination of their thermal stability, which gives reasonable results.

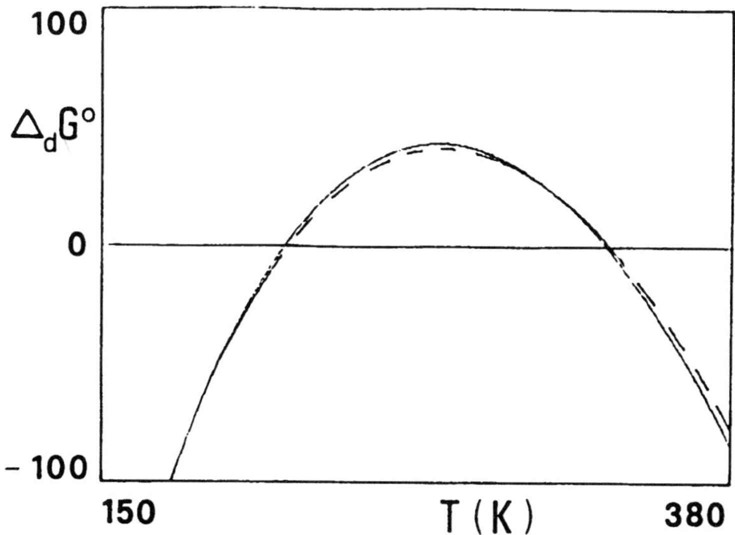

Figure 4. Temperature dependence of denaturation Gibbs energy function $\Delta_d G^\circ$(kJ/mol) for RNAase A at pH 5.0 determined according to the Murphy and Gill model (dashed curve) and calculated from experimental dsc data with Eqs.1-7 (continuous curve). Parameter values used: $\Delta_d C_{p,ap.}$ = 12.3 kJ/mol·K; $\Delta_d C_{p,pol.}$ = -5.6 kJ/mol·K; N_{res} = 124; T_d = 334.5 K; $\Delta_d H^\circ$ = 465 kJ/mol; $\Delta_d C_p$ = 7.0 kJ/mol·K.

When the same approach is applied in the case of RNAase T1, the results are not satisfactory. In the Figure 5 are reported the $\Delta_d G^\circ(T)$ calculated from our experimental data at pH=5.0 [81] (with the same criterion as before for the choice of Δ_dC_p) and the $\Delta_dG°$ evaluated according to Murphy and Gill: the discrepancy is clear. This however is not due to a failure of the model, but rather it confirms that RNAase T1 has a peculiar behaviour, as suggested by the fact that the ratio $\Delta_dH°_{exp}/\Delta_dH°_{v.H.}$ lies in the range 0.75 ÷ 0.85 at all the pHs investigated [81]. The discrepancy arises because the Δ_dC_p obtained experimentally, 6.54 kJ/mol·K, is larger than that calculated summing the polar and apolar contributions, 4.59 kJ/mol·K, as the number of apolar CH groups is very low (see Table 1).

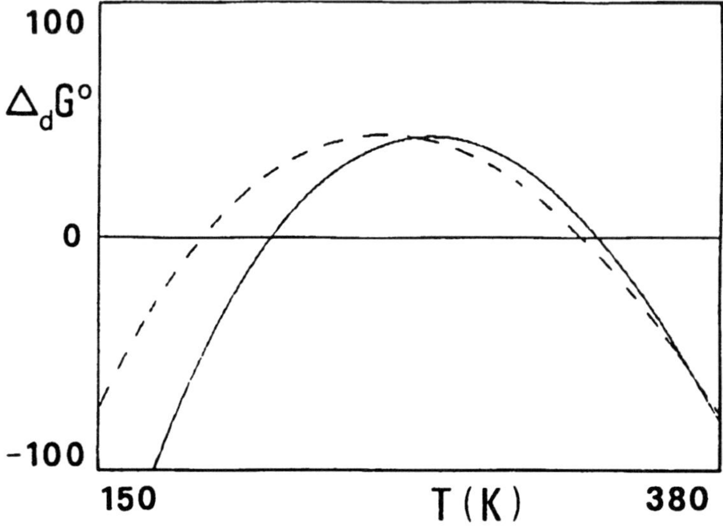

Figure 5. Temperature dependence of denaturation Gibbs energy function $\Delta_dG°$ (kJ/mol) for RNAase T1 at pH 5.0 determined according to the Murphy and Gill model (dashed curve) and calculated from experimental dsc data with Eqs.1-7 (continuous curve). Parameter values used: $\Delta_dC_{p,ap.}$ = 9.15 kJ/mol·K; $\Delta_dC_{p,pol.}$ = -4.56 kJ/mol·K; N_{res} = 104; T_d = 335.1 K; $\Delta_dH°$ = 452 kJ/mol; Δ_dC_p = 6.54 kJ/mol·K.

Further the experimental denaturation enthalpy is greater than that expected for a globular protein of 104 amino acid residues. A different way to calculate the net heat capacity change associated to the protein unfolding from evaluation of the buried accessible surface area, again gives a value of Δ_dC_p for RNAase T1 5.15 kJ/mol·K, lower than the experimental one [101]. Freire et al.[91] on the basis of their dsc measurements of RNAase T1, in presence of GuHCl, have ascribed to kinetic effects the cause of the discrepancy between the actual denaturation process and the one-step transition. But in absence of GuHCl they do not detect kinetic effects.

In heuristic way we have shown that it is possible to simulate this behaviour assuming that two forms are present (not transforming each in the other), that have in common the final denatured state [83]. The presence of a remarkable quantity of other isoform is excluded by amino acid analysis of our sample. The process is schematized as:

$$N \quad \Leftrightarrow \quad D$$

$$N^* \quad \Leftrightarrow \quad D$$

the transformation N $\Leftrightarrow$ N* being hindered or forbidden. It has been suggested that a compact, partially disordered conformation can be present (molten globule) [80,102], that contributes independently to the thermal phenomenon. The structure of RNAase T1 can be suitable for this particular situation (the two disulphide bridges link the extremes of the polypeptide chain). The transformation of the form N in the form N* could be hindered, for instance, by the formation (at the biosynthesis) of a percentage of isomers with the scrambled disulphide bridges. This possibility was observed during the *in vitro* renaturation of reduced RNAase T1 [103]. Another possibility is the very slow cis-trans isomerization of proline [56].

3.2 THE DIMERIC RNAase BS

We tried to apply the Murphy and Gill approach to the dimeric RNAase BS, assuming that the fraction of buried apolar and polar residues can be calculated as for RNAase A. Actually there are some apolar residues, at the interface between the two subunits of RNAase BS, that are not accounted for by the formulas used. However these residues, being very near to the two cysteines 31 and 32, are not exposed to the solvent in the denatured state. The calculated and experimental $\Delta_d H°$ at pH = 5.0 are in good agreement (832 kJ/mol and 863 kJ/mol , respectively). The calculated $\Delta_d C_p$, however, is much higher than the experimental one (14.4 against 10.7 kJ/mol·K). On the other hand, the Eqs. 5-7 cannot be used because all the obtained indications suggest the occurrence of a multistep transition [68,82,83]. For this reason the values determined according to the Murphy and Gill approach must be considered as mediated on the real process, that is not a one-step transition. In a parallel paper we have applied both the model of two independent transitions and that of two sequential transitions, obtaining comparable results with comparable confidence indexes. This is not an artifact and probably expresses the reality of the process. The two "deconvolution" analyses give the same distinct values of Td, (58.4 and 62.3 °C, respectively) for the two transitions. Similarly the determined transition enthalpies are 400±2 kJ/mol and 465±2 kJ/mol, respectively. Keeping in mind that RNAase A shows a two-step denaturation in particular conditions (for example when immobilized on a polymeric matrix) and considering the structure of RNAase BS near the intersubunit links, it is reasonable to assume that the two domains with different thermal stabilities, do not coincide with the two protomers of the enzyme. It is probable that the two domains are formed one by the central part of the dimer, the "core" which contains the two pairs of cysteines 31 and 32, and the other by the peripheral zones of polypeptide arrangement. Work is in progress at our laboratory to confirm the proposed mechanism of thermal unfolding.

Acknowledgement

This work was carried out with the financial support of Italian C.N.R., Target Program "Chimica Fine", (Rome) and of Ministry of University and Scientific and Technological Research.

4. References

1. C.H.W. Hirs, S. Moore, W.H. Stein, *J. Biol. Chem.* **235**, 633, (1960).
2. G. Kartha, J. Bello, D. Harker, *Nature* (London) **213**, 862, (1967).
3. A. Wlodawer in F.A. Jurnak, A.Mc Pherson Eds. <u>Biological Macromolecules &</u> <u>Assemblies</u>, vol.**2**, p. 393, <u>Nucleic Acids and Interactive Proteins</u>, J.Wiley, New York, (1985).
4. A.Wlodawer,.A. Svensson, L. Sjolin and G.L Gilliland, *Biochemistry* **27**, 2705, (1988).
5. M. Rico, M. Bruix, J. Santoro, C. Gonzales, J.L. Neira, J.L. Nieto, J. Herranz, *Eur. J. Biochem.* **183**, 623, (1989).
6. C.B. Anfinsen, E. Haber, M. Sela, F.H. White, *Proc. Natl. Acad. Sci.* **47**, 1309, (1961).
7. C.B. Anfinsen, H.A. Scheraga, *Adv. Protein Chem.* **29**, 205, (1975).
8. T.E. Creighton, D.A. Hillson, R.B. Freedman, *J. Mol. Biol.* **142**, 497, (1980).
9. R.B. Freedman, *Trends Biochem. Sci.* **9**, 438, (1984).
10. H.W. Wyckoff, D. Tsernoglon, A.W. Hanson, J.R. Knox, B.Lee, F.M. Richards, *J. Biol. Chem.* **245**, 305, (1970).
11. T.Y. Tsong, R.P. Hearn, D.P. Wrathall, J.M. Sturtevant, *Biochemistry* **9**, 2666, (1970).
12. G. Barone, F. Catanzano, P. Del Vecchio, D. Fessas, C. Giancola, G. Graziano, in press.
13. R.P. Hearn, F.M. Richards, J.M. Sturtevant, G.D. Watt, *Biochemistry* **10**, 806, (1971).
14. G. D' Alessio, A. Floridi, R. De Prisco, A. Pignero, E. Leone, *Eur. J. Biochem.* **26**, 153, (1972).
15. A.Di Donato, G. D' Alessio, *Biochem. Biophys. Res. Commun.* **55**, 919, (1973).
16. G. D' Alessio, A. Malorni, A. Parente, *Biochemistry* **14**, 1116, (1975).
17. H.Suzuky, L. Greco, A.Parente, B. Farina, R.La Montagna, E.Leone, in <u>Atlas of Protein Sequence and Structure</u> (M. O. Dayhoff Ed.) Vol.**5**, Suppl.2, pag.93, N.B.R. Foundation, Washington, D.C. (1976).
18. S. Capasso, F. Giordano, C.A. Mattia, L. Mazzarella, A. Zagari, *Biopolymers* **22**, 327, (1983).
19. L. Mazzarella, C.A. Mattia, S. Capasso, G.Di Lorenzo, *Gazz. Chim. It.* **117**, 91, (1987).
20. L. Mazzarella, S.Capasso, L. Vitagliano, A. Zagari. This Book.
21. S. Andini, G.D'Alessio, A.Di Donato, L.Paolillo, R.Piccoli, E.Trivellone, *Biochim. et Biophys. Acta* **724** , 530, (1983).

22. L. Greco, G. Misuraca, E. Leone, *Boll. Soc. It. Biol. Sper.* **49**, 1439, (1973).

23. E. Leone, L. Greco, R.K. Rastogi, L. Iela, *J. Reprod. Fert.* **34**, 197, (1973).

24. J. Matousek, *J. Reprod. Fert.* **19**, 63, (1969).

25. M. Tamburrini, G. Scala, C. Verde, M.R. Ruocco, A. Parente, S. Venuta, G. D'Alessio, *Eur. J. Biochem.* **190**, 145, (1990).

26. J. Soucek, A. Hruba, V. Chudomel, J. Dostal, J. Matouse, *Folia Biol.* (Praga) **29**, 250, (1983).

27. J. Soucek, A. Hruba, E. Paluska, V. Chudomel, G. Linderova, *Folia Biol.* (Praga) **27**, 334, (1981).

28. J. Soucek, A. Hruba, E. Paluska, V. Chudomel, J.Dostal, *Folia Biol.* (Praga) **29**, 250, (1983).

29. J. Matousek, *Experientia* **29**, 858, (1973).

30. R. Snanek, J. Matousek, *Folia Biol.* (Praga) **22**, 33, (1976).

31. S.Vescia, D. Tramontano, G. Augusti-Tocco, G. D' Alessio, *Cancer. Res.* **40**, 3740, (1980).

32. C.M. Simm, W.K. Krietsch, G. Isenberg, *Eur. J. Biochem.* **166**, 49, (1987).

33. W.S. Tarnowski, R.L. Kassel, I.M. Mountain, P. Blackburn, G. Wilson, D. Wang, *Cancer. Res.* **36**, 4074, (1976).

34. J. Bartholeyns, P. Baudhuin, *Proc. Natl. Acad. Sci.* **73**, 573, (1976).

35. Cantor and Schimmel, <u>Biophysical Chemistry</u> vol. 3, W.H. Freeman & Company, S. Francisco, (1980).

36. A.M.Crestfield, W.H. Stain, S. Moore, *J.Biol. Chem.* **238**, 2413, (1963).

37. A.M.Crestfield, W.H. Stain, S. Moore, *J.Biol.Chem.* **238**, 2421, (1963).

38. R.Piccoli, M.Tamburrini, G.Piccialli, A.Di Donato, A.Parente,G.D'Alessio, *Proc. Natl. Acad. Sci.* **89**, 1870, (1992).

39. A. Di Donato, P. Galletti, G. D' Alessio, *Biochemistry* **25**, 8361, (1987).

40. N. Quarto, G.F. Tajana, G. D' Alessio, *J. Reprod. Fert.* **80**, 81, (1987).

41. M. Palmieri, A. Carsana, A. Furia, M. Libonati, *Eur. J. Biochem.* **152**, 275, (1985).

42. R.G. Fruchter, A.M. Crestfield, *J. Biol. Chem.* **240**, 3875, (1965).

43. A. Parente, G. D' Alessio, *Eur. J. Biochem.* **149**, 386, (1985).

44. S. Clark, *Int. J. Pept. Protein Res.* **30**, 808, (1987).

45. S.E.Zale, A.M.Klibanov, *Biochemistry* **25**, 5432, (1986).

46. J.L. Casal, T.J. Ahern, R.C. Davenport, G.A. Petsko, A.M. Klibanov, *Biochemistry* **26**, 1258, (1987).

47. T.J. Ahern, J.L. Casal, G.A. Petsko, A.M. Klibanov, *Proc. Natl. Acad. Sci.* **84**, 675, (1987).

48. B.A Johnson, J.M. Shirokawa, W.S. Hancock, M.W. Spellman, L.J. Basa, D.W. Aswad, *J. Biol. Chem.* **246**, 14262, (1989).

49. B.A. Johnson, J.M. Shirokawa, D.W. Aswad, *Arch. Biochem. Biophys.* **268**, 276, (1989).

50. C.N. Pace, *Methods Enzymol.* **131**, 266, (1986).

51. C.N. Pace, G.R. Grimsley, *Biochemistry* **27**, 3242, (1988).

52. C.N. Pace, D.V. Laurents, *Biochemistry* **28**, 2520, (1989).

53. C.N. Pace, B.A. Shirley, J.A. Thomson, in <u>Protein Structure: a Pratical Approach</u>, T.E. Creighton Ed. IRL Press Oxford, (1989).

54. C.N. Pace, D.V. Laurents, J.A. Thomson, *Biochemistry* **29**,2565, (1990).

55. F.G. Walz, S. Kitareewan, *J. Biol. Chem.* **265**, 7127, (1990).

56. T. Kiefhaber, R. Quaas, U. Hahn and F.X. Schmid, *Biochemistry* **29**, 3053, (1990).

57. C.N. Pace, N. Heinemann, H. Hahn, W. Saenger, *Angew. Chemie. Inter. Ed.* **30**, 343, (1991).

58. B.A. Shirley, P. Stanssens, J. Steyaert, C.N. Pace, *J. Biol. Chem.* **264**, 11621, (1989).

59. R. Arni, U. Heinemann, R. Takuoka, W. Saenger, *Acta Crystallogr. B* **43**, 548, (1987).

60. B.A. Shirley, P.Stanssens, U. Hahn, C.N. Pace, *Biochemistry* **31**, 725, (1992).

61. P.L. Privalov, *Adv. Protein Chem.* **33**, 179, (1979).

62. P.L. Privalov, *Adv. Protein Chem.* **35**, 1, (1982).

63. P.L. Privalov, *Biochem. and Mol. Biol.* **25**, 281, (1990).

64. F.Franks, R.H.M. Hatley, H.L. Friedman, *Biophys. Chem.* **6**, 171, (1985).

65. P.L. Privalov, Yu.V. Griko, S. Yu. Venyaminov, V.P. Kutyshenko, *J. Mol. Biol.* **190**, 487, (1986).

66. F.Franks, R.H.M. Hatley, *Cryo-Letters* **7**, 171, (1985).

67. R.H.M. Hatley, F.Franks, *Cryo-Letters* **7**, 226, (1986).

68. G. Barone, P. Del Vecchio, D. Fessas, C. Giancola, G. Graziano. This Book.

69. T.Ackermann in Structural Molecular Biology D.B. Davies,W. Saenger,S.S. Danyluk Eds. Plenum Press New York, (1982).

70. R. Danforth, H. Krakauer, J.M. Sturtevant, *Rev. Sci. Instrum.* **38**, 484, (1967).

71. T.Y. Tsong, R.P. Hearn, D.P. Wrathall, J.M. Sturtevant, *Biochemistry* **9**, 2666, (1970).

72. O.L. Mayorga, W.W. van Osdol, J.L. Lacomba, E.Freire *Proc. Natl. Acad. Sci.* **85**, 9514, (1988).

73. W.W.van Osdol, O.L. Mayorga, E. Freire, *Biophys. J.* **59**, 48, (1991).

74. J.M.Sturtevant, *Ann. Rev. Phys. Chem.* **38**, 463, (1987).

75. B.J. Barisas, S.J. Gill, *Ann. Rev. Phys. Chem.* **29**, 141, (1978).

76. C.Spink, I.Wadsö, in Methods of Biochemical Analysis, Vol. **23**, pag.1, D. Glick Ed., J.Wiley and Sons, New York , (1976).

77. W. Pfeil in Biochemical Thermodynamics, 2nd Ed., M.N.Jones Ed., Elsevier Science Publishers B.V., Amsterdam, (1988).

78. Biological Microcalorimetry, A.E.Beezer Ed., Academic Press Inc., London, (1980).

79. M. Sela, C.B.Anfinsen, *Biochem. Biophys. Acta* **24**, 229, (1970).

80. C.-Q. Hu, J.M. Sturtevant, J.A. Thomson, R.E. Erikson, C.N. Pace, *Biochemistry* **31**,4876, (1992).

81. G. Barone, P. Del Vecchio, D. Fessas, C. Giancola, G. Graziano, P. Pucci, A. Riccio, M. Ruoppolo, *J. Thermal Anal.* **38**, 2791, (1992).

82. G.Barone, P.Del Vecchio, D.Fessas, C.Giancola, G.Graziano, *Thermochim. Acta*, in press, (1993).

83. G. Barone, P. Del Vecchio, D. Fessas, C. Giancola, G. Graziano, A. Riccio in "Proceedings of the International Symposium on Stability and Stabilization of Enzymes" - Elsevier (Dordrecht - NL), (1993).

84. J.F. Brandts, L. Hunt, *J. Am. Chem. Soc.* **89**, 4826, (1967).

85. P.L. Privalov, N.N. Khechinashvili, *J. Mol. Biol.* **86**, 665, (1974).

86. E. Freire, R.L. Biltonen, *Biopolymers* **17**, 463, (1978).

87. A.L. Jacobson, C.L. Turner, *Biochemistry* **19**, 4534, (1980).

88. Y. Fujita,Y. Noda, *Bull. Chem. Soc. Jpn.* **57**,2177, (1984).

89. T.Y. Tsong, R.P. Hearn, D.P. Wrathall, J.M. Sturtevant, *Biochemistry* **9**, 13, (1970).

90. T. Kiefhaber, F.X. Schmid, M. Renner, H.J. Hinz, U.Hahn, R.Quaas, *Biochemistry* **29**, 8250, (1990).

91. I.M. Plaza del Pino, C.N. Pace, E. Freire, *Biochemistry* **31**, 11196, (1992).

92. C. Chothia, *J. Mol. Biol.* **105**, 1, (1976).

93. F.M. Richards, *J. Mol. Biol.* **82**,1, (1973).

94. B. Gavish, E. Gratton, C.J. Hardy, *Proc. Natl. Acad. Sci.* **80**,750, (1983).

95. J. Bello, *Int. J. Peptide Protein Res.* **12**, 38, (1978).

96. A. M. Liquori in <u>Principle of Biomolecular Organization</u> Ciba Foundation Symposium London, (1966).

97. A. M. Liquori, *Quarterly Reviews of Biophysics* **2**, 65, (1969).

98. K.P. Murphy, S.J. Gill, *Thermochim. Acta* **139**,279, (1989).

99. K.P. Murphy, S.J. Gill, *J. Chem. Thermodynam.* **21**,903, (1989).

100. K.P. Murphy, S.J. Gill, *J. Mol. Biol.* **222**, 699, (1991).

101. R.S. Spolar, J.R. Livingston, M.T. Record Jr., *Biochemistry* **31**, 3947, (1992).

102. P.S. Kim, R.L. Baldwin, *Annu. Rev. Biochem.* **51**,459, (1982).

103. C.N. Pace,T.E. Creighton, *J. Mol. Biol.* **188**,477, (1986).

Denaturation of Biological Macromolecules: New Programs for the Deconvolution of DSC Measurements

G. BARONE, P. DEL VECCHIO, D. FESSAS, C. GIANCOLA and G. GRAZIANO
Department of Chemistry, University "Federico II"
Via Mezzocannone, 4 - 80134 - Naples, Italy

1.Introduction

Differential scanning microcalorimetry (dsc), is a useful technique for investigating the conformational transition occurring in the thermal denaturation process of proteins and other biological macromolecules in solution. The power, the advantages and few disadvantages of this methodology are discussed in the previous paper of this book [1]. We wish to discuss here the strategy for analysing the experimental dsc curves and obtain a reliable thermodynamic picture. Usually, in the past, the unfolding of proteins has been assimilated to a reversible all or none process, in other words to a one-step transition between the native and denatured state [2]. But it is now clear that many proteins show calorimetric peaks that cannot be interpreted as the manifestation of this simple mechanism [3,4]. In such cases multistage processes generally occur and it is necessary for obtaining a deep insight on each stage, to try to deconvolve the dsc curves, following some theoretical model. Thus two basic problems must be solved in order to eliminate dangerous misunderstandings: i) to determine if the dsc curve is really well represented by the two-state transition model; ii) to perform a deconvolution analysis of the dsc curve when the unfolding process is not a one-step. For these purposes we have developed and assembled new software tools in our laboratory: TAURO program which allows to establish if the denaturation process is really a two-state transition through the accomplishment of two distinct tests; DEDALUS program which deconvolves calorimetric curves considering that the complex denaturation process can be regarded as the sum of independent two-state transitions; MINOS program which accomplishes the deconvolution according to the sequential transition model.

Some application of these programs will be shown for three Ribonucleases (see the previous paper of this Book [1]).

2. Some general features of the programs

All programs are compiled in the Microsoft Quik Basic 4.5 environment. They run on IBM PC's compatible with VGA Graphic Card. Mathematical coprocessor is desired. RAM memory request depends on the dimensions of the input data files. The input-output file format consists of two files of data in ASCII format, one, containing the

N. Russo et al. (eds.), Properties and Chemistry of Biomolecular Systems, 67–78.
© 1994 *Kluwer Academic Publishers. Printed in the Netherlands.*

experimental or the elaborated data (in kJ mol^{-1}K^{-1}) and the other, all the experimental conditions and information indispensable for the analysis, in a specific order .

In a preceding paper [5] we described the software package THESEUS, that handles the dsc instrument signals bypassing the casual and systematic errors, very common in these experiments, and giving the basis for speculative analyses in the case of complex denaturation processes. The THESEUS program corrects the experimental power-time diagrams, according to the sensitivity-temperature calibration curve of the instruments, eliminates the electronic noise by using different filters, corrects for the baseline and gives a numerical detailed description of the apparent specific heat $\Delta C_p(T)$. By a regression procedure, from the pre- and post-denaturation temperature ranges, it makes possible to extrapolate the respective trends with the temperature into the denaturation range.

The input-output files are automatically produced by the THESEUS program, however the user can produce them with other systems too. The numerical procedures are performed in double precision mode. Numerical errors due to the procedures are almost negligible against the experimental errors of the $\Delta C_p(T)$ function, so they are not considered. Running time of TAURO and MINOS programs is of the order of few seconds, whereas DEDALUS, in the worst case, needs few minutes to perform the best fit. There is no limit number of independent or sequential transitions to take into account. All programs allow zoom, save, print, etc. utilities.

3. The TAURO program

3.1. THE POPULATION FRACTION TEST

To verify that the denaturation process is well represented by the two-state transition model, it is necessary to show that in the whole investigated temperature range the unique two macroscopic states significantly populated are the native one, N, and the denatured one, D. The analysis consists in determining, independently, the population fractions of native and denatured states, f_N and f_D respectively, and checking if their sum is equal to one for each temperature (i.e. no intermediates are present). It must be stressed that the two determinations are independent from each other [6].

The canonical partition function Q(T) of the macromolecule in solution, considering the native state as reference, can be obtained readily from the experimental heat capacity curve:

$$Q(T) = \exp\left[\int_{T_N}^{T} (<\Delta H(T)>_N/RT^2)\, dT\right] \tag{1}$$

where $<DH(T)>_N$ is the excess enthalpy function with respect to the native state, determined through integration of the excess heat capacity function $<\Delta C_p(T)>_N$ (see the following section) and T_N is a temperature at which all protein molecules are in the native conformation. The population fraction of the native state is then calculated as:

$$f_N(T) = 1/Q(T) \tag{2}$$

But it is also possible to consider the canonical partition function Z(T) of the macromolecule in solution, assuming the denatured state as reference, and to calculate it from the relation:

$$Z(T) = \exp\left\{\int_{T}^{T_D} [\,(\Delta_d H - \langle \Delta H (T)_N \rangle)/RT^2\,]\, dT\right\} \tag{3}$$

where $\Delta_d H$ is the change of enthalpy associated to the conformational transition (i.e. the area of the calorimetric peak) and T_D is a temperature at which all protein molecules are denatured. The population fraction of the denatured state is then calculated as:

$$f_D(T) = 1/Z(T) \tag{4}$$

Because $f_N(T)$ and $f_D(T)$ are independently determined, we are in a position to verify that:

$$f_N(T) + f_D(T) = 1 \tag{5}$$

at each temperature, as necessary condition for a two-state process. The use of this test confirmed that the thermal denaturation of RNAase A closely corresponds to a two-state transition in a large pH range (no intermediates are present between native and denatured state [7]). Instead, in the case of RNAase BS (Fig.1a), it is evident that the two-state transition model fails because intermediates are present; finally in the case of RNAase T_1 (Fig.1b) the two population fractions are overestimated, being their sum greater than one (the reasons of this peculiar behaviour have been partially discussed in parallel papers [1,7,8], one being presented in this Book).

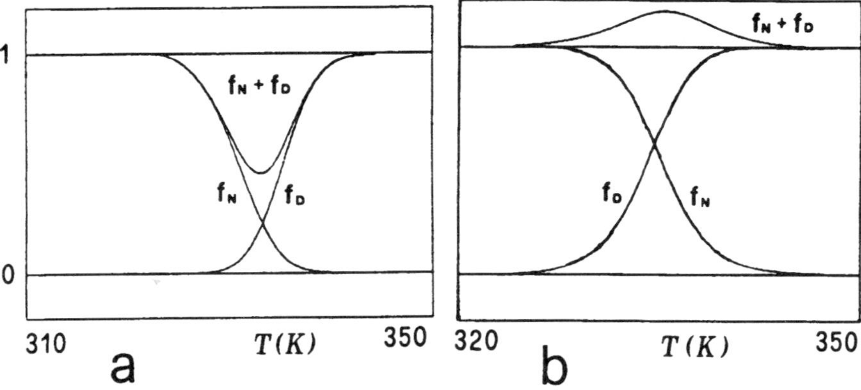

Figure 1. Population fraction test in the case of a) RNAase BS (pH 5.0, acetate buffer; protein concentration 3.8 10^{-5} M) and b) RNAase T_1 (pH 5.0, acetate buffer; protein concentration 1.1 10^{-4} M).

3.2. THE VAN 'T HOFF TEST

The starting hypothesis is that the denaturation process could be correctly described as a two-state transition. We determine the excess heat capacity of the system with respect to native state $<\Delta C_p(T)>_N$ and that with respect to denatured state $<\Delta C_p(T)>_D$, assuming as heat capacity of native state, the linear function obtained by a least square regression of pre-denaturation region, and as heat capacity of denatured state, the linear function obtained by regression of post-denaturation region. So it is possible to calculate the excess enthalpy functions with respect to the native and denatured state, $<\Delta H(T)>_N$ and $<\Delta H(T)>_N$ respectively, by a simple integration:

$$<\Delta H\,(T)>_N = \exp \int_{T_N}^{T} <\Delta C_p\,(T)>_N \; dT \tag{6}$$

and

$$<\Delta H\,(T)>_D = \exp \int_{T_D}^{T} <\Delta C_p\,(T)>_D \; dT \tag{7}$$

where T_N and T_D have the same meaning above expressed. It must be noted that because the D state is the highest in enthalpy scale, the function $<\Delta H(T)>_D$ is always negative and equal to zero only at $T = T_D$. The calorimetric enthalpy, as function of temperature, is then given by [9]:

$$\Delta_d H(T) = <\Delta H(T)>_N - <\Delta H(T)>_D \tag{8}$$

The area of the calorimetric peak is determined as the sum of two contributions and this relation is valid at each temperature. Further, expressing in a suitable manner the equilibrium constant of the process, it is possible to obtain a strictly thermodynamic expression for the van 't Hoff enthalpy change associated to the unfolding transition as [9]:

$$\Delta_d H_v(T) = RT^2[(<\Delta C_p>_N / <\Delta H(T)>_N - (<\Delta C_p>_D / <\Delta H(T)>_D)] \tag{9}$$

It is simple to show that this relation leads for $T = T_d$ to the largely used expression of $\Delta_d H_{vH}(T_d)$ (see Eq.3 of previous paper in this Book). But there is no reason to compare the values of $\Delta_d H$ and $\Delta_d H_{vH}$ only at $T = T_d$. Their equality at this temperature is a necessary but not sufficient condition for asserting that the process is a two-state transition. Instead, if the system is well represented by only two thermodynamic states, $\Delta_d H(T)$ and $\Delta_d H_{vH}(T)$ must be equal, within experimental uncertainties, in the whole temperature range where denaturation occurs. Besides, when this condition holds (i.e. $\Delta_d H(T) = \Delta_d H_{vH}(T)$), it is possible to demonstrate that the system can be described by the two-state transition model [9].

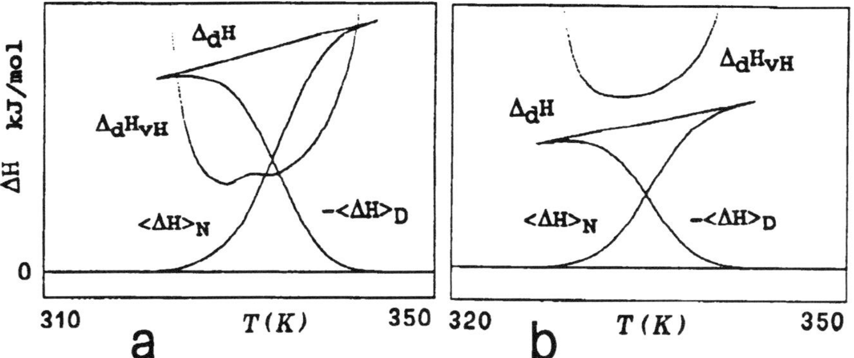

Figure 2. Application of the van 't Hoff criterion test in the case of a) RNAase BS and b) RNAase T1. Same conditions as in Fig.1. The large divergences at the limits of the denaturation temperature ranges are artifacts due to the form of the analytical expression.

The application of this test to RNAase A has shown a good agreement between $\Delta_d H$ and $\Delta_d H_{vH}$, confirming definitely that its denaturation process is a one-step type [7]. But for RNAase BS and RNAase T_1 (Figs.2a,2b) the large differences between the two enthalpy functions demonstrate that the two-state transition model is wrong for describing their thermal unfolding and more complex processes occur.

3.3. THE HEAT CAPACITY PROFILE SIMULATION

When it has been established that the denaturation process is really a two-state transition, the TAURO program allows to simulate the excess heat capacity function for the system under investigation in a large temperature range, showing also the predicted "cold denaturation" of small globular proteins. The canonical partition function Q(T) for a two-state system, assuming the native state as reference, is given by [10]:

$$Q(T) = 1 + K(T) \qquad (10)$$

where K is the equilibrium constant for the process $N \rightarrow D$ and is expressed as function of temperature:

$$K(T) = \exp -(\Delta_d H / R) [(1/T) - (1/T_d) + (\Delta_d C_p / R) [1 - (T_d / T) + \ln (T_d/T)] \qquad (11)$$

where $\Delta_d H$, $\Delta_d C_p$ and T_d are the thermodynamic parameters characterizing the denaturation process and directly determined from the experimental dsc curve. It is possible to determine, $<\Delta H(T)>_N$ $<\Delta H(T)^2>_N$ and f_D according to their statistical mechanical definition from the above partition function and express the excess heat capacity function referred to the native state as [11]:

$$<\Delta C_p(T)>_N = [(<\Delta H^2>_N - <\Delta H>_N^2) / RT^2] + f_D \cdot \Delta_d C_p \qquad (12)$$

that can be put in the simpler form:

$$<\Delta C_p(T)>_N = (\Delta_d H^2 / RT^2] + \{ K(T) / [Q(T)]^2 \} + \Delta_d C_p [K(T) / Q(T)] \qquad (13)$$

This is an analytical expression which allows to simulate a calorimetric curve for a two-state transition. Clearly if the denaturation process of the studied protein molecule closely conforms to a one-step process, the experimental and simulated heat capacity function must agree very well. Indeed this is what happens for RNAase. A as shown in Fig.3, where the "cold denaturation" is also emphasized.

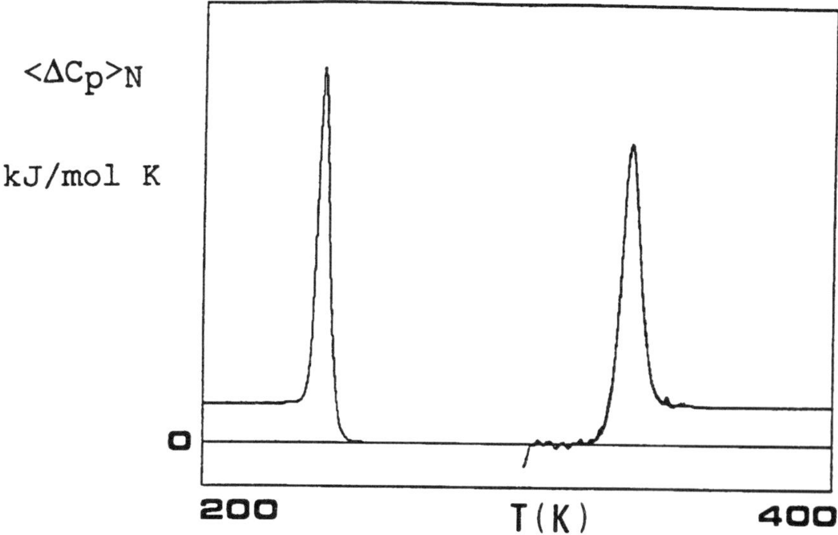

Figure 3. Simulation of the excess heat capacity function for RNAase A. The bold line represents the experimental curve (pH 5.0, acetate buffer; protein concentration 1.8 10^{-4}). The peak on the left describes the predicted "cold denaturation" phenomenon, inaccessible to direct investigation due to freezing of water.

4. Complex denaturation processes

When both tests of TAURO program clearly indicate that the denaturation process is not a two-state transition, it is possible to assume that a multistage mechanism occurs. To analyze in a suitable manner the calorimetric curves we have developed, assembled and implemented two programs: DEDALUS which analyzes the experimental data according to the independent two-state transition model and MINOS which analyzes the experimental data according to the sequential transition model.

4.1 THE DEDALUS PROGRAM

The simplest approach for describing a complex denaturation process is to consider it as a sum of a certain number of independent two-state transitions [12,13]. This hypothesis seems unrealistic, looking at the complexity of the denaturation process of biological macromolecules. The well established existence, in the three dimensional structure of proteins, of more or less independent domains has conferred a particular interest to such an assumption.

The power of this deconvolution approach relies on the possibility to give an analytical expression for the excess heat capacity function referred to the native state $<\Delta C_p(T)>_N$ for a two-state system (Eq.13). Indeed this is the starting point of the deconvolution procedure because, if the denaturation process can be considered as a sum of n independent two-state transitions, the overall excess heat capacity function is simply given by:

$$\left(\Delta C_p (T) \right)_N = \sum_{i=1}^{n} \left(\Delta C_p (T) \right)_{N,i} \tag{14}$$

where each $<\Delta C_p(T)>_{N,i}$ is expressed as in Eq.13. DEDALUS program utilizes as inputs the experimental $<\Delta C_p>_{Exp}$ file, the supposed number of two-state transitions and the associated $\Delta_d H_i$, $\Delta_d C_{p,i}$ and $T_{d,i}$ values. Of course these parameters are unknown and so the operator has to introduce them only in an approximate manner. The program performs a non linear regression using the Levenberg-Marquardt algorithm [14], to adjust them until a best fit between the $<\Delta C_p(T)>_N$ calculated by Eq.14 and the experimental one is reached. Thus, it is possible to determine the thermodynamic parameters characterizing the thermal stability of the independent domains which constitute the overall structure of the macromolecule. Some constraints have been applied to the algorithm (i.e. keeping as positive the sign of the $\Delta C_{p,i}$) in order to obtain physically acceptable values for the parameters. But testing the reliability of DEDALUS program on simulated complex calorimetric curves, it has become clear that the net heat capacity change associated to the unfolding process always complicates the deconvolution work. For this reason the $\Delta_d C_{p,tot}$ has been eliminated using an approximate iterative procedure, based on a reasonable physical basis [15]. Often exothermic processes due to the aggregation of denatured protein molecules are further present in the dsc profiles, sometimes overimposed to the endothermic peak. To analyze even these cases, DEDALUS allows to include, in the deconvolution procedure, gaussian functions for describing the exothermic peak. In Fig.4 the deconvolution of the calorimetric curve of RNAase BS at pH 5.0 is shown; two independent two-state transitions are necessary for fitting well the experimental curve.

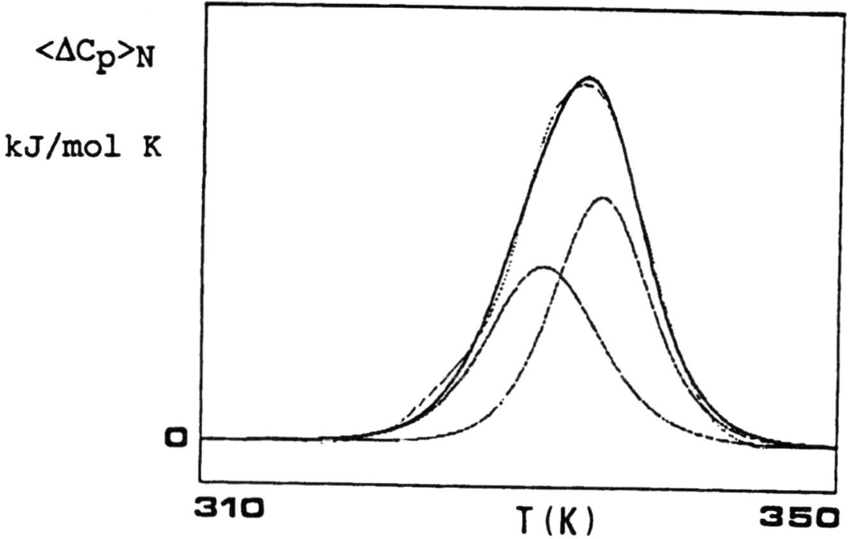

Figure 4. Deconvolution of the excess heat capacity for RNAase BS according to the model of the independent transitions of domains. The continuous curve represents the experimental values, the dashed one are the simulated curves. Same conditions as in Figs. 1a and 2a.

4.2. THE MINOS PROGRAM

MINOS program performs the deconvolution analysis following the procedure suggested by Freire and Biltonen [16,17], which considers the denaturation process as the sequential passage from the native state through a certain number of intermediates to the denatured one:

$$N \overset{K_1}{\leftrightarrow} I_1 \overset{K_2}{\leftrightarrow} I_2 \overset{K_3}{\leftrightarrow} \quad ..... \quad \overset{K_m}{\leftrightarrow} \quad D \tag{15}$$

The canonical partition function Q(T), written considering as reference the native state, is :

$$Q(T) = 1 + K_1 + K_1K_2 + ... + K_1K_2 ...K_m \tag{16}$$

It can be directly calculated from the experimental calorimetric data with a double integration of $<\Delta C_p(T)>_N$ on the temperature and allows to accomplish the deconvolution. The key-function of the procedure is expressed as:

$$\Phi_{N \to 1}(T) = <\Delta H(T)>_N / \{ 1 - [(1 / Q(T)] \} \tag{17}$$

But expliciting the statistical mechanical definitions of the various terms it becomes:

$$\Phi_{N\rightarrow 1}(T) = \Delta_d H_{N\rightarrow 1} + <\Delta H(T)>_1 \tag{18}$$

where $\Delta_d H_{N\rightarrow 1}$ is the denaturation enthalpy change associated to the process $N\rightarrow I_1$ and $<\Delta H(T)>_1$ is the excess enthalpy function referred to the I1 state. Clearly when the temperature is low enough that only the native state is populated, we obtain:

$$\lim_{T\rightarrow T_N} \phi_{N\rightarrow 1}(T) = \Delta_d H_{N\rightarrow 1} \tag{19}$$

Applying this algorithm in a recurrent form it becomes possible to determine all the enthalpy changes which give rise to the complex calorimetric peak. The denaturation temperature for the first transition is determined solving numerically the equation:

$$Q(T_{d,1}) - Q_1(T_{d,1}) = 1 \tag{20}$$

where Q_1 is the partition function of the system assuming the first intermediate I_1 as reference.

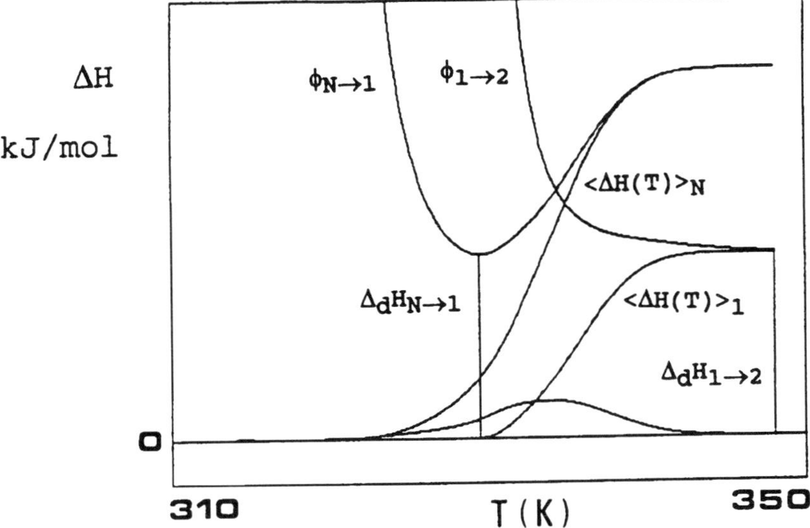

Figure 5. Deconvolution of the excess heat capacity for RNAase BS according to the model of sequential transitions. The experimental curve (arbitrary units) is reported in the lower part of the figure. Same conditions as in Figs.1a and 2a.

Putting this relation in a recurrent form, all the denaturation temperatures can be determined. So the MINOS program can deconvolve the complex process in its constituent transitions and calculate the respective thermodynamic parameters, using only thermodynamic functions. Problems arise because the key-function $\Phi_{N\to 1}$ diverges at low temperatures as pointed out by Filimonov et al.[18]. This divergence can affect the deconvolution procedure especially if the component transitions are separated by few degrees in temperature scale. Indeed in such cases the key-function increases before settling in the true enthalpy value of the intermediate state I_1. For this reason we have established to keep the minimum of the $\Phi_{N\to 1}$ function as the value of $\Delta_d H_{N\to 1}$ and to accomplish a step by step optimization algorithm to overcome this problem. The reliability of MINOS program has been tested on simulated calorimetric curves and the results obtained are satisfactory, even in the worse conditions (i.e. two sequential transitions separated by only two degrees in temperature). In Fig.5 the deconvolution of the calorimetric curve of RNAase BS at pH 5.0 is shown; two sequential transitions are emphasized.

5. Discussion

As it has been shown two different general approaches exist for analyzing dsc curves which are not well represented by the two-state transition model. However the results obtained with them in the case of RNAase BS are very similar [19]. This raises an obvious question: are the two models really different? The answer is yes, they are very different on theoretical grounds [17,20]. The fundamental difference relies on the number of accessible energy states. A sequential transition of n steps generates (n+1) energy states, but a sum of n independent two-state transitions will generate 2^n energy states. If the n two-state transitions are well separated in the temperature scale, only (n+1) energy state will be significantly populated during the unfolding process, and the observed melting curve will be identical to that of a sequential transition. This is the case of RNAase BS, as emphasized by our results [19]. However, when the temperature spacing between transitions is small, the 2^n accessible states will progressively become populated, and the melting curve will be broader than that expected for a sequential transition with the same values of the thermodynamic parameters.
Indeed we have simulated two complex dsc curves: W_0 composed by two independent two-state transitions and W_1 composed by two sequential transitions, using the same thermodynamic parameters and a temperature spacing of only two degrees. The large difference existing between them is evident (Fig.6): W_0 is broader and lower than W_1.
In the first column of Table I are reported the values of the parameters utilized for the simulation of W_0 and in the second column those determined by MINOS deconvolution (i.e. by the wrong model). Clearly MINOS determines quite well the values of the transition enthalpies, but fails regarding the transition temperatures, because the W_0 curve is broader than two sequential transitions.

Table I. Denaturation parameters used to build the W_0 curve according to the independent transition model.

ΔH_1 (kJ/mol)	400.00	379.82
ΔH_2 (kJ/mol)	420.00	423.20
T_{d1} (K)	335.15	334.40
T_{d2} (K)	337.15	338.00

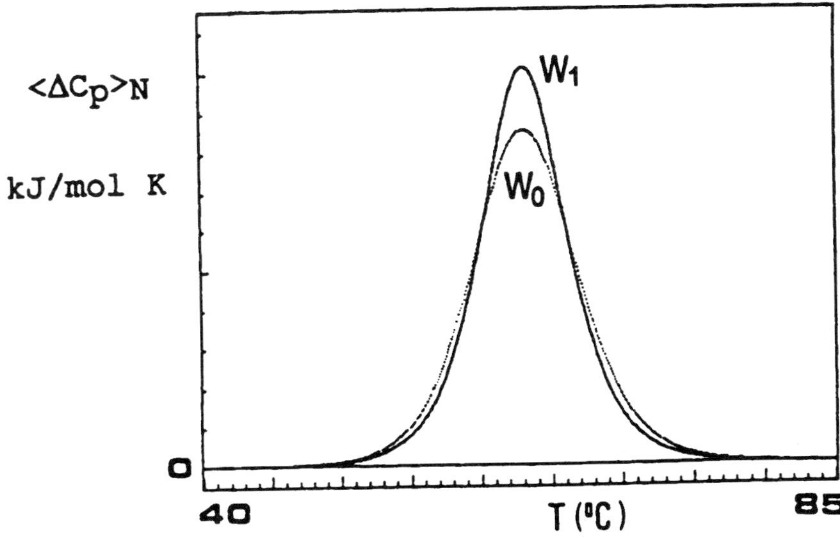

Figure 6. Simulations of the dsc curves according to both denaturation models. Starting values as in Tables I and II. The dashed line is the W_0 curve and the continuous one is the W_1 curve.

Table II. Denaturation parameters used to build the W_1 curve according to the sequential transition model.

ΔH_1 (kJ/mol)	400.00	302.95
ΔH_2 (kJ/mol)	420.00	538.58
T_{d1} (K)	335.15	336.08
T_{d2} (K)	337.15	336.40

In the first column of Table II are reported the parameter values utilized for the simulation of W_1 and in the second column those obtained by DEDALUS deconvolution (i.e. by the wrong model). It is evident that DEDALUS fails in analyzing the W_1 curve, regarding both the enthalpy changes and the transition temperatures, because the peak is too sharp with respect to the sum of two two-state calorimetric functions.

The differences between the two models and the two deconvolution procedures are very great. For this reason it is convenient to analyze carefully the experimental calorimetric curves with both the models, in order to verify which of the two basic mechanisms fits better the investigated denaturation process.

Acknowledgements

This work was carried out with the financial support of Italian C.N.R., Target Program "Chimica Fine", (Rome) and of Ministry of University and Scientific and Technological Research.

References

1. G. Barone, P. Del Vecchio, D. Fessas, C. Giancola, G. Graziano, This Book.
2. P.L. Privalov, *Adv. Protein Chem.* **33**, 179, (1979).
3. P.L. Privalov, *Adv. Protein Chem.* **35**,1, (1982).
4. J.M. Sturtevant, *Ann. Rev. Phys. Chem.* **38**, 463, (1987).
5. G. Barone, P. Del Vecchio, D. Fessas, C. Giancola, G. Graziano, *J. Thermal Anal.* **38**, 2779, (1992).
6. E. Freire, R.L. Biltonen, *Biopolymers* **17**, 463, (1978).
7. G. Barone, P. Del Vecchio, D. Fessas, C. Giancola, G. Graziano, A. Riccio, in Proceedings of the International Symposium on Stability and Stabilization of Enzymes - Elsevier, Dordrecht (NL) (1993).
8. G. Barone, P. Del Vecchio, D. Fessas, C. Giancola, G. Graziano, P. Pucci, A. Riccio, M. Ruoppolo, *J. Thermal Anal.* **38**, 2791, (1992).
9. S.I. Kidokoro, A. Wada, *Biopolymers* **26**, 213, (1987).
10. S.J. Gill, C. H. Robert, J. Wyman, in M.N. Jones Ed., Biochemical Thermodynamics (2nd Edition), Elsevier (1988).
11. T.L. Hill in An Introducti on to Statistical Thermodynamics , Dover, New York (1986).
12. P.L. Privalov, V.V. Filimonov, *J. Mol. Biol.* **122**, 447, (1978).
13. L.H. Chang, S.J. Li, T.L. Ricca, A.G. Marshall, *Anal. Chem.* **56**, 1502, (1984).
14. W.H. Press, B.P. Flannery, S.A. Teukolsky, W.T. Vetterling, in Numerical Recipes, Cambridge University Press (1989).
15. P.L. Privalov, S.A. Potekhin, *Methods Enzym.* **131**, 4, (1986).
16. R.L. Biltonen, E. Freire, *Crit. Rev. Biochem.* **5**, 85, (1978).
17. R.L. Biltonen, E. Freire, *Biopolymers* **17**, 1257, (1978).
18. V.V. Filimonov, S.A. Potekhin, S.V. Matveyev, P.L. Privalov, *Mol. Biol. (U.S.S.R.)* **16**, 551, (1982).
19. G. Barone, P. Del Vecchio, D. Fessas, C. Giancola, G. Graziano, *Thermochimica Acta*, in press (1993).
20. S.V. Matveev, V.V. Filimonov, P.L. Privalov, *Mol. Biol. (U.S.S.R.)* **16**, 1234, (1983).

Solvent Effects on the Molecular Organization and Properties of Biological Systems

R. BARTUCCI, R. GUZZI, P. SAPIA, and L. SPORTELLI
Dipartimento di Fisica, Laboratorio di Biofisica Molecolare and Unità INFM, Università della Calabria, 87036 Arcavacata di Rende, Italy

1. Introduction

The structural organization and overall stability of biological systems is largely determined by the interaction existing between their components and between these components and the aqueous environment, i.e., the solvent. As concerns the latter interaction, there is convincing evidence that the behaviour of membrane components and proteins changes by changing the physico-chemical properties (pH, ionic strength, polarity) of the solvent. In fact, it has been shown by x-ray diffraction and calorimetric investigations that the basic structural arrangement of biological membrane, i.e., the lipid bilayer, in presence of various compounds such as glycerol [1-4], small surface active molecules [5], alcohols [3, 6-9] and ions [10] above a threshold concentration, results completely interdigitated in the gel phase of saturated, symmetrical chain phosphatidylcholines. In this phase, the opposing lipid monolayers interpenetrate each other so that the terminal methyl groups of the lipid chains are located near the interfacial region on the opposite side of the lipid lamellae [for a review see ref. 11].
Moreover, mono-, bi-, and trivalent cations also influence the properties of lipid bilayers. Their action on polar phospholipids is well understood in the limit of the Gouy-Chapman-Stern lipid double layer theory in presence of up to 1 M salt concentration in the dispersion medium [12].
The stability of the native protein structure is determined by the balance of various inter- and intramolecular interactions occurring among different amino acid residues and with the water molecules surrounding the protein. Perturbing agents in solution such as denaturant [13, 14] or stabilizer [15], as well as temperature [16] and pH variations [17] change the balance of these interactions [18-20] and the overall protein structure results weakened or strengthened depending on the perturbing agent.
To get more insight we have investigated by Spin Label Electron Spin Resonance (ESR) and Optical Absorption spectroscopies the effet of different ionic strength values and of glycerol on multilayers of dipalmitoylphosphatidylcholine (DPPC) and of hydro-alcohol mixtures on azurin thermal stability, a metal-protein containing a single copper ion.
The presence of NaCl in the multilayers dispersion medium causes an increase of the molecular orientational order of the lipid chains and a decrease of the fluidity of the bilayer. In other words, the higher ionic strength stabilizes the ordered state of the neutral lipids.
The effect of glycerol on DPPC is quite different: the DPPC/glycerol system is

N. Russo et al. (eds.), Properties and Chemistry of Biomolecular Systems, 79–91.
© 1994 *Kluwer Academic Publishers. Printed in the Netherlands.*

interdigitated as results by comparing the mobility of the lipid chains for dispersions in glycerol and in aqueous buffer. In particular, the motion of the labels 5-, and 16-PCSL is uniformly slow, the anisotropy of the chain motion is reduced and less affected by temperature in the interdigitated system when compared with the normal lipid gel phase. Moreover, the gradient of increasing fluidity towards the terminal methyl end of the chain normally found in a bilayer structure is absent in the interdigitated gel phase induced by glycerol.

The investigation of the thermal stability of azurin in presence of alcohols with different side chain length such as ethanol and n-butanol shows a decrease of the pre- and full denaturation temperatures of the Cu^{++}-N_2SS^* active site of the protein as the molar fraction of the alcohols increases. Further, the down shift of the denaturation temperatures depends on the alcohols side chain length.

2. Materials and Methods

2.1. ESR MEASUREMENTS

Synthetic 1,2-dipalmitoyl, sn-glycero-3-phosphatidylcholine (DPPC) was used as obtained by Fluka. The spin labels 2-(5-carboxypropyl)-4,4-dimethyl-2-tridecyl-3-oxazolidinyl-oxyl (5-NSA) and di-tert-Butyl-Nitroxide (DTBN) were Aldrich products stored at 4 °C in ethanol solution. Phosphatidylcholine spin label with the nitroxide group on the C-n atom of the sn-2 chain, 1-acyl-2-(n-(4-4-dimethyl-oxazolidin-N-oxyl)-stearoyl)-sn-glycero-3-phospho-choline (n-PCSL, with n=5 or 16), was synthesized as described in Marsh and Watts [21]. NaCl and other chemicals used for buffer solutions were C. Erba products of analytical grade purity. Glycerol was from Merk. Distilled water was used throughout.

Multilayers of DPPC were prepared by first mixing the lipids with 1% by weight of the spin labels in chloroform solution, evaporating off the solvent with a stream of dry nitrogen, and then placing them under vacuum overnight. The dried lipids were dispersed in 10 mM phosphate buffer solution (PBS) at pH 8.0 containing 0, 1, 2, or 3 M NaCl to a final concentration of 50 mg/ml by vortex mixing at a temperature of 45 °C, i.e., above the gel to fluid phase transition of DPPC. Samples in glycerol were prepared similarly by adding 25 ml of glycerol to 1 mg of the dry DPPC/spin label mixture. The suspensions were transferred to 1 mm i.d. glass capillaries and centrifuged on a bench centrifuge at 2000 rpm for 15 min. Excess supernatant was removed. The sealed sample capillary tube was accomodated in a standard 4 mm quartz tube containing light silicon oil for thermal stability and was centred in a TE_{102} cavity. Samples were incubated at 4 °C for 24 h, and were then measured immediately, starting at low temperature.

Electron Spin Resonance measurements. The ESR spectra were recorded with a Bruker ER 200D-SRC spectrometer operating at 10 GHz equipped with the ER 4111VT temperature regulation unit (accuracy $\pm$ 0.3 °C) and the ESP 1600 Data System.

First harmonic in-phase absorption ESR spectra were recorded at a modulation frequency of 100 kHz and a modulation amplitude of 1.25 G_{pp} at the microwave power of 10 mW.

The lineshape of the spin labels ESR spectra can be described by the following spin hamiltonian [22-24]:

$$\mathcal{H} = \beta HgS + ITS \tag{1}$$

where the first term represents the Zeeman interaction between the unpaired electron spin $S = 1/2$ with the static magnetic field, $\mathbf{H}$, the second one the hyperfine interaction of the electron spin with the nitrogen nuclear spin $I = 1$ of the nitroxide moiety and β is the Bohr magneton.

In particular, the spectrum of 5-NSA or 5-PCSL is described by the axially symmetric form of the spin hamiltonian (1):

$$\mathcal{H} = \beta \, [g_{\parallel} \, H_z S_z + g_{\perp}(H_y S_y + H_x S_x)] + [\, T_{\parallel} \, I_z S_z + T_{\perp}(I_y S_y + I_x S_x)] \, . \qquad (2)$$

When the anisotropy of the system is large (spin labels in viscous environment or at low temperature) the outer hyperfine splitting, $2T'_{\parallel}$, of the motionally averaged axially symmetric nitrogen hyperfine $\mathbf{T}$-tensor, can be evaluated and related to the degree of molecular order.

The spectrum of DTBN in PBS consists in three sharp lines of equal height centred at g_0 and with hyperfine splitting T_0. The rapid tumbling of the spin probes averages out the anisotropy in the $\mathbf{g}$ and $\mathbf{T}$-tensor elements and the spin hamiltonian (1) assumes the isotropic form:

$$\mathcal{H} = \beta g_0 \mathbf{H S} + T_0 \mathbf{I S} \qquad (3)$$

with

$$g_0 = 1/3 \, \mathrm{Tr} \mathbf{g} \quad T_0 = 1/3 \, \mathrm{Tr} \mathbf{T} \qquad (4)$$

DTBN dissolved in an aqueous lipid dispersion partitions between the aqueous and the fluid lipid phase. The corresponding ESR spectrum is the superposition of two isotropic triplets arising from DTBN in PBS and from the fraction of the label in the fluid hydrophobic environment of lipid bilayers. Since membranes have major viscosity and smaller polarity than water, the spectra coming from the two environments have small differences both in the T_0 and g_0 factors. This leads to a partial spectral resolution being the high field line with $M_I = -1$ only resolved. The measure of the DTBN signal in the aqueous region, H_W, and in the fluid lipid, H_L, allows an extimation of the fraction of the labels in the hydrophobic region of the bilayer by means of the partition coefficient:

$$P_C = H_L/(H_L + H_W). \qquad (5)$$

This partition coefficient gives a measure of the membrane fluidity [23].

2.2. OPTICAL ABSORPTION MEASUREMENTS

Lyophilized azurin from Pseudomonas Aeuriginosa was purchased from Sigma and used without further purification. The optical absorption ratio A_{625}/A_{280} measured at RT was 0.44. Ethanol and n-butanol were from Farmitalia. A 10 mM phosphate buffer solution (PBS) at pH 7.0 was used throughout. The hydro-alcoholic mixtures were prepared and the pH controlled. Protein concentration was checked spectrophotometrically with an extinction cofficient $\varepsilon = 3500 \, M^{-1} \, cm^{-1}$ at $\lambda = 625$ nm. The hydro-alcohol samples were prepared with an ethanol molar fraction, χ_f, ranging between 0 and 0.050, while it was only up to 0.015 for n-butanol due to the alcohol-water miscibility and protein stability.

Thermal denaturation was followed with a JASCO 7850 spectrophotometer equipped with a Haake thermostated bath Mod. D8-G ($\pm$ 0.02 °C). Quartz cuvette of 1 cm optical path were used. Sample temperature was measured directly with a YSI precision thermistor dip in the reference cuvette. The measures started 5 min after the samples had been positioned in the thermostated sample holder at the initial temperature of 50 °C for ethanol-water and 40 °C for n-butanol-water mixtures. The heating rate was of 0.7 °C/min.

3. Results and Discussion

3.1. EFFECT OF HIGH SALT CONCENTRATION ON DPPC-MULTILAYERS.

The spin labels 5-NSA and DTBN have been used to investigate the effect of high salt concentration on the molecular order and fluidity of DPPC-multilayers. This because the spin label 5-NSA, by anchoring its fully dissociated carboxylic group near the positive charge of the choline quaternary ammonium one [25] allows the nitoxide moiety to investigate the order of the polar zone of lipid bilayers with temperature and NaCl concentration increase whilst, the partition of DTBN, between the bulk solution and the fluid hydrophobic core of the bilayers, gives informations on the permeability of this small molecule.

The ESR spectra of 5-NSA are powder like spectra, very similar to that of 5-PCSL (Figure 2) in multilayers. The separation between the outer resonance lines can be calculated and plotted as a function of temperature and NaCl concentration from 0 to 3 M in the dispersion medium (Figure 1A). As can be seen, in absence of the 1:1 electrolyte the pre-, $L_{\beta'} \to P_{\beta'}$, and main-, $P_{\beta'} \to L_\alpha$, DPPC phase transitions occur at the temperatures of T_p= 31.5 °C and T_m= 40.5 °C in agreement with literature data [12, 26, 27]. The pre-transition is more evident than with 5-PCSL due to the location of this spin label into the polar zone. The increase of NaCl concentration in the medium shifts upward T_p from 31.5 to 36.5 °C while, T_m goes up from 40.5 to 41.9 °C. A small decrease of $T'_\parallel$ in the fluid fase of DPPC multilayers is also observed.

Even with some difference, similar results are found using the spin label DTBN (Figure 1B). Without salt in solution T_p occurs at 34.3 °C while, T_m results at 40.5 °C suggesting that the constrains to the rotational motion about the long molecular axis of DPPC molecules, likely imposed by the hydrogen bonds network, should be eliminated before the diffusion of DTBN in the free volume of the lipid bilayers takes place. When the salt concentration in the medium is rised up to 3 M, either the pre- or main transition temperature are shifted upwards to 38.5 and 42.5 °C, respectively. Moreover, the cooperativity of both transitions remains more or less the same in absence and in presence of NaCl.

Since the DPPC multilayers were fully hydrated, the effects observed should be clearly related to the high Na^+ and Cl^- concentration present in the medium, i.e., to the properties of the solvent. The plots in Figure 1 A and B claim an increase in the packing density of the DPPC molecules in the multilayers. In other words, they claim a reduction induced by the salt of the free volume existing in the $L_{\beta'}$ gel phase of the lipid bilayers. The de-hydration of the DPPC polar heads, which is known to induce the increase in packing of the molecules [28], cannot be invoked in this case because the multilayers are prepared in excess of water, i.e., are fully hydrated. From a molecular

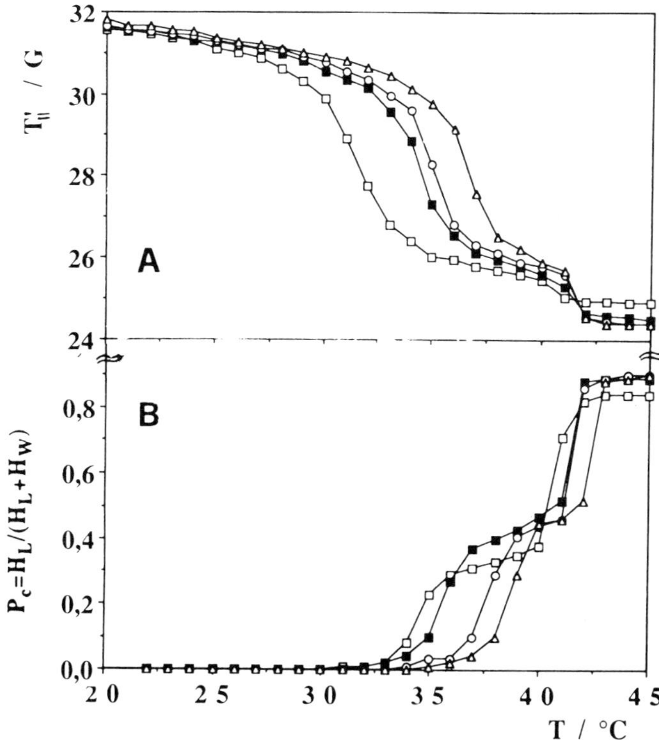

Figure 1 A) T'$_{||}$ values and B) P$_c$ ones vs. temperature of the spin labels 5-NSA and DTBN in DPPC multilayers dispersed in solutions with different contents of salt at the pH value of 8.0: —☐— , no NaCl; —■— , 1 M NaCl; —o— , 2 M NaCl; —△— , 3 M NaCl. The error on T'$_{||}$ is less than the dimension of the symbols.

point of view a mechanism can be suggested to explain the phenomena observed: due to the high radius of curvature of the multilamellar mesophase, the DPPC molecules are tightly packed so that the free volume between the phosphocholine polar heads is very low. In this way, the hydrated Na$^+$ ions cannot reach the -PO$^-$ region to modify, to some extent, the existing hydrogen bonds network.

On the other hand, the Cl$^-$ ions screen the charge on the more exposed -N$^+$(CH$_3$)$_3$ groups and modify the water structure around them. These actions result in the modification of the equilibrium between the hydrophobic, electrostatic and hydration forces, which in turn is restored by an increase of the molecular packing.

This mechanism is supported by ESR measurements [29] performed with DTBN and 5-NSA on DPPC small unilamellar vesicles which show, on the contrary, a down shift of both T$_p$ and T$_m$. This fact suggests that, due to the small radius of curvature of this mesophase, hence the higher free volume in the lipid monolayers, the Na$^+$ ions rearrange the hydrogen bonds network around the -PO$^-$ groups. As a consequence the separation between the DPPC molecules increases and the packing density decreases.

3.2. EFFECT OF GLYCEROL

The ESR spectra of the phosphatidylcholine spin label positional isomers, 5-PCSL and 16-PCSL, incorporated in DPPC dispersions, either in aqueous buffer or in glycerol, are compared in Figure 2A and 2B, respectively. In the gel phase at 0 °C, the spin labels in DPPC/glycerol system remain immobilized on the conventional ESR time scale. On the contrary, some motion, in the slow motion regime takes place for spin labels in DPPC/buffer system leading to a decrease in the outer hyperfine splitting on proceeding towards the therminal methyl end of the chain. Qualitatively this means that there are differences in chain mobility between the two systems which become greater near the methyl terminus of the lipid chain. A larger difference between the ESR lineshapes for both positions of the labels is seen for the two systems at higher temperatures in the gel phase (0-40 °C) (spectra not shown). In fact in glycerol, the spectra of the positional isomer near the end of the chain (16-PCSL) show a spectral splitting still similar to that found for those close to the polar headgroup region (5-PCSL). In other words, throughout the gel phase, the spectra in glycerol are motionally restricted and lie exclusively in the slow motional regime. Instead, in DPPC/buffer system, the spectra exhibit extensive motional narrowing of the hyperfine anisotropy on increasing temperature and on proceeding towards the end of the lipid chain. The chain motions observed in the normal gel phase for DPPC dispersed in aqueous buffer, therefore, display a mobility gradient that is strongly affected by temperature: it is already evident in the $L_{\beta'}$ phase at 0 °C, and it is much more pronunced in the $P_{\beta'}$ phase. This mobility gradient corresponds to the more rapid rotation about the long axis of the lipid chains that takes place in the $P_{\beta'}$ phase (30) and to a degree of chain segmental motion which has been found previously both by ^{2}H-NMR [31, 32] and by spin label ESR [32] for different phosphatidylcholines in the normal non-interdigitated bilayer gel phase. In contrast the fluidity gradient is lost in the interdigitated gel phase induced by glycerol, i.e., the motion of phosphatidylcholines labelled near the terminal methyl end of the chains is restricted to an extent similar to that for positional isomers located much closer to the polar head region. A similar restriction of the chain motion was found in other interdigitated systems also investigated with the spin label ESR technique, both for saturated mixed chain phosphatidylcholines [33, 34] and for acidic phospholipid complexes with polymyxin [3, 35].

For the fluid phase , the spectra from both environments are principally in the fast motional regime and posses axial anisotropy, characteristic of a liquid crystalline L_{α} phase in each case. For the dispersions in glycerol, the spectral anisotropy decreases monotonically with label position down the chain from the polar interface region towards the terminal methyl end. This indicates that chain interdigitation is absent for the fluid phase in glycerol, just as it is for the fluid phase in aqueous buffer. However, for a given label position, the spectral anisotropy is considerably greater for the DPPC dispersions in glycerol than for those in aqueous buffer, when spectra from the two systems are compared. The origin of this effect lies most probably in a somewhat higher chain packing density of the lipid bilayers in glycerol than of those in water, corresponding to the different hydration potentials of these two solvents.

A quantitative comparison between the two systems can be done by means of the temperature dependence of the outer hyperfine splitting, $2T'_{||}$, of the spin labels. The plots of $2T'_{||}$ vs. temperature of 5- and 16-PCSL in DPPC/glycerol and of 5-PCSL

DPPC/buffer dispersions are given in Figure 2C. In the normal gel phase, i.e., DPPC/buffer, the pre-transition is evident between the $L_{\beta'}$ and $P_{\beta'}$ gel phases as a drop at ca. 25-30 °C, and the chain melting phase transition as another step drop at ca. 40 °C. In the interdigitated gel phase, i.e., DPPC/glycerol, however, the values of $2T'_{\parallel}$ are not only greater for both the 5-PCSL and 16-PCSL labels than for the 5-PCSL label in the noninterdigitated phase, but also the values remain uniformly high throughout the gel

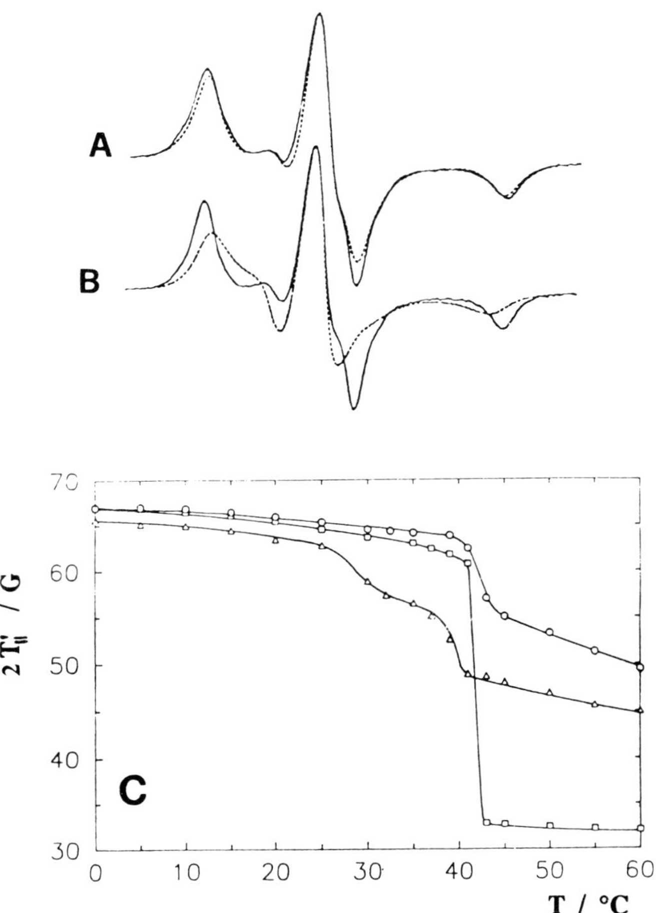

Figure 2 ESR spectra of the spin labels A) 5-PCSL and B) 16-PCSL in dipersions of DPPC in aqueous buffer (----) and in glycerol (——). Spectra recorded in the lipid gel phase at 0 °C. C) Temperature dependence of the outer hyperfine splitting, $2T'_{\parallel}$, of 5-PCSL in dispersions of DPPC in aqueous buffer ($\triangle$) and in glycerol ($\bigcirc$), and of 16-PCSL in glycerol ($\square$).

phase, with no indication of any pretransitional behaviour. Appreciable changes in the values of $2T'_{\parallel}$ are seen only at the chain melting transition which is found to occur at 42 °C, in agreement with other measurements [1, 3]. The lack of the pretransition involving the $L_{\beta i}$ phase is in agreement with previous findings, not only with DPPC/glycerol [1, 3], but also with interdigitated phosphatidylcholine systems induced either by short-chain alcohols [4] or by ions [10]. As already evidenced from the spectra, the present findings with DPPC/glycerol system suggest that a uniformly tight packing of the interdigitated chains is maintained at all temperatures in the $L_{\beta i}$ phase, up to the transition at the normal bilayer fluid phase, L_α.

3.3. EFFECT OF HYDRO-ALCOHOL MIXTURES ON AZURIN THERMAL DENATURATION

Azurin is a bacterial protein which possesses a single "blue" copper type I Cu^{++} center with peculiar spectroscopic features [36, 37]. Its function is of electron transfer [38]. The Cu^{++} ion is located near the molecular surface, about 7.5 Å dccp, but buried in a very hydrophobic pocket shielded from the solvent. It is coordinated by one S from Cys-112, one S* from Met-121 and two N from His-117, and -46 residues in a distorted tetrahedral array. One edge of a histidine ligand is accessible to the solvent [39]. The intense optical absorption at 625 nm due to the $S p\pi \rightarrow d_x 2_{-y} 2$ charge transfer transition occurring in the Cu^{++}-N_2SS^* active site of azurin has been used by us to follow the thermal disruption of the metal binding site of the protein with ethanol and n-butanol molar fraction increase in solution.

From the plot in Figure 3B we can observe that, in pure PBS, the thermal denaturation curve at $\lambda=625$ nm is characterized by a linear OD reduction between 67 and 76 °C, hereafter refered to as pre-denaturation step, before an abrupt reduction, corresponding to the full denaturation of the Cu^{++}-N_2SS^* active site at $T_d \cong 80$ °C, takes place.

This result suggests a gradual weakening of the hydrogen bonds network present on the protein surface which propagates toward the metal binding site of azurin. In other words, there is a coupling between the solvent fluctuations and the protein interior through the protein matrix. This hypothesis is supported by a similar result observed looking at the OD variation of the absorption band at $\lambda=280$ nm essentially coming from the absorption of the unic tryptophan residue of azurin (Figure 3A).

In fact, a pre-denaturation step is also observed at about 66 °C, which is followed by a step-wise OD variation at about 79 °C regarding the total unfolding of the protein.

Due to the similar behaviour of the two thermal denaturation curves and our interest on the properties of the Cu^{++}-N_2SS^* active site we have followed the effect of cosolvents looking at the band at $\lambda=625$ nm.

In Figure 3 C and D the reference denaturation curve at $\lambda=625$ nm is compared with those recorded in the presence of increasing amount in solution of ethanol and n-butanol, respectively. As can be seen, the effect of both organic cosolvents is to lower the temperature either of the pre-denaturation step or of the disruption of the Cu^{++}-N_2SS^* site. However, from the plots it results that ethanol, in solution up to 0.032 molar fraction, produces the same effect on T_d. In fact, in all three cases T_d results down shifted from 80 to about 68 °C while, the pre-denaturation region ranges now between 68 and 74 °C. A more marked effect is observed in presence of ethanol $\chi_f = 0.040$. In fact, the denaturation temperature results further down shifted to 72 °C. n-butanol shows a denaturating capability higher than ethanol also at the lower molar fraction. By comparing, for example, the results obtained at the same alcohol molar fraction,

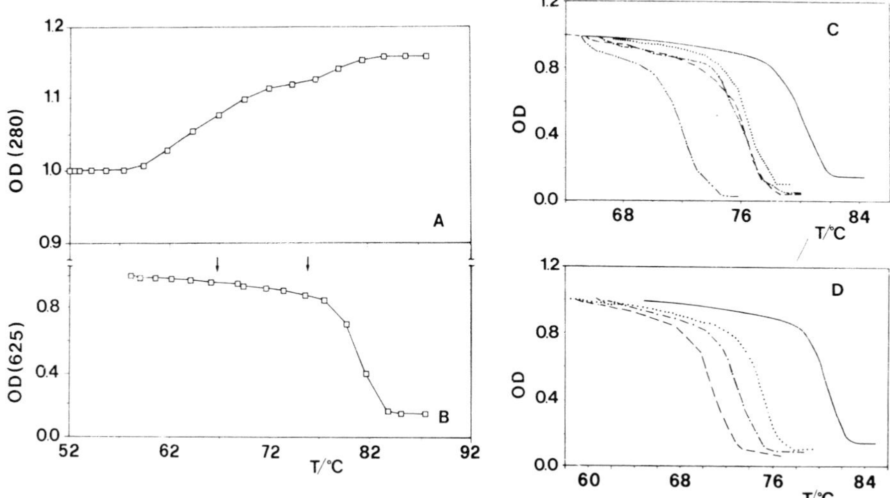

Figure 3 Normalized OD variation vs. temperature at A) λ= 280 and B) 625 nm of azurin in PBS. The heating rate was of 0.7 °C/min. The arrows in B indicate the pre-denaturation region. Normalized OD variation vs temperature at λ= 625 nm of azurin in C) ethanol- PBS mixtures with (——) 0.0, ($\cdots\cdots$) 0.015, (– · – · –) 0.023, (– – – –) 0.032 and (– ·· –) 0.040 molar fractions and in D) n-butanol-PBS mixtures with (——) 0.0, ($\cdots\cdots$) 0.007, (– · – ·) 0.010 and (– – –) 0.015 alcohol molar fractions.

χ_f=0.015, it results that T_d is 76 °C for the water-ethanol mixture, while it is 71 °C for the water-n-butanol one. Clearly the chain length of the two alcohols is at the origin of the difference observed. This result is similar to those found for other proteins such as RNase A and lysozyme [14, 40, 41], which show a down shift of the thermal denaturation temperature depending both on the alcohols side chain length and their molar fraction.

The overall effect observed on the active site, hence on the protein structure is, at least, the sum of two: one coming from the variations of the solvent properties   and the other from the increase of the thermal molecular motion. It is very difficult to separate quantitatively the two contributions. Neverthless, something can be said on the possible effect firstly of organic additives and then of temperature.

At RT, the non polar amino acid residues on the azurin surface exposed to the solvent are surrounded by water molecules forming well defined structures, such as clathrates, so that the structure of the protein is stabilized [41]. The addition of increasing amount of cosolvents decreases the water activity. Of course, the clathrates stability reduces and the alcohol molecules gain the solvation shell of the protein. This fact gives rise to the disruption of the clathrate structures which is followed by the destabilization of the protein structure as suggested by the 8 % loss of the OD at 625 nm in the presence of the highest alcohols molar fraction [42].

At high temperature, the solvation around the macromolecule is much more random even in pure water and the protein is destabilized; the thermal energy increases the global breathing motion [43] of the protein. As the temperature is increased and the properties of the solvent changed due to the alcohols, the breathing motion increases to such an extent as to cause the disruption of the Cu^{++} coordination sphere at lower temperature if compared with pure water. Of course, the dynamics of formation and breaking of hydrogen bonds which couples the molecular protein conformations and water fluctuations results modified, also. Time-dependent thermal denaturation of azurin (not reported) in different alcohol-water mixtures, followed at 75 °C, have shown that the kinetic of the denaturation process is biphasic. This result can be interpreted in two different ways: i) denaturation of two predominant different conformational substates assumed by the protein at room temperature in the presence of alcohols; ii) the action of the cosolvent molecules is essentially directed toward the protein region where the Cys-112 is located. The simultaneous action of solvent and temperature first removes the sulfur atom of this residue from the Cu^{++}-N_2SS^* active site of azurin with consequent rapid disappearing of the $S\pi \rightarrow d_{x^2-y^2}$ charge transfer band of the protein at $\lambda=625$ nm. The residual optical absorption, likely due to crystal field d-d transitions of remaining ligands, disappears slowly with temperature giving rise to the second decay process.

At present, the second hypothesis seems more plausible. In fact, although two possible structures have been suggested for crystallized azurin [36], in solution and at RT they should be no more observable due to the large atomic positional fluctuations.

Conclusion. The thermal denaturation of azurin, in presence of cosolvents, has allowed to get interesting informations both on the protein structure and protein-solvent interaction. The results of the present work can be synthesized as follows: i) the two thermal denaturation steps observed in pure water are related to the shell structure of azurin and to its interaction with the solvent. In pure water, the temperature increases the fluctuations of the interfacial hydrogen bonds network and, at the end, disrupts the coordination sphere around the Cu^{++} ion; ii) the effects of the alcohols on azurin are similar to those observed with other proteins, i.e., lowers the temperature of the pre-denaturation step and that of the full denaturation of the Cu^{++}-N_2SS^* active site. These effects depend on the side chain length of the alcohols used. iii) The target of the alcohols is the region of the azurin outer shell where the Cys-112 is located, furnishing the ligand sulfur atom responsable of the CT-band at $\lambda=625$ nm. The simultaneous action of alcohol and temperature in the thermal denaturation process first moves the Cys sulfur atom from the coordination sphere of the metal ion and, after, the nitrogens and the sulfur atom of methionine, giving rise to the biphasic decay.

Acknowledgements
Thanks are due to CNR and MURST for research grants.

References

1. R. McDaniel, T. McIntosh, and S. Simon, "Nonelectrolyte Substitution for Water in Phosphatidylcholine Bilayers", *Biochim. Biophys. Acta* **731**, 97-108, (1983).
2. T. O'Leary, and I. Levin, "Raman Spectroscopic Study of an Interdigitated Lipid Bilayer Di-palmitoylphosphatidylcholine Dispersed in Glycerol", *Biochim. Biophys. Acta* **776**, 185-189, (1984).
3. J.M. Boggs, and R. Rangaraj, "Phase Transitions and Fatty Acid Spin Label Behavior in Interdigitated Lipid Phase Induced by Glycerol and Polymyxin",

Biochim. Biophys. Acta **816**, 221-233, (1985).

4. J.A.Veiro, P. Nambi, L.C.Herold, and E. Rowe, "Effect of n-Alcohols and Glycerol on the Pretransition of Dipalmitoylphosphatidylcholine" *Biochim. Biophys. Acta* **900**, 230-238, (1987).

5. T. McIntosh, R. McDaniel, and S. Simon, "Induction of an Interdigitated Gel Phase in Fully Hydrated PC Bilayers", *Biochim. Biophys. Acta* **731**, 109-114, (1983).

6. S. Simon, and T. McIntosh, "Interdigitated Hydrocarbon Chain Packing Causes the Biphasic Transition Behavior in Lipid:Alcohol Suspensions", *Biochim. Biophys. Acta* **773**, 169-172, (1984).

7. E. Rowe, "Thermodynamic reversibility of Phase Transitions. Specific effects of Alcohols on Phosphatidylcholines", *Biochim. Biophys. Acta* **813**, 321-330, (1985).

8. P. Nambi, E. Rowe, E., and S. McIntosh, "Studies of the Ethanol-Induced Interdigitated Gel Phase in Phosphatidylcholines using the Fluorophore 1,6-Diphenyl-1,3,5-Hexatriene", *Biochemistry* **27**, 9175-9182, (1988).

9. B. G. Tenchov, H. Yao, and I. Hatta, "Time-Resolved X-Ray Diffraction and Calorimetric Studies at Low Scan Rates. I. Fully Hydrated Di-palmitoyl-phosphatidylcholine (DPPC) and DPPC/Water/Ethanol Phases", *Biophys. J.* **56**, 557-568, (1989).

10. B.A. Cunningham, W. Tamura-Lis, L.J. Lis, and J.M. Collins, "Thermodynamic Properties of Acyl Chain and Mesophase Transitions for Phospholipids in KSCN", *Biochim. Biophys. Acta* **984**, 109-112, (1989).

11. J.L. Slater, and C.H. Huang, "Interdigitated Bilayer Membranes", *Prog. Lip. Res.* **27**, 325-359, (1988).

12. G. Cevc, and D. Marsh, <u>Phospholipid Bilayers: Physical principles and Models</u>, J. Wiley & Sons, New York, 1987.

13. A. Cupane, D. Giacomazza, F. Madonia, P.L. San Biagio, P.L., and E. Vitrano, in: <u>Development in Biophysical Research,</u> (A. Borsellino, P. Omodeo, R. Strom, A. Vecli, and E. Wanke, Eds.), pp. 269-277, New York, 1980 .

14. P.H. Von Hippel, and K.Y. Wong, "On the Conformational Stability of Globular Proteins", *J. Biol. Chem.* **240**, 3909-3923, (1963).

15. L. Sportelli, A. Desideri, and A. Campaniello, "Isotopic Effect on the Kinetics of Thermal Denaturation of Ceruloplasmin", *Z. Naturforsch.* **40 C**, 551-554, (1985).

16. R. Jaenicke, "Protein Structure and Function at Low Temperatures", *Phil. Trans. R. Soc. Lond.* **B 326**, 535-553,(1990) .

17. E.T.Adman, G.W. Canters, H.A.O. Hill, and N.A. Kitchen, "The Effect of pH and Temperature on the Structure of the Active Site of Azurin from Pseudomonas Aeuriginosa", *FEBS Letters* **143**, 287-292, (1982).

18. A.J. Doig, and D.H. Williams, "Is the Hydrophobic Effect Stabilizing or Destabilizing in Proteins ?", *J. Mol. Biol.* **217**, 389-398, (1991) .

19. G. Némethy, W.J. Peer, and H.A. Scheraga, "Effect of Protein-Solvent Interactions on Protein Conformation", Ann. Rev. Biophys. Bioeng. **10**, 459-497, (1981).

20. P.L. Privalov, and S.A. Potekhin, "Scanning Microcalorimetry in Studying Temperature-Induced Changes in Proteins", *Methods in Enzimology* **131**, 1-51, (1986).

21. D. Marsh, and A. Watts, in: <u>Lipid-Protein Interactions</u> (P.C. Jost, and O.H. Griffith, Eds.) Vol. **2**, pp. 53-126, Wiley-Interscience, New York, 1982.

22. P.C. Jost, and O.H. Griffith, in: <u>Spin Labeling.Theory and Applications,</u> (L.J. Berliner, Ed.), pp. 454-523, Acad. Press, New York, 1976.

23. D. Marsh, in: <u>Membrane Spectroscopy</u>, (E. Grell, Ed.), pp. 51-142, Springer, New York, 1981.
24. P.L. Nordio, in: <u>Spin Labeling. Theory and Applications.</u> (L.J. Berliner, Ed.), pp. 2-52, Acad. Press, New York, 1976,.
25. A. Sanson, M. Ptak, J.L. Rigand, and C.M. Gary-Bobo, "An ESR Study of the Anchoring of Spin-Labeled Stearic Acid in Lecithin Multilayers", *Chem. Phys. Lipids* **17**, 435-444, (1976).
26. R. Bartucci, N. Gulfo, and L. Sportelli, "Effect of High Electrolyte Concentration on the Phase Transition Behaviour of DPPC Vesicles: a Spin Label Study", *Biochim. Biophys.Acta* **1025**, 117-121, (1990).
27. E. Sackmann, in: <u>Biophysics</u> (W. Hoppe, W.Lohmann, H. Markl, H. Ziegler, Eds.), pp. 425-460, Springer, New York,1983.
28. D. Chapman, W.E. Peel, B. Kingston, and T.H. Lylley, "Lipid phase transitions in model membranes. The effect of ions on phosphatidylcholine bilayers", *Biochim. Biophys.Acta* **464**, 260-267, (1977) .
29. P. Sapia, and L. Sportelli, "Effects of High Ionic Strength on 1,2-dipalmitoyl-sn-glycero-3-phosphatidylcholine Multilamellar and Vesicle Mesophases: a Comparative Electron Spin Resonance Study", *Colloid and Surfices*, (1993), in press.
30. D. Marsh, "Molecular Motion in Phospholipid Bilayers in the Gel Phase: Long Axis Rotation", *Biochemistry* **19**, 1632-1637, (1980).
31.J.H. Davis, "Deuterium Magnetic Resonance Study of the Gel and Liquid Crystalline Phase of Di-palmitoylphosphatidylcholine", *Biophys. J.* **27**, 339-358, (1979).
32. M. Moser, D. Marsh, P. Meier, K.H. Wassmer, and G. Kothe, "Chain Configuration and Flexibility Gradient in Phospholipid Membranes. Comparison between Spin Label Electron Spin Resonance and Deuterium Magnetic Resonance and Identification of New Conformations", *Biophys. J.* **55**, 111-123, (1989).
33. J.M. Boggs, and J. Mason, "Calorimetric and Fatty Acid Spin Label Study of Subgel and Interdigitated Gel Phases Formed by Asymmetric Phosphatidylcholines", *Biochim. Biophys. Acta* **863**, 231-242, (1986).
34. J.M. Boggs, R. Rangaraj, and A. Watts, "Behavior of Spin Labels in a variety of Interdigitated Lipid Bilayer", *Biochim. Biophys. Acta* **981**, 243-253, (1989).
35. P. Kubesh, J.M. Boggs, L. Luciano, G. Maass, and B. Tummler, "Interaction of Polymixin B Nonapeptide with Anionic Phospholipids", *Biochemistry* **26**, 2139-2149, (1987).
36. E.I. Solomon, J.W. Hare, D.M. Dooley, J.H. Dawson, P.J. Stephens, and H.B. Gray, "Spectroscopic Studies of Stellacyanin, Plastocyanin and Azurin. Electronic Structure of Blue Copper Sites", *J. Am. Chem. Soc.* **102**, 168-178, (1980).
37. G.E. Norris, B.F. Anderson, and E.N. Baker, "Blue Copper Proteins. The Copper Site in Azurin from Alcaligenes Denitrificans", *J. Am. Chem. Soc.* **108**, 2784-2785, (1988).
38. I.A. Fee, "Copper Proteins-Systems Containing the Blue Copper Center", in: <u>Structure and Bonding</u>, (D.J. Dunitz, P. Hemmerch, R.H. Holm, J.A. Ibers, C.K. Jorgensen,J.B. Neilands, D. Reinen, and R.J.P. Williams, Eds.) Springer-Verlag, Berlin-Heidelbeg-New York, Vol. **23**, pp.2-60, (1975).
39. E.T. Adman, and L.H. Jensen, "Structural Features of Azurin at 2.7 Å Resolution", *Israel J. of Chem.***21**, 8-12, (1981) .

40. R.G. Biringer, and A.L. Fink, "Methanol-stabilized Intermediates in the Thermal Unfolding of Ribonuclease A", *J. Mol. Biol.* **160**, 87-116, (1982).

41. J.F. Brandts, and L. Hunt, "The Thermodynamics of Protein Denaturation. III. The Denaturation of Ribonuclease in Water and in Aqueous Urea and Aqueous Ethanol Mixtures", *J. Am. Chem. Soc.* **89**, 4826-4838, (1967).

42. R. Guzzi, and L. Sportelli, "Interaction of Azurin with Alcohols: An ESR Optical Absorption and Fluorescence Emission", *J. Inorg. Biochem.* **45**, 39-45, (1992).

43. J.L. Finney, B.J. Gellatly, I.C. Golton, and J. Goodfellow, "Solvent Effects and Polar Interactions in the Structural Stability and Dynamics of Globular Proteins", *Biophys. J.* **32**, 17-33, (1980).

Superstructural Informations in the Base Sequences of Nucleic Acids

G. BENEDETTI, P. DE SANTIS, M. FUA' , S. MOROSETTI, A. PALLESCHI ,
M. SAVINO and A. SCIPIONI
*Dipartimento di Chimica and Dipartimento di Genetica e Biologia Molecolare ,
Università di Roma "La Sapienza" P.le A.Moro 5,00185 Roma , Italy*

1. Introduction

In the recent past, the prevalent concept of the nucleic acids was restricted to the biological functions of conserving and carryng the genetic informations whose control was fully delegated to proteins. Such concept, however, is being in rapid evolution in that DNA and RNAs are now considered to represent a relevant part of the management of the complex physico-chemical transformations in living matter.
In fact, proteins are involved in a large variety of biological functions as a result of their structural polymorphism and conformational versatility due to the large chemical dispersion of the 20 different amino acid side groups which strongly modulate the conformational features of the polypeptide chain. In contrast, in the case of the nucleic acids, because of the low chemical variance and the rather homogeneity of the intramolecular forces, base pairing and stacking interactions, a strong structural degeneration of the polynucleotide chain should be expected. In fact, the perfect base pairing of the DNA double helix, restricts the structural variance in the biological conditions, to relatively slow modulations of the the B DNA structure along the helical axis which produce superstructures, but with a number of important consequences in its properties and biological functions.
In the case of RNAs whose biological roles are currently increased in number and importance, the incomplete base pairing should give rise to a large numbers of structures; these are, however, characterized by assemblage of tracts of double helix directed by the optimization of their stacking and by maximization of the tertiary base pairing under topological constraints.

2. Theoretical prediction of sequence dependent DNA Superstructures

There are at present few doubts that the intrinsic superstructural features of DNA have biological implications. The concept of sequence directed DNA curvature is in fact

93

N. Russo et al. (eds.), Properties and Chemistry of Biomolecular Systems, 93–108.
© 1994 *Kluwer Academic Publishers. Printed in the Netherlands.*

promoting a rapidly increasing interest about its physical origin and biological meaning. In fact, some functional regions of DNA as promoters, enhancers, origin of replications and transcriptions, sites of recombination and in general regulative regions are currently found to be curved.

Essentially, two models were proposed to explain the origin of the curvature in DNA: one is based on the nearest-neighbor differential interactions within the nucleotide steps which result in a sequence of wedges between base pairs whose integration along DNA bends the helix axis . The other model identifies the origin of bends at the conformational transitions between a non canonical structure characterizing poly A sequences and the standard B-DNA; such a model, implicitly, recognizes the existence of determinant distal and cooperative factors at the origin of the DNA curvature .

This paper illustrates achievements and prospects of a theoretical nearest-neighbor model of DNA curvature which we proposed few years ago, based on the local deviations from the standard B-DNA of the different dinucleotide steps conformations as obtained by minimizing their conformational energy [1-8].

It should be noted that among the different conformational and structural variations observed in the dinucleotide steps, only the angular deviations, (see Fig.1), are relevant for DNA bending and in particular the roll (ρ) and tilt (τ) angles , where the only average value of the twist angles (Ω) seems to be effective .

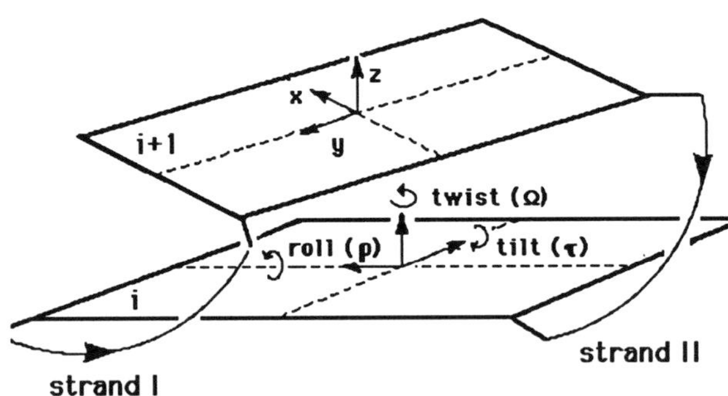

Figure 1. Orientational parameters of the dinucleotide step.

The relevant superstructural parameters are conveniently represented by the modulus and the relative phase of the local curvature per turn $\underline{C}(n)$ along the sequence number n

$$\underline{C}(n) = v/g \sum_{n-g/2}^{n+g/2} {}_{S}\, d(s)\, \exp(i\, w(s))$$

n, the sequence number; v, the average B DNA periodicity; g, the integration grid, usually, a period of the double helix or its multiple ; $d(s) = \rho - i\, \tau$, the deviation of the dinucleotide step at position s of the sequence from the canonical B DNA in terms of the roll and tilt angles and w(s), the corresponding phase angle with respect to the first dinucleotide step evaluated by summing the relative twist angles (Ω) .

This form is particularly suitable for the integration of the local deviations along a tract of DNA which results in the writhing of the DNA axis. In fact, using the moduli of the curvature as bend angles, and the phase increments as torsional angles of the segmental chain of the local helical axes, it is easy to obtain the 3-dimensional writhing of a DNA fragment.

The complex representation of the curvature corresponds to the linear term of the Taylor series of the pertinent matrix transformations which is justified by the smallness of the local deviations as already reported.

In addition, it contains the invariance of the modulus of curvature under the interchange of the two strands sequences, (in the correct 5'-3' direction) according the symmetry of the double helix, as well as the condition of coherence of the angular deviations sequence, d(s) with the periodicity of B-DNA. In fact, it is easy to realize that, independently on the approximations of the model, only the harmonic term of d(s) with frequency n, contributes to the curvature; as a consequence the addition of a constant to d(s) does not change the curvature although it changes the dinucleotide step structures.

We found that the theoretical values of the angular deviations ,reported below in matrix form,were in satisfactory agreement with the local structures of double helical dodecamers solved by X-ray crystal analysis (considering the plausible distortions due to the crystal packing) [4-6].

d, Ω	A	T	G	C
T	(8.0, 0.0), 35	(-5.4, -0.5), 36	(6.8, 0.4), 35	(2.0, -1.7), 36
A	(-5.4, 0.5),36	(-7.3, 0.0),35	(1.0, 1.6),36	(-2.5, 2.7),35
C	(6.8, -0.4),35	(1.0, -1.6),36	(4.6, 0.0),34	(1.3, -0.6),33
G	(2.0, 1.7),36	(-2.5, -2.7),35	(1.3, 0.6),33	(-3.7, 0.0),33

d=(ρ,-τ) and Ω in degrees.

The moduli of curvature, calculated in the case of ten-fold periodical biosynthetic DNAs, correlate very well with the gel electrophoresis retardation factors, R, quantified as the ratio between the apparent and the real chain lengths ; it is very interesting that the model correctly predicts the dramatic change of electrophoretic mobility of the two pairs of multimeric ten-fold synthetic DNAs investigated by Hagerman , which are different only for the polarity of the sequence [9].

Superstructures of biologically relevant DNA tracts were investigated and generally found to account very satisfactorily for the electrophoretic manifestations and for the electron microscopy visualizations as well as, in some cases, for the biological function.

Fig.2 illustrates the curvature profile, modulus and phase, along the sequence of the circular DNA 1453 bp TRP1ARS1 of yeast.

The prominent maximum corresponds to the replication origin which was experimentally found as a strongly curved DNA tract [10].

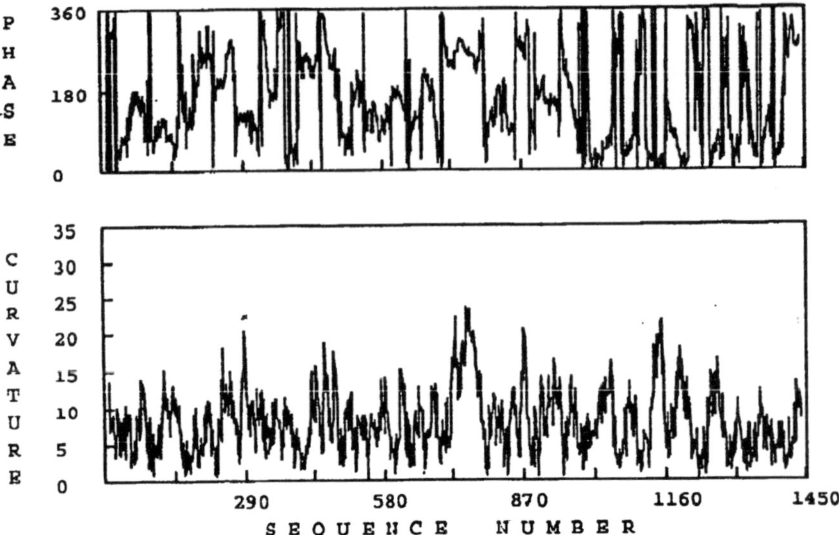

Figure 2. Curvature profile, modulus and phase, averaged on 21 bp, of the circular DNA 1453 bp TRP1ARS1 of yeast.

3. Dispersion of Curvature and Gel Electrophoresis Retardation

At present, the most sensitive tool for detecting the presence of curvatures in DNA tracts, as well as for localizing the main curvature sites, is represented by the polyacrylamide gel electrophoresis. In fact, curved DNA tracts are characterized by lower mobility than that of straight DNAs with the same molecular weight. The retardation factor quantifies such a property; it increases with the acrylamide concentration in the gel and decreasing temperature.

The site of the main curvature along the sequence of a DNA tract is currently localized with the Wu and Crothers permutation assay [11]; it corresponds to localize the minimum of a retardation plot of cyclically permuted DNA tracts , obtained by single restrictions from the tandem dimer of the tract considered.

We recently investigated the origin of the gel electrophoresis retardation in terms of curvature and found that it seems to be related to the central dispersion of the curvature, σ^2, namely to the angular dispersion of the writhing helical axis . Such a quantity is a measure of the average activation energy required for straightening the DNA axis if a simple Hooke's harmonic potential is adopted. As a consequence, it should be proportional to the logarithm of the ratio between the electrophoretic mobilities of the straight and the curved DNA fragment, practically, to the logarithm of the retardation, R [7].

In fact, Fig. 2 illustrates the correlation between log R and σ^2 calculated using matrix transformations and the theoretical local angular deviations, d(s), in the case of about

500 multimeric double helical oligomers investigated by various authors and different for sequence, periodicity [9,10,11,12,15, 20, 21] and molecular weight (50 - 200 bp). Further, we reproduced the trend of the permutation gel electrophoresis of DNA tracts investigated by different authors for their biological relevance [3,5-8].
The results are very convincing and offer an easy theoretical alternative to the experimental technique. In any case, the good agreement obtained further proves the validity of our theoretical model for predicting the intrinsic DNA superstructures.
The most critical test of the model is represented by the striking results obtained in predicting the gel electrophoresis retardation changes due to point mutations in a tract of SV40 DNA recently investigated by Milton et al.[12] .
Fig. 5 illustrates the case when each one of the bases along the 60 bp curved region of the SV40 tract (1836 - 2008) is substituted by Cytosine.
The general pattern of agreement is strikingly good taking into account that the model predicts the gel electrophoresis retardation changes induced in a tract of 173 bp by a single base substitution [7,13]. It is noteworthy that the large majority of artificially induced point mutations, decreases the retardation and then the related curvature, suggesting that this superstructural feature is plausibly biologically selected.
As a general conclusion we want to stress that our theoretical model, whilst based on a simple nearest-neighbor interaction hypothesis and on the linear response theory, is capable to translate the deterministic fluctuations of base sequences in pieces of information on DNA superstructure according with the experimental manifestations. Such a general pattern of agreement between theoretical and experimental data, convincingly proves the physical origin of the curvature and allows, adopting the theoretical calculations with large confidence, as a valid alternative to the experiments, the investigation of the implications of the superstructural features on the biological functions of DNA.

4. DNA Curvature in Recognition Mechanisms by Proteins: Intrinsic and Induced Curvature

The analysis of DNA-protein interactions is an essential step towards elucidating molecular mechanisms for genic regulation. Proteins are in fact, capable of recognizing and amplifying as well as inducing curvature in biologically relevant DNA tracts. The sequence dependent polymorphism of DNA allows the recognition of structural and superstructural features by proteins which plausibly precedes their intimate interactions with the specific base pairs along the double helix.
In fact, as a consequence of the intrinsic curvature, the relative amplitudes of the DNA grooves are proportionally and periodically changed with respect to the canonical B DNA. This produces a differential electrostatic field along the curved DNA tracts, changes in hydration, and, finally, in affinity towards proteins. Therefore, the curvature could, represent a first step of DNA recognition by proteins. The cooperation of short range forces could afterwards, result in an amplification of the curvature as monitored, in several cases investigated, by the parallel amplification of the permutation gel electrophoresis assays performed in the presence and in absence of protein. A similar mechanism probably holds for the nucleosome formation on DNA.

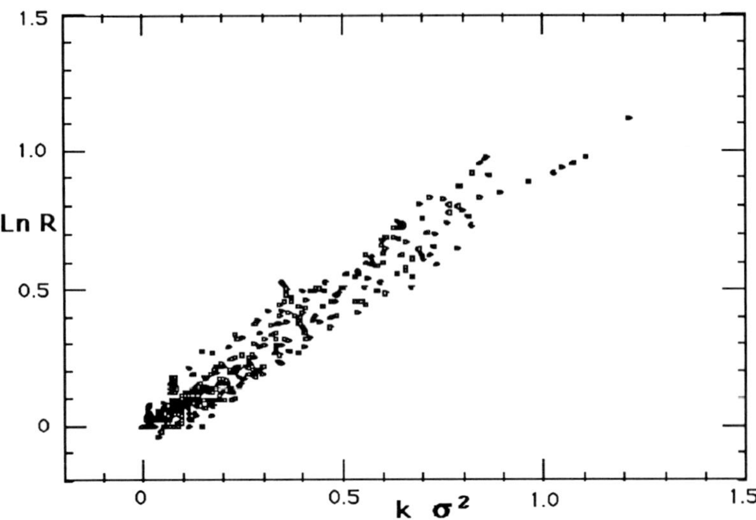

Figure 3. Correlation between gel electrophoretic retardation, R, and the dispersion of curvature, σ^2, for about 500 multimeric double helical oligonucleotides with different sequences, periodicities and lengths.

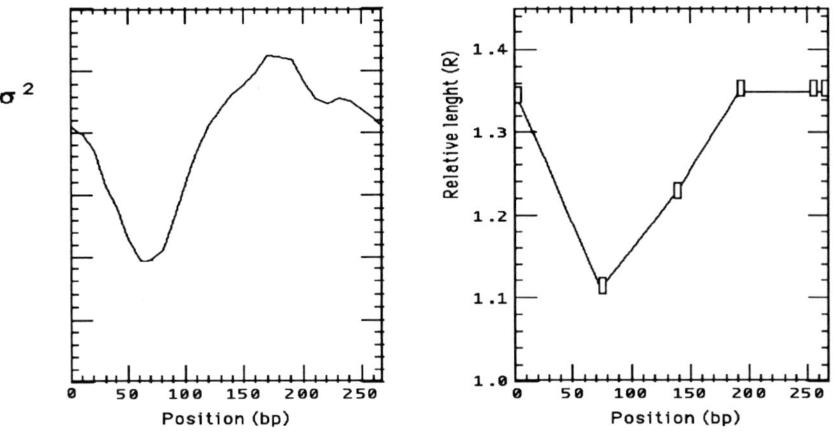

Figure 4. Comparison between experimental and theoretical permuted gel electrophoresis for the DNA tract 2534-2801 of SV40. ▯ represents the restriction enzymes positions.

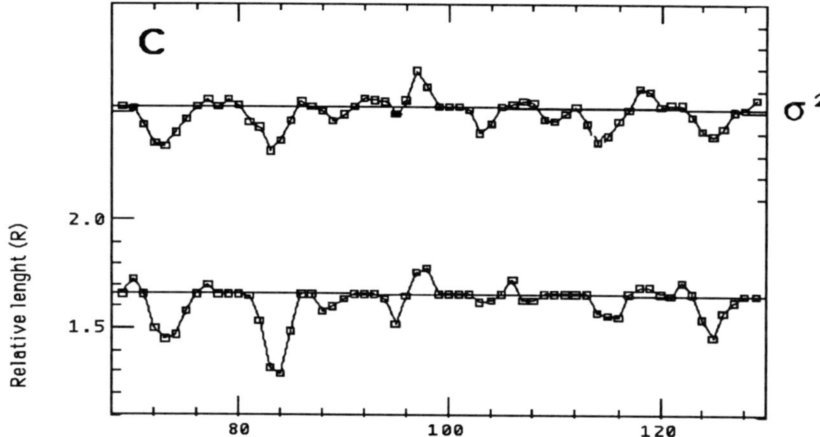

Figure 5. Experimental retardation (lower) in comparison with the theoretical σ^2 diagram (upper), when each one of the bases on the 60 bp curved region of the SV40 tract (1836 - 2008) is substituted by Cytosine.

In fact, often DNA interacts with proteins wrapping around the proteic core, as in nucleosome, and such nucleoproteic superstructures promote the next events of the replication, transcription or recombination. Therefore, the intrinsic curvature of DNA appears to be of general biological importance, containing informations which are recognized and amplified by the protein binding at a functional level.

The amplification and induction of curvature by interaction with proteins introduces the concept of bendability which is only in part sequence dependent involving the protein mojety. It is however currently known as the intrinsic anisotropic potential flexibility of DNA.

We analyzed this point adopting our theoretical model of the sequence depend curvature and assuming a principle of minimum deformation energy of the intrinsic superstructure necessary to obtain the induced superstructure.

On such a basis we have predicted the superstructures induced in DNA tracts by circularization or nucleosome formation.

Furthermore, the predicted nucleosome positioning and phasing along DNAs investigated is found in general agreement with the experiments .

Circularization of a DNA tract is the result of stochastic and deterministic motions of the double helix. DNA characterized by in phase bends have, in fact, a higher probability of cyclization with respect to a straight DNA with the same length.

We tried to simulate the circularization process starting from linear DNA tracts and constraining the ends to a correct ring closure.

We used conjugate gradient optimization method under the condition of minimum deformation energy evaluated in a simple harmonic approximation as proportional to the average of the square of the curvature deviation namely the dispersion of the local curvature as for the model of an elastic cylindric bar deformation [14)].

If $\underline{C}_0(n)$ and $\underline{C}(n)$ are the starting and final local curvature (per turn of the double helix), the deformation potential energy E is assumed as:

$$E = \langle\, (\underline{C}(n) - \underline{C}_O(n)\,) \mid (\,\underline{C}(n) - \underline{C}_O(n)\,) \rangle$$

Fig. 6 illustrates the curvature diagrams,modulus and phase, of the linear and the corresponding circular form of a DNA tract 169 bp investigated by Drew and Travers [15].
Fig 7 shows the experimental cleavage probability by DNAse I, averaged on both the strands, of the circular, with respect the linear, form. The typical trend mirrors the periodical deformation of the small groove which is generally contracted where it faces into the cycle then inhibiting the cleavage reaction. However, in some tracts of the sequence, a strange increase of the cleavage probability of the cyclic form is observed; it is very interesting that, accordingly,the curvature of these tracts results higher in the linear than in the circular form as indicated by the crossing of the corresponding profiles in Fig. 6 .

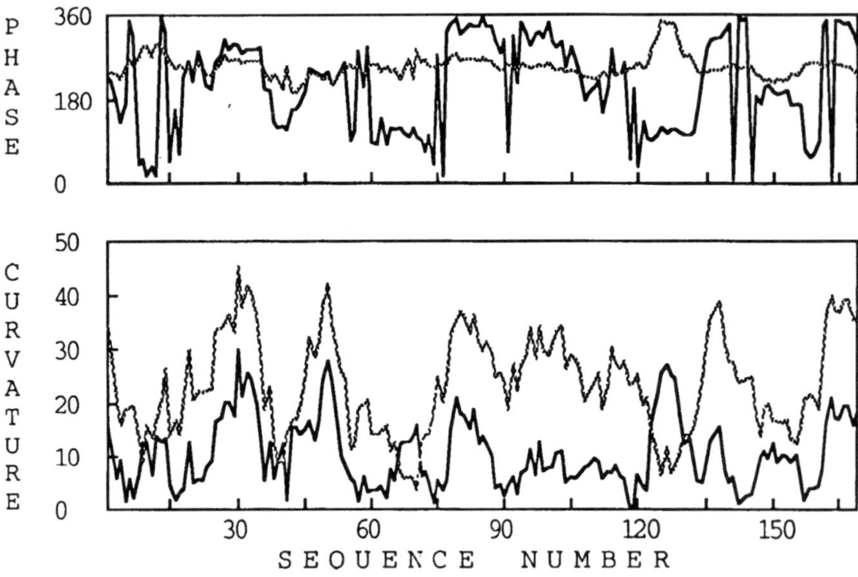

Figure 6. Curvature diagrams, averaged on 10 bp, of the linear (full lines) and the corresponding circular form (dot lines) of a DNA tract 169 bp investigated by Drew and Travers [15].

It is interesting that, in spite of the relatively low curvature which characterize such DNA tract, the same results were obtained for different cyclic permuting sequences strongly suggesting a dominant deterministic circularization process, in account of the high cooperativity of the structural transformations in a topological domain. In order to reduce the calculation times, an alternative and new mathematical approach was attempted: it corresponds to find the deformations of a uniform cyclic DNA under the force field of the differential interactions along the sequence.

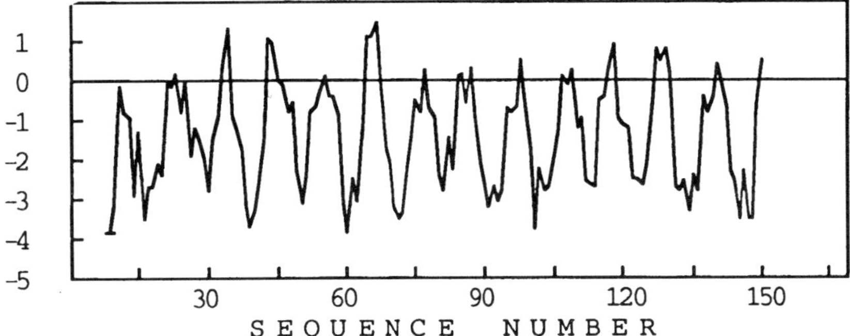

Figure 7. Logarithm of the differential cleavage probability by DNAse I (circular-linear) from the reference [15] but averaged on both the strands.

Such deformations are given in the curvature complex representation by a Fourier series, with variational amplitudes and phases. These are easily obtained by minimizing the deformation energy E with respect to the Fourier coefficents using a stepeest descent method. The amplitude of the first harmonic was set equal to zero in order to mantain fixed the cyclic form.

In fact such a condition, assures the persistence of the cyclization for small deformations of the structure: in fact, only the first harmonic provides changes of the cycle to concerted local curvature which may affect its integrity; it corresponds, in the linear approximation of the response theory, to the integration of the effects of a periodical perturbation on a periodical structure which is non zero only when the periods coincide. Further, the amplitude of the zero order was fixed in order to assure the topological invariance of the cyclic DNA with the modulus equal to $360°$ divided the nearest integral number of DNA turns and the phase equal to that of the average local curvature profile of the linear DNA.

Practically, the method should require iterative calculations of the harmonic components within a prefixed resolution power. However, taking advantage from the theorem of the minimum quadratic average deviation of two functions, given in terms of Fourier series, and using the Parseval's formula, it is possible to analitically solve the problem of finding the most stable superstructure of a circular DNA in terms of that of the corresponding linear DNA [14].

Thus, adopting the harmonic approximation of the deformation energy, the curvature function of the cycle, $\underline{C}'(n)$, results:

$$\underline{C}'(n) = \underline{C}(n) - A0 - A1 + A'0 + A'1$$

where A0 and A1 are the complex components of the zero and the first harmonics of the Fourier series of the intrinsic curvature, whereas A'0 and A'1 are the corresponding harmonics of the cyclic DNA, as defined before.

We have tried such a method of minimum harmonic deformation energy in the case investigated by Drew and Travers and found exactly the same result as previously obtained, but in two seconds of a personal computer, which is extraordinarily shorter.

The success of such a method prompted us to its extension to   problems where the writhing of a chain is changed under the action of a force field as in the protein folding and for what concerns this paper, in the nucleosome formation along DNA sequences.

5.   Prediction of nucleosome positioning

Nucleosome positioning along a DNA sequence is defined by two parameters: the translation, marking where the histone octamer dyad axis is placed and the orientation of DNA relative to the direction of curvature. Many authors agree that little is known about the translation, whereas the rotational parameter appears to involve structures where the A/T rich minor grooves face in towards the protein core.

Adopting the model of the minimum harmonic energy deformation, calculations were tried to predict the nucleosome formation along DNAs. We have designed a computer program to predict the nucleosome positioning along DNAs , by localizing the minima of the deformation energy function required to distort recurrent tracts 145 bp of the sequence in a nucleosomal superstructure; the values of the distortion energy are given in terms of that of an equivalent tract of Poly dA-Poly dT which is well known to be unable to form nucleosome.

Single nucleosome positions were succesfully predicted for some DNA tracts investigated in different laboratories; particularly interesting the prediction in the case of Alu-repeat of ribosomal DNA in Drosophila which was experimentally confirmed by Mirzabekov [16].

The possibility of finding the virtual positioning of nucleosomes along DNA sequences by localizing the minima of the distorsion energy function, (evaluated at a rate of a nucleosome per second, of a personal computer) provides a suitable theoretical method to predict the translation and the rotational parameters of a set of nucleosomes along DNAs, as well as their phase relations.

Thus, we extended the model to the more complex case of the   formation of several nucleosomes on long DNA tracts [14].

Fig. 7 illustrates the case of the circular DNA 1453 bp TRP1ARS1 of a yeast plasmid. The minima of the distortion energy obtained, represent the virtual positions of the nucleosome dyad axis along the DNA sequence.

The obvious exclusion condition of nucleosomes at distances less than 160 bp, allows an optimal assembly of seven nucleosomes. This appears to be closely related with the experimental distribution of the nucleosome positions obtained by Thoma [17] for the reconstituted minichromosome, in particular the ARS1 region appears to be nucleosome free.

6. RNAs   Secondary and Tertiary Structures from the Sequence

Wide classes of RNA molecules (for example tRNAs and rRNAs) are involved in a structural role rather than in carrying information [18-19].

Moreover, in the last years the role of these molecules are extended to the management of information through structural interactions; see for example the maturation process and the completely new functional ability exhibited by the ribozymes [20].

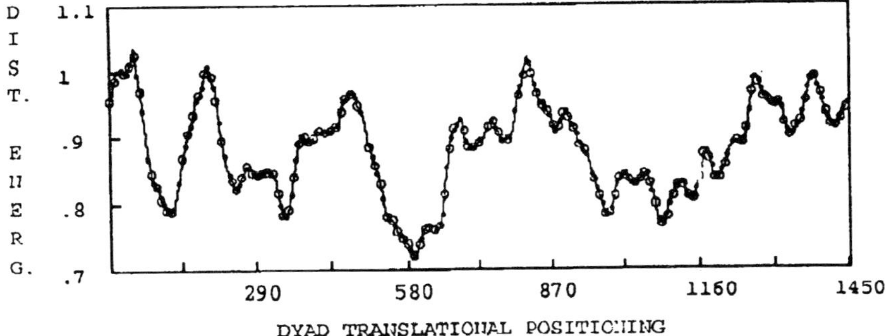

Figure 8. Deformation energy of the circular DNA 1450 bp TRP1ARS1 of a yeast plasmid. The minima represent the virtual positions of the nucleosome dyad axis along the DNA sequence.

The separation between the two biological roles results to be only an oversimplified model. This behavior can be important in the field of biotechnological applications and it can suggest new hypotheses on the origin of life [21].

The importance of the role of secondary and tertiary RNA structure in biological processes has grown in the last years [18, 20]. It is now generally assumed that primary sequence carries the information required for its actual three-dimensional folding. Predicting secondary structure first and then proceeding on to tertiary structure can be supposed to be a fruitful, if not infallible, approach. The main free energy contributions derive from hydrogen bonds in AU, GC and GU base pair and stacking interactions among them. The problem resides in the optimization of these interactions under geometrical and topological constraints: the molecule must fold without crossing itself or forming knots.

The free energy evaluation is based on the favorable stacking interactions and the destabilizing loop terms as obtained by different authors [22] and considered as additive free energy contributions. With the increase in length and number of the determinated nucleic acid sequences, there has been a growing need for algorithms that can efficiently search for the more probable secondary structures. Several methods have been developed to this aim [23 and reference therein]. Some of them predict only one optimal free energy structure for each nucleotide sequence. Alternate equivalent and suboptimal free energy structures are not identified despite their possible biological significance, and the free energy error inherent in these evaluation methods.

Moreover they generally require a computational time proportional to N^3, and memory requirements of the order of N^2, where N is the number of nucleotides in the sequence. The experimental (enzimatic and chemical) data, if considered, are generally introduced into the programs as rigid constraints: discarding or imposing some helical regions.

We developed an algorithm able to select a set of optimal free energy structures, with computer time requirement proportional to N^2, and with the possibility of introducing a gradual competition between the experimental data and the free energy content of the structure [23]. Our method is based upon the use of an original mathematical process rising from the convolution theorem, which is applied to a discrete function whose terms are tetradimensional complex vectors representing the bases. This allows to obtain a list of all possible helical regions that can be derived from a given sequence. The topologically compatible helices are collected to give the possible secondary structures. The growth of the number of possible structures caused by the increase of the number of considered helical regions, requires a "pruning" criterion. It consists in the exclusion of the alternative substructure if it has an evaluation function worse than the previous one deprived of the incompatible helical regions. The evaluation function takes into account the free energy and the experimental data when available. The methods were applied to Tetrahymena thermophila rRNA IVS fragment [24], the first molecule exhibiting self-splicing ability [25], and enzymatic activity [20], with the aim to select a set of the deepest free energy RNA secondary structures under constraints of model hypotheses and experimental evidences. The secondary structures obtained are characterized by the close proximity of self-reactions sites (imposed via model hypothesis) and account for double mutations experiments, and differential digestion data; they are shown in Fig.9 and 10.

In a following step we folded the obtained secondary structures, in low resolution tertiary models, which kept up the proximity of the catalytic sites also in the space [26].

Most of the building rules were derived from consideration of the experimental structures of tRNA [27] and X-ray fiber diffraction [28] as it follows:

a) RNA duplexes were assumed to have A-form RNA helix conformation;

b) the sites directly or indirectly involved in self-splicing reaction are located in convenient positions for transesterification reaction;

c) the helical axes of parallel strands are 22 Å distant ;

d) two helical regions joined by single strand(s) are colinearly stacked when possible; otherwise they are placed in adjacent position following c) criterion;

e) single bulged bases were stacked within helix;

f) a distance of 6 Å was allowed between two adjacent nucleotides in single strand regions.

Following these rules we realized the whole three-dimensional model of the best free energy secondary structure of IVS. The stereoview of the model is shown in Fig. 11

The proposed tertiary folding seems to provide for a better explanation to the transesterification mechanisms and moreover it is in good agreement with the experimental data (activity of mutants, enzymatic cleavages, phylogenetically conserved regions).

Functionally homologous RNA sequences can substantially diverge in primary sequence but it can be reasonably assumed that they are related in higher degree structure. As a consequence the common features in secondary and tertiary structures are more important than the consensus of base sequences. Therefore, the most powerful approach in functionally related sequences appears to finalize the search to the common secondary structures rather than to compare the primary sequences. The problem to find such structures, and simultaneously to satisfy as far as possible the free energy minimization criterion, is considered in two aspects. First: a quantitative measure of the folding consensus among secondary structures is defined translating each structure into a linear representation and using the correlation theorem to compare them. Second: an algorithm for the parallel search for secondary structures according to the free energy minimization criterion, but with a filtering action on the basis of the folding consensus measure is developed [29].

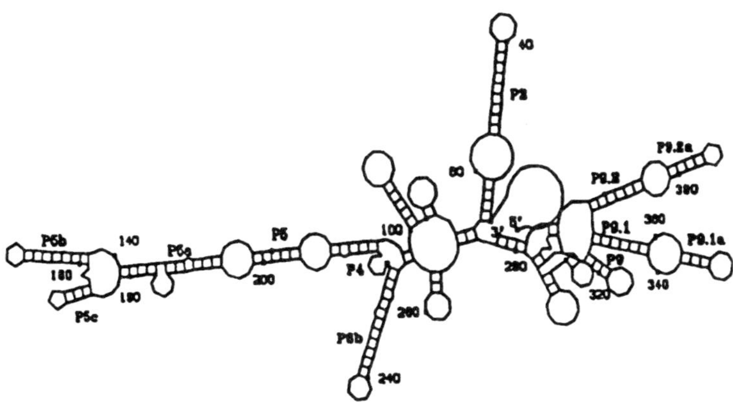

Figure 9. The IVS minimum free energy secondary structure obtained with experimental data and 1-15 cyclization model hypotheses imposed. The enzymatic cleavages are shown.

Figure 10. The mutant U14G minimum free energy secondary structure for the 1-19 cyclization reaction.

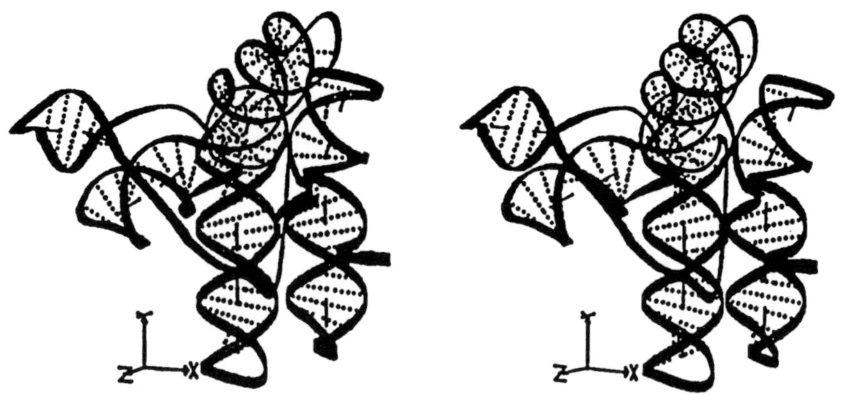

Figure 11. Stereoview of our proposal of three-dimensional folding.

The method has been tested on groups of RNA sequences, different in origin and in functions, for which proposals of homologous secondary structures based on experimental data exist. Its application also allows the "transfer" of experimental data available for one sequence, to a functionally related and therefore homologous one. A result is shown in Fig.12.

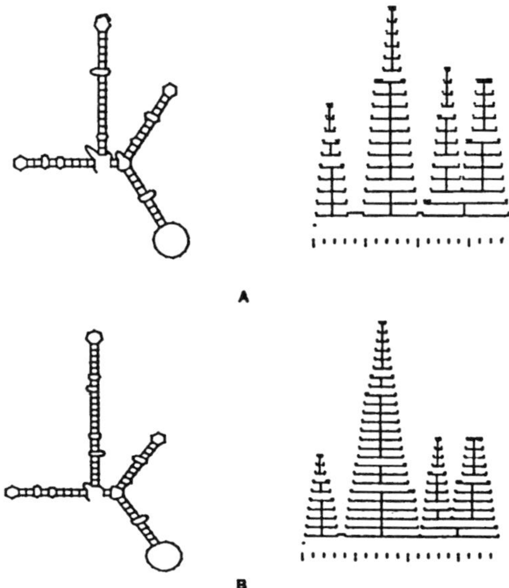

Figure 12. Human (A) and rat (B) U3 snRNAs. Stem and loop (left) and mountains (right) representations of the homologous better free energy secondary structures are show.

7. Concluding remarks

The implication of DNA curvature in mechanisms which govern biological processes
such as replication , transcription and chromatin organization, appears to be a complex
problem involving static and large scale dynamic effects and certainly requires further
investigations.
However, the theoretical model we adopted, which is based on the integration of the
local deviations from the canonical B DNA structure due to differential interactions
along the sequence as evaluated by conformational energy calculations, appears to be
suitable for translating the sequence fluctuations in superstructural elements of DNA. It
allowed the prediction of superstructures of a large set of biosynthetic periodical.
DNAs as well as of biologically relevant tracts of natural DNAs and their electrophoretic
manifestations in excellent agreement with the experiments. Beside this reliability, it
shows a surprisingly high sensitivity as demonstrated by the good results obtained in
the prediction of the electrophoretic changes observed in a tract of SV40 DNA after
extensive point mutations.
 The results obtained, adopting a simple harmonic model to evaluate the deformation
energy, are very promising; they allow the theoretical prediction of fine effects of
nuclease activity along DNA sequences as well as the 'a priori' positioning and phasing
of nucleosomes. This indicates that the intrinsic superstructures of DNA are the main
determinant of the induced superstructures so providing the possibility to predict the
virtual recognition sites of regulative proteins.
The important ability of DNA to filter and productively integrate the slightly different
stereochemistry of the nucleotide steps giving rise to sequence dependent
superstructures, is not yet quite evaluated, but it is easy to predict that it could get an
insight into the molecular mechanisms of management of the still latent informations
encoded in DNA.
In the case of RNAs structures the usefulnes of the model resides in its ability in
obtaining a wide set of alternative secondary structures also for long sequences,among
which selecting that biological significant on the basis of experimental evidence.
Two possible applications can be suggested: searching for a universal model for
selfsplicing RNAs, as class I introns, and then designing the minimal sequence which is
able to preserve the essential features of the threedimensional structure and ,as a
consequence, the working mechanism of the ribozymes; predicting the secondary
structures which can account for alternative proteins obtained by transcription under
different biological conditions .
Finally, a more ambitious task can be the proposal of secondary structures that can be
involved in interactions with proteins in the still unknown ribosome structure.

References

1. P. De Santis, S. Morosetti, A. Palleschi and M. Savino, in "Structures and
    Dynamics of Nucleic Acids, Proteins and Membranes", (E.Clementi and S. Chin
    eds.), pp 31, (1986).

2. P. De Santis, S. Morosetti, A. Palleschi, M. Savino, and A. Scipioni , in "Biological and Artificial Intelligence System", (E. Clementi and S. Chin eds.), pp 143, (1988).

3. P. De Santis, A. Palleschi, M. Savino, and A. Scipioni, *Biophys. Chem.* **32**, 305, (1988).

4. S. Cacchione, P. De Santis, D. Foti, A. Palleschi, and M.Savino, *Biochemistry* **28**, 8706, (1989).

5. P. De Santis, A. Palleschi, M. Savino, and A. Scipioni, *Biochemistry* **29**, 9269, (1991).

6. D. Boffelli, P. De Santis, A. Palleschi and M. Savino, *Biophys. Chem.* **39**, 127, (1991).

7. P. De Santis, A. Palleschi, M. Savino, and A. Scipioni, *Int. J. Quantum Chem.* **42**, 1409, (1992).

8. D. Boffelli, P. De Santis, A. Palleschi, G. Risuleo and M. Savino, *FEBS Lett.* **300**, 175, (1992).

9. P.J. Hagerman, *Nature* **321**, 449, (1986).

10. F. Thoma, *Biochim. et Biophys. Acta* **1130**, 1, (1992).

11. H.M. Wu and D.M. Crothers, *Nature* **308**, 509, (1984).

12. D.H. Milton, M.C. Casper and R.F. Gesteland, *J. Mol. Biol.* **186**, 773, (1990).

13. P. De Santis, A. Palleschi and M. Savino, *Biophys. Chem.* **42** , 147, (1992).

14. P. De Santis, M. Fua', A. Palleschi and M. Savino, *Biophys. Chem.* (in press).

15. H.R. Drew and A.A. Travers, *J. Mol. Biol.* **186**, 773, (1985).

16. A. Mirzabekov, (private communication).

17. F. Thoma, *J. Mol. Biol.* **190**, 177, (1986).

18. L.P. Eperon, J.P. Estibeiro, & I.C. Eperon, *Nature* **324**, 280-282, (1986).

19. G. Stoffler, et al. In Ribosomes: Structure, Function and Genetics (G. Chambliss, G. Craven, J. Davies, K. Davis, L. Kahan, & M. Nomura eds), pp. 171-205, University Park Press, Baltimore MD, (1980).

20. A.J. Zaug, M.D. Been, & T.R. Cech, *Nature* **324**, 429-433, (1986).

21. J. Maddox, *Nature* **342**, 609-613, (1989).

22. S.M. Freier, R. Kierzek, J.A. Jaeger, N. Sugimoto, M.H. Caruthers, T. Neilson, and D.H. Turner, *Proc. Natl. Acad. Sci. USA* **83**, 9373-9377, (1986).

23. G. Benedetti, P. De Santis, and S. Morosetti, *Nucleic Acids Research* **17**, 5149-5161, (1989).

24. G. Benedetti, P. De Santis, and S. Morosetti, *J. Biomol. Struct. Dyn.* **7**, 1269-1277, (1990).

25. T.R. Cech, *Science* **236**, 1532-1539, (1987).

26. G. Benedetti, P. De Santis, and S. Morosetti, *J. Biomol. Struct. Dyn.* **8**, 1045-1055, (1991).

27. S.-H. Kim, in Transfer RNA: Structure, Properties, and Recognition Cold Spring Harbor Monograph Series 9A eds P. Schimmel, D. Soll, and J. Abelson, (Cold Spring Harbor Laboratory, Cold Spring Harbor, NY), 83-100, (1979).

28. S. Arnott, *Progr. Biophys. Mol. Biol.* **21**, 267-319, (1970).

29. G. Benedetti, and S. Morosetti, *Eur. J. Biochem.* **202** , 241, (1991).

DNA Superstructures: Relevance on Physicochemical Properties and in Recognition Mechanism with Proteins

S. CACCHIONE, P. DE SANTIS, L. FANTUZZI, L. LEONI, B. SAMPAOLESE, M. SAVINO and A. TUFILLARO.
Dipartimento di Genetica e Biologia Molecolare. Dipartimento di Chimica. Centro per lo Studio degli Acidi Nucleici del CNR. Università di Roma "La Sapienza", P.le A. Moro 5, 00185 Roma.

1. Introduction

DNA curvature is now considered of great importance in determining its physicochemical properties as well as in regulating specific interactions with proteins. The idea of deviations of B DNA double helix from a straight line was firstly advanced by Trifonov, who observed a harmonic component of the base distribution in eukaryotic nucleotide sequences and associated to this component a geometrical meaning, mainly the wedge of the AA dinucleotide [1]. This hypothesis received experimental support from the observation that some DNAs migrate more slowly than normal DNA molecules, having the same molecular weight, on polyacrylamide gel electrophoresis [2]. In the last few years, numerous research groups have spent strong efforts to correlate DNA superstructural features to the nucleotide sequence as well as to develop methods to experimentally characterize curved, with respect to straight, DNAs [3]. DNA superstructural features have resulted of great relevance in the control of important biological processes such as transcription [4] and chromatin organization [5]. Taking advantage of the theoretical method developed by De Santis and coworkers [6,7], which derives DNA curvature from the nucleotide sequence, and of the rapidly increasing number of known sequences of DNA regulative regions derived via genetic engineering, it is now possible to predict the superstructural features of these sequences and to try a correlation with their biological role. Furthermore, on the basis of theoretical curvature profiles of the investigated DNA tracts, it is easier to plan experimental approaches to verify the theoretical prediction as well as to evaluate the physicochemical properties of the considered DNAs. We want to report in the present paper two examples of biologically significant sequences studied according to this approach. The first one refers to the sequences which regulate transcription in the pea multigenic *rbc*S family, which code for the small subunit of ribulose bisphosphate carboxylase, a key enzyme in photosynthesis. A very satisfactory correlation between the theoretically predicted superstructural features of these sequences and their physicochemical properties, such as electrophoretic mobility, circular dichroism and cyclization probability, was found. The second system is a DNA fragment about 500 bp long, containing a highly curved sequence derived from the protozoa *Crithidia fasciculata* [8].

109

N. Russo et al. (eds.), Properties and Chemistry of Biomolecular Systems, 109–126.
© 1994 *Kluwer Academic Publishers. Printed in the Netherlands.*

The influence of this curved sequence in nucleosome positioning has been predicted on the basis of DNA distortion energy, finding also in this more complex system a surprisingly good agreement between theoretical prediction and experimental evaluation.

2. Materials and methods

2.1. OLIGOMERS SYNTHESIS AND MULTIMERIZATION

The LREs, 62 bp long Oligodeoxyribonucleotydes, were synthesized on a Biosearch DNA synthesizer and purified on a 15% polyacrylamide gel in the presence of 7M urea, followed by high salt (0.1 M ammpnium bicarbonate) elution from a Sphadex G-50 column. Oligomer sequences were purified by using the chemical sequencing technique of Maxam and Gilbert. Their sequences are as follows:

LRE-3A -170 ACACAAAATTTCAAATCTTGTGTGGTTAATA -140

LRE-3.6 -170 ACACACAACTTTTCAATCTTGTGTGGTTAAT -140

LRE-3A -139 TGGCTGCAAACTTTATCATTTTCACTATCTA -109

LRE-3.6 -139 ATGGCTGCAAAGTTTATCATTTCACAATCTA -109

Oligonucleotides were phosphorylated and complementary sequences were annealed as previously described [9]. Multimers were obtained by ligation with T4 DNA ligase as previously reported [9].

2.2. RETARDATION MEASUREMENTS

Ligated products were run on a non denaturing 8% polyacrylamide gels. In the measurements of the complexes with spermine, this polyamine was present in the gel as well as in the running buffer.

2.3. CIRCULAR DICHROISM SPECTRA

Circular dichroism (CD) spectra were obtained with a Jasco model J 500 A dichrograph. The CD spectra were obtained in a 1 cm pathlenght quartz cell where the oligonucleotides concentration fall within 0.4-0.5 OD units/ml. The spectra showed a good reproducibility ruling out relevant effects on CD features due to the slight turbidity of spermine-DNA complexes solutions, at the highest neutralization ratio.

2.4. CYCLIZATION KINETICS

DNA fragments, 410-3A and 410-E9, were obtained from pUC-3A/E9 and pBR325-rbcS-E9 clones kindly provided by N-H. Chua. The two DNAs were restricted with HindIII (3A) and HinfI and HindIII (E9). Protruding ends were filled with the Klenow enzyme in the presence of all four dNTPs. The resulting fragments were ligated to octameric and dodecameric *Hind*III linkers, and then extensively cut with *Hind*III, in order to have two fragments of the same length and with the same protruding ends. The two 410 DNAs were then cloned in pUC18. The sequences were verified by the dideoxy sequencing approach, modified for sequencing of double-stranded DNAs 0.5 µg of 410-3A and 410-E9 were treated with calf intestinal alkaline phosphatase and 5'-labelled by incubation with 50 µCi γ^{32}P ATP and Polynucleotide Kinase. Cyclization reactions were carried out in 10 mM Tris pH 7.5, 50 mM NaCl, 10 mM $MgCl_2$, 5 mM dithithreitol, 0.25 mM ATP at 20 °C. DNA concentration was set at 100 ng/ml to minimize dimer formation. After ligase addition, at a final concentration of 0.5-1 U/ml, aliquots were removed and quenched with EDTA at 50 mM and heating at 65 °C for 10 min. Samples were run on 4% agarose (NuSieve 3:1, FMC). Gels were dried and DNA was visualized by autoradiography. Quantity was obtained by scanning densitometry.

2.5. NUCLEOSOMES PREPARATION AND ASSEMBLY

The 524 bp DNA containing the 211 bp highly curved sequence from *C. fasciculata* was prepared by cleaving with *Hind* III the pPK201/CAT plasmid, kindly provided by E. Di Mauro, and 5' labelled with γ^{32}P-ATP and Polynucleotide Kinase according to standard procedures. Nucleosomes were prepared by cleavage with Micrococcal Nuclease of chromatin extracted from chicken erithrocyte nuclei, and separation of mononucleosome with 5-20% sucrose gradient ultracentrifugation. Modified nucleosomes, namely nucleosomes lacking NH_2 terminal histone domains, were obtained from normal nucleosomes by digestion with immobilized trypsin. Nucleosomes assembly on the 524 bp DNA fragment was performed by adding a constant aliquot of nucleosome (normal or trypsinized) to a varying amount of 524 bp DNA at 0.8 M NaCl, newly nucleosomes formed on the fragment were stabilized lowering, step by step, NaCl concentration by adding small aliquots of the same buffer. The 524 bp DNA 5'-end labelled at *Hind* III restriction site, reconstituted with normal and trypsinized nucleosomes, has been treated with Exonuclease III (2 U/µl) in the presence of 3 mM $MgCl_2$ for 10' at 30 °C. After DNA purification, the molecular weights of the DNA fragments were measured on denaturing 6% polyacrylamide gel electrophoresis.

3. Superstructural features of DNA regulative regions of *rbcS* genes and their relevance on gel electrophoretic mobility, cyclization probability and circular dichroism spectra

Curvature of the regulative regions of three pea genes which code for the small subunit of the enzyme ribulose-1,5-bisphosphate carboxylase *rbcS-3A, rbcS-E9,* and *rbcS-3.6* was studied. These three genes represent an interesting system in order to evaluate the possible correlation between superstructural features of upstream regulative regions and DNA transcription. In fact, although the coding sequences have a high degree of homology, the transcription of *rbcS-3A* has been estimated about 40% of the total mRNA, while *rbcS-3.6* and *rbcS-E9* together account for only 7% [10]. It is worth noting that the upstream regulative DNA sequences have high chemical homology, suggesting that their different behaviour could depend on their different superstructural features. The theoretical curvature profiles of the three sequences from -410 to the start site of transcription, derived with the method of De Santis et al [6,7], are shown in Fig. 1, reported as the modulus of the curvature vector averaged each 21 bp. The profiles appear complex and characterized by local motifs, corresponding to biologically significant DNA tracts. In particular, the sequences from -40 to the start site of transcription that are specifically recognized by RNA polymerase, show a very similar curvature profile for the three genes. On the contrary, the curvature profiles of the sequences from -170 to -110, named "Light responsive elements" (LRE) for their relevance in light regulation, are largely different, although their sequences differ in only five bases. We assayed both the difference in the overall curvature of the whole sequences and the different local curvature of the LREs. We have compared *rbcS-3A,* and *E9* in the case of the curvature of the entire fragment and *rbcS-3A* and *3.6* in the case of LRE, since they show the largest differences.

3.1. CORRELATION BETWEEN THEORETICAL AND EXPERIMENTAL EVALUATION OF THE SUPERSTRUCTURES OF THE REGULATIVE DNA TRACTS OF *rbcS-3A* AND *E9* GENES FROM MEASUREMENT OF GEL ELECTROPHORETIC RETARDATION AND CYCLIZATION PROBABILITY

The correlation between the curvature of large DNA fragments and their electrophoretic behaviour is now possible using the equation which connects the curvature dispersion with R, the retardation, namely the ratio between the apparent and the real DNA molecular weight [11]. The two predicted values of R for the 410 bp long DNA fragments 410-3A and 410-E9 were respectively of ≈ 1.20 and 1.00 (straight DNA). The experimental R values derived from a 8% polyacrylamide gel in standard conditions, are $R_{410\text{-}3A} = 1.15$ and $R_{410\text{-}E9} = 1.00$, in satisfactorily good agreement with the predicted values, taking into account that for long DNAs, thermal fluctuations surely decrease the sequence intrinsic curvature. While the electrophoretic mobility is the more generally used method to experimentally measure DNA curvature, this can be also correlated with the probability of cyclization. Ring closure reactions of linear DNAs with complementary ends are catalyzed *in vitro* by the enzyme DNA ligase. Joining of cohesive ends, the substrate for ligase closure, requires large fluctuations in the molecular dimensions, since both ends of the molecule must come close together for the ligation. The cyclization equilibria are therefore expected to be quite sensitive to the DNA mechanical rigidity as well as to sequence directed DNA curvature. Using the distortion energy function, evaluated adopting a simple harmonic model [12], it is possible to calculate the energy cost of the cyclization of different DNA sequences with the same length in comparison with that of the equivalent tract of polydA-polydT, a polydeoxynucleotide, that has been reported to be rigid on account of its inability to form nucleosomes. The theoretical curvature of 410-3A and E9, that we reported previously as stereoviews [9] strongly suggest their different cyclization probability (see Fig. 2). In fact the cyclic forms of the two sequences, shown in the same figure, correspond to largely different distortion energies, namely the necessary energy to cyclize 410-E9 is about two times that of 410-3A. The theoretical predictions were found in satisfactorily good agreement with measurements of the kinetics of cyclization

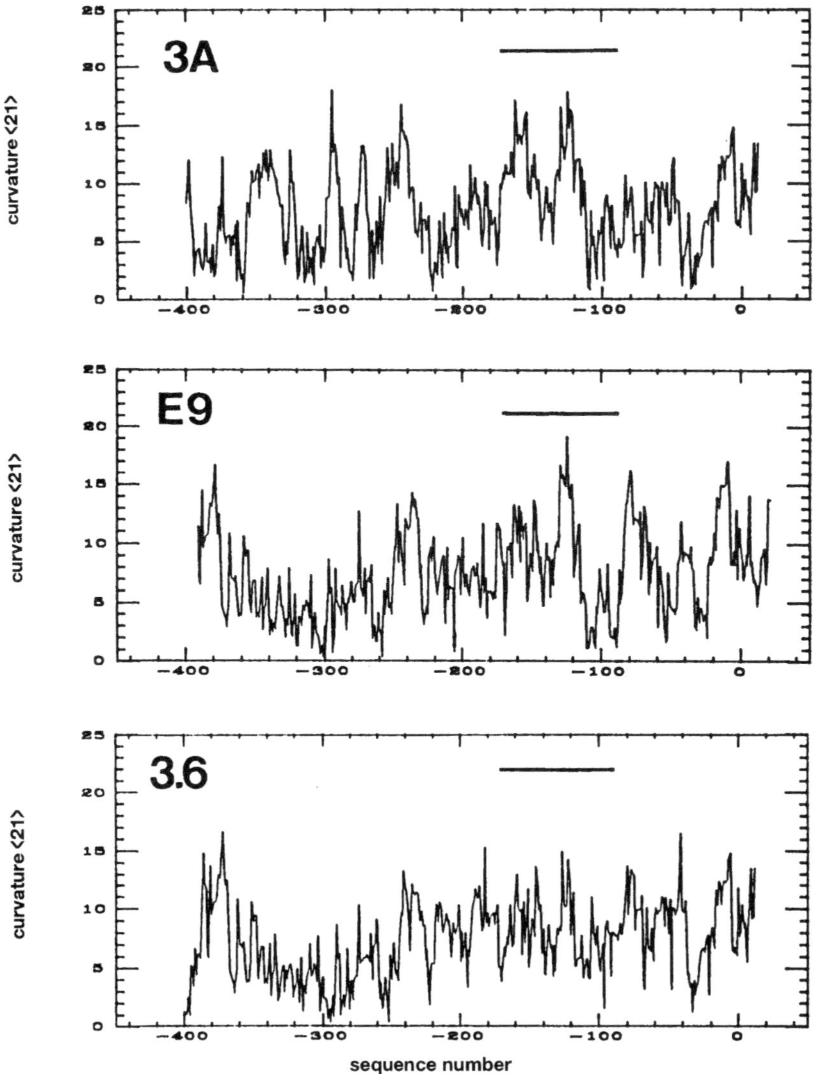

Figure 1. Theoretical curvature profiles of the regulatory regions from -410 to +10 from the genes *rbcS-3A*, *rbcS-E9*, *rbcS-3.6*. The bars indicate the LREs sequences.

of the two fragments. In Fig. 3a the ligation products of 410-3A and 410-E9 at increasing ligation times are shown; the different cyclization kinetics for the two fragments are clearly visible. In Fig. 3b the disappearance of the linear forms is reported as a function of time. After 3 minutes of reaction about 70% of 410-3A is in the circular form, whereas only 40% of 410-E9 has cyclized, in good agreement with theoretical prediction.

3.2. CORRELATION BETWEEN THEORETICAL AND EXPERIMENTAL EVALUATION OF THE CURVATURE OF THE TWO 62 BASE PAIRS LREs OBTAINED BY GEL ELECTROPHORESIS RETARDATIONS OF IN PHASE MULTIMERS

In the study of the curvature of the 62 bp long DNA tracts corresponding to the LREs of the two genes *rbcS-3A* and *rbcS-3.6*, using gel electrophoresis retardation, we have taken advantage of the possibility to amplify the curvature of the sequence by its multimerization. In fact it has been previously shown that if the connected units have the phase of B DNA (multimers of 10.5), the electrophoretic retardation are strongly enhanced [13]. Theoretical evaluation of retardation of the multimers of the two LREs predict a higher retardation for *rbcS-3A* LRE, R = 1.25, than for *rbcS-3.6*, R = 1.15, on account of its more defined superstructure. The experimental measurements of the two multimers retardation (see Fig. 4) result in satisfactorily good agreement: respectively R= 1.22± 0.02 for 3A, and R = 1.18 ±0.02 for 3.6. With the aim to evaluate the influence of DNA counterions on the electrophoretic mobility, similar electrophoretic measurements were carried out in the presence of the polycation, spermine, known to strongly interact with DNA. In the presence of the polycation the difference between the retardations increases, suggesting different interactions between spermine and the two LREs. It is reasonable to suggest that this behaviour could derive from the different superstructural features of the LREs, taking into account the high chemical homology of the two sequences.

3.3 SPECIFIC INTERACTIONS BETWEEN CURVED DNA SEQUENCES AND THE POLYCATION SPERMINE

This model has been confirmed by the CD spectra of the two LREs and of their complexes with spermine. CD spectra of the two LREs in Tris 1mM at pH=7.5, show that the two DNA elements are in the standard B conformation; minor differences with respect to the Cd spectra of eukaryotic mixed sequences DNAs in the same salt conditions can be attributed to the effect of different nucleotide sequences on DNA structural parameters [14]. The addition of spermine to the LRE solutions, either in the case of *rbcS-3A* LRE or *rbcS-3.6* LRE, increases the positive band and decreases the negative one, although with different trends (Fig. 5a and 5b). This different behaviuor can be more easily evaluated in Fig. 6a and 6b, where the ellipticities at 280 nm and their derivative, are reported versus the neutralization ratios, namely the ratio between spermine amino groups and DNA phosphates (r_{spm}).These plots show that a larger neutralization ratio is necessary in the case of *rbcS-3.6* with respect to *rbcS-3A* to obtain the same increase of the positive band. Furthermore the LRE of *rbcS-3A* show a sort of double transition, which deserves further investigation.

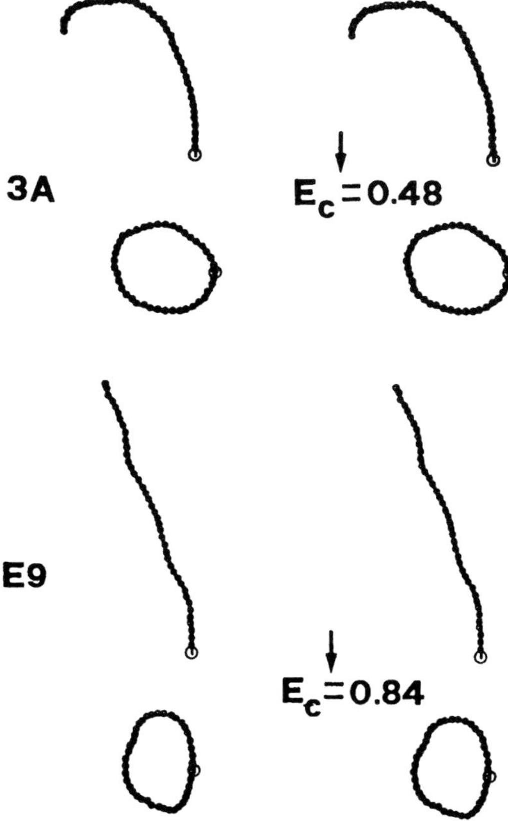

Figure 2. Stereoviews of the superstructures of the linear and circular forms of 410-3A and 410-E9. The predicted energies of cyclization are reported in arbitrary units.

It is tempting to suggest that the larger curvature of the *rbcS-3A* LRE, theoretically predicted and experimentally confirmed, could increase DNA local electrostatic field, increasing the local association constant with polycations. While further investigations are surely necessary, however, from these findings, the study of the physico-chemical properties of different nucleotide sequences on account of their different superstructural features appears very intriguing from the theoretical, as well as from the experimental, point of view.

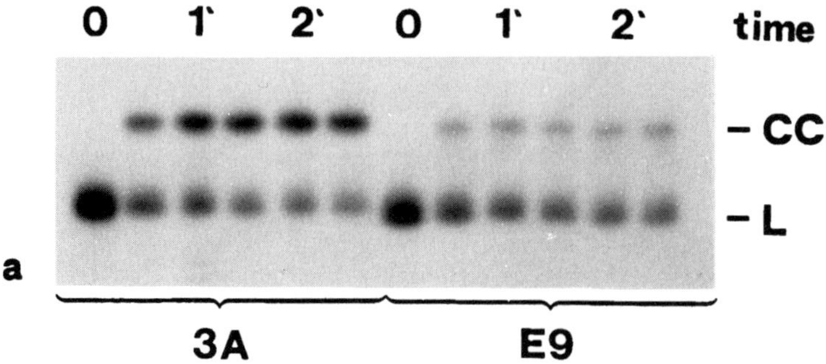

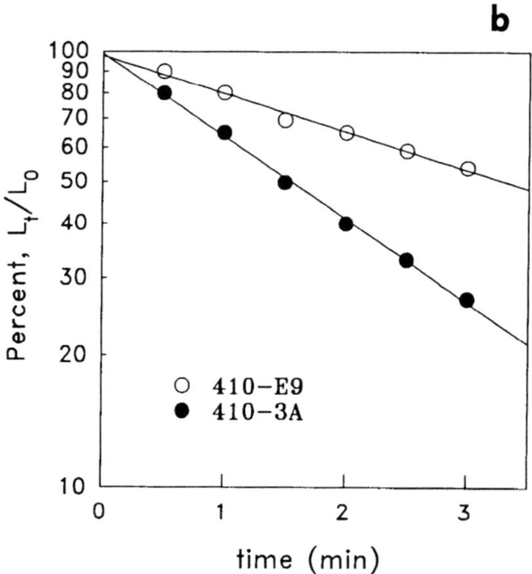

Figure 3. a) Autoradiogram of a 4% agarose gel showing progressive ligation of the linear form of 410-3A and 410-E9 (L), to form covalently closed circles (CC). b) Circularization kinetics of 410-3A and 410-E9, plotted as the fraction of linear DNA remaining at the various times.

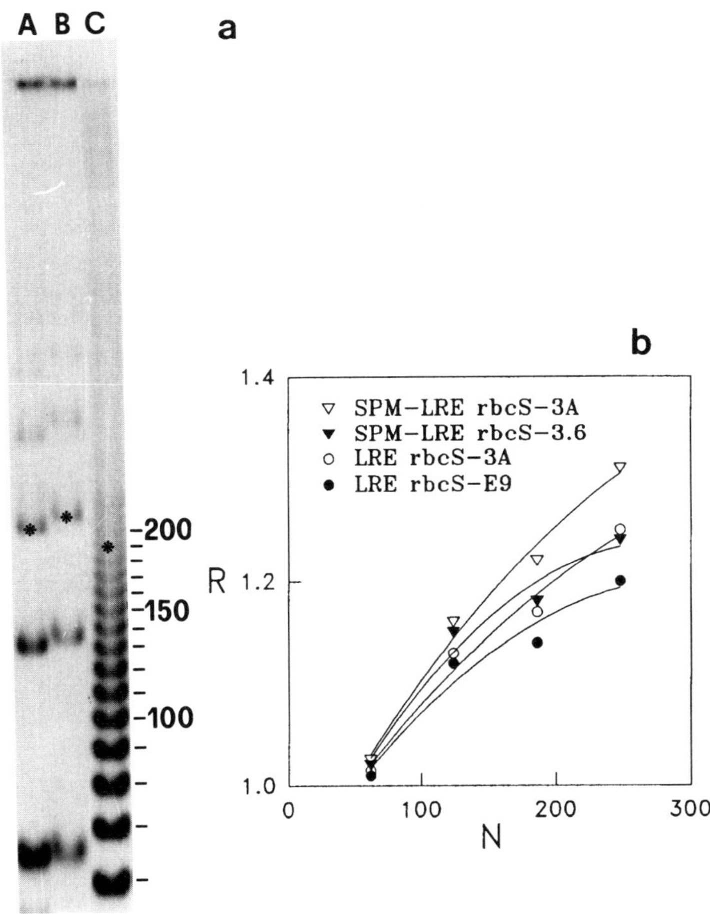

Figure 4. a) Autoradiogram of the multimers series (derived from the *rbcS-3A* and *rbcS-3.6* LRE sequences) run on a non-denaturing 10% polyacrylamide gel at room temperature. Lane A, *rbcS-3.6* LRE; lane B, *rbcS-3A* LRE; lane C, multimeric series of a decamer with a normal electrophoretic behaviour. b) The retardation values (R) are plotted as function of N (the actual chain length).

4. DNA Superstructural features as determinant in theoretical prediction of nucleosome positioning

It is now generally accepted that DNA sequence preference can detemine nucleosome positioning. It is worth remembering that the idea of DNA base sequence was firstly introduced by Trifonov, who found a harmonic periodicity of DNA base sequence in chromatin [1]. Numerous research reports, on nucleosome positioning on specific DNA sequences *in vitro*, have shown that histone octamer recognizes specific DNA sequences [15]. More recently, the most exciting work describing nucleosomal sequence preference has involved sequences that have the actual or potential property to curve the B DNA double helix. From the research of Travers and coworkers [16] on cloned nucleosomal sequences, has emerged the concept that in nucleosomes the wider minor grooves of GC base pairs prefer to face away from the surface of the octamer proteins while the narrower minor grooves of the AT base pairs prefer to face toward the octamer. The DNA curvability is intrinsic to the nucleotide sequence and is not induced by protein/DNA interactions, as shown by the finding that the relationship of major and minor grooves (rotational setting) in a constrained small circle of naked DNA is conserved, when the linearized DNA is wrapped around the histones of nucleosomes [17, 18]. On these basis De Santis and coworkers [12, 19] recently developed a theoretical methods to derive nucleosomes virtual positioning along a nucleotide sequence from the minima of DNA distosrion energy.

The obtained results correlate very successfully with nucleosome positioning in many different systems, mainly consisting of one nucleosome on short DNA tract. Furthermore, taking into consideration the phase relationships between different nucleosomes, the theoretical method appears promising also in the study of more complex systems. Here we report the results obtained in the study of nucleosomes positioning on a DNA fragment 524 bp long, containing a highly curved tract of 211 bp. On this fragment we succeeded to organize either one or two nucleosomes and thus to study the relationship between their positioning.

4.1. NUCLEOSOMES POSITIONING ALONG THE DNA FRAGMENT 524 BP, CONTAINING THE HIGHLY CURVED SEQUENCE FROM *C. fasciculata*

The 524 bp DNA fragment, containing the 221 bp curved sequence from *C. fasciculata*, represents a very valuable system to study the influence of DNA curvature on DNA physico-chemical propeties and specific interactions with proteins.

Furthermore, it is a very good test of theoretical methods to derive DNA superstructures from the sequence.

In fact, due to the fact that the curvature of the 211 bp tract is very high and in phase, the fragment visualized by electron microscopy by Griffith et al. [8], looks like a circle with a tail. Other analyses such as gel retardation and cyclization probability gave results in agreement with the very high curvature of this DNA tract. Theoretical prediction of the superstructural features of the fragments carried out by De Santis et al. results in fairly good agreement with experiments, as shown in Fig. 7 where the stereoviews of the superstructure of the fragment are reported; they look pratically equal to electron microscopy visualization. Recently the positioning of one nucleosome on the fragment was carried out by Di Mauro and coworkers [20], who found the nucleosome strongly positioned on the curved DNA tract. We have studied the positioning of two nucleosomes on the same fragment, with the aim to explore the role of nucleosome-nucleosome interactions and to correlate experimental evaluation and theoretical prediction in this more complex system. Nucleosomes were assembled on the fragment

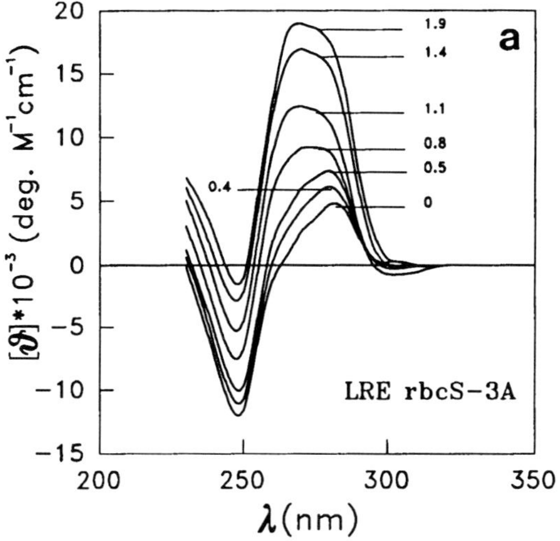

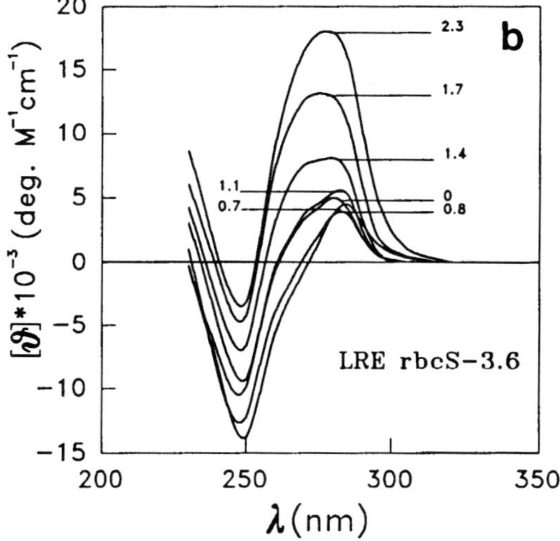

Figure 5. CD spectra of the complexes between spermine and (a), the LRE of the *rbcS-3A* gene, and (b), the LRE of the *rbcS-3.6* gene, in Tris pH 7.5 1 mM at increasing values of r_{spm}.

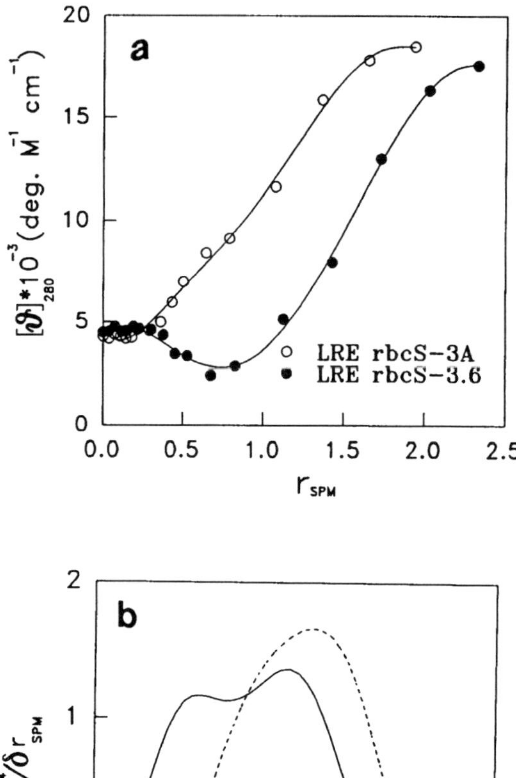

Figure 6. a) Molar ellipticities $[\theta]$ (at = 280 nm) of spermine-LRE complexes reported in Figure 5, versus r_{spm}. b) $\delta [\theta] / \delta r_{spm}$ of the molar ellipticities reported in (a) versus r_{spm}.

by exchange with mononucleosomes from erythrocyte nuclei, at different DNA concentrations, as reported in detail in the section Materials and Methods.

Nucleosomes positions were derived from the stops of the DNA cleavage due to the exonucleasic activity of the enzyme Exo III. The molecular weight of the fragments were obtained by polyacrylamide gel electrophoresis and allowed us to derive the positions of nucleosomes borders as shown in Fig. 8. We considered two different types of nucleosomes: normal nucleosomes and nucleosomes lacking the NH_2- terminal domains of histone octamer. In fact we have previously shown [21, 22] that modified nucleosomes are unable to self interacting, thus they should be useful to explore the virtual nucleosome positions derived from theoretical analysis, relaxing nucleosome-nucleosome interactions. The obtained results are reported in Fig. 9. The densitometric profiles of the gel electrophoresis allow us to evaluate the nucleosomes positions. These are reported in the same figure, as arrows indicating the nucleosome dyad axis along the 524 bp fragment. It is worth noting that near the most populated positions, other positions are present at ± 10 bp, which probably represent a Boltzmann distribution. From these results it appears that the first nucleosome, with the highest association constant, is organized on the *C.fasciculata* sequence with the dyad axis at 370 or 390 bp, in the case of normal as well as modified nucleosomes.

The second nucleosome shows a different behaviour in the two case. While the normal one occupies only one position with the dyad axis at 120 bp, the nucleosome lacking the NH_2 terminal domains of histones, appears to occupy three different positions at 120 bp, about 170 bp and 230 bp. The first and the third one, best resolved, have the same frequency. The concept that nucleosome positioning derives in both cases from DNA intrinsic distortion energy emerges clearly from the comparison between the experimental positioning and the distortion energy profile, calculated according to the theoretical method [12], as shown in Fig. 10. The deepest minimum corresponds fairly well to the first assembled nucleosome in both cases. The difference of 20 bp between the two equivalent positions of the dyad axis probably depends on the minimum width. The other minima of the distortion energy profile at 100, 190 and 225 bp correlate satisfactorily with the positions occupied by the second nucleosome in the case of modified octamer, as previously shown. The slight discrepancy can be reasonably attributed to the width of the minima.

The second normal nucleosome, due to the interactions with the first one, is able to occupy only the first and more distant minimum.

This behaviour could depend on the necessity to curve the linker DNA; this, in fact, should be easier for longer linkers. From the reported results it is possible to conclude that:

a) DNA superstructural features are dominant in specific interactions with histon proteins.

b) The theoretical method, which derives distortion energy from the DNA sequence, allows us to predict nucleosomes positioning in very good agreement with experimental results.

c) Nucleosome-nucleosome interactions are very relevant in the organization of systems with many nucleosomes. However the actual nucleosome positions belong to the set of virtual theoretically derived positions.

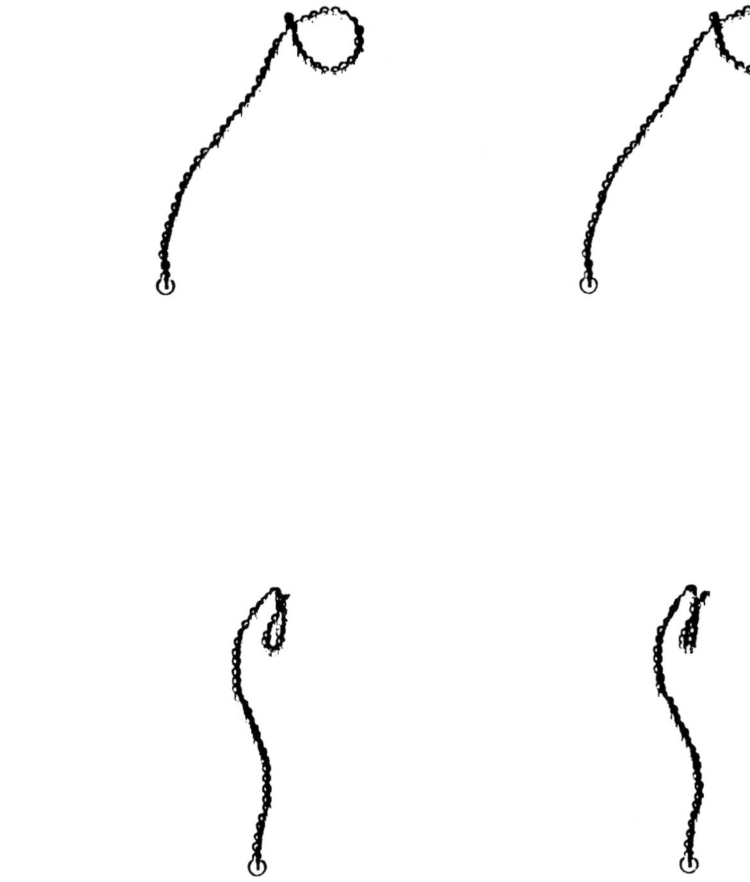

Figure 7. Two stereoprojections of the superstructure of the 524 bp DNA fragment, containing the highly curved 211 bp sequence from *C.fasciculata*. The superstructure of the DNA was obtained from the nucleotide sequence with the method of De Santis et al [6,7].

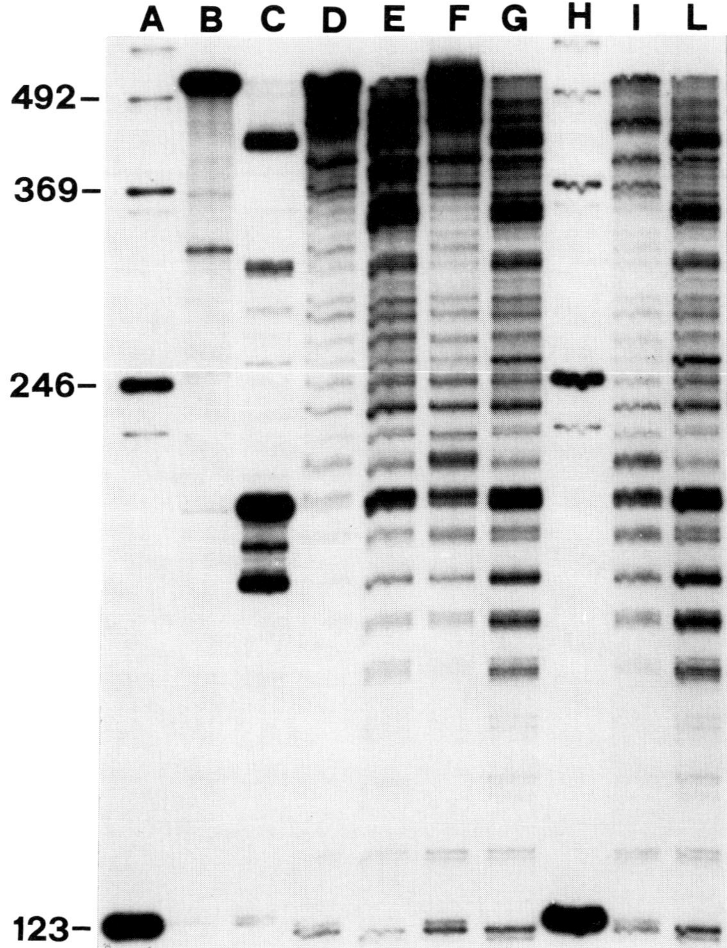

Figure 8. Localization of nucleosomes on the 524 bp DNA fragment. After reaction with Exonuclease III, the labelled DNA was run on a 6% denaturing polyacrylamide gel. Lanes A, H, molecular weights scale; lane B, naked DNA; lane C, naked DNA after reaction with Exo III; lanes D, E, F, a costant amount of normal nucleosomes complexed with increasing DNA quantities (2,4,6 µg); lanes E, G, L, as in (D) (F) (I) except that nucleosomes lacking histones NH$_2$-terminal domains were used.

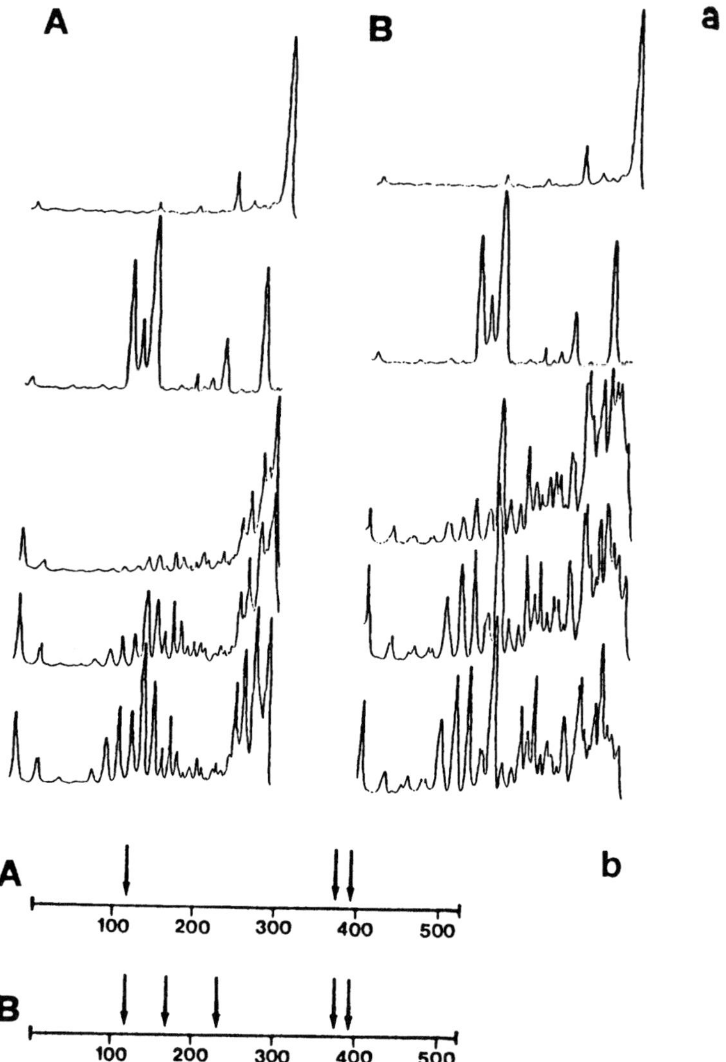

Figure 9. a) Densitometric profiles of the gel electrophoresis shown in Fig. 8. (A) Normal nucleosomes; (B) modified nucleosomes. The positioning of nucleosomes on the 524 bp DNA fragment was derived from the molecular weights determination. b) Positioning of nucleosomes on the 524 bp fragment. (A) Normal nucleosomes; (B) modified nucleosomes. The nucleosomes positions are referred to the dyad axis.

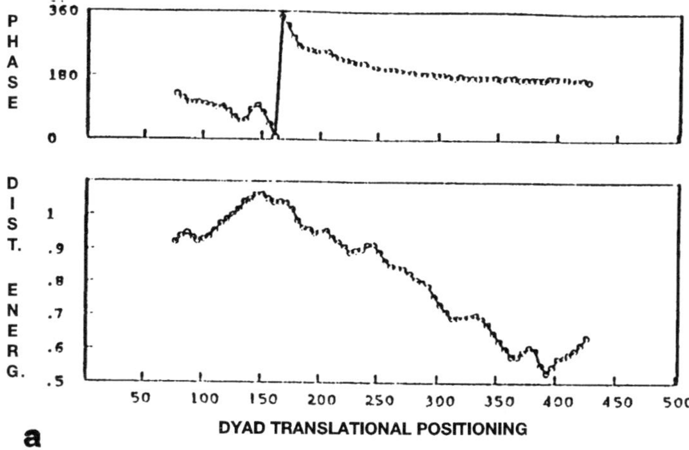

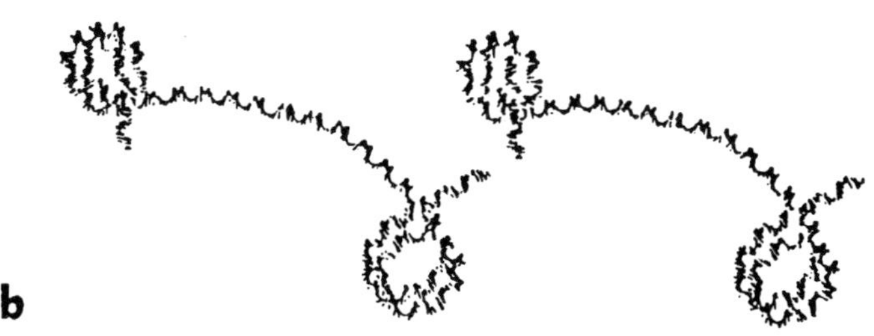

Figure 10. a) Distortion energy profile of the 524 bp DNA fragment containing the highly curved sequence from *C.fasciculata*, and the corresponding phase of the nucleosome dyad axis, according to De Santis et al [12]. b) Stereoprojection of the predicted most stable assembly of two nucleosomes on the 524 bp fragment.

Acknowledgements

This research has been partially supported by the Fondazione Cenci-Bolognetti - Istituto Pasteur, and by CNR Comitato per le Biotecnologie e la Biologia Molecolare.

References

1. E.N. Trifonov, *Nucleic Acids Res.* **8**, 4041, (1980).
2. S.C. Marini, S.D. Levene, D.M. Crothers & P.T. Englund, *C. S. H. Symp. Quant. Biol.* **47**, 279, (1983).
3. A.A. Travers, *Cell* **60**, 177, (1990).
4. L. Bracco, D. Kotlarz, A. Kolb, S. Diekmann & H. Buc, *EMBO J.* **8**, 4298, (1989).
5. M. Grunstein, *Annu. Rev. Cell Biol.* **6**, 643, (1990).
6. P. De Santis, A. Palleschi, M. Savino & A. Scipioni, *Biophys. Chem.* **32**, 305, (1988).
7. D. Boffelli, P. De Santis, A. Palleschi, A. Scipioni & M. Savino, *Int. J. Quant. Chem.* **42**, 1409, (1992).
8. C.H. Hsieh & J.P. Griffith, *Cell* **52**, 535, (1988).
9. S. Cacchione, P. De Santis, D. Foti, A. Palleschi & M. Savino, *Biochemistry* **33**, 8706, (1989).
10. C. Kuhlemeier, P.J. Green & N.-H. Chua, *Annu. Rev. Plant Physiol.* **38**, 221, (1987).
11. P. De Santis, A. Palleschi, M. Savino & A. Scipioni, *Biochemistry* **29**, 9269, (1990).
12. P. De Santis, M. Fuà, A. Palleschi & M. Savino, *Biophys. Chem.*, in press, (1992).
13. S. Cacchione, M. Savino & A. Tufillaro, *FEBS Letters* **289**, 244, (1991).
14. L. Fairall, M. Stephen & D. Rhodes, *EMBO J.* **8**, 1809, (1989).
15. A.A. Travers & A. Klug, *Phil. Trans. R. Soc. Lond.* **317**, 537, (1987).
16. S.C. Satchwell, H.R. Drew & A.A. Travers, *J. Mol. Biol.* **191**, 659, (1986).
17. H.R. Drew, A.A. Travers, *J. Mol. Biol.* **186**, 773, (1985).
18. F. Thoma, *J. Mol. Biol.* **190**, 177, (1986).
19. D. Boffelli, P. De Santis, A. Palleschi & M. Savino, *Biophys. Chem.* **39**, 127, (1991).
20. G. Costanzo, E. Di Mauro, G. Salina & R. Neri, *J. Mol. Biol.* **216**, 363, (1990).
21. P. Forte, L. Leoni, B. Sampaolese & M. Savino, *Nucleic Acids Res.* **17**, 8683, (1989).
22. M. Buttinelli, L. Leoni, B. Sampaolese & M. Savino, *Nucleic Acids Res.* **19**, 4543, (1991).

Redox and pH-Sensitive Polymer-Grafted Membranes: Thermodynamic Characteristics in Drug Delivery

M. CASOLARO
*Department of Chemistry, University of Siena, Piano dei Mantellini 44,
53100 SIENA, Italy*

1. Introduction

The electrostatic effects due to changes of protonation can affect the conformational properties of the chain segments of flexible polyions grafted in the membrane channels. Water-soluble functional polymers incorporating carboxyl acid groups [*poly(acrylic acid)* PAA, and *poly(N-acryloyl glycine)* P1] or the simpler nicotinamide moiety [*poly(3-carbamoyl-1-(p-vinylbenzyl)pyridinium chloride)* CVPy] have recently been studied as stimulus-responsive polymers, contracting and lengthening reversibly as function of the external conditions, such as pH and redox activity [1, 2].

PAA   [-CH2-CH-]$_x$
              COOH

P1   [-CH2-CH-]$_x$
              C=O
              NH-CH2-COOH

CVPy   [-CH2-CH-]$_x$
              CH2
              N+ Cl$^-$
              CONH2

When these polymers are grafted onto matrices (*cellulose*, CL; *polyurethane*, PU; *polyethersulphone*, PES) the polymer chains have reduced mobility but chain extension can be easily varied by changing solvent composition [3]. A composite structure, having "*chemical valve*" function (Figure 1), can be used in drug delivery systems to control the permeation of solutes [4, 5].

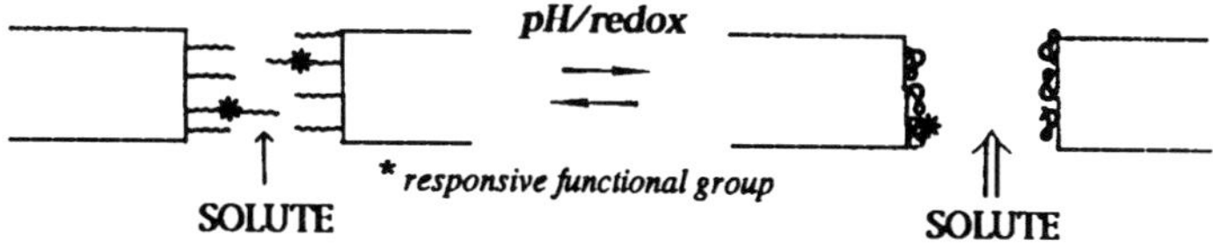

Figure 1. The membrane valve mechanism.

N. Russo et al. (eds.), Properties and Chemistry of Biomolecular Systems, 127–141.
© 1994 *Kluwer Academic Publishers. Printed in the Netherlands.*

It consists of porous membranes onto which the vinyl polymer has been graft-polymerized by the method of plasma-initiated polymerization [5, 6]. The principle of the chemomechanical expansion and contraction of the grafted chains is based on the specific reversible ionization of suitable functional groups on the polymer by alternating addition of *base (oxidizer)* or *acid (reducer)*, whereby the former produces an electrostatic repulsion between fixed charges giving rise to a significant conformational change in the grafted polymer. Since the ionized carboxylic group or the oxidized pyridinium ring carries negative or positive charges respectively, they can assume extended conformation. The pores in the membrane are obstructed by polymer chains and the apparent solute permeability is low. Low pH or reduced states bring the polymer chains back to the conformation of a collapsed coil. The composite membrane thus exhibits high solute permeability.

Our main interest in this system was the protonation thermodynamics of chain segments and their ability to form stable complex species with heavy metal ions in relation to conformational properties of the graft chains [4, 7]. The solution properties of free and grafted polymers were compared in order to elucidate the effect of the solid support on thermodynamic characteristics.

2. pH-Sensitive Polymers

2.1. THERMODYNAMICS OF PROTONATION

The two carboxylic acid polymers were studied in aqueous media in the free and grafted states. The methods used to study thermodynamic behaviour in the heterogeneous phase were limited to *potentiometry*, titration *calorimetry* and *FT-IR spectroscopic* techniques [8, 9]. Potentiometry gives information on the equilibrium constant and on the types and amount of titrable carboxylic groups present on the surface. Unlike the free polymers, the acid-base titration curves of the grafted systems always display large hysteresis loops between forward and backward titrations [8-10]. This occurs when the titrant (H^+/OH^-) addition time is short. Figure 2 shows a typical plot of pH against mequiv H^+ for P_1 grafted on polyurethane.

Similar behaviour was observed with the cellulose as support. The stepwise addition of titrant does not allow a condition of thermodynamic equilibrium, at least for of less time intervals. Moreover, the magnitude of the hysteresis loop also depends on two factors: the time lag and the amount of grafted polymer. High concentrations of the latter result in increased distance between the two curves. By increasing the time necessary to reach equilibrium pH at each step of titrant addition, the two curves becomes closer but do not overlap in thermodynamic equilibrium for many hours. The shape of the titration curve with H^+ approaches that of equilibrium more quickly. The *basicity constants (logK)* of the protonation reaction $-COO^- + H^+ \Leftrightarrow -COOH$ were evaluated and compared to those of the free polymers (Table 1).

In the two cases, polyelectrolyte behaviour was slightly different, depending on the hydrophilic quality of the support. Equilibrium conditions were reached more quickly with decreasing hydrophilicity of the support.

The *logK°* and the *(n-1)* terms of the modified Henderson-Hasselbalch equation increased with increasing hydrophilic character.

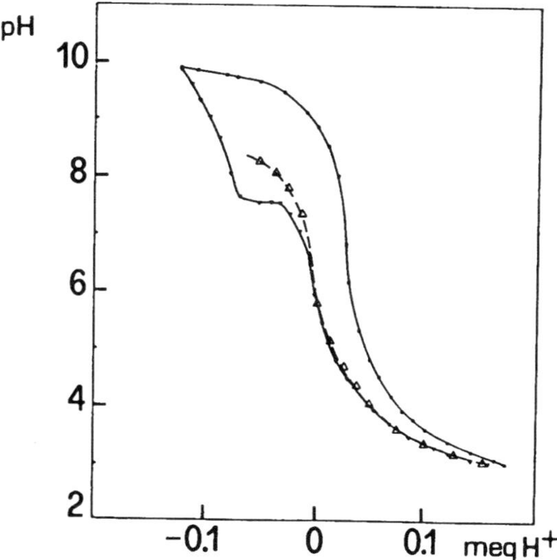

Figure 2. Potentiometric titration curves in 0.1M NaCl at 25°C for *P1-g-PU* (5wt% graft content); 30 minutes for each titration point: upper curve *titration* with 0.1M NaOH, lower curve *backtitration* with 0.1M HCl. (Equilibrium curve is the dotted line).

Table 1. Basicity Constants for Free and Grafted Polymers at 25° in 0.1M NaCl.

System	logK° [a]	(n-1) [a]	Time to reach equilibrium pH/hrs
PAA-free	*5.32*	*0.96*	*fast*
PAA-g-CL	5.79	0.86	16-20
PAA-g-PU	5.54	0.75	8-12
P1-free	*4.32*	*0.50*	*fast*
P1-g-CL	4.72	0.44	16-20
P1-g-PU	4.47	0.14	2-10

[a] $\log K = \log K° + (n-1)\log[(1-\alpha)/\alpha]$

This is the usual trend observed in the polyelectrolyte domain, for acid and basic polymers [11, 12].

To ascertain the influence of the substrate, calorimetric investigation of stepwise (incremental) protonation of ionized graft PAA chains on PU was performed at slow titrant delivery rates [13]. The chemical reaction revealed a slightly endothermic effect, decreasing with increasing charge neutralization. Figure 3 shows the characteristic trend of the thermodynamic functions ($-\Delta G°$, $-\Delta H°$ and $\Delta S°$) for free and grafted PAA. Unlike the free polymer [14], the $-\Delta H°$ values of the grafted PAA approached zero as

the degree of protonation α tended to one [10]. The presence of the PU substrate
affected the energetics of the protonation process. This process may be offset for PAA
chain segments, elements of which have relative freedom of rotation, on account of
changes in chain conformation caused by repulsion of negatively charged COO⁻groups.

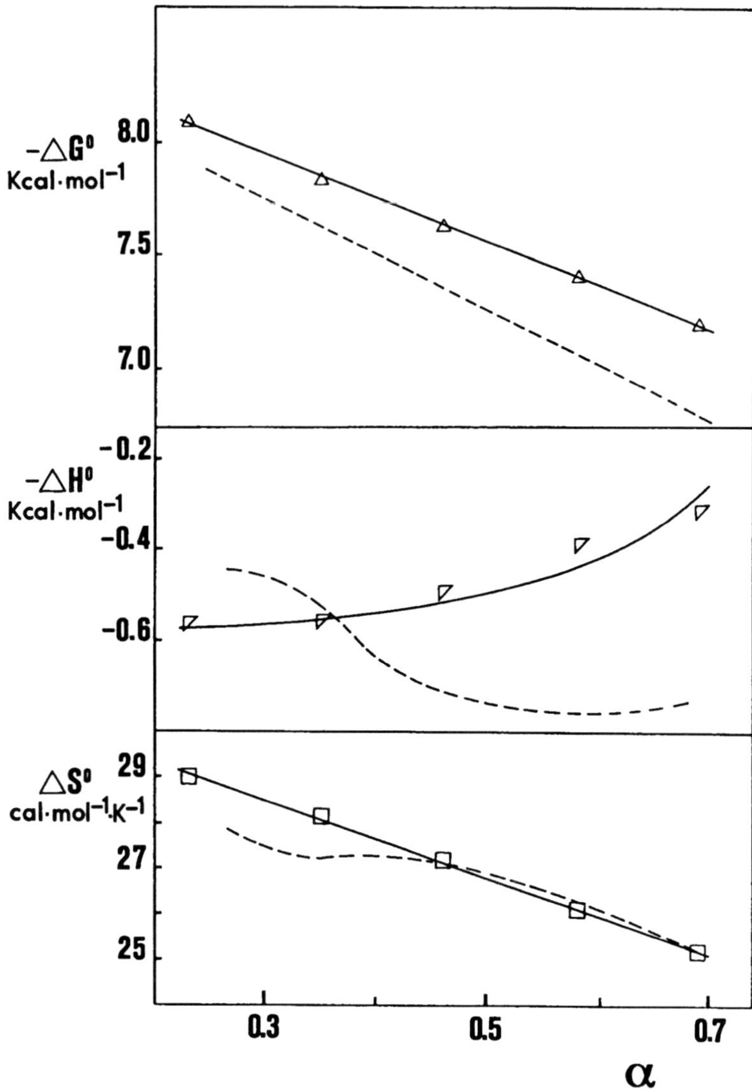

Figure 3. Thermodynamic functions vs degree of protonation a for PAA in the free
(----) and PU-grafted (——) states (25°C in 0.1M NaCl).

In the grafted system, the ionization of PAA differs from the similar process undergone by statistical coils in solution. The PAA chains fixed at one end to the PU surface ensure a degree of stability towards conformational changes. A close similarity was previously found in some porous carboxylic cationites with crosslinked methacrylic acid [15]. Further evidence of the role played by the support was clearly shown by FT-IR spectroscopic analysis at low and high pHs. Unlike the single band at 1728 cm^{-1} for free P_1 at pH 2 [1], the P_1-g-CL system shows two bands in the C=O stretching region [8] (Figure 4). The splitting at 1735 cm^{-1} and 1719 cm^{-1} was probably due to an interaction between the COOH and some functional groups of cellulose. At higher pHs no differences were observed, since the carboxylate groups are prevented from interacting with the matrix by electrostatic effects between ionized grafted chains.

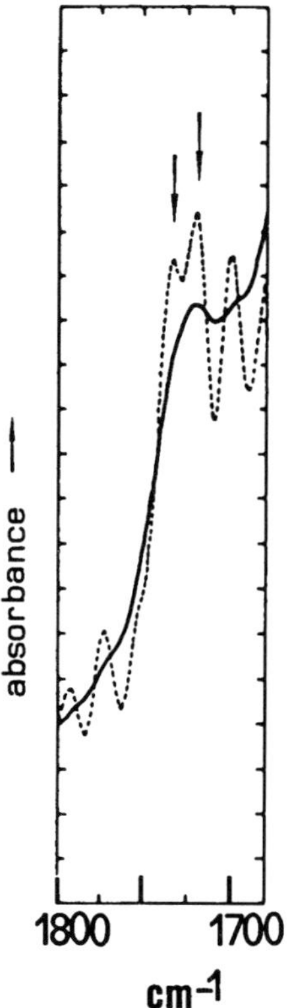

Figure 4. ATR/FT-IR *deconvoluted* difference spectrum at pH 2 for P_1-g-CL.

2.2. THERMODYNAMICS OF COPPER(II)-COMPLEX FORMATION

The two vinyl polymers formed complexes with copper(II) ions in aqueous media in the free and grafted forms, the latter requiring longer times to reach equilibrium pH for each step of titrant addition. Compared to the protonation process, equilibrium conditions were always reached in shorter time in the presence of Cu^{2+} ions. The metal ions, in fact, enhance the splitting-off of protons from the COOH on the grafts, since the macromolecule is in a compact coil. This facilitates the acid-base neutralization process.

Logβ (*stability constant*) was calculated from the potentiometric data using the dependence of the previously calculated logK values and the modified Henderson-Hasselbalch equation [16]. The refined logβ values fit *CuL₂* (*L⁻ is the monomer unit of the graft polymer*) stoichiometry in a wide range of pHs (Table 2). The logβ for the PAA-g-PU/Cu(II) system is close to that reported for the free PAA [17]. This suggests that the matrix does not participate in the coordination. PAA-g-CL showed higher logβ values. The difference in coordination behaviour may be due to the participation of the hydroxyl groups in the coordination by virtue of their more hydrophilic nature. Unlike the simple CuL^+ stoichiometry found in solution [12], the polymer P_1 grafted onto cellulose showed CuL_2 stoichiometry similar to that of PAA, but its logβ value was lower, even though the same cellulose support was present.

Table 2. Stability Constants of Copper(II)-Complex Species for Free and Grafted Polymers at 25°C in 0.1M NaCl.

System	Reaction	pH-range	logb
Cu(II)/PAA-g-PU	$Cu^{2+} + 2L^- \Leftrightarrow CuL_2$	3.1-5.5	7.4
Cu(II)/PAA-g-CL	$Cu^{2+} + 2L^- \Leftrightarrow CuL_2$	4.0-5.7	9.3
Cu(II)/PAA-Free	$Cu^{2+} + 2L^- \Leftrightarrow CuL_2$	----	*7.48*
Cu(II)/P₁-g-CL	$Cu^{2+} + 2L^- \Leftrightarrow CuL_2$	3.2-4.8	7.9
Cu(II)/P₁-Free	$Cu^{2+} + 2L^- \Leftrightarrow CuL^+$	*2.8-4.0*	*3.30*

The longer lateral chain length in P_1 is sterically unfavourable for coordination with copper(II) ion. The free polymer is unable to form a high stability bond, since the evaluated logβ value involves coordination of a single monomer unit with the amido and the carboxylic groups as coordination sites.

2.3. SOLUTE PERMEABILITY

2.3.1. *Polyoxyethylene*

The permeation of solutes through the porous membranes and permeability control on changes of pH showed typical features. *Polyoxyethylene* (POE) having a *molecular weight* in the range *1,000* to *80,000* was used in the preliminary permeation experiments through porous cellulose (*0.2 μm* pore size) membranes at the two different pHs and several PAA graft contents [5] (Figure 5). The two main results were the permeation control of POE permeation by pore size and the effect of pH on conformational changes in the grafted polymer chains.

The PAA grafted polymer was ionized at pH 7, this causing chain extension for the electrostatic repulsion and preventing POE permeation. Solute permeation was allowed at pH 2 by chain contraction.

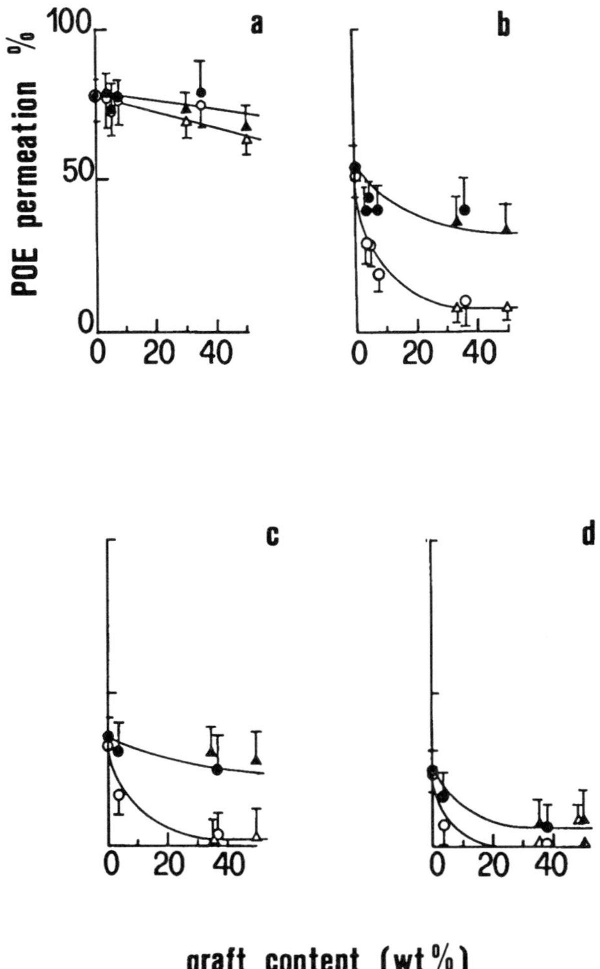

Figure 5. Permeability of *POEs* through *PAA-g-CL* at pH 7 (*lower curve*) and pH 2 (*upper curve*) [molecular weight of POE: (a) *1,000;* (b) *6,000*; (c) *20,000*; (d) *80,000*].

2.3.2. *Insulin*

Insulin was selected as testing drug for permeability control. The amount of drug permeated increased always linearly with time [5, 8]. Moreover, permeation through the grafted cellulose membranes was lower than through the native cellulose. While in the latter case the permeation was not affected by pH changes, the presence of the grafts produced permeability changes that strongly depend on the pH and amount of grafted polymer. Higher degrees of grafting led to larger differences in permeability at the two pHs considered. The amount of insulin permeating with time and its pH-dependence for the porous cellulose grafted with P1 [8] is reported in Figure 6. Permeation through a non-grafted membrane was independent of pH, with significant differences with respect to the grafted system.

These "chemical valves", being pH-sensitive, can be used as a *self-regulating insulin delivery* system. Since the enzyme GOD (*Glucose Oxidase*) catalyzes the conversion of

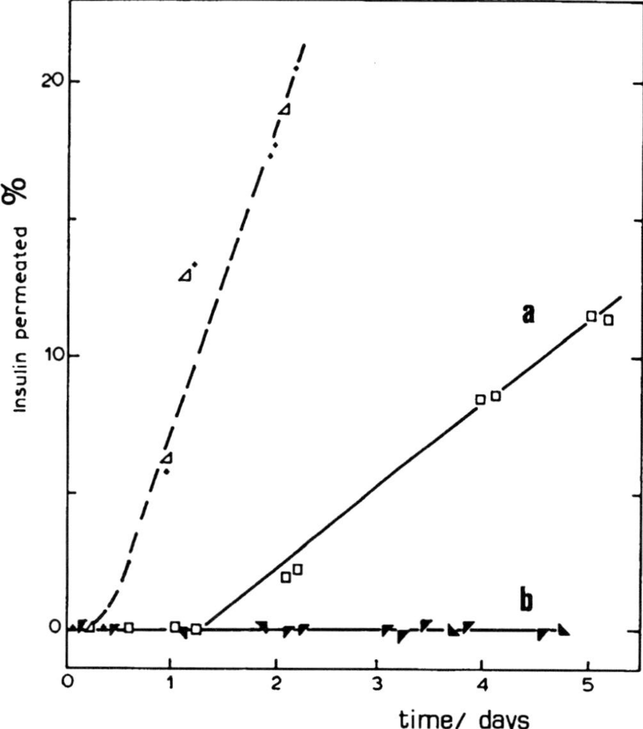

Figure 6. Percentage of insulin permeated through P1-g-CL at pH 8.6 (b) and pH 2.7 (a). [*Broken line shows the performance of a non-grafted membrane under the same conditions*].

glucose to *gluconic acid*, a glucose-sensitive membrane system containing immobilized GOD on the graft chains was developed [5, 8, 18] (Figure 7).

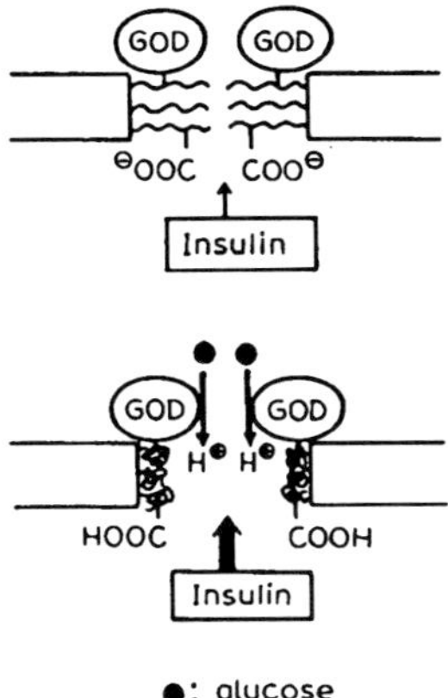

Figure 7. Principle of controlled release system for *insulin* (GOD: glucose oxidase).

When glucose diffuses into the porous membrane it is converted to gluconic acid, causing a decrease in pH within the microenvironment [19]. The COO⁻ groups, having a higher logK value than gluconic acid [20] (logK 3.6), protonate, causing the rod-like chains to coil and interact with the matrix. The pores open, allowing the faster passage of insulin molecules. Data on the permeation experiments for P_1-grafted and GOD-immobilized cellulose, with glucose additions, are reported in Figure 8 [8] .

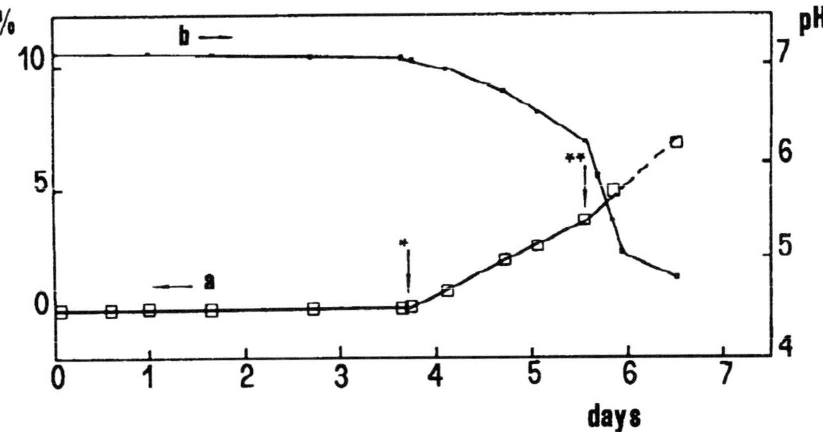

Figure 8. a: Percentage of insulin permeating through P_1-grafted and GOD-immobilized cellulose membrane (*0.01M Tris-HCl buffer solution;* 110 wt% grafts content; 3wt% GOD content). b: *pH-change profile with glucose addition (* 50 mg/100 ml; ** 400 mg/100 ml).*

It can be seen that the release rate was negligible for several days. Insulin flowed only upon addition of glucose, and the permeation slopes increased with increasing glucose concentration. Changes in pH were also observed in solution; below pH 5 insulin aggregation was enhanced. This phenomenon can be avoided with an increase in buffer strength [8] . The quantity of graft with its COO^- groups plays a fundamental role in the change in pH for medical applications of this *"artificial pancreas"* [18]. Since the buffering strength of human blood is closer to 0.01M than 0.1M Tris-HCl, it is necessary to have more carboxylate groups in the graft in order to avoid pH drop. These groups, themselves acting as buffering agents against H^+ ions, prevent a significant decrease in pH.

3.   Redox-Sensitive Polymers

3.1. THERMODYNAMIC CHARACTERISTICS

The oxidized form of the polymer *CVPy* is only freely soluble in water. The behaviour of its aqueous solution was studied with respect to the action of *hydroxide ions* on the quaternary salts of the *nicotinamide* residue [21-23]. The characteristic feature of the polymer solution was the appearence of the yellow colour on addition of OH^- ions. The latter affect the molecular structure of the nicotinamide residue as happens for simple pyridinium salts, several intermediate compounds of which are believed to form in the reduced state [24]. The electrostatic effect of positive charges was detected by a simple *viscometric titration* of the polymer. The reduced viscosity (η_{sp}/C) showed a decreasing pattern on addition of OH^- ions (Figure 9).

This means that the oxidized form of the CVPy was hydrated and in extended conformation. The presence of the positive charge on the quaternary nitrogen causes electrostatic repulsion between neighbouring units of the macromolecule. The stepwise addition of base leads to a decrease in the number of positive charges on the macromolecule, reflected by a lower η_{sp}/C. The polymer solution was yellow and non-ionic in character. This hydrophobic character was revealed by soap-like behaviour. The linear decrease in viscosity reached   its minimum at a molar ratio $OH^-/(CVPy)_{unit}$ close to one.

Unlike the potentiometric data, calorimetric titrations also revealed two definite end-points (Figure 10) that suggest chemical reactions in two-step mechanisms involving the nicotinamide moiety.

A low exothermic effect was revealed also for a copolymer containing acrylamide units together with the nicotinamide residue. The larger exothermic effect beyond 1:1 stoichiometry may be related to the incipient yellow solid phase separation appearing beyond the second end-point. Precipitation of polymers in water is often accompanied by considerable heat release associated with the build-up of water clusters [25]. *Spectrophotometric* analysis showed that the process was reversible.

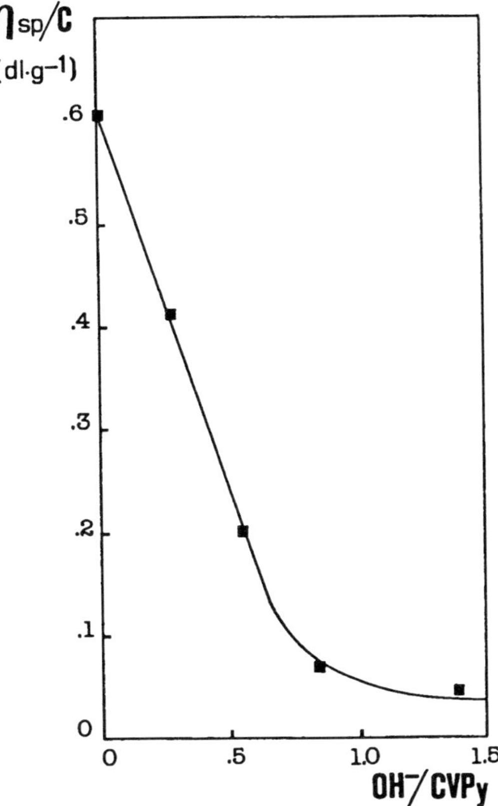

Figure 9. Viscometric titration of *CVPy* with sodium hydroxide solution at 25°C in 0.1M NaCl.

The absorption maximum at *265 nm* can be assigned to the pyridinium ion of the nicotinamide residue. The addition of base would be expected to increase the ratio of the intensity of the long wavelength peak (*350 nm*) to that of the shorter. This was found to occur in both forward and backward addition of OH⁻ and H⁺. The spectral behaviour of the polymer CVPy was quite similar to that observed for aqueous solution of simple quaternary pyridinium halide upon treatment with base [22]. The spectral changes suggest that the process is reversible.

3.2. WATER PERMEATION

The construction of a system in which changes in length of the graft polymer chains are produced by alternating chemical action of reducing and oxidizing agents can also be used to control the flow rates in solvent separation processes under pressure. It has been shown [26] that the water flow rate can easily be controlled by pH-sensitive membranes with straight pores.

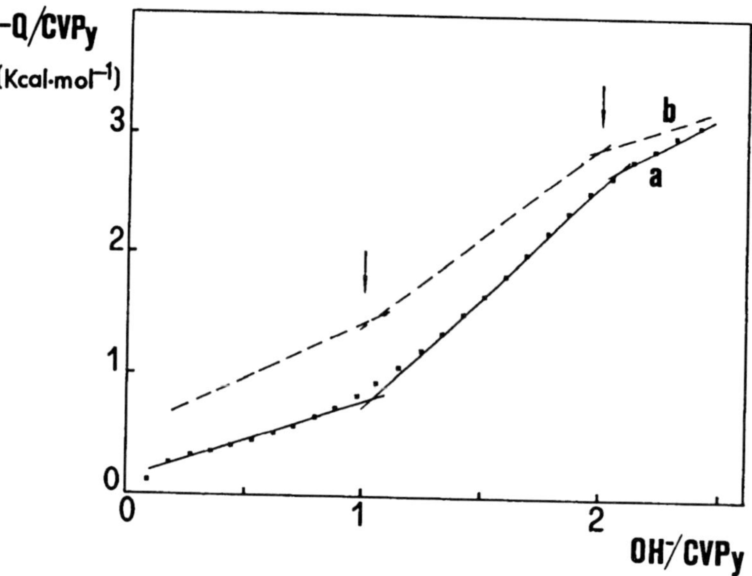

Figure 10. Calorimetric titration data of: *a CVPy* homopolymer; *b CVPy* copolymerized with acrylamide (*AAm*) in a 4:6 *CVPy:AAm* molar ratio (0.1M NaCl and 25°C).

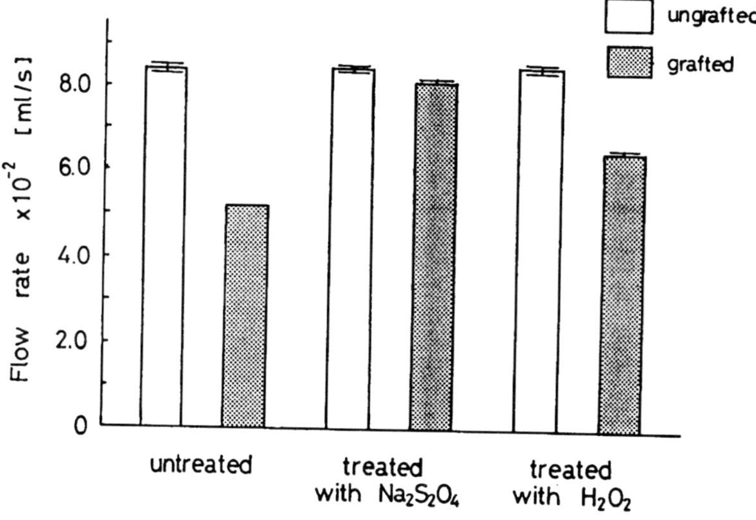

Figure 11. Redox-dependent water permeation through the *CVPy-grafted polyethersulphone membrane.*

Furthermore, they demonstrated the "chemical valve" function of *CVPy* grafted on porous *polyethersulphone* membrane (PES, 0.8 μm pore size) by permeation of water containing *reducer* ($Na_2S_2O_4$, 20 mM) or *oxidizer* (H_2O_2, 50 mM) species [27]. Water permeation through the grafted membrane was lower than through the corresponding native membrane. Moreover, for the grafted membrane the water flow rate was dependent on the presence of oxidizing or reducing species (Figure 11).

The presence of the latter enhanced water permeation whereas the oxidizing species did not allow such a low flow rate as in the membrane originally grafted with CVPy. The positive charges present in the oxidized form enhanced chain extension and limited pore size, slowing the flow. The presence of reducing species made the polymer chains collapse in a coil. The viscosity data reported above showed that the coils were uncharged. The on-off switch of this system was not fully reversible under redox action, because crosslinkings of nicotinamide residues probably occurred during the reduction process [28].

4.   Conclusions

Composites made of polyelectrolytes grafted on the surface of porous membranes are of special interest for the construction of systems for the purification and separation of substances. The grafted polyelectrolyte assumes extended conformations in the ionized form, acting as a barrier to substances permeating through the pores. The gate opens when external signals modify the forces of interaction on the charged grafted chains. These forces can be of different nature, but electrostatic forces are the main ones.

A decrease in charge causes the extended macromolecule to coil; the size of the coil also depends on hydrophobic forces within the polymer itself and the magnitude of polymer-matrix interaction.

A highly hydrophobic nature in the neutral form and a highly hydrophilic nature in the ionized form of grafts is required if they are to act as effective permeation valves.

The nature of the matrix influences the reversible ionization process and is reflected in the kinetics of protonation. The results of this work show that weak polymer-matrix interactions have a faster response in the protonation thermodynamics. It can therefore be expected that hydrophilic polymers grafted on hydrophobic materials undergo more rapid ionization than those linked to a hydrophilic matrix. The protonation of the ionized carboxylate groups on the rod-like chain was faster than the kinetics of ionization of the COOH groups buried inside macromolecular coil. This may be attributed to the gradual neutralization of substrate-interacting carboxylic groups by hydroxide ions that slowly penetrate the coil domain. On the other hand, the charged polymer with its extended chains is more available to the protonation process. The uptake of protons only involves the group to be protonated and is less complicated by the substrate.

This comparative study suggests that similar thermodynamic behaviour can be expected for the redox graft polymer.

In studying the thermodynamics of graft polymers it is important to compare the results with the corresponding soluble analogous.

Acknowledgement

This work was partially supported by MURST (Italy, 60% funds) and by a Grant for an International Joint Research Project from NEDO (Japan).   The author expresses its gratitude to Dr. Yoshihiro Ito (Dept. of Polymer Chemistry, Kyoto University) for providing redox polymer samples and data on water permeation.

References

1. R. Barbucci, M. Casolaro and A. Magnani, *Makromol. Chem.* **190**, 2627, (1989).
2. K. Ishihara, M. Kobayashi and I. Shionohara, *Makromol. Chem., Rapid Comm.* **4**, 327, (1983).
3. R.S. Ross and P. Pincus, *Macromolecules* **25**, 2177, (1992).
4. Y. Osada. *Adv. Polym. Sci.* **82**, 1, (1987).
5. Y. Ito, M. Casolaro, K. Kono and Y. Imanishi, *J. Controlled Release* **10**, 195, (1989).
6. W.R. Gombotz and A.S. Hoffman, *CRC Critical Reviews in Biocompatibility* **4**, 1, (1987).
7. M. Casolaro, C. Roncolini, R. Barbucci and A. Magnani, in *Proceedings on the "4th Macromolecule-Metal Complexes, IUPAC-MMC IV"*, Siena (Italy),152,(1991).
8. R. Barbucci, M. Casolaro and A. Magnani, *J. Controlled Release* **17**, 79, (1991).
9. R. Barbucci, M. Casolaro and A. Magnani, *Clinical Materials* **11**, 37, (1992).
10. M. Casolaro and R. Barbucci, *Colloids and Surfaces*, submitted.
11. M. Casolaro and R. Barbucci, in "Thermodynamics of Protonation and Cu(II)-Complex Formation with Polyelectrolytes Containing Different Functional Groups", 2nd International Symposium on Polymer Electrolyte, Ed. B.Scrosati, Elsevier Appl. Sci., London, 311, (1990).
12. R. Barbucci, M. Casolaro and A. Magnani, *Coord. Chem. Rev.* **120**, 29, (1992).
13. M. Casolaro, F. Tempesti and E. Busi, in "Applied Thermodynamics", The Development of Science for the Improvement of Human Life, First Kyoto-Siena Symposium, Ed. F.Casprini and R.Barbucci, Siena, 83, (1992).
14. V. Crescenzi, F. Quadrifoglio and F. Delben, *J. Polym. Sci.: Part A2*, **10**, 357, (1972).
15. A. Chernova, V.S. Yurchenko, O.A. Pisarev and G.V. Samsonov, *Polym. Sci.*, **20**, 417, *USSR*, (1978).
16. R. Barbucci, M. Casolaro, M. Nocentini, S. Corezzi, P. Ferruti and V. Barone, *Macromolecules* **19**, 37, (1986).
17. H.P. Gregor, L.B. Luttinger and E.Loebl, *J. Phys. Chem.* **59**, 34, (1955).
18. M. Casolaro and R. Barbucci, *Int. Journ. Art. Organs* **14**, 732, (1991).
19. G.W. Albin, T.A. Horbett, S.R. Miller and N.L. Ricker, *J. Controlled Release* **6**, 267, (1987).
20. A.E. Martell and R.M. Smith, in "Critical Stability Constants", Plenum Press, N.Y., (1974).
21. A.G. Anderson and G. Berkelhammer, *J. Org. Chem.* **23**, 1109, (1958).
22. R.M. Burton and N.O. Kaplan, *Arch. Biochem. Biophys.* **101**, 139, (1963).

23. D.C. Dittmer and J.M. Kolyer, *J. Org. Chem.* **28**, 2288, (1963).
24. U. Eisner and J. Kuthan, *Chem. Rev.* **72**, 1, (1972).
25. M. Morcellet, C. Loucheux and H. Daoust, *Macromolecules* **15**, 890, (1982).
26. Y. Ito, S. Kotera, M. Inaba, K. Kono and Y. Imanishi, *Polymer* **31**, 2157, (1990).
27. S. Nishi, Y. Ito and Y. Imanishi, in <u>Proceedings on the "36th Polymer Symposium</u>", Kobe (Japan), 56, (1990).
28. T. Endo and M. Okawara, *J. Polym. Sci.: Polym. Chem. Ed.* **17**, 3667, (1979).

Iron-sulfur Proteins: Part II Valence-specific Assignment in Oxidized Hipip through ^{1}H NMR Spectroscopy

S. CIURLI[§], C. LUCHINAT[§], and A. SCOZZAFAVA[#]
Contribution from §*Institute of Agricultural Chemistry, University of Bologna, Viale Berti Pichat 10, 40127 Bologna, Italy.*
#*Department of Chemistry, University of Florence, Via Gino Capponi 7, 50121 Florence, Italy*

1. Introduction

High potential iron-sulfur proteins (HiPIP) are a small class of iron-sulfur proteins found in several photosyntetic bacteria [1]. HiPIP contain a [4Fe-4S] cluster [2] which is bound to the protein backbone as exemplified in Figure 1 for C. *vinosum* HiPIP [3-5]. It has been proposed that these metalloproteins may act as electron carriers that shuttle electrons from the bc_1 complex to the bacteriochlorophyll contained in the photosynthetic reaction center [6]. This function is usually carried out by cytochrome c_2, but some degree of complementarity has been found between cyt c_2 and HiPIP in many species of bacteria, thus suggesting this hypothesis [6]. The redox process involves the $[Fe_4S_4]^{3+/2+}$ couple [7]. The redox property peculiar to HiPIP is the high potential at which they transfer electrons ($E_O' = +50 \div +450$ mV) [8] as compared to the much more widespread ferrodoxins ($E_O' = -250 \div -650$ mV) [9,10], containing an analogous [4Fe-4S] cluster, but in which the redox couple $[Fe_4S_4]^{2+/1+}$ is involved. A large collection of spectroscopic data has been gathered on these proteins, in an effort to understand both the structural, electronic, and magnetic factors that characterize these clusters, and the relationships between these factors and the functional properties. In part I, [11] we have shown how ^{1}H NMR represents a formidable tool for investigating these systems by monitoring the $\beta-CH_2$ resonances of the cysteinyl residues bound to the paramagnetic cluster.

Hyperfine shifts and their temperature dependence can be rationalized on the basis of magnetic coupling operative among the iron atoms [11], leading to a description of the electronic structure of these delocalized systems at room temperature which is complementary to that obtained through other spectroscopic techniques at low temperature. At this point we wish to review briefly the information obtained on HiPIP through Mössbauer, Electron Paramagnetic Resonance (EPR), Electron-Nucleus Double Resonance (ENDOR), Resonance Raman (RR), and X-ray crystallography, and to show how, using advanced 1D and 2D techiniques, NMR is able to provide a unique piece of information by specifically identifying the valence state of each iron ion and its position in the cluster core with respect to the protein frame.

143

N. Russo et al. (eds.), Properties and Chemistry of Biomolecular Systems, 143–157.
© 1994 *Kluwer Academic Publishers. Printed in the Netherlands.*

S. CIURLI ET AL.

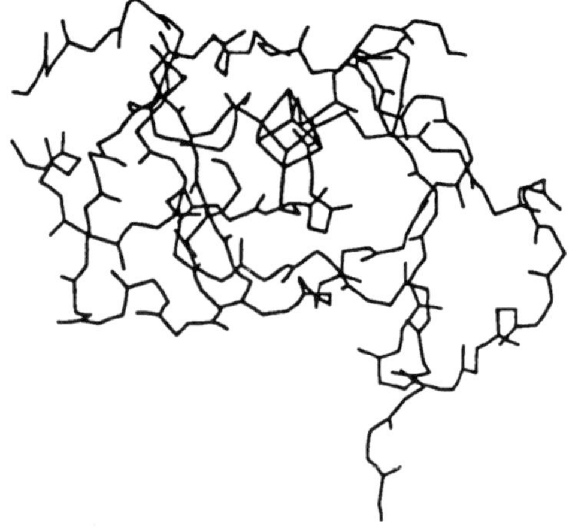

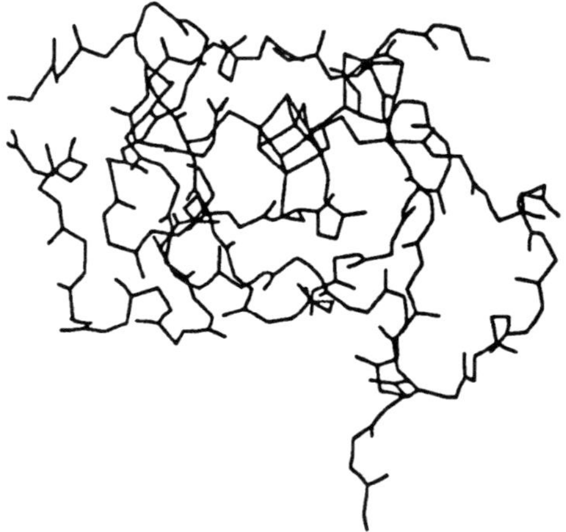

Figure 1. Stick-bond stereo-view of HiPIP from *C. vinosum* showing the cluster core bound to the protein backbone. The structure was retrieved from the Brookhaven National Laboratory Protein Data Bank, and elaborated using the graphic program package SYBYL (Tripos Associates, Inc., 1699 South Hanley Road, St. Louis, MO 63144, USA).

2. Spectroscopic Background

Early Mössbauer studies [12-15] on oxidized HiPIP isolated from C. *vinosum* suggested [13-15] the hypothesis of the existence of *two couples* of iron ions, the (Fe^{3+} -Fe^{3+}) pair, having a positive hyperfine constant, and the (Fe^{3+} -Fe^{2+}) pair, with a negative hyperfine constant. The total spin of the cluster, $S=1/2$, would then be a result of the antiferromagnetic coupling between the smaller subspin S_{12}, associated with the ferric pair, and the larger subspin S_{34}, associated with the mixed-valence pair [15]. The values for $| S_{34}, S_{12} >$ that would be consistent with the values of the hyperfine constants are $| 9/2, 4 >$ and $| 7/2, 3 >$ [16-20]. This model is consistent with early EPR studies conducted on oxidized HIPIP from C. *vinosum* [21-23]. The unique and quite complex EPR spectrum was originally interpreted as arising from two major components, one axial and the other rhombical, equally weighted, both of which having a $S=1/2$ ground state well isolated from higher excited states [21-23]. A third, minor component was considered was arising from an impurity. A model was thus proposed, in which the two main components of the EPR spectrum originated from two species generated by the hopping of one electron between the two iron ions constituting the

mixed-valence pair, a process for which t $\approx$ 10^{-7} ÷ 10^{-8} s [23]. In a more recent investigation [24], the two major components of the EPR spectrum of C. *vinosum* HiPIP have not been found to be in a 1:1 ratio, but in a ratio larger than 10:1, with the axial component being the prevalent species. Moreover, the signal previously attributed to an impurity has been interpreted as arising from inter-protein cluster interactions, with its intensity depending on the ionic strengh of the solution. In a more recent Mössbauer study conducted on the HiPIP II isolated from E. *halophila*, the high similarity of the observed hyperfine constants with the corresponding constants found for C. *vinosum* HiPIP [25] is evidenced, but, in contrast with the latter protein, the EPR spectrum of E.*halophila* HiPIP II consist of a single axial component similar to the prevalent species observed in C. *vinosum* HiPIP [25]. Single crystal ENDOR spectroscopy could in principle permit a more precise determination of the hyperfine constants than that obtained from Mössbauer studies. This technique, when applied to model complexes containing the $[Fe_4S_4]^{3+}$ core, has in fact allowed researchers to distinguish the individual iron ions, and to accurately determine their hyperfine constants [26, 27]. However, ENDOR spectra of oxidized C. *vinosum* HiPIP resulted to be broad and poorly structured because of the multiple orientations taken up by the protein in the sample [28]. This disorder has thus prevented the identification of more than *two* types of iron ions [28]. ENDOR studies on single crystals of HiPIP would then be desirable in the near future. Likewise, Resonance Raman spectroscopy has been utilized to study the cluster core structure of several HiPIP [29-31]. Even in this case, however, the D_{2d} distortion of the T_d ideal symmetry is consistent with *two* types of iron sites. Further distortions, that would indicate a lower geometry, thus leading to the differentiation of the four iron ions, would require a major resolution than that presently attainable with this technique. Reduced HiPIP contain the $[Fe_4S_4]^{2+}$ core, which is formally constituted by two Fe^{3+} and two Fe^{2+} ions. Mössbauer data are consistent with the presence of two mixed-valence pairs, antiferromagnetically coupled to give an $S=0$ ground state [14, 15, 32]. The reduced state will not be further discussed here. Several theoretical models have been proposed to rationalize the spectroscopic observations [33] described above, including models based on simple Heisenberg coupling assumptions, which account very well for the NMR properties of HiPIP [11, 19, 20].

3. Crystallographic studies

The molecular structures of C. *vinosum* HiPIP [3-5] and of E. *halophila* HiPIP I [34] have been solved by X-ray crystallography. In both cases the [4Fe-4S] cluster has as a cubane-like conformation and is bound to the protein by four cysteine residues. The cubane cluster is slightly distorted, as a tetragonal compression along a S_4 rotation axis results in an approximately D_{2d} symmetry. The main issues that will be discussed here are the possible relationships between the protein structure and the redox properties of the cluster, specifically, the value of the electrochemical potential and the distribution of the valence in the oxidized cluster. For the latter, it is important to compare the atomic distances in the $FeS_3(SCys)$ units, because they are responsive to the iron oxidation state and decrease with increasing Fe^{3+} character [35]. In Table I are reported the interatomic distances in the $[Fe_4S_4(SCys)_4]^{1-}$ moiety found in oxidized HiPIP from C. *vinosum* and E. *halophila* I. The cysteine residues and the iron ions are numbered according to the order in which they are found in the aminoacid sequence of the proteins [36], and the complete numbering scheme is shown in Figure 2. It is noteworthy to mention that there is a close structural correspondence between the cysteines coordinating the cluster in the two proteins, the pattern being (Cys I)-$(X)_2$-(Cys II)-$(X)_{15}$-(Cys III)-$(X)_{12}$-(Cys IV) for C. *vinosum* HiPIP and (Cys I)-$(X)_2$-(Cys II)-$(X)_{13}$-(CysIII)-$(X)_{15}$-(Cys IV) for E. *halophila* HiPIP I [36]. The most interesting parameter to be evaluated, in the effort of determining the iron oxidation state, is the Fe-S(Cys) distance. In fact, accurate crystallographic determinations of this parameter in model complexes indicate an approximate distance of 2.3 Å for Fe^{3+}-SR and of 2.5 Å for Fe^{2+}-SR, R being an alkyl group of a thiolate ligand [35]. As it is evident from the values presented in Table I, there are hardly any differences between the two cluster cores, and, within each cluster core, among the various Fe-S(Cys) distances. This may be a consequence of the low resolution of the crystal structure of the protein, but, more important, of the fact that, in proteins, the rigidity of the backbone can impose distances which are very different from those observed in model complexes, for which there are no constraints imposed by the thiolate ligands. The only interesting distance involves the Fe(2)-S(Cys II) in E. *halophila* HiPIP I: this distance of 2.41 Å is much larger than those involving Fe(1), Fe(3), and Fe(4), which would indicate a somewhat reduced ferrous state for Fe(2). In any case, these parameters are not useful to firmly establish a valence specific assignment based on X-ray crystallography, and, consequently, to infer any structure-function relationships. In this respect, it has been shown that the hydrogen bonding pattern to the sulfur ions of the core is nearly identical in the two cases [34], so that the difference in redox potential between HiPIP I from E. *halophila* (E_0' = 110 mV, [36]) andHiPIP from C. *vinosum* (E_0' = 356 mV, [37]) cannot be simply due to differences in the number of hydrogen bonds [38]. The possibility exists that very small, undetectable variations in bond distances and angles determine the charge distribution within the cluster core and the overall potential. On the other hand, the anisotropic electrostatic field exerted by the protein on the cluster core, may be responsible both for the large redox potential range found in HiPIP and for the valence distribution within the cluster core. The relevance of the latter contribution can be ascertained by molecular dynamics calculations, performed after the electron distribution within the cluster core and the framing of this core in the protein matrix is established. We will see now how this goal has been accomplished using ^{1}H NMR spectroscopy, which, taking advantage of the paramagnetic ground state and of the overall antiferromagnetism of oxidized HiPIP cluster core, exploits the extreme sensitivity of the isotropic shift to very small differences in the structural and electronic features of the individual iron ions.

Table 1. Interatomic distances (Å) in $[Fe_4S_4(SCys)_4]^{1-}$ clusters found in oxidized HiPIP from *C. vinosum* and *E. halophila*.

Bond Type	*C. vinosum* HiPIP	*E. halophila* HiPIP I
Fe(1)-Fe(2)	2.57	2.64
Fe(1)-Fe(3)	2.63	2.66
Fe(1)-Fe(4)	2.64	2.65
Fe(2)-Fe(3)	2.57	2.65
Fe(2)-Fe(4)	2.61	2.66
Fe(3)-Fe(4)	2.54	2.67
Mean	2.59	2.66
Fe(1)-S(Cys I)	2.17	2.18
Fe(2)-S(Cys II)	2.19	2.41
Fe(3)-S(Cys III)	2.25	2.09
Fe(4)-S(Cys IV)	2.17	2.02
Mean	2.20	2.18
Fe(1)-S(1)	2.34	2.22[a]
Fe(1)-S(2)	2.32	2.22[a]
Fe(1)-S(3)	2.24	2.22[a]
Fe(2)-S(1)	2.23	2.22[a]
Fe(2)-S(2)	2.16	2.22[a]
Fe(2)-S(4)	2.25	2.22[a]
Fe(3)-S(2)	2.24	2.22[a]
Fe(3)-S(3)	2.14	2.22[a]
Fe(3)-S(4)	2.36	2.22[a]
Fe(4)-S(1)	2.27	2.22[a]
Fe(4)-S(3)	2.22	2.22[a]
Fe(4)-S(4)	2.15	2.22[a]

[a] Imposed equal in the X-ray structure refinement.

4. The valence-specific assignment of the iron ions in oxidized HiPIP

The application of NMR spectroscopy to iron-sulfur proteins dates back to the early seventies [39-42]. In recent years, however, major advancements have occurred in the understanding of the NMR spectra of HiPIP, specifically of the temperature dependence of the isotropically shifted signals [11,43-45]. Advanced 1D and 2D NMR spectroscopies have been utilized to study the oxidized HiPIP from *C. vinosum* [46-50] and the oxidized HiPIP II from *E. halophila* [51-53].

These studies have taken advantage of the paramagnetism of the polymetallic cluster, which allows the detection of the isotropically shifted signals of protons which are scalarly or dipolarly coupled to the $[Fe_4S_4]^{3+}$ core [54]. In the previous article of this series [11] we have discussed the theoretical background which allows the interpretation of the 1D NMR spectra of these proteins. Here we will focus the procedure which we have used to perform the valence specific assignment, and the conclusions that we have reached regarding the electron distribution within the cluster core. The cysteine numbering scheme is shown in Figure 2.

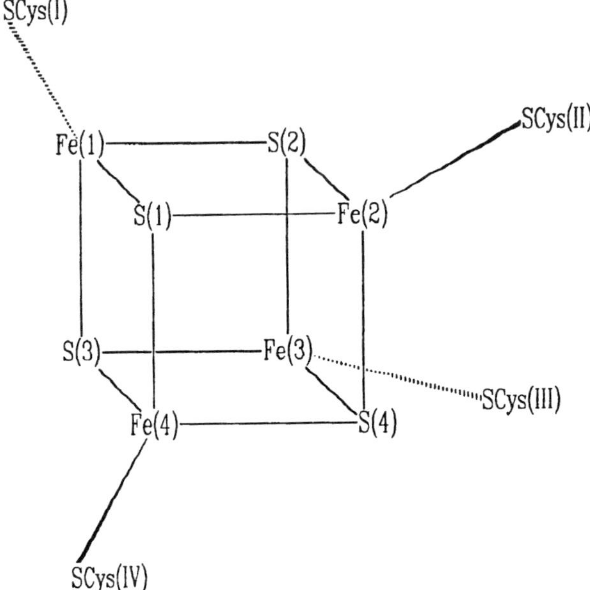

Figure 2. Close-up view of the iron-sulfur cluster in HiPIP, showing the numbering scheme followed throughout this paper.

4.1. OXIDIZED HiPIP FROM *C. VINOSUM*

In the ^{1}H NMR spectrum of this protein, shown in Figure 3A, seven hyperfine-shifted signals appear downfield of the intense diamagnetic region, and two upfield [46-50].

[1]H NOE (Nuclear Overhauser Enhancement) and COSY (COrrelation SpectroscopY) experiments performed on these signals have revealed strong through-space and through-bond connectivities between signals *A-B, C-D, F-G,* and *Y-Z,* and they have thus been assigned to the β-CH_2 protons of the cluster-bound cysteine residues [47,48]. Signal E shown dipolar and scalar connectivities to *C* and *D*, and it has been assigned to the α-CH proton of the cysteine to which signals *C* and *D* also belong [48].

The temperature dependence of the eight signals of the β-CH_2 protons is characterized by the shift of signals *A, B, C, D* (and *E*) to higher fields with increasing temperature, and by the opposite trend for signals *F, G, Y* and *Z* [48]. On the basis of the information drawn from Mössbauer spectroscopy, this temperature dependence behavior has been interpreted as a consequence of the sign of the hyperfine constants observed for the mixed-valence pair (negative, consistent with a larger subspin S_{34}) and for the ferric pair (positive, consistent with a smaller subspin S_{12}) [15]. Thus, the downfield signals *A, B, C,* and *D* belong to cysteines bound to the mixed-valence pair, while the signals *F, G, Y,* and *Z* must then belong to cysteines bound to the ferric pair. A coupling scheme has been proposed which accounts for the shift and temperature dependence of these signals [11, 43, 45, 48]. To place the cluster core within the protein frame, that is to establish wich cysteine residue is bound to the ferric ions and which to the mixed-valence iron ions, we have utilized the information coming from the tree-dimensional crystal structure of the protein [3-5], which indicates that some Cys β-CH_2 or α-CH protons lie very close to protons belonging to the aromatic rings of residues surrounding the cluster. In spite of the fast nuclear relaxation rate characterizing these protons, we have been able detect ^{1}H NOE connectivities in 1D and 2D NMR experiments.

The difference NOE spectrum obtained by irradiating signal F, reveals signals w_1 and w_2 which, on the basis of 2D EXSY (EXchange SpectroscopY), NOESY (Nuclear Overhauser Enhancement SpectroscopY) and TOCSY (TOtal Correlation SpectroscopY) spectra, have been assigned to ring NH protons of the two, deeply buried, triptophan residues Trp 76 and Trp 80 (Figure 4A-C).

The crystal structure of the protein [3-5] indicates that Cys II Hβ1 is the only Cysβ-CH2 proton lying close to a Trp NH proton, namely the NH of Trp 76, thus allowing the assignment of signal *G* to Cys II Hβ2. We then concluded that Cys II is bound to one iron ion belonging to the ferric pair, on the basis of the temperature dependence of signals *F* and *G*. The presence of NOE and NOESY correlations between the signals of the aromatic ring of Trp 76 and a signal corresponding to signal E in the reduced protein (Figure 4B), and the short distance between the α-CH proton of Cys IV and the aromatic ring of Trp 76 observed in the crystal structure, allowed the assignment of signals *C, D* and *E* to β-CH_2 and α-CH of Cys IV. From the temperature dependence behavior of these signals, we then concluded that Cys IV is bound to an iron ion belonging to the mixed-valence pair. A NOESY cross peak in the reduced protein, (Figure 4D) is observed between signal *b*, corresponding to signal, *A* of the oxidized protein, and a signal belonging to a scalar pattern typical of a Phe residue (Figure 4E).

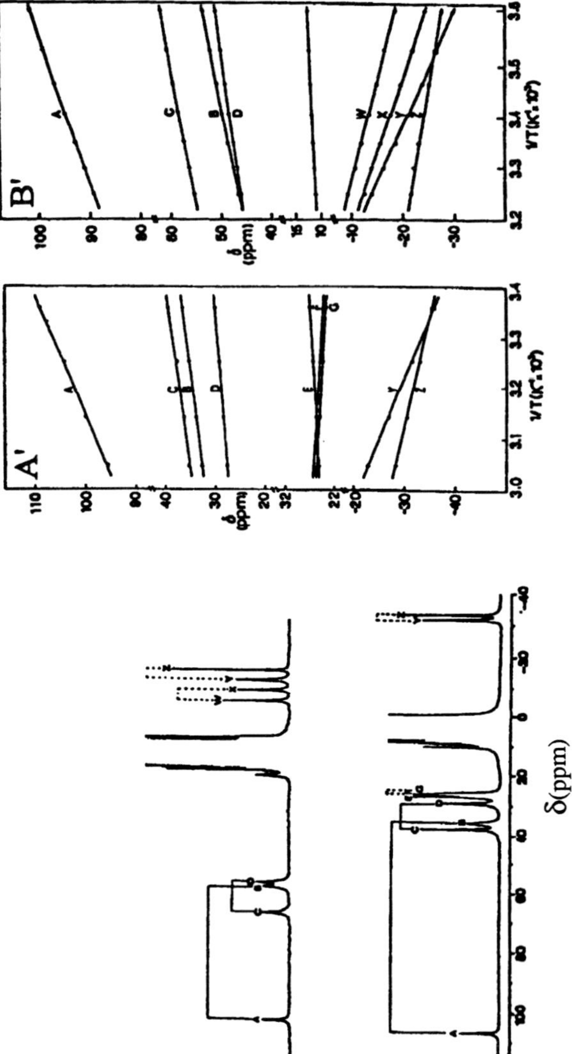

Figure 3. 600 MHz 300 K ^{1}H NMR spectra of oxidized high potential iron-sulfur proteins. A: *C. vinosum* HiPIP; *E. halophila* HiPIP II. The temperature dependence of the isotropically shifted signals are shown in A' and B', respectively.

The fact that, in the crystal structure, one of the β-CH$_2$ protons of Cys III is the only cysteine methylene proton lying close to a ring proton of Phe 66, allowed the assignment of signals A and C as β-CH$_2$ protons of Cys III. Thus, Cys III is bound to one iron ion belonging to the mixed-valence pair, on the basis of the temperature dependence of signals A and B. Finally, the last cysteine, Cys I, must then be bound to an iron ion of the ferric pair, and the signals corresponding to its β-CH$_2$ protons are Y and Z. This is further confirmed by the presence of NOE connectivities between signal Z and signals typical of NH protons, and by the fact that Cys I Hβ1 lies very close to the NH protons of Cys I itself and of Ala 44.

4.2. OXIDIZED HiPIP II FROM *E. HALOPHILA*

The 1D NMR spectrum of oxidized *E. halophila* HiPIP II, shown in Figure 3B [52], presents four downfield shifted and four upfield shifted signals which, on the basis of their temperature dependence, short relaxation times, and NOE measurements have been assigned to four couples of methylene protons of the iron-bound cysteine residues, the pairs being *A-B, C-D, W-X, Y-Z* [52]. The downfield signals move to higher fields with increasing temperature, and are thus assigned to cysteines bound to mixed-valence iron ions, while the upfield signals are characterized by an opposite behavior, and must correspond to cysteines bound to ferric ions on the basis of the Mössbauer parameters observed for this protein [25]. This NMR spectrum was sufficiently different from the corresponding spectrum observed for *C. vinosum* HiPIP to prompt an investigation of the electron distribution in this protein. The framing of the cluster core inside the HiPIP from *C. vinosum* [49] was made possible by the availability of this crystal structure [3-5]. The corresponding X-ray structure is not known for *E. halophila* HiPIP II, but a crystallographic study has been conducted on the HiPIP I from the same microorganism [34], a protein which has a 65% homology with HiPIP II [36]. Moreover, the aromatic residues around the cluster are all conserved in the two sequences. This homology has been exploited by performing molecular dynamics (MD) calculations in order to predict the tridimensional structure of *E. halophila* HiPIP II in solution [53]. This approach has followed the successful development of the parameters describing the [4Fe-4S] cluster in *C. vinosum* HiPIP [55].

Using these parameters, a model for the solution structure of HiPIP II has been obtained, thus providing a set of key inter-proton distances, subsequently used in the NMR study to frame the valence distribution of the cluster core in the protein backbone [53]. This is the first time that MD calculations and NMR spectroscopy complement each other in providing structural information on a paramagnetic protein for which the crystal structure is not available. The difference NOE spectrum which is obtained upon irradiation of signal Z shown two signals at 6.30 and 5.58 ppm. These signals have been assigned to aromatic protons of a Trp residue through the TOCSY (Figure 5) and NOESY (Figure 6) spectra. Only one methylene proton, belonging to Cys I, can be found in the vicinity of the aromatic ring of a triptophan, namely Trp 45, in the model obtained by MD calculations. This allows the assignment of signals Y and Z to Cys I, which is then bound to a purely ferric ion. By saturating signals C and D, a common

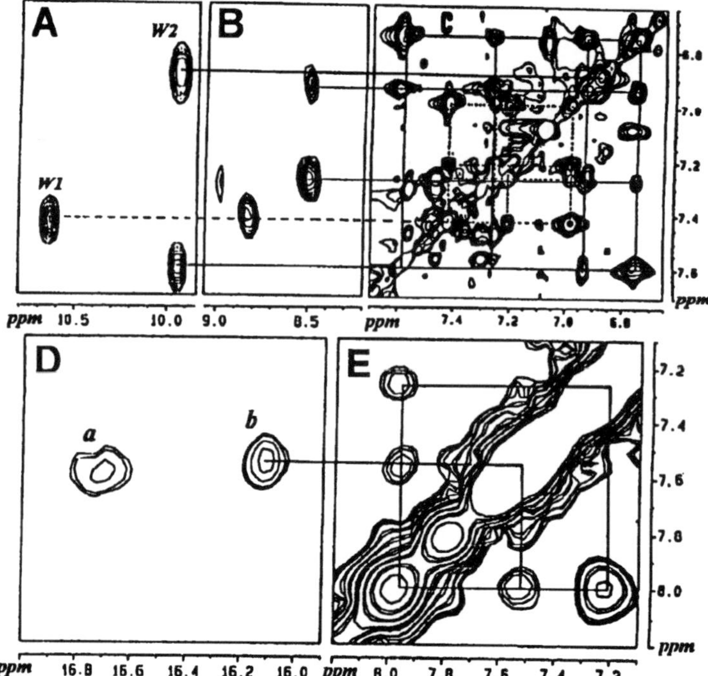

Figure 4. 600 MHz 300 K NOESY (A,B, and D), TOCSY (C), and COSY (E) spectra of the aromatic region of reduced *C. vinosum* HiPIP. The TOCSY spectrum shows the spin systems of Trp 76 (-------) and Trp 80 (- - - -). The COSY spectrum reveals the scalar pattern of Phe 66 (----). The NOESY spectrum allows detection of dipolar connectivities from Trp 80 and Trp 76 to w_1 and w_2 NH protons, respectively (A), from Phe 66 to signal b (D), and from the signal at 8.52, corresponding to signal E in the oxidized protein, and the aromatic protons of Trp 76.

NOE is observed at 8.20 ppm, which is part of a NOESY and TOCSY pattern of a Phe residue. The uniqueness of Phe 60 in the vicinity of Cys IV, immediately suggests that *C* and *D* belong to this cysteine residue, which is also bound to a ferric ion. As a consequence, Cys II and Cys III are bound to the mixed-valence pair. The NOE difference spectrum obtained saturating signal X shows two signals in the aromatic region, at 6.80 and 6.42 ppm, which are also dipolarly and scalarly coupled. This observation suggests that W and X belong to Cys II because the alternative Cys III cannot give dipolar coupling signals in the aromatic region being far away from any aromatic residues.

5. Conclusions and Perspectives

We have shown how NMR spectroscopy can be more successful than Mössbauer, EPR, ENDOR, RR and X-ray crystallography in identifying and characterizing the *four non-*

equivalent iron ions in HiPIP, and how the framing of the cluster in the protein can be accomplished. A summary of the results obtained so far is given in Table II. The first observation that can be drawn from the present results is that the valence distribution within the cluster cores is different in the case of *C. vinosum* HiPIP and *E. halophila* HiPIP II. An illustration of this fact is given in Figure 7. Only the iron ion bound to Cys III maintains a mixed-valence character, and the iron ion bound to Cys I remains ferric in both proteins. On the other hand, the iron ion bound to Cys II is ferric in the case of *C. vinosum*, but is part of the mixed-valence pair in *E. halophila* HiPIP II, while the iron ion bound to Cys IV behaves conversely, having a mixed-valence in *C. vinosum* HiPIP and a ferric valence in *E. halophila* HiPIP II. A consequence of this behavior could be that it may not be important, for the electron transfer process, where the excess electron density is removed away upon oxidation of the reduced HiPIP, and that only the overall redox potential is functionally controlled by the protein backbone. If this is true, the electron distribution may be simply modulated by very small variations in bond distances and angles of the cluster core.

Similar studies are available for oxidized HiPIP isolated from *R. gelatinosus* [56, 57], *R. globiformis* [58] and *E. vacuolata* [59]. From such studies, it has been evidenced the existence of two main types of electron distribution: *C. vinosum* and *R. gelatinosus* HiPIP have one orientation of the cluster within the protein, whereas *R. globiformis* and *E. halophila* HiPIP II have another orientation, the HiPIP II isolated from *E.vacuolata* representing an intermediate situation.

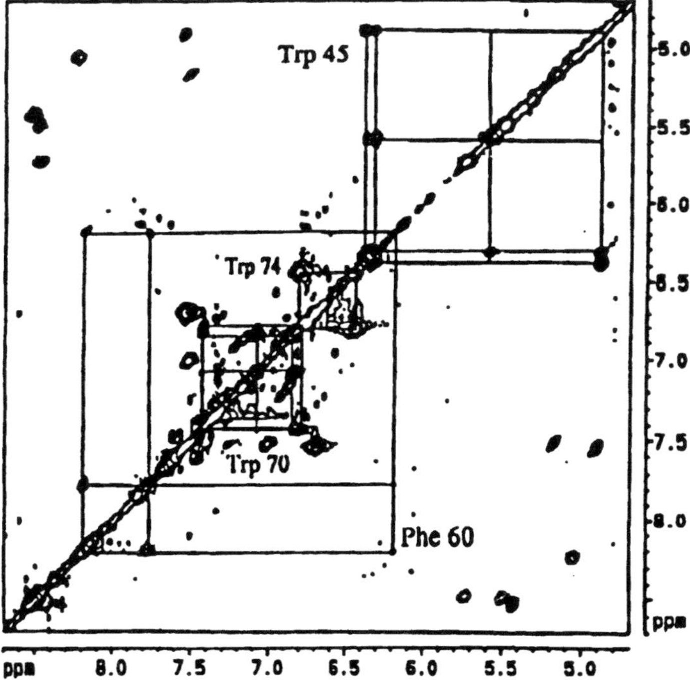

Figure 5. 600 MHz, 300 K ^{1}H NMR TOCSY spectrum (aromatic region) of the oxidized HiPIP II from *E. halophila*.

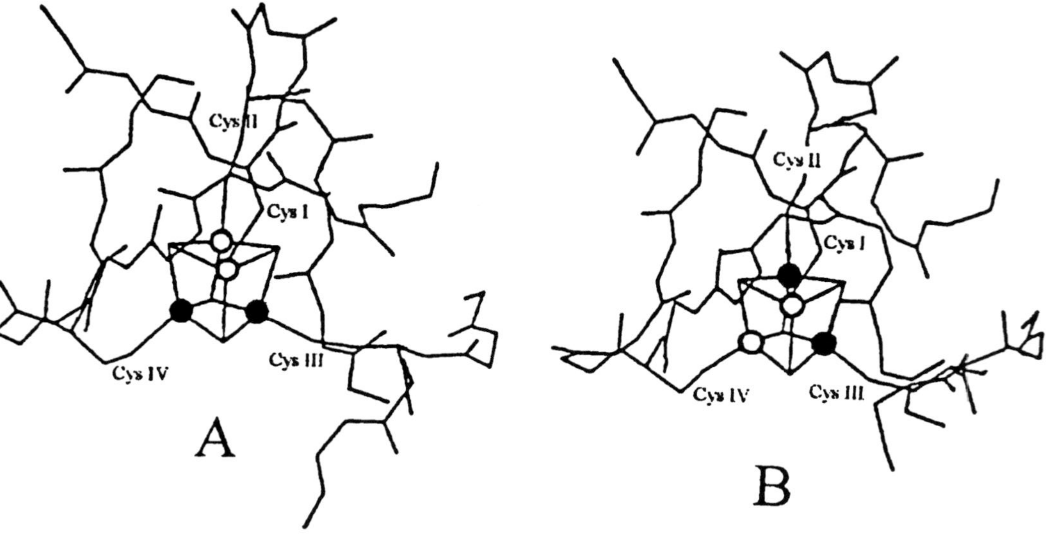

Figure 6. Geometric arrangement of the cluster and of the sorrounding residues in *C. vinosum* HiPIP derived from crystallographic data (A) and *E. halophila* HiPIP II derived from MD calculations (B). White circles evidence Fe^{3+} ions and black circles $Fe^{2.5+}$ ions as indicated by NMR.

Table II. Summary of the results obtained on HiPIP

Iron Pair	Valence Type	S	A	Resonating Magnetic Field	Cysteine coordination in *C. vinosum* HiPIP	Cysteine coordination in *E. halophila* HiPIP
3,4	Mixed Valence	9/2 (or 7/2)	< 0	Increase with T	Cys III Cys IV	Cys II Cys III
1,2	Ferric	4 (or 3)	> 0	Decrease with T	Cys I Cys II	Cys I Cys IV

Molecular dynamics calculations are currently in progress on all these proteins in the effort to determine the local and overall electrostatic potential exerted on the iron ions and on the whole cluster core by the protein surroundings. These calcualtions should allow the rationalization of the electron distributions found in HiPIP and of the influence of the protein bakbone on the establishment of redox potentials.

References

1. R.G. Bartsch, *Methods Enzymol.* **53**, 329-340, (1978).
2. C.W.IR. Carter, in <u>Iron-Sulfure Proteins,</u> W. Lovenberg, Ed., Academic Press, New York; Vol. **3**, 157-204, (1977).
3. C.W.,Jr. Carter, J. Kraut, S.T. Freer, N.-H. Xuong, R.A. Alden, R.G. Bartsch, *J. Biol. Chem.* **249**, 4212, (1974).
4. C.W.,Jr. Carter, J. Kraut, S.T. Freer, R.A. Alden, *J. Biol. Chem.* **249**, 6339, (1974).
5. S.T. Freer, R.A. Alden, C.W.Jr. Carter, J. Kraut, *J. Biol. Chem.* **250**, 46, (1975).
6. R.G. Bartsch, *Biochim. Biophys. Acta* **28**, 1058, (1991).
7. C.W.,Jr. Carter, J. Kraut, S.T. Freer, R.A. Alden, L.C. Sieker, E. Adman, L.H. Jensen, *Proc. Natl. Acad. Sci. U.S.A.* **69**, 3526, (1972).
8. T.E. Meyer, C.T. Przysiecki, J.A. Watkins, A. Bhattacharyya, R.P. Simondsen, M.A. Cusanovich, G. Tollin, *Proc. Natl. Acad. Sci. U.S.A.*, **80**, 6740, (1983).
9. D.C. Yoch, R.P. Carithers, *Microbiol. Rev.* **43**, 384, (1979).
10. F.A. Armstrong, S.J. George, A.J. Thomson, M.G. Yates, *FEBS Lett.* **234**, 107, (1988).
11. L. Banci, I. Bertini, F. Briganti, C. Luchinat, A. Scozzafava, Topics in <u>Molecular Organization and Engineering</u>, W.N. Lipscomb, and J. Maruani, eds., Kluwer Academic Publishers, Dordrecht, Vol. **8**, pp.; (1991).
12. T.H. Moss, A. J. Bearden, R.G. Bartsch, M.A. Cusanovich, *Biochemistry* **7**, 1591, (1968).
13. M.C.W. Evans, D.O. Hall, C.E. Johnson, *Biochem. J.* **119**, 289, (1970).
14. D.P.E. Dickson, C.E. Johnson, R. Cammack, M.C.W. Evans, D.O. Hall, K.K. Rao, *Biochem. J.* **139**, 105, (1974).
15. P. Middleton, D.P.E. Dickson, C.E. Johnson, J.D. Rush, *Eur. J. Biochem.* **104**,

289, (1980).

16. L. Noodleman, *Inorg. Chem.* **27**, 3677, (1988).

17. J.M. Mauesca, B. Lamotte, G.J. Rius, *Inorg. Biochem.* **43**, 251, (1991).

18. L. Banci, I. Bertini, F. Briganti, C. Luchinat, A. Scozzafava, M. Vicens Oliver, *Inorg. Chem.* **30**, 4517, (1991).

19. L. Banci, I. Bertini, C. Luchinat, *Struct. Bonding* **72**, 113, (1990).

20. L. Banci, I. Bertini, F. Briganti, C. Luchinat, *New J. Chem.* **15**, 467, (1991).

21. T.H. Moss, D. Petering, G. Palmer, *J. Biol. Chem.* **244**, 2275, (1969).

22. W.D. Phillips, M. Poe, C.C. McDonalds, R.G. Bartsch, *Proc. Natl. Acad. Sci. U.S.A.* **67**, 682, (1970).

23. B.C. Antanaitis, T.H. Moss, *Biochim. Biophys. Acta*, **405**, 262-279, (1975).

24. W.R. Dunham, W.R. Hagen, J.A. Fee, R.H. Sands, J.B. Dunbar, C. Humblet, *Biochim. Biophys. Acta* **253**, 1079, (1991).

25. I. Bertini, A.P. Campos, C. Luchinat, M. Teixeira, submitted

26. G. Rius, B. Lamotte, *J. Am. Chem. Soc.* **111**, 2464, (1989).

27. J.M. Mauesca, B. Lamotte, G. Rius, *J. Inorg. Biochem.* **43**, 251, (1991).

28. R.E. Anderson, G. Anger, A. Petersson, Ehrenberg, R. Cammack, D.O. Hall, R. Mullinger, K.K. Rao, *Biochim. Biophys. Acta* **376**, 63, (1975).

29. R.S. Czernuszewicz, M.A. Macor, M.K. Johnson, A. Gewirth, T.G. Spiro, *J. Am. Chem. Soc.* **109**, 7178, (1987).

30. J.-M. Moulis, J. Meyer, M. Lutz, *Biochemistry* **23**, 6605, (1984).

31. G. Backes, Y. Mino, T.M. Loehr, T.E. Meyer, M.A. Cusanovich, W.V. Sweeney, E.T. Adman, J. Sanders-Loehr, *J. Am. Chem. Soc.* **113**, 2055, (1991).

32. V. Papaefthymiou, M.M. Millar, E. Münck, *Inorg. Chem.* **25**, 3010, (1986).

33. L. Noodleman, D.A. Case, *Adv. Inorg. Chem.* **38**, 423, (1992).

34. D.R. Breiter, T.E. Meyer, I. Rayment, H.M. Holden, *J. Biol. Chem.* **266**, 18660, (1991).

35. M.J. Carney, G. Papaefthymiou, K. Spartalian, R.B. Frankel, R.H. Holm, *J. Chem. Soc.* **110**, 6084, (1988), and references cited therein.

36. C.T. Przysiecki, T.E. Meyer, M.A. Cusanovich, *Biochemistry* **24**, 2542, (1985).

37. I.A. Mizrahi, T.E. Meyer, M.A. Cusanovich, *Biochemistry* **19**, 4727, (1980).

38. R.P. Sheridan, R.C. Allen, J.W.Jr. Carter, *J. Biol. Chem.* **256**, 5052, (1981).

39. M. Poe, W.D. Phillips, C.C. MacDonald, W. Lovenberg, *Proc. Natl. Acad. Sci. U.S.A.* **65**, 797, (1970).

40. M. Poe, W.D. Phillips, J.D. Glickson, C.C. MacDonald, A. San Pietro, *Proc. Natl. Acad. Sci. U.S.A.*, **68**, 68, (1971).

41. W.D. Phillips, M. Poe, C.C. MacDonald, R.G. Bartsch, *Proc. Natl. Acad. Sci. U.S.A.* **67**, 682, (1970).

42. W.R. Dunham, G. Palmer, R.H. Sands, A.J. Bearden, *Biochim. Biophys. Acta* **253**, 373, (1971).

43. L. Banci, I. Bertini, C. Luchinat, *Struct. Bonding* **71**, 113, (1990).

44. C. Luchinat, S. Ciurli, "NMR of Polymetallic Systems in Proteins" in <u>Biological Magnetic Resonance,</u> L.J. Berliner and J. Reuben Eds., Plenum Press, New York, (1992) in press.

45. L. Banci, I. Bertini, F. Briganti, C. Luchinat, *New. J. Chem.* **15**, 467, (1991).

46. D.G. Nettesheim, T.E. Meyer, B.A. Feinberg, J.D. Otvos, *J. Biol. Chem.* **258**, 8235, (1983).

47. J.A. Cowan, M. Sola, *Biochemistry* **29**, 5633, (1990).
48. I. Bertini, F. Briganti, C. Luchinat, A. Scozzafava, M. Sola, *J. Am. Chem. Soc.* **113**, 1237, (1991).
49. I. Bertini, F. Capozzi, S. Ciurli, C. Luchinat, L. Messori, M. Piccioli, *J. Am. Chem. Soc.* **114**, 3332, (1992).
50. (a) D.G. Nettesheim, S.R. Harder, B.A. Feinberg, J.D. Otvos, *Biochemistry* **31**, 1234, (1992). (b) J. Gaillard, J.P. Albrand, J.-M. Moulis, D.E. Wemmer, *Biochemistry*, in press.
51. R. Krishnamoorthy, J.L. Markley, M.A. Cusanovich, C.T. Przysiecki, *Biochemistry* **25**, 60, (1986).
52. L. Banci, I. Bertini, F. Briganti, C. Luchinat, A. Scozzafava, M. Vicens Oliver, *Inorg. Chim. Acta* **180**, 171, (1991).
53. L. Banci, I. Bertini, F. Capozzi, P. Carloni, S. Ciurli, C. Luchinat, M. Piccioli, submitted.
54. I. Bertini, C. Luchinat, in <u>NMR of Paramagnetic Molecules in Biological Systems</u>, Benjamin/Cummings, Menlo Park, California, (1986).
55. L. Banci, I. Bertini, P. Carloni, C. Luchinat, P.L. Orioli, *J. Am. Chem. Soc.* in press.
56. L. Banci, I. Bertini, F. Briganti, C. Luchinat, A. Scozzafava, M. Vicens Oliver, *Inorg. Chem.* **30**, 4517, (1991).
57. I. Bertini, F. Capozzi, C. Luchinat, M. Piccioli, M. Vicens Oliver, *Inorg. Chim. Acta* **483**, 198-200, (1992).
58. I. Bertini, F. Capozzi, C. Luchinat, M. Piccioli, *Eur. J. Biochem.* in press.
59. L. Bianci, I. Bertini, S. Ciurli, S. Ferretti, C. Luchinat, M. Piccioli, submitted.

Chlorophyll a Molecular Organization and Photoreactivity

P. COSMA[§], A. AGOSTIANO, L. CATUCCI[§], A. CEGLIE, G. COLAFEMMINA,
A. MALLARDI[§], G. PALAZZO, M. TROTTA[§] and M. DELLA MONICA.
Dipartimento di Chimica-Università degli Studi di Bari
[§] *CNR-Centro Studi Chimico-Fisici sull'Interazione Luce-Materia 4, Traversa Re David
200, 70126 Bari (Italy)*

1.Introduction

The Chlorophyll a (Chl a) is the main pigment in plant photosynthesis, whose relevance
resides in its functional duality. The Chl a, indeed, acts as antenna, collecting and
funneling light, and as photoreactive center, where the energy transduction takes place.
The different roles played by this molecule are closely related to its molecular
organization due to stereospecific interactions involving the Mg atom and the C=O
group of the Chl a with water or side chains (N-H, S-H, O-H) of natural aminoacids
[1].
For this reason, the problem of pigment aggregation and interaction with water is of
great interest in the *in vitro* studies of photosynthesis, whose important focus is the
characterization of the photoreactive form of the Chlorophyll a [2, 3].
Currently the work on Chl a aggregation and photochemistry is focussed on its aqueous
solutions with organic solvents [4-11]. Chl a aggregation, in a wide range of
composition of the water/polar solvent mixtures, has been the subject of investigation in
our laboratories [12-15].
The first part of this paper summarizes the results of our researches on the different
factors affecting the Chl a aggregation. The influence of the different functional groups
of the solvents, the role played by the water content of the solution, the relevance of the
hydrophobic effect and the correlation with the excess properties of the medium will be
discussed.
Although the existence of an efficient energy transfer between different Chl a aggregates
in water/polar solvent mixtures has been demonstrated [16], the study of Chl a
photochemistry is not an easy task because of the lack of compartmentalization, which is
essential for the stabilization of the charge separated state and vectorial electron transfer
in *in vivo* photosynthesis.
The second part of the paper reports the attempts to immobilize the photoreactive Chl a
at the interface between different phases. Reverse micelles were chosen as microreactors
in which the trapping of the Chlorophyll could supposedly be possible by the use of
water/organic solvent binary mixtures [17, 18].
The results of NMR studies on this system, indicating the impossibility of Chl a to be

159

N. Russo et al. (eds.), Properties and Chemistry of Biomolecular Systems, 159–174.
© 1994 *Kluwer Academic Publishers. Printed in the Netherlands.*

trapped inside the reverse micelles [19], shifted our attention on a new family of water-in-oil microemulsions formed by lecithin in n-heptane. Upon water addition the microemulsion is transformed into transparent gel [20].

The last part of this note reports the results obtained by introducing a photoreactive form of Chl a at the interface between the water and the lipidic phase. The photoinduced electron transfer catalyzed by the Chlorophyll was verified in presence of suitable electron donors and acceptors solubilized in the different phases of the microemulsion [21].

2. Binary Mixtures

The behavior of the Chl a in solution is governed by the inter and intra molecular interactions of its functional groups.

Namely, the Mg atom of the porphyrin ring provides the site of binding for the nucleophylic moiety of solvents, such as the oxygen of water, ketones or alcohols. The carbonyl groups in C-13[1], C-13[3], C-17[3] act in addition as a nucleophyl in the hydrogen bonding to the solvents or to a second Chl a molecule. Finally, the hydrophobic phytil chain allows the Chl a solubilization in the hydrocarbon constituent of the medium (Fig. 1). In other words in the binary mixtures the polar component favors the aggregation of porphirinic rings of Chl a, while the hydrophobic component tends to keep the phytilic chains far apart. The formation of various solvated and aggregated species is a multistep

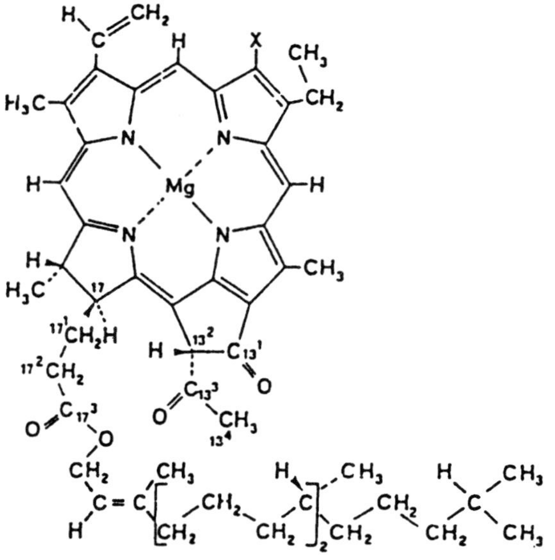

Figure 1. Chlorophyll a molecule.

equilibrium process, whose position depends more largely upon the nature and the composition of the solvent and upon Chl a concentration.

The existence of monomeric and aggregated forms of Chl a in dry or wet nonpolar medium has been widely investigated [22-24]. The variations in the absorption and fluorescence spectra of the Chl a with the composition of binary polar solvent mixtures with water have been reported by early researchers [4, 5, 25]. Only recently attempts were made to relate the spectroscopic observations to Chl a aggregation as a consequence of its hydration behaviour [12, 16, 26].

In many of the pure polar solvents examined Chl a was present as a monomer, with the nucleophylic part of the solvent linked to the Mg. Its characteristic spectrum exibits the red S1 transition with 4 distinct satellite peaks and a strong Soret transition in the blue, with two satellite peaks also. Fig. 2 shows the spectrum of a solution 5×10^{-6} M of Chl a in pure acetone and in presence of various amounts of water. (Note that dimer formation has been reported in literature [4, 27] at high concentration in methanol and ethanol in contrast with our findings).

Fig. 3 shows the changes in the absorption intensity and, in the position of the main red band of a solution 5×10^{-6} M Chl a in acetonitrile (ACN) reported as function of the water molar fraction (χ_{H2O}). Up to a χ_{H2O} of 0.66 only small red shifts in the band position are observed. In the same range of χ_{H2O} the Adsorptive Stripping Voltammetry

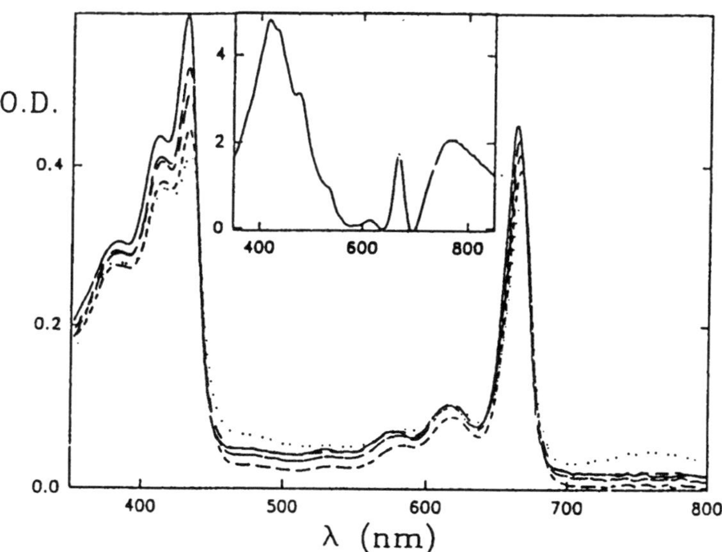

Figure 2. Absorption spectra of a 5×10^{-6} M Chl a solution in acetone/water at different water amount. ───── Pure acetone; ---- 10%; ─ ─ ─ 20 %; ─ . ─ . ─ 30%; ············· 50% H_2O. Inset: absorption spectrum of 5×10^{-5} M Chl a solution in acetone/water 50%.

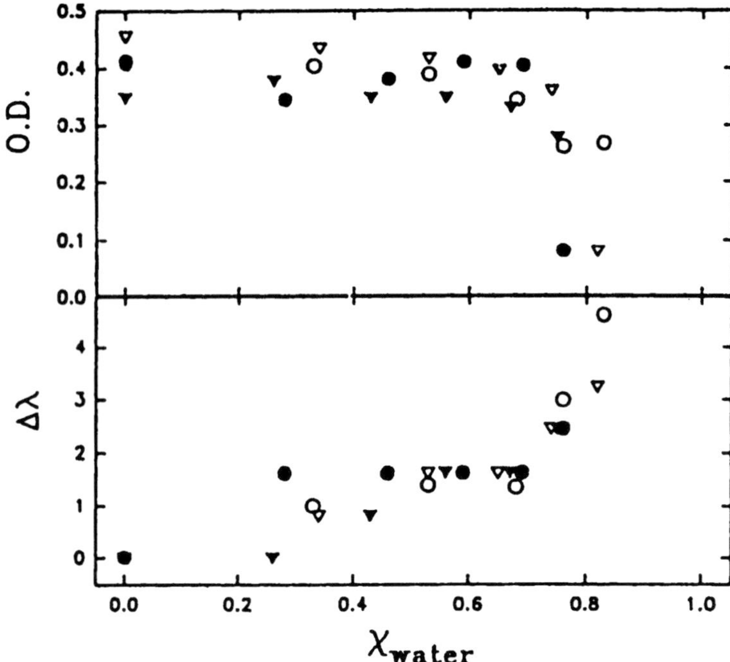

Figure 3. Dependence of the optical absorption shift (Dl) of the main red peak for different solvents: (—) ACN, (—) Acetone, (•) dimethyl formamid, (o) ethilic alcohol.

shows the presence of two peaks, indicative of at least two solvation Chl a complexes with distinctly different adsorption/desorption kinetics.

The spectroscopic and electrochemical data concordly suggest the exchange of solvates with the water replacing the other solvents at the Mg atom [12, 26].

A steeper decrease in the intensity of the peak maximum and the asymmetric broadening of the red region characterize the spectra of samples of Chl a in polar solvents with χ_{H2O} exceeding 0.66.

The presence of a unique peak in the corresponding voltammograms indicates a complete conversion of the Chl a-organic solvent complex to the hydrated Chl a complexes [12, 14, 25, 28].

More information on the nature of the Chl a species can be obtained from the Circular Dichroism (CD) and fluorescence data reported in Fig. 4 and 5 respectively.

It is well known that although the Chl a contains three asymmetric carbon atoms, only a weak value of ellipticity is recorded in its solution in pure polar solvent, where it is present as a monomer. Samples of 5 x 10^{-6} M of Chl a in acetone and ethanol with χ_{H2O} exceeding O.66 show a complex CD spectrum characterized by the splitting into two components of the Soret band and a great enhancement in the value of the ellipticity

(Fig.4). These data account for the coupling of the oscillators associated with the corresponding transitions in the different molecules. This characterizes a largely overlapping dimer in which a water molecule links the Mg of one Chl $\underline{a}$ with the C-17^3 of a second one [22, 23, 29].

The presence of a dimeric form of Chl $\underline{a}$ is further supported by the fluorescence data reported in Fig. 5, in which the relative fluorescence efficiency, as function of the solvent composition, is shown for Chl $\underline{a}$ solutions in all the polar solvent examined.

The quenching of the fluorescence parameters observed in all the solvent for χ_{H2O} exceeding 0.66 can be interpreted in terms of a resonant energy transfer between monomeric and dimeric Chl $\underline{a}$ [6, 8, 13, 25]. The expression for the dependence of the fluorescence quantum efficiency on the solvent composition has been calculated assuming that four water molecules are involved in the equilibrium conversion from the monomer to the dihydrate dimer of Chl $\underline{a}$. The quadratic dependence on the concentration C of the pigment given in the equation of Fig. 5 is consistent with the general theory of energy transfer. The probability of finding the acceptor in the nearest neighbour position is proportional to C^{n-1}, where n is the number of bodies involved in the process. The monomer to dimer transfer is indeed a coherent "three body process"[16]. The oligomer formation in the samples of Fig. 5 is indicated by the deviation from the values expected from the theoretical numerical fit.

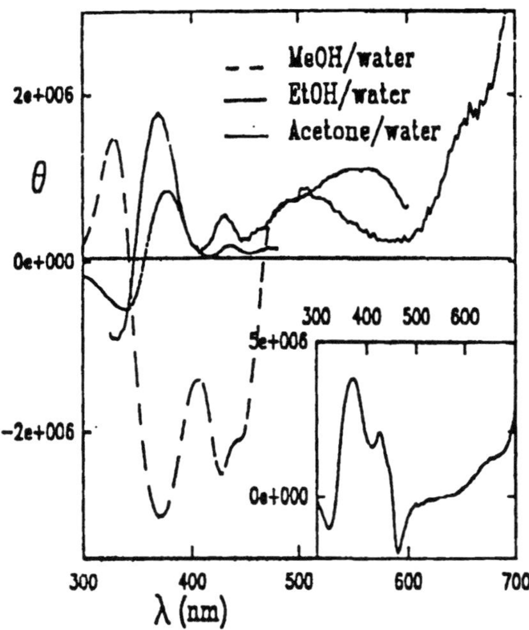

Figure 4. CD spectra of 5 x 10^{-6} M solutions of Chl $\underline{a}$ of various water/organic solvent mixtures at fixed composition 50%.vol. Inset: CD spectrum of oligomeric aggregate in water/ACN 50% vol. All the data are expressed as deg·cm^2/dmol.

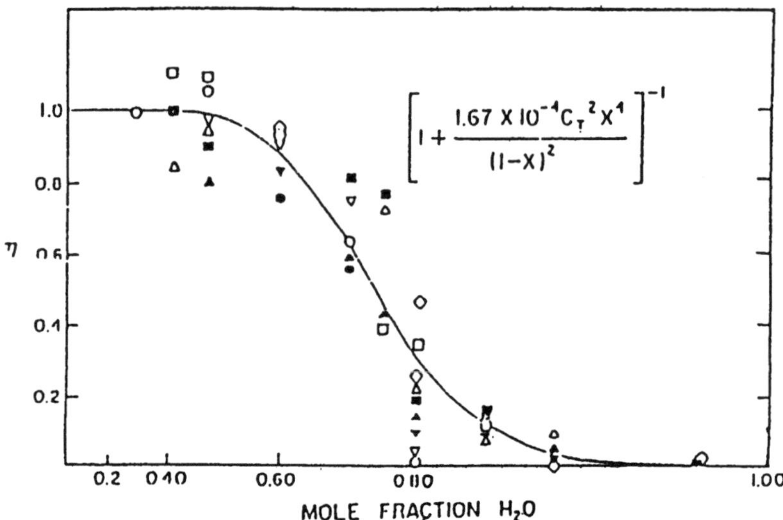

Figure 5. Concentration and solvent dependence of relative fluorescence efficiency. Life-time measurement: (•) 5 x 10^{-5} M Chl a in acetone; (■) 5 x 10^{-6} M Chl a in ACN; (▲) 2 x 10^{-6} M Chl a in ethanol. Relative fluorescence intensity measurements: (o) 5 x 10^{-5} M Chl a in acetone; (x) 5 x 10^{-6} M Chl a in acetone; (▼) 5 x 10^{-7} M Chl a in acetone; ( ) 5 x 10^{-6} M Chl a in ACN; (Δ) 2 x 10^{-6} M Chl a in ethanol. Quantum yield: (∇) 5 x 10^{-5} M Chl a in acetone.

The Chl a concentration and the functional group of the polar solvent play a relevant role in determining Chl a aggregation. The inset in Fig. 2 shows, at higher Chl a concentration (up to 5 x 10^{-5} M) in acetone containing the 50 % of water, a new peak absorbing around 745 nm. This peak is assigned to an oligomeric form of the hydrated Chl a. The same band is present in the spectrum of a solution of ACN with the same water amount but at a Chl a concentration ten times smaller.

The structure of the 745 nm absorbing Chl a form is still controversial. One hypothesis, based on Small-Angle Neutron Scattering experiments [30], assumes the formation of a cilindrical reverse micelle with a single water molecule cross-linking the magnesium, the keto and esteric carbonyl group of two chlorophylls [3]. In a second hypothesis, based on the X-ray crystallographic structure of the ethyl-chlorophyllide, two water molecules are involved in the bridge connecting the Mg atom and the esteric carbonyl of one Chl a to the C-13 keto group of a second Chl a molecule [3]. The CD spectrum (inset of Fig. 4) in water/ACN containing the oligomeric Chl a as a prevalent species does not evidence relevant differences in shape compared to the CD spectra of solutions containing the dimer. This observation is consistent with the structure of the oligomer as a repetition in the space of the dimeric minimal unit.

The high value of the ellipticity, indicating a strong coupling between neighbouring molecules, seems to rule out the possibility of formation of a cylindrical micelle with the porphyrin rings at the separation surface, whose value of ellipticity is expected to be extremely low.

When the percentage of organic solvent in water does not exceed 1%, the absorption spectra of the Chl a show the appearing in time of a new species absorbing at 713 nm.(Fig.6).

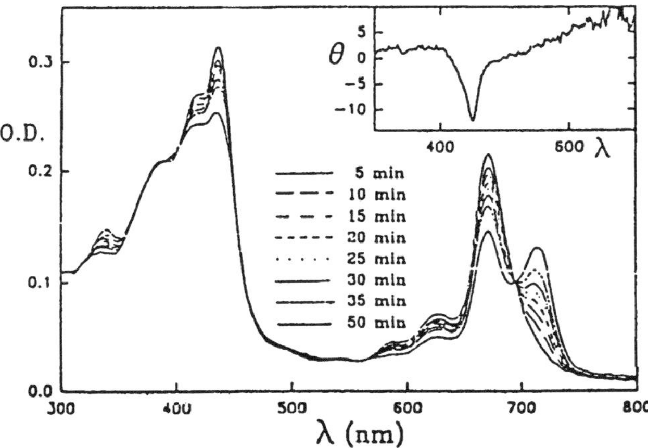

Figure 6. Formation in time of the 713 nm absorbing species. The inset shows its CD spectrum in the 300-700 nm region.

This is the product of an equilibrium conversion between different Chl a aggregates, as indicated by the presence of an isosbestic point at 695 nm [15]. The CD spectrum in the inset of Fig.6 shows the presence of a unique, non conservative negative term at 450 nm, indicative of the formation of a Chl a aggregate characterized by a geometry that allows only the coupling between non-degenerate excited states of the molecules [31]. These data are consistent with a structure formed by Chl a molecules with the phytilic chains surrounded by the organic solvent and anchored by hydrophobic interactions in the inside of the "micelle-like" aggregate. The macrocycle heads should be the interface with the water bulk [15]. A very large splitting but a negligible oscillatory strength can be predicted for the above described geometry, and the only observable CD signal should arise from non-conservative terms in the expression of the rotational strength of the aggregate caused by coupled oscillator interactions with non-degenerate exited states of other Chl a molecules. It is interesting to note that similar effects are observed in synthetically prepared Chl a dimers and in the presence of interactions between monomeric Chl a and proteins [32]. In Fig. 7 is summarized the behaviour of Chl a in the binary mixture water/ACN .

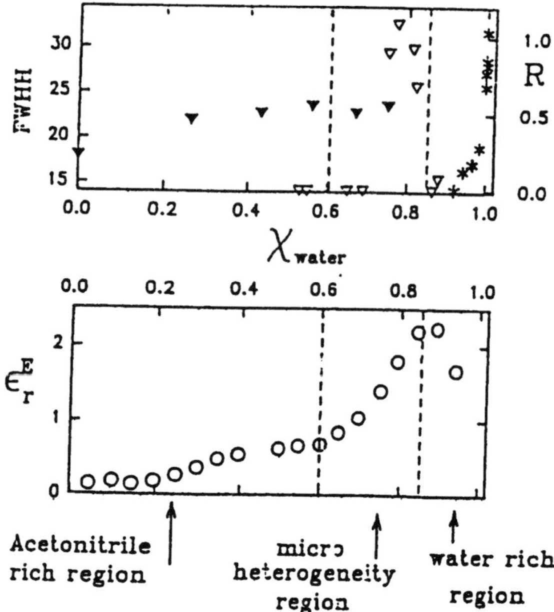

Figure 7. Dependence of various parameters on the water molar fraction. See text for explanation.

For this purpose three different parameters are used: Full Width at Half Height (FWHH) ($\blacktriangledown$), the absorbance ratio between the oligomer and 660 nm peaks (∇) and the absorbance ratio between the 713 nm and 660 nm peaks ($*$). For χ_{H2O} up to 0.8 (acetonitrile rich region) the formation of the dimer can be followed through the increase of the FWHH. Above $\chi_{H2O} \approx 0.5$ (microheterogeneity region) a peak at ≈ 745 nm formation appears, due to the oligomers. Increasing χ_{H2O} the relative amount of the oligomers increases and then decreases passing through a maximum at $\chi_{H2O} \approx 0.75$ and disappearing at $\chi_{H2O} > 0.9$ In Fig. 5 this is showed by the variation of the absorbance ratio of the ≈ 745 nm and ≈ 660 nm peaks.

Finally for $\chi_{H2O} \geq 0.9$ (water rich region) the species absorbing at 713 nm is formed and the increase in its relative amount can be followed by the increase of the absorbance ratio between 713 nm and 660 nm peaks.

It is surprising to note the behaviour of the parameters in Fig. 6 which closely reminds that of the excess dielectric constant as a function of the water molar fraction in ACN reported by Moreau and Douheret [33]. The absence of relevant changes in the value of the dielectric constant up to a water molar fraction of 0.60 indicates how in this region the solvent mixture behaves essentially as the organic solvent. The presence of a unique monomeric form of Chl $\underline{a}$ is consequently expected in this area. Increasing amounts of water result in the increasing of the value of the excess dielectric costant up to maximum at $\chi_{H2O}=0.8$, defining a "microheterogeneous region" in which the progressively stronger water-water interactions take place. This can account for the occurring of the Chl $\underline{a}$ aggregation described so far. Water molar fractions exceeding 0.85 characterize the so-called "water rich region" where the ACN is assumed to fill the cavities of the aqueous lattice, without breaking its structure. The hydrophobic effect can consequently lead to a micellar reorganization of the Chl $\underline{a}$ to minimize the interaction of the phytol chain with the water [33].

In conclusion it is possible to assume that the existence of the Chl $\underline{a}$ in different dominions of aggregation can be modulated through changes in the bulk properties of the solvent and of the Chl $\underline{a}$ concentration.

3. Compartimentalized Systems

3.1 REVERSE MICELLES

In order to obtain a system able to stabilize the charge separated states of an electron donor-acceptor couple it is necessary to introduce it in immobilized compartimentalized environments.

One possibility could be to introduce species that are able to donate electrons (dimer) in a microemulsion: the multicomponent AOT/isooctane/water-acetone system [17].

The structure modification in the AOT/isooctane/water system caused by acetone addition has been studied by the analysis of self-diffusion coefficient D [19].

Confinement of molecules within closed domains dramatically influences their translational mobility over supradomain distances. The three limiting cases are: a water-in-oil droplet (reverse micelles) structure with D_{water} ($\approx D_{droplet}$) $<< D_{oil}$ where D_{oil} is of the same order of magnitude of D of neat oil or oil solution. An oil-in-water droplet structure would give D_{oil} ($\approx D_{droplet}$) $<< D_{water}$, D_{water} being of the same order of magnitude as D of neat water or relevant aqueous solution. The third case is a bicontinuos structure with high values of both D_{water} and D_{oil} [34].

The self-diffusion coefficients of water, isooctane and acetone were obtained by using the FT Pulsed Gradient Spin Echo NMR technique [35].

Acetone is soluble both in water and in isooctane. Supposing a fast exchange between two sites, the organic bulk and the reverse micelles, one can write:

$$D_{Ac}^{obs} = (1-p^{mic})\, D_{Ac}^{free} + p^{mic}\, D_{Ac}^{mic}$$

where p^{mic} is the molar fraction of acetone in the aggregate. p^{mic} can then be calculated from self diffusion coefficients. In the inset of Fig. 8 are reported the p^{mic} values in function of different [Ac]/[H2O] ratios.

In the AOT/isooctane/water system the $D_{water}/D^{neat}_{water}$ value = 0.03, with D^{neat}_{water} = 2.3x10^{-9} m^{2}s^{-1} confirm that the coarse microstructure is always a water-in-oil droplet also in the case of acetone addition ($D_{water}/D^{neat}_{water}$ in the 0.09-0.05 interval). In Fig. 8 the $D_{water}/D^{neat}_{water}$ values increase as [Acetone]/[H2O] increases. Assuming spherical shape for the aggregates, this means smaller reverse micelles, a variation of the curvature of the interface [36,37] and, generally, implies the presence of a cosurfactant, now acetone.

From the inset of Fig. 8 it could be noticed that increasing acetone content, pAc^{mic} decreases from 0.70 to 0.55. If one considers all the acetone in the water pool, is the calculation of the water enrichment in the micellar core respect to the injected water-acetone mixture straightforward. $pAc^{mic}= 0.55$ corresponds to an acetone-water core 75% ca. in water. But if acetone behaves as a cosurfactant it means a further acetone partition between surfactant palisade and internal core. This, again, implies a further water enrichment of the water pool.

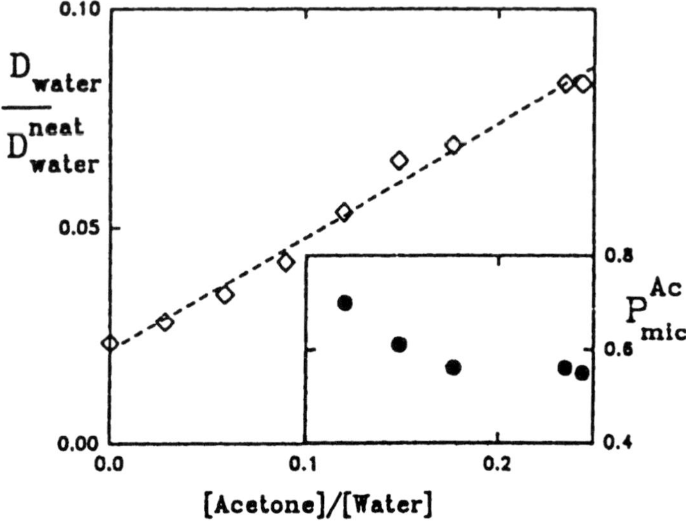

Figure 8. $D_{water}/D^{neat}_{water}$ values at 25°C vs [Acetone]/[H2O] ratio for the AOT/isooctane/water-acetone. The inset shows the calculated pAc^{mic} as acetone concentration increases.

In AOT systems the pAc^{mic} decreases with the increasing Chl a concentration indicating some acetone going into organic bulk. This rules a competition between acetone and Chl a to penetrate the surfactant palisade or to interact with the aggregate. All the AOT-acetone microemulsions retain a water-in-oil droplet microstructure ($D_{water}/D^{neat}_{water}$ ≈ 0.1). Increasing Chl a content, at constant acetone concentration, Chl a molecules penetrate into outer palisade expelling some acetone.
Chl a in isooctane is always present as oligomers [17,19,38]. AOT-acetone microemulsions never show Chl a, for any concentration, in an oligomeric form. The dichroic spectrum of Chl a in AOT-acetone microemulsions does not show any excitonic coupling peculiar of the dimer presence [19]. This means that Chl a interacts so strongly with micelle that it is unable to form any kind of aggregate or/and, at the same time, the interaction induces degradation of chlorophyll.
Chl a is degradated in acid solutions but in AOT-acetone microemulsions is insensitive to an acid water pool [19]. This implies that Chl a is not inside the micelle and strongly interacts with the surfactant palisade [38]. The interaction could take place in the outher

regions of the surfactant layers (accordingly with the competition, stated above, with acetone to penetrate) or in the inner palisade side.

3.2 LECITHIN ORGANOGELS

As already mentioned, solutions of lecithin in organic solvents can be transformed into highly viscous gel-like solutions by addition of a small amount of water [20]. The Chl a has been introduced in these microemulsions and its behavior has been extensively studied as function of the Chl a concentration [21].

The visible absorption spectrum of 1×10^{-4} M Chl a in n-heptane shows a peak in the red region at 745 nm peculiar of oligomeric forms of hydrated chlorophyll. Upon addition of lecithin to a final concentration of 0.1 M, the absorption spectrum is modified showing the typical feature of the monomeric Chl a (maxima at 435 and 666 nm). No changes are observed in the spectrum when water is added to the solution in the critical amount (0.2 M) to obtain its gelation.

An increase in the Chl a concentration up to 1×10^{-3} M does not lead to any changes in the spectra recorded in n-heptane (Fig. 9a) and n-heptane containing lecithin (Fig. 9b). Instead the addition of water to form gels induces a shift in the position of the peak to a value of 670 nm indicative of Chl a dimer formation. (Fig. 9c). The analysis of the infrared (IR) spectra of Fig. 9A gives a more detailed picture. The Chl a in n-heptane is characterized by a split of the ester carbonyl peak into two components at 1740 and 1725 cm^{-1}. This splitting can be interpreted considering that the C-17^3 and C-13^3 ester carbonyls are differently involved in the hydrogen bond between the two Chl a molecules of the minimal unit of the oligomer (dimer). The keto-carbonyl peak at 1632 cm^{-1} is shifted of 60 cm^{-1} respect to the free ester-carbonyl functional group. This shift provides evidence for the hydrogen bond of the C-13^1 ketone-carbonyl to the water molecule coordinated by the central magnesium atom of the nearby Chl a (39). Skeletal C=C and C=N absorption maxima at 1605 cm^{-1} are also evident (Fig. 9A). When lecithin is added to the solution of Chl a in n-heptane, the spectrum shows an intense band at 1680 cm^{-1} of the keto-carbonyl function. Moreover the keto-ester band is shifted to 1745 cm^{-1} and does not show any splitting (Fig. 9B). This pattern is the characteristic of non coordinated functionalities. Upon gel formation the keto-ester peak splits again into two components at 1730 and 1710 cm^{-1} respectively. A broad ketone-carbonyl band centred at 1665 cm^{-1} is also present (Fig. 9C).

Again we have seen that changes in the solvent composition and Chl a concentration modulates the formation of the hydrate aggregates. Oligomers are formed in pure n-heptane while in lecithin gels the pigment is organized in a different form, probably a dimer. At lower chlorophyll concentration the formation of the dimer does not occur.

The photoreactivity of the Chl a dimer in the lecithin gel has been demonstrated by performing the reaction of the water splitting (Hill reaction). The following equation:

$$H_2O + 2DCPIP \xrightarrow{\;(Chl\ a.2H_2O)_{2*}\;} 1/2\ O_2 + 2DCPIHP$$

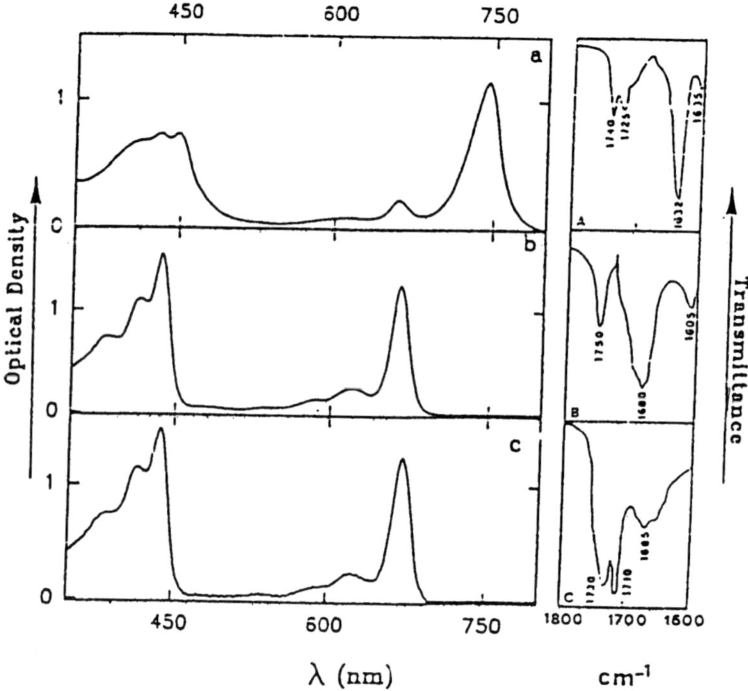

Figure 9. Visible and IR spectra for Chl a in lecithin gels. See text for explanation.

shows the reduction of the 2,6-dichlorophenol-indophenol (DCPIP), the oxidation of the water and the catalytic role of the dimer.

Under continuous illumination, the occurrence of the Hill reaction in the gel is revealed by the bleaching of the absorption maximum of the DCPIP at 590 nm.

In Fig. 10 is shown the absorption decrease with time for a gel containing 1×10^{-3} M Chl a and 1×10^{-4} M DCPIP.

No changes in the absorption are shown by a lecithin sample containing the same amount of DCPIP but a Chl a concentration ten times smaller. We can assume that lecithin gels provide an excellent membrane mimetic system where vectorial electron transfer reactions can be performed.

4. Conclusions and perspective

The main achievement of the work of our lab, above described, is that we are able to modulate the formation of Chl a dimers.

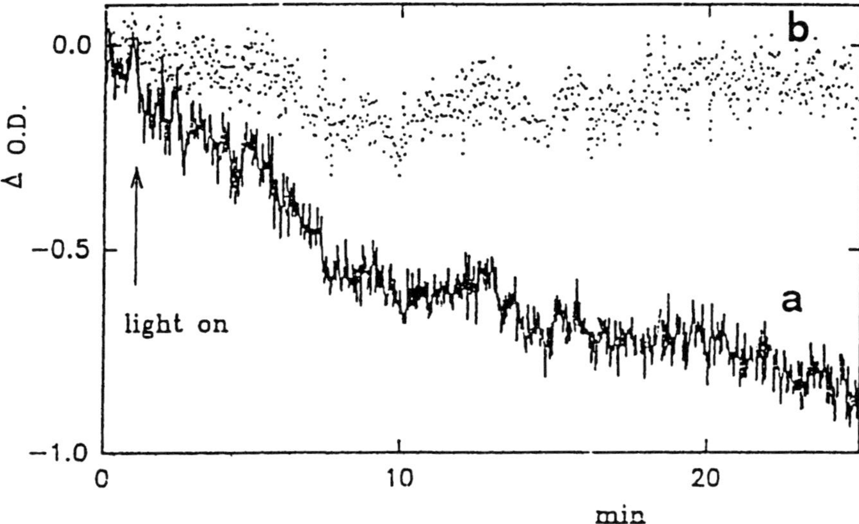

Figure 10. Variation of optical density (ΔOD) for Chl a 1 x 10^{-3} M (a) and 1 x 10^{-4} M (b).

These are the photoreactive species we are most interested in and using immobilized systems we can realize a vectorial electron transfer.
We want to search for more compartimentalized systems such as normal micelles in which electron acceptors and donors are separated in different domains.
Moreover the next step will be the introduction of other components of the photosynthetic chain in the lecithin gels and further characterize the electron transfer.

References

1. G. Forti *Photosynthesis* ed. J. Amesz, <u>New Comprehensive Biochemistry</u> **15**, Elsevier, Amsterdam 1, (1987).
2. A. J. Hoff <u>Light reaction path of photosynthesis</u> Springer-Verlag, Berlin, Heidelberg, New York, 80, (1982).
3. H. C. Chow, R. Serlin and C.E. Strouse, *J. Am. Chem. Soc.* "The crystal and molecular structure and absolute configuration of ethyl chlorophyllide a dihydrate. A model for the different spectral forms of chlorophyll a", *J. Am. Chem. Soc* **97**, 7230, (1975).
4. A. Hochapfel, R. Journeaux and R. Viovy, "Existence et nature des formes agrégées de la chlorophylle a dans les solvants binaires. I. -Etude des spectres d'absorption", *J. Chim. Phys* **66**, 1467, (1969) .
5. R. Journeaux, A. Hochapfel and R. Viovy, *"Existence et nature des formes agrégées de la chlorophylle a dans les solvants binaires. II. -Etude des spectres de fluorescence"*, *J. Chim. Phys* **66**, 1474, (1969).

6. A. Agostiano, K.A. Butcher, M.S. Showell, A.J. Gotch and F.K. Fong, "Dipole-dipole transfer between acetone solvates of chlorophyll a and chlorophyll a dihydrate dimers in water/acetone mixtures. A model for P680 sensitized excitation", *Chem. Phys. Lett.* **137**, 37, (1987) .

7. F.K. Fong, M. Kosunoki, L. Galloway, T.G. Mathews, F.E. Litle, A.J. Hoff and F.A. Brinkman, "Elementary reconstitution of the water splitting light reaction in photosynthesis. 1. Time-resolved fluorescence and electron spin resonance studies of Chlorophyll a dihydrate photoreaction with water in nonpolar solution",*J. Am. Chem. Soc* **104**, 2759, (1982).

8. A.J. Alfano, F.E. Lytle, M.S. Showell and F.K. Fong, "Excited singlet-state lifetimes of hydrated chlorophyll aggregates", *J. Chem. Phys.* **82**, 758-764, (1985).

9. A.J. Alfano, M.S. Showell and F.K. Fong, "Triplet-state decay kinetics of hydrated chlorophyll complexes", *J. Chem. Phys.* **82**, 765, (1985).

10. A.J. Alfano and F.K. Fong, "Elementary reconstitution of the water splitting light reaction in photisynthesis. 2. Optical double resonance study of (Chl a·2H$_2$0)n two-photon interactions in nonpolar solutions", *J. Am. Chem. Soc.*, **104**, 2767, (1982).

11. P. Maly, R. Danielus and R. Gadonas, "Picosecond absorption spectroscopy of chlorophyll a dihydrate", *Photochem. Photobiol.* **45**, 7, (1987).

12. A. Agostiano, P. Cosma and M. Della Monica, "Spectroscopic and electrochemical characterization of chlorophyll a in different water+organic solvent mixtures", *Bioelectrochem. and Bioenerg.***298**, 311, (1990).

13. A. Agostiano, M. Caselli and M. Della Monica, "Polarographic and wavelenght-selected fluorescence excitation studies of chlorophyll a aggregation in water containing trace amounts of acetone", *Bioelectrochem. and Bioenerg.* **23**, 301, (1990).

14. A. Agostiano, M. Caselli, M. Della Monica, A.J. Gotch and F.K. Fong, "Polarographic studies of photocatalytic chlorophyll. Formation of chlorophill a dihydrate dimer and oligomer in water-acetone binary solvent mixtures", *Biochem. Biophys. Acta* **936**, 171, (1988).

15. A. Agostiano, M. Della Monica, G. Palazzo and M. Trotta, "Chlorophyll a autoaggregation in water rich regions", submitted to *Biophys. Chem.*

16. A. Agostiano, K.A. Butcher, M.S. Showell, J.L. You, A.J. Gotch and F.K. Fong, *Research* "Resonant energy transfer between bulk chlorophyll a and chlorophyll a dihydrate dimers in water/acetone mixtures. A model of sensitized excitation in plant photosynthesis", *Progress in Photosynthesis* **1**, 19, (1987).

17. A. Agostiano, M. Caselli and M. Della Monica, "Use of mimetic membrane system in photosynthesis studies: different aggregations and hydration forms of Chl a in reverse micelles" *J. Surface Sci. Technol.* **5**, 21, (1989).

18. G. R. Seely, X. C. Ma, R. A. Nieman and D. Gust, "Association of chlorophyll with inverted micelles of dodecylpyridinium iodide in toluene" *J. Phys. Chem* **94**, 1581, (1990).

19. A. Ceglie, G. Colafemmina, M. Della Monica and G. Palazzo, "NMR study of AOT microemulsion with acetone in the presence of Chl a: distribution of acetone and role of chlorophyll" *Colloids and Surfaces* . In press.

20. R. Scartazzini and P. L. Luisi, "Organogels from lecithins" *J. Phys. Chem* **92**, 829, (1988).

21. A. Agostiano, M. Della Monica and A. Mallardi, "Chlorophyll a dimer photoreactions in lecithin organogels" *J. Photochem. Photobiol.* **13**, 241, (1992).

22. C. Houssier and K. Sauer "Circular dichroism and magnetic circular dichroism of the chlorophyll and protochlorophyll pigments" *J. Am. Chem. Soc.* **92**, 779, (1970).

23. M. D. Kandelaki, A. G. Volkov, V. V. Shubin and L. Boguslavsky, Chlorophyll-water interaction during oxygen photoevolution at the octane-water interface" *Bioch. Biophys. Acta* **893**, 170, (1987).

24. M. R. Wasielevski in Light reaction path of photosynthesis Springer-Verlag, Berlin, Heidelberg, New York, **35**, p. 234, (1982).

25. R. Vladikova, D. Kafalieva, V. Kolev and V. Spasov, *Conference proceeding of 5th international seminar on energy transfer in condensed media* "Time-decay fluorescence studies of chlorophyll a in different solvents" Praga, p. 22, (1985).

26. A. Agostiano, P. Cosma and M. Della Monica, "Formation of chlorophyll a photoreactive dimers in alcoholic mixtures: spectroscopic and electrochemical study" *J.Photochem. Photobiol.* **58**, 201, (1991).

27. S. S. Brody and M. Brody, "Spectral characteristics of aggregated chlorophyll and its possible role in photosinthesis" *Nature* **189**, 547, (1961).

28. A. Agostiano, M. Caselli and M. Della Monica, "Adsorption stripping voltammetry of different solvated species of chlorophyll a in water + acetone mixtures" *J. Electroanal. Chem.* **249**, 89, (1988).

29. V. Fidler and A. D. Osborne, "Strong circular dichroism of chlorophill a in 50:50 ethanol-water solution" *J. C. S. Chem. Comm.* **7**, 1056, (1980).

30. D. L. Worcester, T. J. Michalski and J. J. Katz, "Small-angle neutron scattering studies of chlorophyll micelles: models for bacterial antenna chlorophyll" *Proc. Natl. Acad. Sci* **83**, 3791, (1986).

31. I. Tinoco Jr. and C. R. Cantor in Methods of biochemical analysis **18** ed. D. GLIck, Interscience Publishers, John Wiley and Sons, New York, London, Sydney and Toronto.

32. S. G. Boxer, "Model reactions in photosynthesis" *Biochem. Biophys. Acta* **726**, 265, (1983).

33. C. Moreau and G. Douheret, "Thermodynamic and physical behaviour of water + acetonitrile mixtures. Dielectric properties" *J. Chem. Thermodynamics* **8**, 403, (1976).

34. J.O. Carnali, A. Ceglie, B. Lindman and K. Shinoda, '"Structure of microemulsions in the brine/Aerosol OT/Isooctane system at the Hydrophile-Lipophile Balance temperature studied by the self-diffusion technique" *Langmuir* **2**, 417, (1987).

35. B.Lindman and P.Stilbs, in Microemulsions S. Friberg and P. Bothorel Eds, CRC Press , 119, (1987).

36. B. W. Ninham, E. F. Evans and S.J. Chen, "Role of oils and other factors in microemulsion design" *J. Phys. Chem.* **88**, 5855, (1984).

37. E.F. Evans, D. J. Mitchell and B. W. Ninham, "Oil, water, and surfactant: properties and conjectured structure of simple microemulsions" *J. Phys. Chem.* **90**, 2817, (1986).

38. P. Brochette, T. Zemb, P. Mathis and M. P. Pileni, "Photoelectron transfer from Chlorophyll to Violagens in reverse micelles. Unusual interfacial effect on the reaction" *J. Phys. Chem.* **91**, 1444, (1987).
39. K. Ballschmiter and J. J. Katz, "An infrared study of chlorophyll-chlorophyll and chlorophyll-water interactions" *J. Am. Chem. Soc.* **91**, 2661, (1969).

Biocompatible Catalysis of Polyene Polymers by Manganese[III] Porphyrins

A.G. COUTSOLELOS and M.J. TORNARITIS
*Laboratory of Bioinorganic and Coordination Chemistry Department of Chemistry,
University of Crete, P.O. Box 1470, Heraklion GR-711 10, Crete, Greece.*

1. Introduction

Modelling of monooxygenase enzymes such as cytochrome P-450 was the origin of
the development of a new field in the domain of catalytic oxidations in liquid phase:
olefin epoxidation and alkane hydroxylation was mediated by high valent
oxometalloporphyrins. Iodosylbenzene, hypochlorites, amine N-oxides,
alkylhydroperoxides or hydrogen peroxide have been used as oxygen donors to
manganese or iron-porphyrin complexes in order to achieve a complete modelling of
the short cycle of cytochrome P-450 (see Scheme 1, and references [6-9], while the
long catalytic cycle of the enzymes has been successfully mimicked by Tabushi and
coworker [10]. The epoxidation reaction of double bonds, is highly relevant to the
catalytic transformation of any polymeric material with carbon-carbon double bonds.
The polyepoxides, polyalcoholic precursors, can also be transformed finally to small
molecular weight alcohols [11]. This is an easy way to transform vague materials to
raw materials, which happens during their auto degradation in nature.
In this study a complete transformation of polyene-polymers to corresponding
polyepoxides has been performed when NaOCl or PhIO is used as oxygen donor.
$KHSO_5$ though was however, less efficient as oxygen donor. ^{1}H-NMR spectroscopy
permitted the monitoring in the polymer evolution, while UV-visible spectroscopy the
porphyrin evolution. IR-spectroscopy was used as a complementary identification
method. Achieving catalytic, stereospecific, and regioselective olefin epoxidation with
synthetic metalloporphyrin models of cytochrome P-450 [12] is a topic of continuing
research interest. Iodosylbenzene, PhIO, has been used as a single oxygen atom
transfer reagent in a wide variety of studies, usually with the aim of generating reactive
intermediates analogous to those occurring naturally in metalloenzyme catalyzed
reactions [13].

Abbreviation: Por, unspecified (2⁻) porphyrinate (2⁻); TPP, 5, 10, 15, 20 - tetraphenyl
porphyrinate (2⁻).

N. Russo et al. (eds.), Properties and Chemistry of Biomolecular Systems, 175–185.
© 1994 *Kluwer Academic Publishers. Printed in the Netherlands.*

2. Experimental

2.1. PHYSICAL MEASUREMENTS

^{1}H-NMR: Spectra were recorded on a Varian FT-80. An aliquot (0.3 ml) was taken from the reaction mixture and the solvent removed. Deuterated chloroform (0.3 ml) with tetramethylsilane as internal reference, dissolved the residual and its ^{1}H-NMR was obtained.

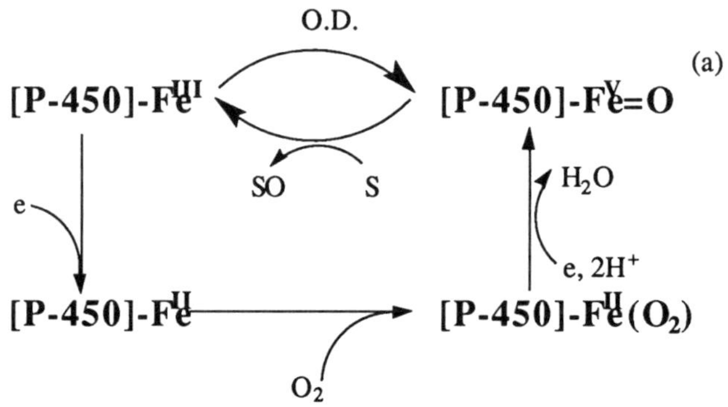

Scheme 1: Long (with O$_2$ and electrons) and short (with oxygen donors)
catalytic of cytochrome P-450.

UV-visible-data: Electronic absorption spectra were recorded on a Perkin-Elmer 330 (or Lambda 6) spectrophotometer using 5×10^{-3} mol. l^{-1} toluene solutions, or an ORIEL diode-array visible spectrophotometer. The experiments were repeated in microscale and in a UV-visible cell at room temperature.
IR-data: Samples were prepared as 1% dispersions in CsI pellets or Nujol mulls. The IR spectra were recorded on a Perkin-Elmer 298 spectrophotometer, or on an FT-IR 1760 series Perkin-Elmer spectrophotometer.

2.2. PREPARATION OF COMPLEXES

Metalloporphyrins were usually synthesized as described by Alder et al [14] with recent modifications (for Mn(TPP)Cl: see ref 12). Mn(TPP)OAc was prepared by DMF a metallation procedure [15] from Mn(OAc)$_2$ followed by alumina column chromatography. *Cis*-polybutadiene,*cis*-polyisoprene, *trans*-polyisoprene and benzyldimethyltetradecylammonium chloride were obtained from Aldrich and used without purification. Sodium hypochlorite and potassium hydrogenpersulfate were obtained from Aldrich Sodium hypochlorite was titrated by the iodometric method. Iodosylbenzene was prepared according to Saltzmann et al [16].

2.2.1. Epoxidation with iodosylbenzene.

In a two neck round bottle (50 ml) polyene-polymer (0.15g~2.78 mmole) was dissolved in methylene chloride (14 ml). Mn(TPP)Cl or Mn(TPP)OAc (0.025 g ~0.0355 mmol) was added and the solution cooled to 0° C. Iodosylbenzene (0.6 g) was introduced under argon and the reaction mixture was stirred for 6 hours. An additional amount of iodosyl benzene (0.6 g ~ 3 mmol) was added and the mixture was stirred at room temperature for 18 hours.

2.2.2. Epoxidation with sodium hypochlorite.

In a two neck round bottle (50 ml) polyene-polymer (0.21 g, 4 mmole) was dissolved in methylene chloride (10 ml). To the solution were added manganese acetato-porhyrin (0.036 g, 0.051 mmole), benzyldimethyltetradecylammonium chloride (0.04 g, 0.0099 mmole) and pyridine (0.1ml~1.2mmole). Sodium hypochlorite (40 ml) was introduced to the round bottle under argon. The reaction mixture was stirred for half an hour at room temperature and a portion of the organic layer taken for ^{1}H-NMR. Seven hours after the reaction started, the two layers were separated, the organic layer was dried with sodium sulfate anhydrous, the solvent was evaporated and from the residual the ^{1}H-NMR spectrum was obtained.
Experiments under the same conditions but without the catalyst for both of the above two epoxidation reactions have been carried out. The efficiency of these experiments was also checked by their ^{1}H-NMR spectra.
The experiments were repeated in microscale under exactly the same conditions in a UV-visible cell at room temperature. In a typical experiment a portion of the reaction mixture without oxygen donor and without polymer was added to the cell and was diluted with methylene chloride. After the introduction of the oxygen donor in the above solution a series of 20-25 spectra was taken every 30 or 60 seconds for the PhIO and NaOCl, respectively. The same experiments were performed as above but the polymer was dissolved in the reaction mixtures before the introduction of the oxygen donor.

2.2.3. Epoxidation with potassium hydrogenpersulfate.

In order to make a comparative study hydrogenpersulfate was used as an oxygen donor. For this experiment a 0.3 M phosphate buffer pH 7.0 is prepared by dilution of KH_2PO_4 (16 g) and Na_2HPO_4 (25.9 g) in 1000 mL of H_2O. For the epoxidations 25 mg of polyene-polymers (*cis*-polyisoprene, *trans*-polyisoprene and *cis*-polybutadiene), were used: CH_2Cl_2, 10 ml; 4-t-Bu-pyridine, 75 mmol; CBDTA 0.025 mmol; Mn(TPP)X, 3 $\eta = \mu$mol; $KHSO_5$, 7400 μmol (diluted in *ca.* 12 mL of 0.25 M phosphate buffer pH 7.0).

3. Results and Discussion:

3.1. ^{1}H-NMR

The catalytic conversion of polyene-polymers to the corresponding polyepoxides was achieved in few hours when iodosylbenzene or sodium hypochlorite was used as the oxygen donor. In contrast, when $KHSO_5$, was used, only a partial transformation could be observed. The above epoxidation reaction could be the first step for the degradation of polymers containing double bonds. Two spectroscopic methods were used to monitor the reaction; ^{1}H-NMR and UV-visible spectrophotometry. By the first technique, the conversion of the polyene-polymer to polyepoxide was followed continuously with time whereas the latter allowed to monitor all changes in the catalyst. The ^{1}H-NMR spectra (Figure 1) obtained from the epoxidation reaction of *cis*-polyisoprene with iodosylbenzene clearly shows the signals of the reacting polymer and the products. The peaks at 5.3 and 2.2 ppm correspond to -CH- and -CH$_2$ groups respectively of the polyolefin (spectrum 1). The multiple peaks at the area of 7 to 8 ppm correspond to the signals of the phenyl ring hydrogens of benzene iodine, the product of iodosylbenzene. The peaks at 3.1 and 1.8 ppm are attributed to the hydrogen signals of the groups -CH- and -CH$_2$ of the polyepoxide. The ^{1}H-NMR spectrum 3 shows only the peaks of the products indicating that the reaction was complete.

Table 1. ^{1}H-NMR data of the catalytic conversion of the polyene-polymers (in CDCl$_3$).

	Signal of protons (ppm)		
Polymer	CH	CH$_2$	CH$_3$
cis-polyisopren	4.75	2.1	1.69
trans-polyisoprene	4.75	2.1	1.69
polyepoxide product	2.68	1.69	1.24
cis-polybutadiene	5.3	2.2	
polyepoxide product	3.1	1.8	

The ^{1}H-NMR spectrum of the experimental run under the same conditions but withoutthe catalyst clearly shows that the reaction didn't proceed. In the 7-8 ppm region the signals observed were only 2-3% of those when the reaction was carried out with the catalyst (spectrum 3). Only the protons belonging to cis-polybutadiene, the starting material, were observed (iodosylbenzene is not soluble in chloroform or in methylene chloride).

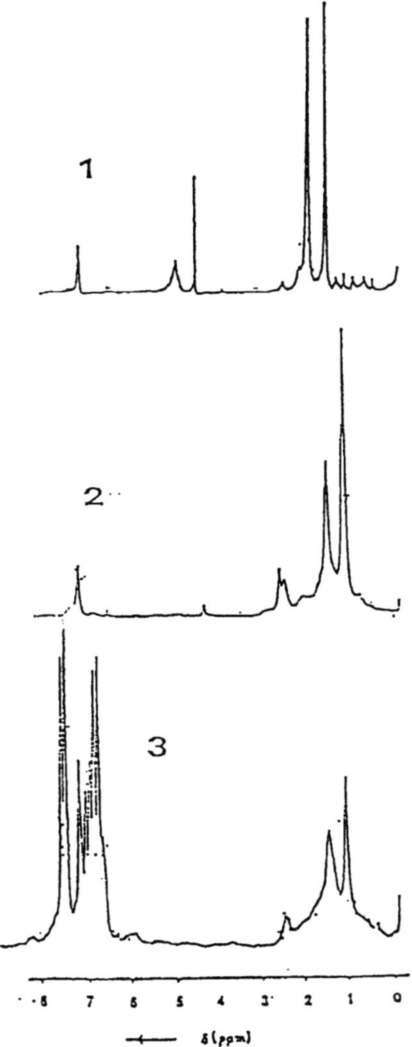

Figure 1: ^{1}H-NMR spectra monitoring the epoxidation reaction where polyene was the *cis*-polyisoprene. Spectrum (A) shows the *cis*-polyisoprene (Aldrich commercial). Spectra (B) and (C) show the polypepoxide product in the reaction mixture of the epoxidation reaction with NaOCl and PhIO, respectively (*CH$_2$Cl$_2$).

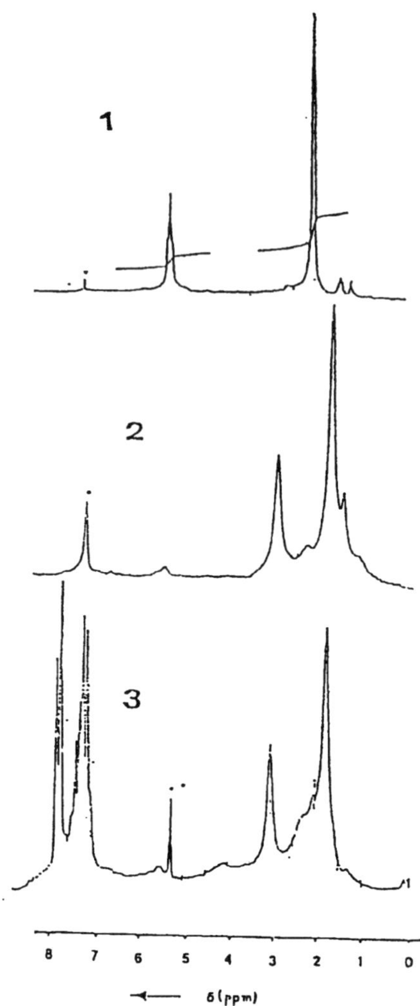

Figure 2: ^{1}H-NMR spectra monitoring the epoxidation reaction where polyene was the *cis*-polybutadiene. Spectrum (A) shows the cis-polybutadiene. Spectrum (A) shows the *cis*-polybutadiene (Aldrich commercial). Spectra (B) and (C) show the polyoxide product in the reaction mixture of the epoxidation reaction with NaOCl and PhIO, respectively (*CHCl$_3$, **impurity in CDCl$_3$).

With NaOCl as oxygen donor, a "conventional" method developed by Meunier et al. [10], 20% of the polymer substrate was converted to polyepoxide in only half an hour as shown by the ^{1}H-NMR spectra (Figure 2). In addition to the peaks corresponding to the signals of the hydrogens of the polymer, the starting material, and the polyepoxide product, very small peaks from the manganese porphyrin catalyst and the phase transfer catalyst are visible in this spectrum. The spectrum recorded 7 hours after the reaction, shows that 85% of the starting polymer was converted to polyepoxide [3], when the catalyst was omitted. No polyepoxide product was detected by ^{1}H-NMR, even after a period of 24 hours.

3.2. UV-VISIBLE

The typical UV-visible absorption spectrum of Mn(TPP)Cl or Mn(TPP)OAc shows peaks at 463 nm (Soret band) and 525, 577, 613 nm (annex bands), in methylene chloride. Figures 3 and 4 show the time-resolved UV-visible spectra of the catalyst Mn(TPP)Cl or Mn(TPP)OAc with PhIO and NaOCl as oxygen donors. In each case the annex bands of the starting material were replaced as the reaction proceeded by novel bonds, which were different for each of the oxygen donors. This shows that the intermediate species [8] cannot be manganese linked to oxygen via a double bond since this intermediate should have given identical UV-visible spectra for each of the oxygen donors. When PhIO or when NaOCl were the donor and Mn(TPP)OAC the catalyst, these changes in the spectra of the porphyrin were observed with or without the polymer. When NaOCl was the oxygen donor and Mn(TPP)Cl the catalyst, changes in the spectrum of the original porphyrin were only observed with the addition of polymer. Moreover, the changes in the UV-visible spectra of the porphyrin cannot be due to formation of a manganese porphyrin complex (linked to double bond via the oxygen atom) since the reaction was observed even in absence of the polymer.

Table 2. UV-visible data of time resolved spectra for the catalytic reaction by (TPP)MnCl and (TPP)MnOAc.

Porphyrin	Oxygen donor	Polymer	Annex bands (λ,nm)			
Mn(TPP)Cl	PhIO	none	524	575	611	
Mn(TPP)OAc	PhIO	none	524	572	607	
Mn(TPP)Cl	NaOCl	none	----	----	----	----
Mn(TPP)OAc	NaOCl	none	516	568	606	660
Mn(TPP)Cl	PhIO	yes	530	577	611	
Mn(TPP)OAc	PhIO	yes	524	573	607	
Mn(TPP)$_n$Cl	NaOCl	yes	511	574	611	664
Mn(TPP)OAc	NaOCl	yes	510	564	610	650
Mn(TPP)Cl	none	none	525	578	614	
Mn(TPP)OAc	none	none	527	577	613	

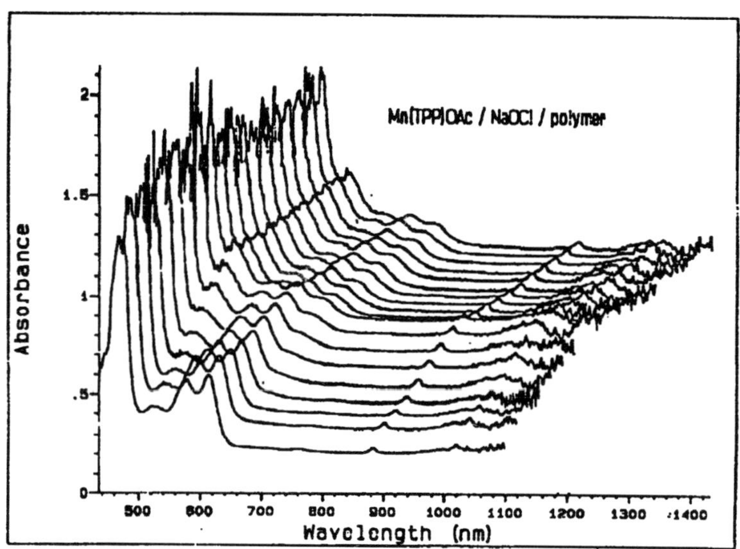

Figure 3. Time resolved spectra of *cis-polyisoprene* epoxidation reaction by NaOCl in CH$_2$Cl$_2$, interval time 120 seconds (first spectrum without oxygen donor).

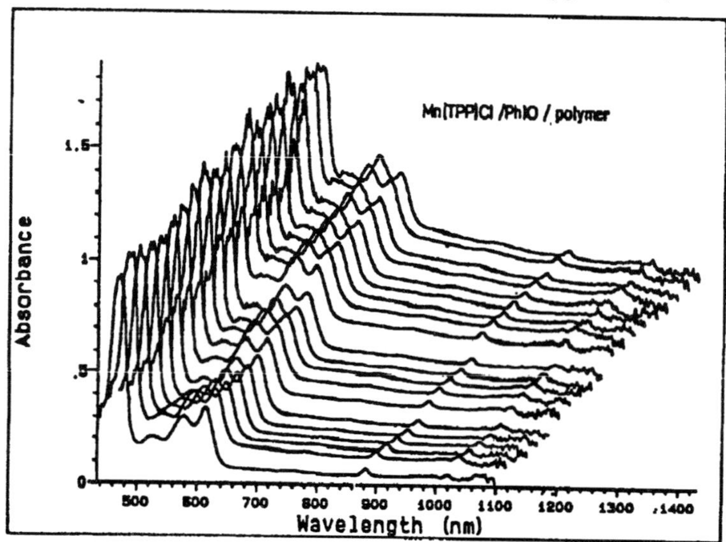

Figure 4. Time resolved spectra of *cis-polyisoprene* epoxidation reaction by PhIO in CH$_2$Cl$_2$, interval time 120 seconds (first spectrum without oxygen donor).

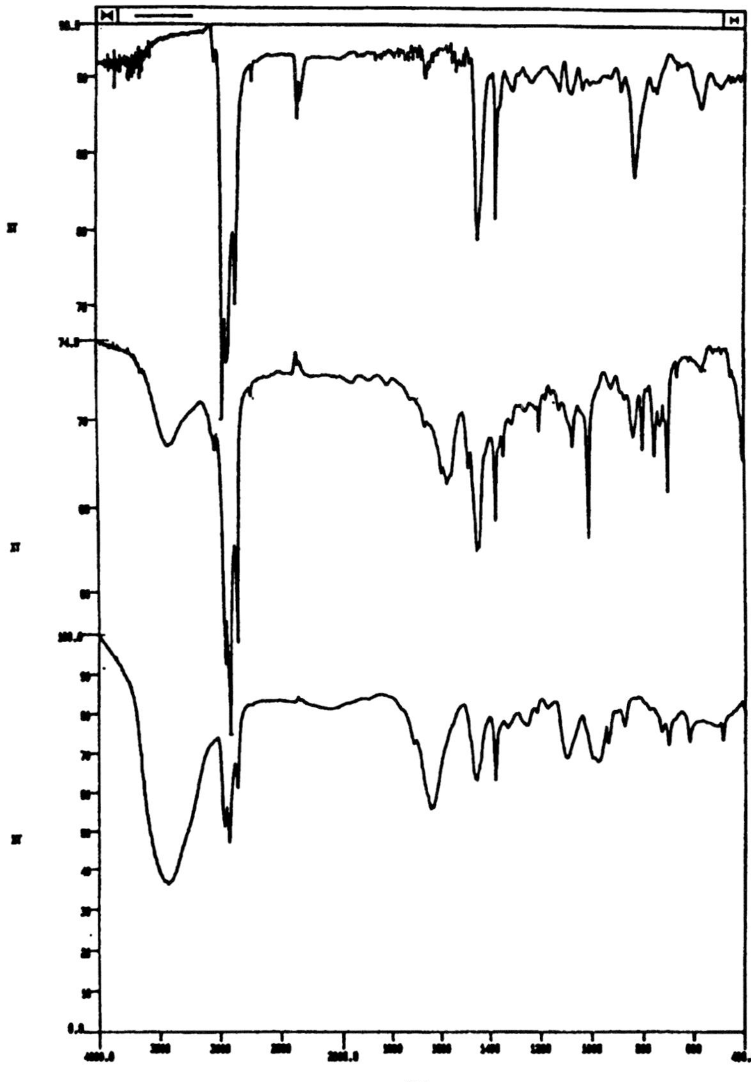

Figure 5. FT-IR spectra, (1) *cis*-polyisoprene, (2) reaction mixture of epoxidation of *cis*-polyisoprene without oxygen donor and (3) with oxygen donor.-

Thus we conclude that the only species which could give the spectrum was the complex of the porphyrin connected to the oxygen donor. That is, this species dominated the spectra by its long life time. These conclusions are confirmed by the lack of changes in the soret band area of the spectra during the progress of the reaction.

The catalytic transformation of the cis-polybutadiene to polyepoxide can be followed easily by ^{1}H-NMR and UV-visible spectroscopy. A 100% transformation has been observed in the case where PhIO is the oxygen donor and 85% after 7 hours when NaOCl was used. Preliminary experiments in our laboratory with *cis*-polyisoprene (natural rubber) or *trans*-isoprene as substrate, show the complete transformation of the substrates to polyepoxides within few hours.

3.3. INFRA-RED

The infrared spectroscopy used in the case of the epoxidation of natural rubber with sodium hydrochlorite, substantiated the conclusions drawn from ^{1}H-NMR and UV-visible spectroscopy. IR spectrum 3, taken from the organic layer of the reaction mixture shows significant differences from IR spectrum 2 taken from the control experiment without the oxygen donor. The latter spectrum is quite similar to the IR spectrum 1 taken from natural rubber dissolved in methylene chloride. Spectrum 1 and 2 clearly show a peak at 845 cm^{-1} typical of a carbon-hydrogen bond of a cis $R_2C=CHR$. The lack of this peak in spectrum 3 and the appearance a peak at 890 cm^{-1} characteristic of oxiranes, indicated the complete transformation of the polyolefin to polyepoxide.

4. Conclusion

The above epoxidation of the polyene polymers has very little cost and perhaps it is a way towards solving the pollution problem of plastic materials. The different oxygen donors show a different activity than what we attribute probably to the different intermediates that can be formed.

5. Acknowledgement

This work was supported by a grant from the Greek General Secretary of Research and Technology. We thank Mr. V. Davoras for preparation of starting materials.

References

1. P.R. Ortiz de Montellano, Ed. <u>Cytochrome P-450, Structure, Mechanism, and Biochemistry;</u> Plenum: New York, (1986).
2. M. Pasdeloup, B. Meunier, *New J. Chem.* **13**, 797, (1989).
3. M. J. Tornaritis, A.G. Coutsolelos, *J. of Polym. Sci.,* in press 1992.
4. Patent No 920100361/18.8.92
5. G. Pratviel, J. Bernadou, M. Ricci, B. Meunier, *Biochem. Biophys. Res. Comm.* **160**, 1212, (1986).

6. T.J. McMurry, J.T. Groves, in "Cytochrome P-450: Structure, Mechanism and Biochemistry"; Plenum, New York, 1-28, (1986).

7. B. Meunier , *Bull. Soc. Chim. Fr.*, 578-598, (1986).

8. T.C. Bruice, *Ann. N.Y. Acad. Soc.*, 83-98, (1986).

9. D. Mansuy, *Pure Appl. Chem.* **59**, 759-770, (1987).

10. I. Tabuchi, M. Kodera, M. Yokoyama, *J. Am. Chem. Soc.* **107**, 4466-4473, and references therein, (1985).

11. A.G. Coutsolelos et al. unpublished results.

12. Recent reviews:

  (a) P.R. Ortiz de Montellano, *Acc. Chem. Res.* **20**, 289, (1987).

  (b) J.H. Dawson, M. Sono, *Chem. Rev.* **87**, 1255, (1987).

13. (a) J.T. Groves, T.E. Nemo, *J. Am. Chem. Soc.* **105**, 6243, (1983).

  (b) J.T. Groves, R.S. Myers, *J. Am. Chem. Soc.* **105**, 5791, (1983).

  (c) T.G. Traylor, W.-P. Fann, D. Bandyopadhyay, *J. Am. Chem. Soc.* **111**, 8008, (1989).

  (d) D. Ostovic, T.C. Bruice, *J. Am. Chem. Soc.* **111**, 6511, (1989).

  (e) C.L. Hill, B. Schardt, *J. Am. Chem. Soc.* **102**, 6375, (1980).

  (f) J.T. Groves, W.J.Jr. Kruper, R.C. Haushalter, *J. Am. Chem. Soc.*, **102**, 6375.

  (g) R.B. Hopkins, A.D. Hamilton, *J. Chem. Soc. Chem. Commun.* 171, (1987).

  (h) C.-M. Che, W.-H. Leung, C.-H. Poon, *J. Chem. Soc. Chem. Commun.* 173, (1987).

  (i) C.-M. Che, W.-K Cheng, *J. Chem. Soc. Chem. Commun.* 1443, (1986).

  (j) J.F. Kinneary, J.S. Albert, C.J. Burrows, *J. Am. Chem. Soc.* **110**, 6124, (1988).

  (k) W. Nam, J.S. Valentine, *J. Am. Chem. Soc.* **112**, 4977-4979, and references therein, (1990).

14. A.D. Adler, F.R. Longo, F. Kampas, J. Kim, *J. Inorg. Nucl. Chem.* **32**, 2443, (1970).

15. R.D. Jones, D.A. Summerville, F. Basolo, *J. Am. Chem. Soc.* **100**, 4416, (1978).

16. H. Saltzman, M. Sharefkin, *J. of Org. Synth.* **43**, 60-61, (1963).

Engineering of Bovine Seminal Ribonuclease: Expression of the Secreted Recombinant Protein

A. DI DONATO, V. CAFARO, M. DE NIGRIS, G. MINOPOLI and G. D'ALESSIO
Department of Organic and Biological Chemistry, University of Naples Federico II, Via Mezzocannone 16, 80134 Napoli, Italy

1. Introduction

Bovine seminal RNAase (BS-RNAase) is a singular ribonuclease of the pancreatic RNAase superfamily, with unusual structural, catalytic, and biological properties [1]. It is the only dimeric protein in the superfamily, with intersubunit disulfide bridges [2]. In native BS-RNAase the majority of the subunits interchange their N-terminal segments, while in the remaining dimers each N-terminal segment folds onto its respective subunit [3]. It is homotropically regulated in the second, rate determininig step of the catalytic reaction, with negative and positive cooperativity [4]. It is selectively deamidated; this produces a characteristic isoenzymic composition, made up of α_2, $\alpha\beta$ and β_2 isoenzymes, produced by the progressive transformation of β subunits into α subunits through selective deamidation at Asn^{67} [5]. It is a RISBASE, i.e. a *R*ibonuclease with *S*pecial *B*iological *A*ctions, including its antitumor, anti-spermatogenic and immunosuppressive actions [1]. All these properties are lost when the enzyme is monomerized, although the monomeric derivative is fully and catalytically active [1, 4].
A definition at the molecular level of the structural determinants underlying these diverse biological properties may help to shed light on the molecular mechanisms of such activities. To reach this goal, protein engineering can be a very effective approach, as it should allow to verify the effects of appropriate mutations on the catalytic and the "special" biological activities of the enzyme .
The first, essential steps in this direction are the cloning and the expression of the DNA coding for the protein. Expression of ribonucleases has been cumbersome probably because of their toxicity and/or because of difficulties in forming the correct intra- and inter-subunit disulfide bridges. So far, only few examples are available in the literature of ribonucleases expressed in heterologous systems [6-9]. We thus focussed our attention on the possibility of using secretion expression systems, which could help avoid both toxicity and folding problems.
Here we report a successful procedure for expressing secreted recombinant BS-RNAase.

187

N. Russo et al. (eds.), Properties and Chemistry of Biomolecular Systems, 187–192.
© 1994 *Kluwer Academic Publishers. Printed in the Netherlands.*

2. Experimental procedures

BS-RNAase was purified from seminal vesicles as described by Tamburrini *et al.* [10]. cDNA coding for BS-RNAase subunit sequence was obtained by ligating a synthetic double-strand oligonucleotide, coding for residue 1-46, to cDNA from clone 19-17G3, coding for the sequence 47-124 of BS-RNAase, with a procedure that will be described elsewhere (de Nigris et al., manuscript submitted). The prokaryotic secretion vector pIN-III-OmpA3, containing the strong signal sequence of the OmpA protein, a major outer membrane protein of *Escherichia coli*, was provided by Dr. M. Inouye (University of Medicine and Dentistry of New Jersey).
The prokaryotic expression vector pT7-7 was provided by Dr. G. Ciliberto (University of Naples). Plasmid pUC18, RF-M13mp18, and *E. coli* strain JM101 were purchased from Boehringer; *E. coli* strain BL21(DE3)pLysE from AMS Biotechnology. Polymerase Chain Reaction (PCR) experiments were performed using the reagent kit from Cetus-Perkin Elmer, with the appropriate synthetic oligonucleotide primers. Site-directed mutagenesis was carried out using Muta-Gene Kit from Bio-Rad, based on the mutagenesis method described by Kunkel [11]. Immunoblots were performed using anti-BS-RNAase polyclonal antibodies prepared as described [12]. The Gene-Clean kit for elution of DNA fragments from agarose gel was obtained from Bio 101. Enzymes, including restrictases and reagents for DNA manipulation, were from Promega Biotech. Bacterial cultures, plasmid purifications and transformations were performed according to Sambrook *et al.* [13]. Double-strand DNA was sequenced with the dideoxy method of Sanger *et al.* [14], carried out with a Sequenase Sequencing Kit (U.S.Biochemical Corporation) with deoxynucleotide triphosphates purchased from Pharmacia. RNAase activity on yeast RNA was assayed with the method of Kunitz [15]. SDS-PAGE was performed according to Laemmli [16].

3. Results

3.1. CONSTRUCTION OF THE EXPRESSION VECTOR

The cDNA coding for BS-RNAase subunit chain, cloned into vector pUC18 (de Nigris et al., manuscript submitted), was excised using Eco RI and Hind III restriction endonucleases and inserted into vector pIN-III-ompA3 between Eco RI/Hind III poly-linker sites (see Figure 1). From this vector a fragment was excised with Xba I, spanning the sequence coding for the ompA signal sequence fused upstream to the sequence coding for BS-RNAase subunit chain and inserted into RF-M13mp18 Xba I poly-linker site. A deletion mutagenesis was carried out, to eliminate the region between the triplet coding for the last residue of the signal peptide and the first codon of BS-RNAase subunit sequence. The mutated DNA sequence was excised from RF-M13mp18 using Xba I restriction endonuclease and subjected to another mutagenesis round to convert the Xba I endings into Nde I, at 5', and into Sal I at 3', respectively, and to remove the ribosome binding sequence (*rbs*) derived from vector pIN-III-OmpA3 (see Fig. 1). This double mutagenesis was carried out directly on the DNA fragment by a PCR procedure, using the mutagenic oligonucleotides as primers. The resulting fragment, after trimming with Nde I and Sal I restriction endonucleases, was eventually inserted into the pT7-7 vector, a convenient sy- stem for the controlled expression of genes coding for toxic products [17,18], previously cut with the same enzymes, and

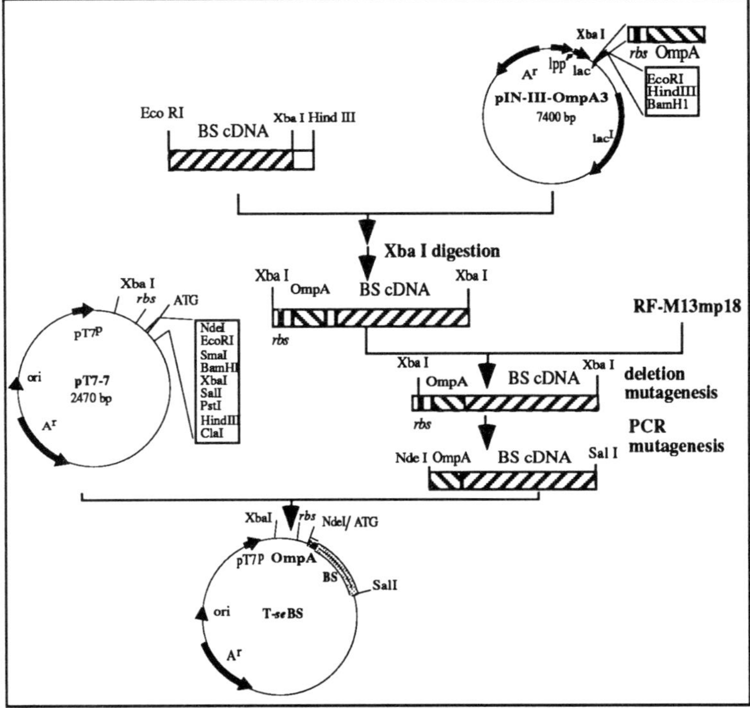

Figure 1. Strategy for the construction of vector p*T-seBS* for expression and secretion of recombinant bovine seminal RNAase.

used to transform JM101 competent cells. Selection of a recombinant clone, *T-seBS*, containing the BS-RNAase coding sequence, was achieved by direct sequencing of recombinant clones.

3.2. EXPRESSION OF SECRETED BS-RNAase

Plasmid DNA from clone *T-seBS* was used to transform BL21(DE3)pLysE cells that express the T7 RNA polymerase under the control of the IPTG inducible lacUV 5 promoter [17]. Fresh recombinant colonies were used to inoculate liquid cultures in TB medium, grown at 37 °C under vigorous shaking to an absorbancy value of 0.7 at 600 nm. Then IPTG was added to 0.25 mM final concentration and bacterial growth was continued, at 30 °C, overnight. Cells were then harvested, and the periplasmic fraction was obtained by osmotic shock as described [9]. Upon SDS-PAGE analysis of induced cell periplasms, carried out under non-reducing conditions, a band was detected after RNAase staining (Fig. 2A), comigrating with BS-RNAase isolated from seminal vesicles, and present only in the IPTG-induced cells. This protein band was found to react with anti-BS-RNAase antibodies after blotting of the gel on a nitrocellulose membrane (Fig. 2B). These results indicated that secreted recombinant BS-RNAase had the same molecular size as natural BS-RNAase, and was immunologically undistinguishable from it. It should be added that a band, approximately comigrating with monomeric BS-RNAase, staining for RNAase activity, and reacting with anti-BS-RNAase antibodies, was also detected.

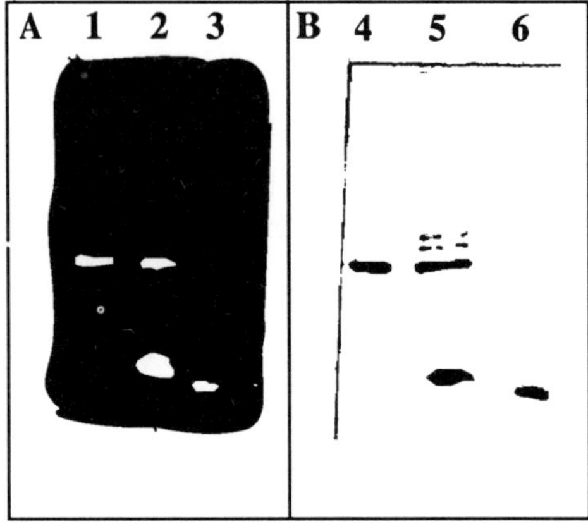

Figure 2. Polyacrylamide gel electrophoresis in sodium dodecyl sulfate in the absence of 2-mercaptoethanol. **A.** RNAase activity staining; **B.** immunoblot treated with anti-BS-RNAase antibodies. Lanes A1 and B4, BS-RNAase purified from seminal vesicles. Lanes A2 and B5, periplasmic fraction from induced cultures. Lanes A3 and B6, monomeric carboxymethylated BS-RNAase.

3.3. PARTIAL PURIFICATION OF SECRETED RECOMBINANT BS-RNAase

The periplasmic fraction (30 ml) from 250 ml culture of BL21(DE3)pLysE cells, transformed with p*T-seBS* and IPTG-induced, was dialyzed against 0.1 M Tris-Cl, pH 8. The dialyzed material, after addition of NaCl to 0.2 M final concentration, was loaded (in 9 separate runs) on a Mono-S column equilibrated with 0.1 M Tris-Cl, pH 8 containing 0.2 M NaCl. After 30 min the NaCl concentration was raised to 0.36 M, resulting in elution of a single peak. Peak fractions from individual cromathographic runs were combined and an aliquot was analyzed by SDS-PAGE under reducing and non-reducing conditions. Fig. 3 shows that the material eluted from the Mono S columns contains a protein mixture enriched in dimeric recombinant BS-RNAase. Moreover, upon reduction, the protein band corresponding to dimeric BS-RNAase comigrated with monomeric BS-RNAase (see Fig. 3A, lanes 2 and 3, and 3C, lanes 9 and 11). It should be noted that a band with a molecular size slightly higher than monomeric BS-RNAase, and active as a ribonuclease, was present in the protein fraction not retained by the Mono S column (see Figure 3B, lane 5). This material likely represents unprocessed pro-BS-RNAase subunit chain.

An estimate of the amount of the dimeric protein present in the Mono S eluate indicated that about 100 µg can be obtained from 1 liter of bacterial culture. As for the putative pro-BS-RNAase subunit chain, it can be estimated that an amount of about 1 mg is secreted into the periplasmic space.

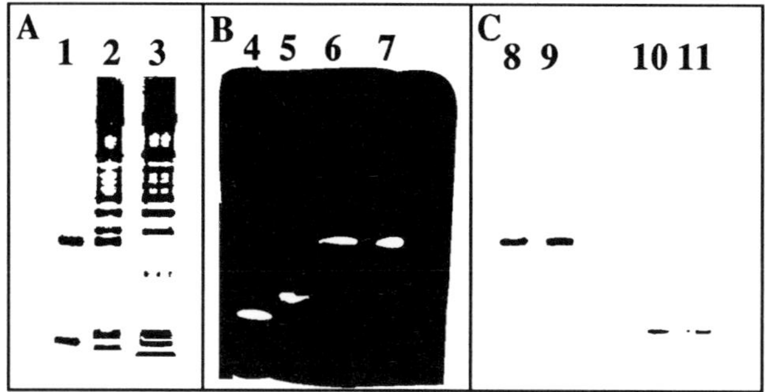

Figure 3. Polyacrylamide gel electrophoresis in sodium dodecyl sulfate in the presence (lanes A3, C10 and C11), and in the absence (all other lanes) of 2-mercaptoethanol. **A.** Silver nitrate staining; **B.** RNAase activity staining; **C.** immunoblot treated with anti-BS-RNAase antibodies. Lanes A1, B7, C8, and C10, BS-RNAase purified from seminal vesicles. Lanes A2, A3, B6, C9 and C11, material eluted from the Mono S column with 0.36 M NaCl (see text). Lane 5, material not retained from the Mono S column (see text). Lane B4, monomeric carboxymethylated BS-RNAase. This material is also present in lane A1.

4. Conclusions

The results described here show that the production of secreted recombinant BS-RNAase can be obtained. Yields are low; however, this seems to be due to the uncomplete processing of the pro-BS-RNAase subunit chain. This may be likely due to the clogging of the translation/secretion machinery, given the high level of transcription which can be obtained under the control of the strong T7 promoter [17]. Further studies are in progress for the optimization of the system.

References

1. G. D'Alessio, A. Di Donato, A. Parente, and R. Piccoli, *Trends Bioch. Sci.* **16**, 104, (1991).
2. G. D'Alessio, M.C. Malorni, and A. Parente, *Biochemistry* **14**, 1116, (1975).
3. R. Piccoli, M. Tamburrini, G. Piccialli, A. Di Donato, A. Parente, and G. D'Alessio, *Proc. Natl. Acad. Sci.* USA **89**, 1870, (1992).
4. R. Piccoli, A. Di Donato, and G. D'Alessio, *Biochem. J.* **253**, 329, (1988).
5. A. Di Donato, and G. D'Alessio, *Biochemistry* **20**, 7232, (1981).

6. G. M. McGeehan, and S.A. Benner, *FEBS Lett.* **247**, 55, (1989).

7. D. Schultz, and R. Baldwin, *Prot. Science* **1**, 910, (1992).

8. R. Shapiro, J. W. Harper, E. A. Fox, H. W. Jansen, F. Hein, and E. Uhlman, *Anal. Biochem.* **175**, 450, (1988).

9. R. Quaas, Y. McKeown, P. Stanssen, R. Frank, H. Blöcker, and U. Hahn, *Eur. J. Biochem.* **173**, 617, (1988).

10. M. Tamburrini, R. Piccoli, R. De Prisco, A. Di Donato, and G. D'Alessio, *Ital. J. Biochem.* **35** 22, (1986).

11. T.A. Kunkel, J.D. Roberts, and R.A. Zabour, *Methods Enzymol.* **154**, 367, (1987).

12. R. Piccoli, A. Ciardiello, F. De Nigris, F. Fiore, L. Grumetto, R. Mastronicola, G. Russo, N. Russo, C. Verde, A. Di Donato, A. Parente, T. Pietropaolo, and G. D'Alessio, in <u>Structure, Mechanism and Function of Ribonucleases</u> (C.M. Cuchillo, R. Llorens, M.V. Nogues, X. and Pares, eds) Universitat Autonoma de Barcelona, Bellaterra, Spain, (1991).

13. J. Sambrook, E.F. Fritsch, and T. Maniatis, <u>Molecular Cloning. A laboratory Manual</u> 2nd Edition, Cold Spring Harbor Laboratory Press, Cold Spring Harbor, NY, (1989).

14. M. Kunitz, *J. Biol. Chem.* **164**, 563, (1946).

15. F. Sanger, S. Nicklen, and A.R. Coulson, *Proc. Natl. Acad. Sci.* U.S.A., **76** 5653, (1977).

16. U. Laemmli, *Nature* (London) **227**, 680, (1970).

17. F.W. Studier, A.H. Rosemberg, J.J. Dunn, and J.W. Dubendorff, *Methods Enzymol.* **185**, 60, (1990).

18. S. Tabor, in <u>Current Protocols in Molecular Biology</u> suppl.11, F. M. Ausubel *et al.* eds., Geene Pub. Assoc. & Wiley-Interscience, New York, (1990).

Membrane Operation in Biochemical Processing

E. DRIOLI, L. GIORNO, L. DONATO, R. MOLINARI, A. BASILE*
University of Calabria, Department of Chemistry, Chemical Engineering Section, I-87036 Arcavacata di Rende (CS) - Italy
**Research Institute on Membranes and Chemical Reactors, CNR, I-87036 Arcavacata di Rende (CS) - Italy*

1. Introduction

The scientific and technological progresses in membrane processes during the past two decades have been huge. The existing and potential industrial applications of membrane operations are the widest if compared with any other available separation technology. Pressure driven membrane operations by themselves have already been successfully applied practically in all industrial sectors: from textile to pharmaceutical industry, from leather to agrofood, from automobile to water industry. Biomedical and biotechnological applications, membrane sensors, artificial organs, advanced components in analytical instruments, are other examples of well established membrane systems.

The fundamental studies carried out particularly in transport phenomena, also clarify, at molecular level, the mechanisms which control selectivities and permeabilities of the used polymeric membranes. A large number of membrane operations and hybrid membrane processes are today used worldwide. Pressure driven membrane processes, electrodialysis, membrane reactors can be indicated as basic unit operations. In Table 1 some membrane processes, developed at industrial level, are summarized. The combination of a selective mass transfer across a membrane with a chemical reaction, typical in various biological processes, is one of the most promising field for future progresses. This combination can be realized by integrating any traditional reactor with the most appropriate membrane operation or by using the membrane itself as a reactor [1]. Synthetic polymeric or ceramic membranes provide an ideal support for catalysts and biocatalysts immobilization due to a wide available surface per unit volume and for offering possibilities to various new immobilization procedures. Enzymes are retained in the reaction site, do not pollute the products and can be continuously re-used. Immobilization has also been proven to enhance enzymes stability. Moreover, provided that membranes of suitable cut-off are used, chemical reactions and physical separation of catalysts (and/or substrates) from the products may take place in the same unit. Selective removal of products through membranes yields effective conversions in the case of thermodynamically unfavorable reactions. Substrate partition at the membrane/fluid interface can be used to improve the selectivity of the catalytic reaction towards the desired products with minimal side reactions. Membranes are also attractive to retain in the reaction volume expensive cofactors often required to carry on some

193

N. Russo et al. (eds.), Properties and Chemistry of Biomolecular Systems, 193–204.
© *1994 Kluwer Academic Publishers. Printed in the Netherlands.*

Table 1. Main membrane separation processes: operating principles and applications

Separation processe	Membrane type	Driving force	Method of separation	Purpose
Microfiltration	Symmetric microporous membrane 0.1 to 10 µm pore radius	Hydrostatic pressure difference 0.1 to 1 bar	Sieving mechanism due to pore radius and absorption	Sterile filtration clarification
Ultrafiltration	Asymmetric microporous membrane 1 to 10 µm pore radius	Hydrostatic pressure difference 0.5 to 5 bar	Sieving mechanism	Separation of macromolecular solutions
Reverse osmosis	Asymmetric "skin type" membrane	Hydrostatic pressure 20 to 100 bar	Solution-diffusion mechanism	Separation of salt and microsolutes from solutions
Dialysis	Symmetric microporous membrane 0.1 to 10 µm pore size	Concentration gradient	Diffusion in convection free layer	Separation of salt and microsolutes from macromolecular solutions
Electrodialysis	Cation and anion exchange membranes	Electrical potential gradient	Electrical charge of particle and size	Desalting of ionic solution
Gas separation	Homogeneous or porous polymer	Hydrostatic pressure concentration gradient	Solution-diffusion mechanism	Separation from gas mixture
Supported liquid membrane	Symmetric microporous membrane with adsorbed organic liquid	Chemical gradient	Solution-diffusion via carrier	Separation and concentration of metal ions and biological species
Membrane distillation	Microporous membrane	Vapour pressure gradient	Vapour transport into hydrophobic membrane	Ultrapure water, concentration of solutions
Pervaporation	Dense membrane	Chemical gradient	Solution-diffusion mechanism	Separation of organic and aqueous-organic solutions

enzymatic reactions. Controlled addition of one of the reactants supplied to the reaction zone, possibility of controlled sequential reactions are other interesting possibilities. In Tables 2, 3 and 4 some of the membrane bioreactors explored for various purposes and reported in the literature, are summarized.

Enzyme membranes will also contribute to the growth of new areas for industrial biochemistry such as the non-aqueous enzymology. It has been shown that most enzymes are able to operate in organic solvents containing no or little water [2]. It has also been clarified that the change in solvents dramatically affects key catalytic

Table 2. Membrane reactors for pharmaceutical applications

Reaction	Membrane Reactor	Purpose
Production of ampicillin and amoxycillin (penicillin amidase)	Entrapment in cellulose triacetate fibers	Production of antibiotics
Conversion of fumaric acid to L-aspartic acid (E. coli with aspartase)	Entrapment in polyacrylamide gel	Pharmaceutical and feed additive, raw material for aspartame
Conversion of L-aspartic acid to L-alanine (P. dacunhae)	Entrapment in polyacrylamide gel	Pharmaceutical additive
Conversion of cortexolone to hydrocortisone and prednisolone (C. lunata/C. simplex)	Entrapment in polyacrylamide gel	Production of steroids
Conversion of acetyl-D,L-amino acid to L-aminoacid (aminoacylase)	Ionic binding to deae-sephadex	Production of L-amino acid for pharmaceutical use
Synthesis of tyrosine from phenol, pyruvate and ammonia (tyrosinase)	Entrapment in cellulose triacetate membrane	Production of L-amino acid for pharmaceutical use
Hydrolysis of a cyano-ester to ibuprofen (lipase)	Gelification on UF capillary membrane	Production of anti-inflammatories
Hydrolysis of 5-p-HP-hydantoine to D-p-HP-glycine (hydantoinase/carbamilase)	Entrapment in UF polysulfone membrane	Production of cephalosporin
Dehydrogenation reactions (NAD(P)H dependant enzyme systems)	Confination with UF-charged membrane	Production of chemical substances
Hydrolysis of DNA to olygonucleotides (DNase)	Gelification on UF capillary membrane	Production of pharmaceutical substances

characteristics of enzymes including thermostability, enantioselectivity, substrate specificity and regioselectivity. The use of organic solvents as reaction media is of particular interest because: a) various substrates have a low solubility in water; b) some side reactions as hydrolysis or isomerization can be avoided in non-aqueous media; c) microbial contaminations can be minimized. Higher enzyme concentration might be however necessary for increasing reaction rate, often slower in organic solvents. The influence of the solvent on enzyme life time, specificity, activity, etc., has to be studied at the moment case by case. Hydrophobic and hydrophilic polymeric membranes, ceramic membranes, in their various configurations are used for realizing biphasic aqueous/organic reactors where the membranes act as ideal surface of reaction separating the aqueous and non aqueous phase.

Some membrane bioreactors have been investigated in our laboratories. Production of fatty acids and glycerine from renewable resources such as vegetable oils and synthesis

Table 3. Membrane reactors for biomedical applications

Reaction	Membrane reactor	Purpose
Hydrolysis of hydrogen peroxide (bovine liver catalase)	Entrapment in cellulose tryacetate membrane	Liver failure
Hydrolysis of inulin to glucose and levulose (inulase)	Entrapment in sodium-alginate gel	Care of liver disease
Oxidation of bilirubin (bilirubin oxidase)	Immobilization on sepharose	Prevention of jaundic in newborn
Degradation of heparin (heparinase)	Immobilization on sepharose	Heparin neutralization
Hydrolysis of arginine and asparagine (arginase and asparaginase)	Entrapment in polyuretane membrane	Care and prevention of leukemia and cancer
Hydrolysis of blood proteic toxins (trypsin, pronase)	Entrapment in polyuretane membrane	Removal of blood toxins in dialytic patients
Insulin secretion (Langerhans islets)	Polyamide UF membrane	Bioartificial pancreas
Hydrolysis of whey proteins (trypsin, chymotrypsin)	UF polysulfone membrane	Production of peptides for medical use

of glycerides from fatty acids and glycerine, have been performed by using lipase enzyme immobilized within the membrane pores or on the top of the membrane skin as an enzymatic gel [3] [4]. Bacterial and mammalian cells have been effectively immobilized by means of membranes [5]. Cells of *S. solfataricus*, a thermophilic and acidophilic microorganism of about 1 μm diameter, have been successfully entrapped in the wall of polysulfone ultrafiltration capillary membranes in the fiber formation process [6]. Protein/lactose solutions (such as whey) can be processed with these immobilized cell membranes to yield a protein rich, lactose poor retentate (useful to the dairy industry) and a glucose rich permeate (useful as a sweetener). Islets of Langerhans (the cellular aggregates responsible for pancreatic insulin secretion) have been immobilized in the shell of capillary membrane reactors in "tube-and-shell" configuration to simulate pancreatic function and protect the islets against the attack of immunocompetent proteins in the host blood [7].

Two examples of Enzyme Membrane Reactors studied in our laboratories are below reported in detail.

2. Kinetic Behavior of Hydantoinase in a Membrane Segregated Enzyme Reactor with Tube-and-Shell Configuration

The hydantoinase studied catalyzes the conversion of 5-p-hydroxy-phenylhydantoine (5-p-HPH) to the corresponding N-carbamyl. The enzyme catalytic activity was preliminarily assayed in solution and, afterwards, under immobilization conditions, either by entrapment within flat membranes or by segregation in hollow fiber. Enzyme activity was measured in terms of International Units:

IU = μmol/min;

enzyme specific activity = IU/g$_{(crude\ enzyme\ solution)}$.

Table 4. Membrane reactors in Agro-Industry

Reaction	Membrane Bioreactor	Purpose
Hydrolysis of lactose to glucose and β-galactose (β-galactosidase)	Axial-annular flow reactor	Delactosization of milk or whey for human consumption
Hydrolysis of starch to maltose (α-, β-amilase, pullulanase)	CSTR with UF membrane	Production of syrups 42 DE and HFCS
Fermentation of all fermentable sugars (yeast)	CSTR with UF membrane	Brewing industry
Anaerobic fermentation (S. Cerevisiae)	CSTR with UF membrane	Production of alcohol
Hydrolysis of pectines (pectinase)	CSTR with UF membrane	Production of bitterness and clarification of fruit juice and wine
Removal of limonene and naringin (β-cyclodextrin)	CSTR with UF membrane	Production of bitterness and clarification of fruit juice
Hydrolysis of K-casein (endopeptidase)	CSTR with UF membrane	Milk coagulation for dairy products
Hydrolysis of collagen and muscle proteins (protease, papain)	CSTR with UF membrane	Tenderization of meat and particular beef
Conversion of glucose to gluconic acid (glucose oxidase/catalase)	Packed bed reactor	Prevention discoloration and off-flavor of egg produt during storage
Hydrolysis of triglycerides to fatty acids and glycerol (lipase)	UF capillary membrane reactor	Production of food, cosmetics, emulsificant
Hydrolysis of cellulose to cellobiose and glucose (cellulase/β-glucosidase)	Asimmetric hollow fiber reactor	Production of ethanol and protein

The optimal reaction conditions (pH, T, K_m) were investigated on the enzyme free in solution (Figs. 1, 2, 3). Whole microbial cells containing hydantoinase enzyme were, also, immobilized by entrapment in flat sheet polysulfone (PS) membranes, during the membrane formation process by phase inversion [8]. A comparative analysis of the enzyme-loaded membranes activity was carried out using membranes containing

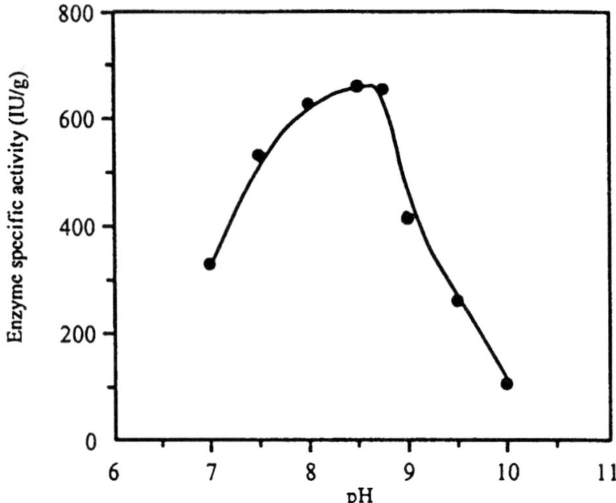

Figure 1. Effect of pH on enzyme activity.
reaction conditions: 0.1 M phosphate buffer, 50
mM 5-p-HPH, T = 40 °C.

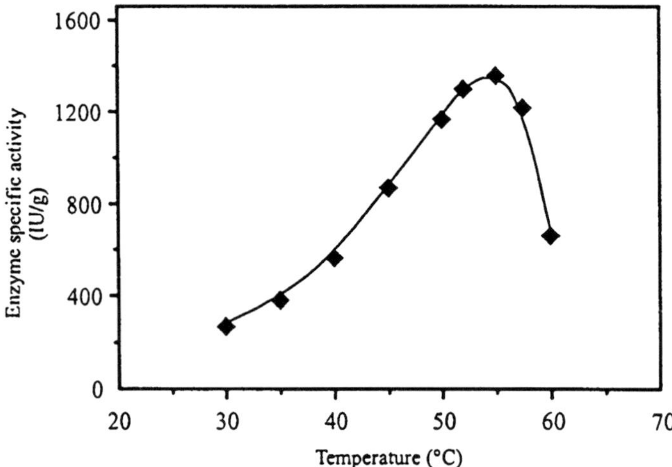

Figure 2. Effect of temperature on enzyme activity.
Reaction conditions: 0.1 M phosphate buffer, 50 mM
5-p-HPH, pH = 8.50.

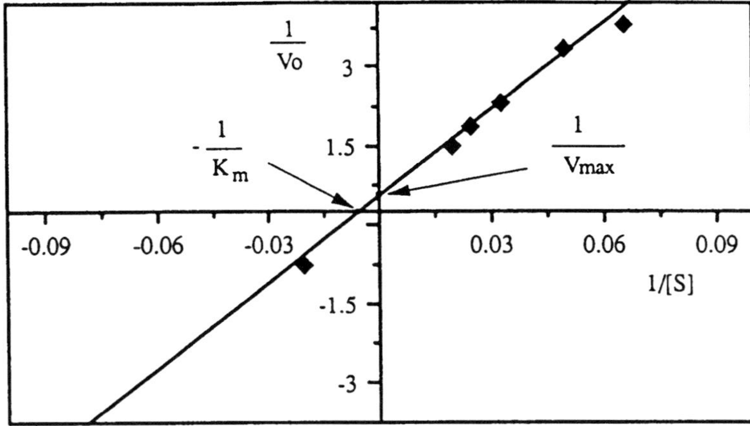

Figure 3. Lineweaver-Burk plot of experimental data for hydantoinase. Reaction conditions: 0.1 phosphate buffer, 0.68 mg/ml enzyme, pH = 8.50, T = 40 °C. K_m = 142.85 mM, V_{max} = 2.86 mM/min.

Table 5. Residual activity of enzyme-loaded membranes (DMF was used as solvent).

PS (%)	PVP (%)	IU/g (Free biomass)	IU/g (Immobilized biomass)	Residual activity (%)
25	5	144	18.18	12.6
20	10	144	53.61	37.0
18	12	144	114.28	79.0
16	14	144	98.00	68.0

different amount of polymer (PS) and copolymer (polyvinilpyrrolidone, PVP) (Table 5). Due to the partial solubility of the substrate in the aqueous phase, the reaction rate was strongly limited by the substrate transport through the membrane. To overcome this problem, a different enzyme immobilization technique was attempted, leading to the realization of a Membrane Segregated Enzyme Reactor (MSER), where the enzyme, still in solution, could be confined in a well defined region of reaction space. The reactor essentially consists of hollow fiber membranes, the enzyme being continuously flushed into the fibers lumen (Fig. 4). A number of experimental tests were carried out to assay both the activity and the stability of the segregated enzyme. As shown in Fig. 5, after 10 hours 1 M of 5-p-HPH was converted, the enzyme maintaining about 20% of activity. The optimal operating conditions for hydantoinase catalysis were found to be pH = 8.50 - 8.75, T = 55 °C, K_m = 142.85 mM. However, all the experiments were performed at

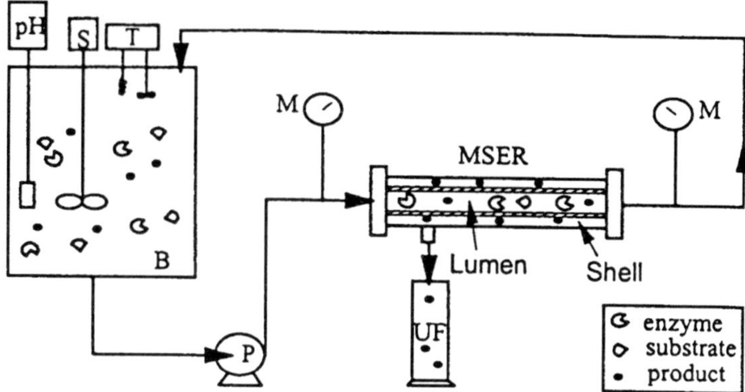

Figure 4. Membrane segregated enzyme system. B=batch, M=manometer, P=peristaltic pump, pH=pHmeter, UF=ultrafiltered reservoir, S=stirrer, T=thermostat.

T = 40 °C because the enzyme exhibited a great stability at this temperature. The biomass entrapped in the flat sheet membranes (PS 18% and PVP 12%) showed a residual activity of 79% as compared to that of biomass in solution. The enzyme confined in the fibers' lumen exhibited an activity decay when continuously reused. The ultrafiltration membrane system permitted a complete separation and recovery of the reaction product.

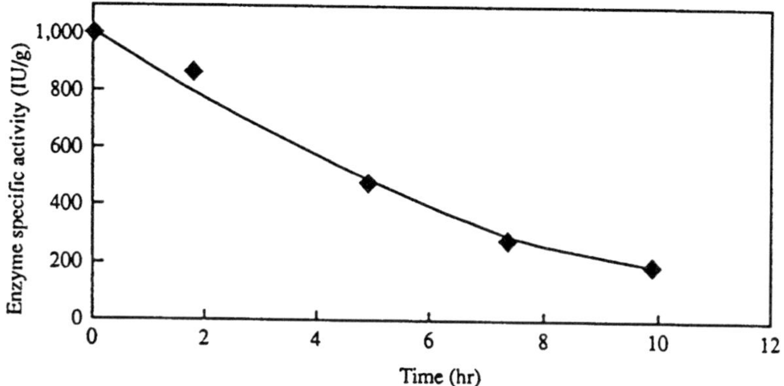

Figure 5. Activity of segregated enzyme vs reaction time. Reaction conditions: phosphate buffer 0.1 M, pH = 8.50, T = 40 °C, 3.6 mg/ml crude enzyme, 0.2 M 5-p-HPH.

3. Enzymatic Hydrolysis of Dna in Enzyme Membrane Reactors

The enzymatic hydrolysis of DNA was studied by using deoxyribonuclease I (EC 3.1.21.1.) enzyme from bovine pancreas. It degrades DNA molecules to a mixture of oligonucleotides having a free 5'-phosphate terminal group [9].

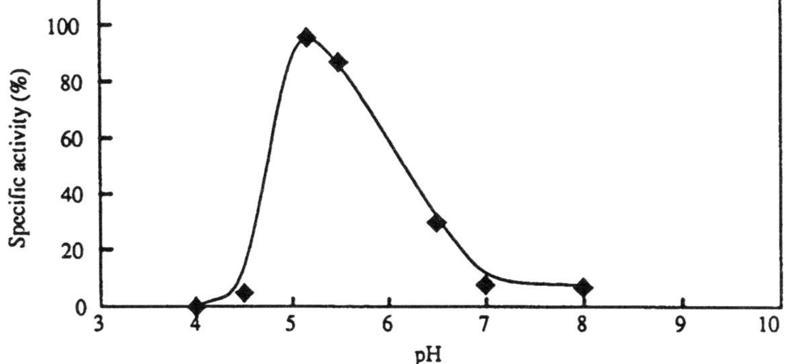

Figure 6. Specific activity of DNase vs pH. reaction conditions:
100 mM sodium acetate buffer; Mg^{2+} = 10 mM; T = 25 °C; DNA
= 40 µg/ml; DNase = 100 K/ml.

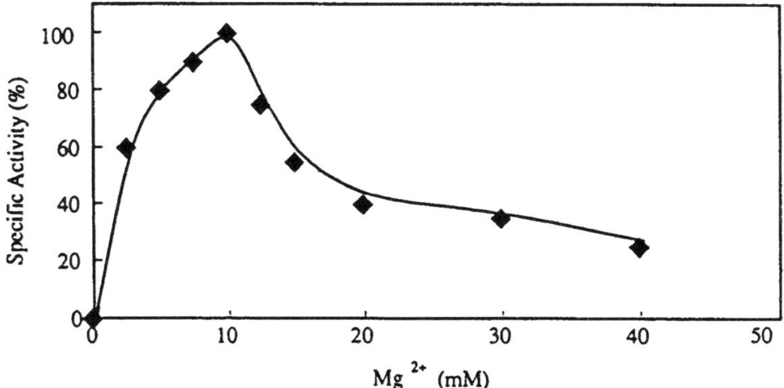

Figure 7. Specific activity of DNase vs Mg^{2+}. Reaction conditions:
100 mM sodium acetate buffer; pH = 5; T = 25 °C; DNA = 40
µg/ml; DNase = 100 K/ml.

Initially, catalytic properties of free DNase have been investigated; subsequently, a
comparative analysis of the kinetic behavior with respect to the immobilized enzyme has
been carried out.
The experimental results show that DNase in immobilized form, maintains catalytic
activity.
DNase I was from Sigma, molecular weight = 31,000 Daltons, specific activity = 2,000
Kunitz units/mg. One Kunitz unit is the amount of enzyme that produces a
ΔA_{260nm} of 0.001 per minute per ml at pH 5 at 25 °C, using DNA as the substrate [10].

The enzyme is activated by bivalent metal ions as Mn2+ and Mg2+. The DNase activity
was measured as a function of pH. As shown in Fig. 6, the optimal pH value is 5.00.
The catalytic behavior of the enzyme as a function of the cofactor Mg2+ concentration
was investigated. As reported in Fig. 7, the highest activity was obtained at 10 mM of
Mg2+. In summery, operating at T = 25 °C, DNase I from bovine pancreas shows a
maximum of its activity at pH = 5.00; Mg2+ = 10 mM in a sodium acetate buffer of 100
mM.

The K_m and V_{max} values were estimated in according to the Lineweaver-Burk plot. By
using the optimal reaction conditions the following results were obtained:

$K_m = 33.4$ µg/ml; $V_{max} = 6.83$ mg/ml·min.

The influence of the enzyme storage conditions on the enzyme activity has also been
studied. DNase solutions were incubated at 4 °C and 25 °C. The enzyme activity was
tested in the time, for both the temperatures at two different concentration values (50 and
200 K/ml) collecting samples from the incubated solutions. As shown in Figs. 8 and 9 it
is convenient to store the DNase solutions at 4 °C and at high concentration.

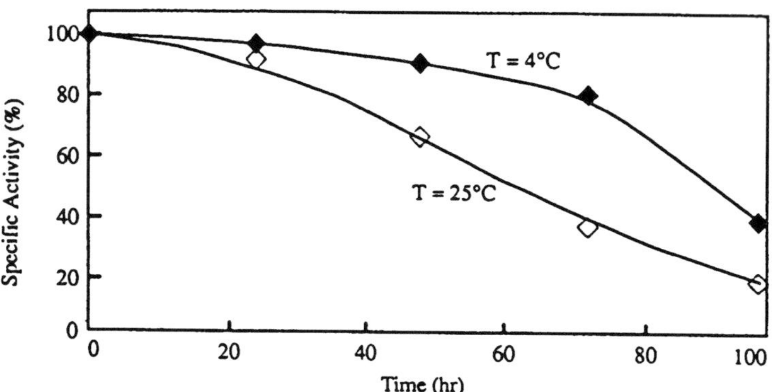

Figure 8. Specific activity of DNase stored at concentration of
50 K/ml. The activity test was carried out in the optimal
conditions (see text).

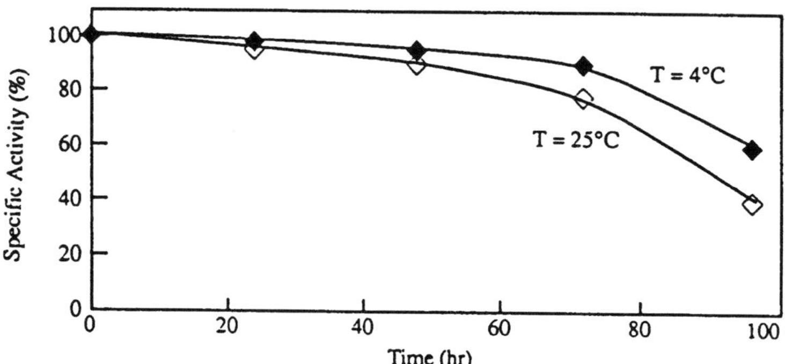

Figure 9. Specific activity of stored DNase at a concentration of
200 K/ml. The activity test was carried out in the optimal
conditions (see text).

The DNase has been then gelled on the inner surface of capillary membranes made of polyammide (PA) realizing an Enzyme Membrane Reactor. Activity and stability of immobilized enzyme was tested.
It was observed that the immobilized DNase looses 45% of the catalytic activity with respect to the same amount of enzyme free in a batch reactor, while increases its stability (Fig. 10).
Due to the increased stability of the immobilized enzyme, it is possible to reuse it for more than 14 hours at a practically constant specific activity. The membrane system also gives the possibility to separate the oligonucleotides produced and to recover them in the pure form as the chemical reaction goes on.

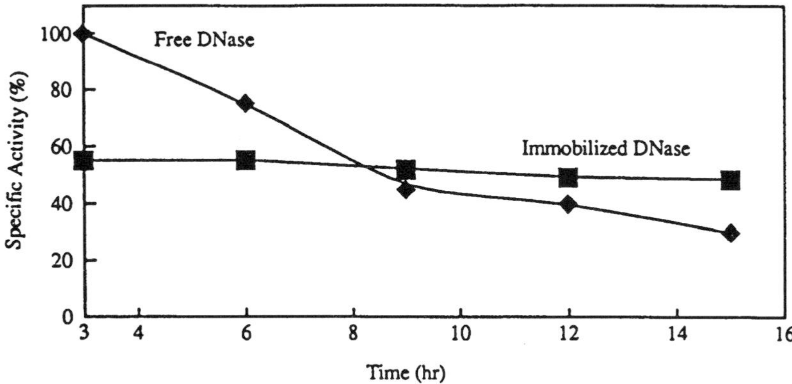

Figure 10. Specific activity of free and immobilized DNase vs reaction time using the same initial enzyme amount (1 mg). Reactions were carried out in the optimal conditions (see text).

4. Conclusion

The experimental results obtained in the described studies well confirm the potentiality of membrane operations in solving significant problems typical of biochemical processes. Enzyme stability can be enhanced immobilizing it on a membrane, and products separations simultaneously the reactions can be realized. Studies are in progress for a better understanding of specific interactions of enzyme with membrane surface and for explaining the observed increasing of life time and the kinetics of the immobilized enzyme.

Acknowledgements

This work was performed as a part of the research project "Processi a Membrana nell'Industria Farmaceutica", sponsored by Tecnofarmaci S.p.A., Pomezia (Roma).

References

1. E. Drioli, G. Iorio, G. Catapano, in <u>Handbook of Industrial Membrane Technology</u>, Mark C. Porter (Ed), Noyes Publ., Park Ridge, NJ, p. 401, (1989).

2. A.M. Klibanov, "Asymmetric Transformations Catalyzed by Enzymes in Organic Solvents", *Ac. Chem. Res.* **23**, 114, (1990).

3. R. Molinari, E. Drioli, "Conversione di oli vegetali in un Reattore Enzimatico a Membrane", *Acqua Aria*, 685, (1987).

4. F.X. Malcata, H.R. Reyes, H.S. Garcia, C.G. Hill, Jr. and C.H. Amundson, "Kinetics and Mechanisms of Reactions Catalysed by Immobilized Lipases", *Enzyme Microb. Technol.* **14**, 426, (1992).

5. G. Catapano, M. Filosa, G. Iorio, E. Drioli, A. Gambacorta, M. De Rosa, M. Demma, "Capillary Polisulfone Membranes with entrapped whole cells: influence of cell loading and filler molecular weight on membrane mechanical and kinetic performance", *J. Molecular Cat.* **58**, 277, (1990).

6. M. De Rosa, A. Gambacorta, E. Esposito, E. Drioli, S. Gaeta, "Immobilized Thermophylic microbial cells in cellulose acetate membranes", *Biochimica* **67**, 517, (1980).

7. F. Crucitti, R. Bellantone, G.B. Doglietto, C.P. Lombardi, G. Catapano, G. Iorio, E. Drioli, in <u>Argomenti di Endocrino Chirurgia</u>, Monduzzi (Ed), p. 231,(1987).

8. P. Radovanovic, S.W. Thiel, S.T. Hwang, "Formation of asymmetric polyulfone membranes by inversion precipitation. Part II. The effects of casting solution and gelation bath compositions on membrane structure and skin formation", *J. Membrane Sci.* **65**, 231, (1992).

9. M. McDonald, in <u>Methods in Enzymology</u>, S.P. Colowich and N.O. Kaplan (Eds), Ac. Press, New York, NY, p. 2, (1976).

10. M. Kunitz, *J. Gen. Physiology* **33**, 349, (1950).

The Conformational Properties of (-)-Dolastatin 10, a Powerful Antineoplastic Agent.

P. FANTUCCI, and E. MATTIOLI
Department of Inorganic, Metallorganic and Analytical Chemistry, Via Venezian 21, 20133 Milano, Italy

T. MARINO and N. RUSSO
Department of Chemistry, University of Calabria, I-87030 Arcavacata di Rende, Cosenza, Italy

1. Introduction

(-)- Dolastatin 10 has been known for a long time as a powerful antineoplastic agent. Its absolute configuration has been recently object of careful analysis [1] , while indications concerning its preferred conformational structure are scarse. We present in this note results obtained from Molecular Mechanics (MM) and Molecular Dynamics (MD) calculations, carried out with the aim of elucidating the basic conformational features of the title compound.

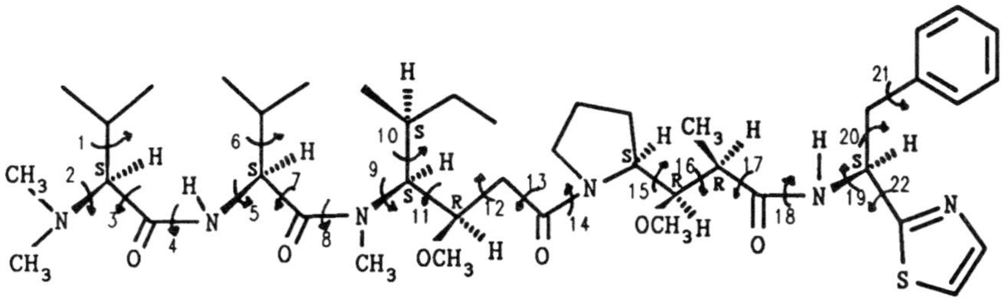

Figure 1. (-)- Dolastatin 10

N. Russo et al. (eds.), Properties and Chemistry of Biomolecular Systems, 205–209.
© 1994 *Kluwer Academic Publishers. Printed in the Netherlands.*

2. Molecular Mechanics study of (-)- Dolastatin 10

The skeleton of the (-)- Dolastatin 10 molecule was built joining fragments in their standard geometries [2] . The chirality of the nine asymmetric centres was carefully checked using special features offered by PIMMS program [3]. A first optimization carried out using a quite complete potential (including third order corrections and cross terms) gave a minimum conformation with energy of 172.5 Kcal/mol. This molecular geometry was used in the following second phase of the study. A further investigation of the preferred conformation is motivated by the well known fact that optimization carried out with gradient methods can only reach the local minima. In order to overcome these difficulties, a quite complete analysis of the Potential Energy Surface (PES) was carried out using a recently proposed algorithm for random sampling [4] . This procedure is especially useful for molecules characterized by a high number of torsional degrees of freedom and can allow to study the combined effect of all the rotation axes, or the effect of individual rotations, leading to the identification of structures suitable as starting points for MM optimization or MD simulations. More than 50000 conformations were generated with the random sampling in the conformational torsional space defined by the 22 free rotation axes of the molecule (see Fig.1). According to the RANCONF procedure [4], the generated conformations are selected when their energy is not higher than a threshold ΔE with respect to the lowest conformer found. Moreover the i-th conformation, characterized by angles τ^i_x (k=1,.......22), can be discarded if $|\tau^i_x - \tau^j_x| < \Delta E$ ($\forall$k), j being one of the already selected conformations. In the first batch of random sampling the thresholds ΔE and $\Delta \tau$ were fixed at 30 Kcal/mol and 25°, respectively. It was found that, using such thresholds, no acceptable conformations could be selected. In fact, most of the generated conformations are hindered by intramolecular contacts, giving rise to high energy. A better strategy was designed, based on random sampling in a subspace identified by a sequence of individual rotations. This can help in identifying the axes which cause intramolecular contacts for small rotations and, therefore, which critically determine the high energy regions of the PES. The procedure gave evidence that only 9 axes (1,2,3,4,5,6,19,20,21) correspond to easy rotations, which identify the section of the PES on which the random search followed by standard geometry optimization can be carried out. This leads to the location of 10 new minima within an energy gap of about 10 Kcal/mol, but all lying at an energy higher than the initial minimum. Finally, the largest set of rotation axes which can be operated simultaneously , has been identified by a random sampling applied to PES sections of increasing dimensionality, spanned by combinations of axes of type 1, 1+2, 1+2+3 .... The final section of the PES is identified by the axes 1, 2, 3, 4, 5, 6, 7, 8, 13, 17, 18, 19. A number of randomly generated conformations (of the order of 10^5) subjected to the selection thresholds ΔE=50 Kcal/mol and $\Delta \tau$=25°, produced 52 accepted molecular forms, and those of lowest energy were subjected to optimization.

Among the 12 possible local minima falling within a range of 10 Kcal/mol (see Tab. 1), a minimum of about 2 Kcal/mol lower in energy than the initial structure was found (see Fig. 2).

Table 1. Relative energies of the 12 lowest minima.

Conformer	Potential Energy	ΔE
A	170.2	0.0
B	172.5	2.3
C	172.7	2.5
D	173.5	3.3
E	173.6	3.4
F	175.3	5.1
G	176.0	5.8
H	176.1	5.9
I	177.2	7.0
L	177.9	7.7
M	179.3	9.1
N	180.1	9.9

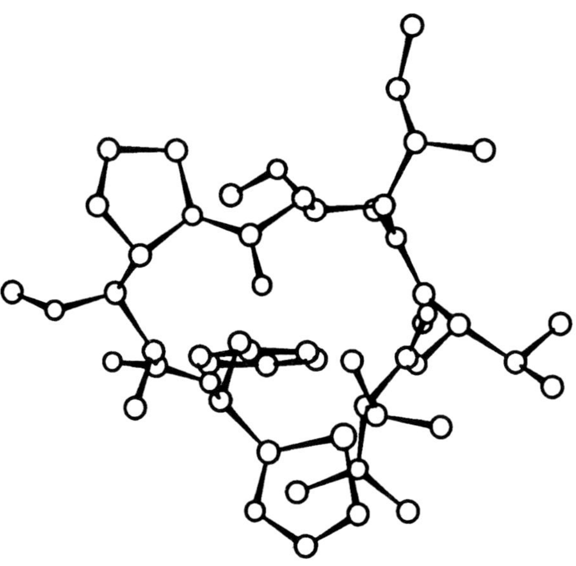

Figure 2. Global minimum of (-)- Dolastatin 10 obtained from MM calculations.

3. Molecular Dynamics study of (-)- Dolastatin 10

MD calculations at constant temperature (300°K), covering a time range of about 100 ps (time-step = 1 fs) were carried out for the molecule both in gas phase and in presence of more than 150 water molecules, distributed within a sphere of diameter equal to 20 A. The concentration of the H_2O molecules was carefully chosen not to be too high in order to avoid an undue constraint for the hydrophobic part of the molecule. At the same time, such a concentration of water molecules is quite sufficient to simulate the most relevant hydration effects. For gas and solution phases, the starting point of the MD simulation was chosen to be the lowest lying minimum.

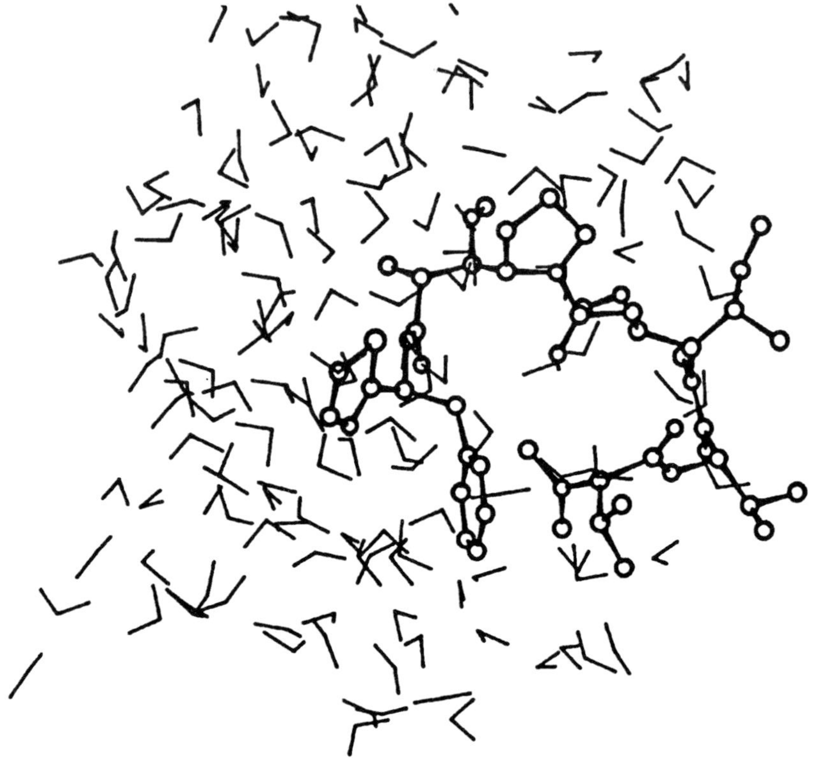

Figure 3. Stucture of (-)- Dolastatin 10 in presence of water, obtained at a low energy conformation in the long time interval of MD simulation.

An equilibration carried out at 300°K for 8000 steps was followed by the 100 ps simulation. In the time vs. potential energy plot of the gas phase MD a few most stable conformations were identified which, after optimization, originated minima all coincident among them and with the global one. This confirms that also the MD shaking of the initial structure cannot give a conformer more stable than that identified by the analysis of the PES described in previous sections .

4. Conclusions

The results of the MM and MD study can be summarized in the following few points: i) The random search of the PES of (-)- Dolastatin 10 helps in identifying 12 distinct local minima .
ii) Since the global minimum is about 2.3 Kcal/mol more stable than the next higher conformer, one can conclude that the statistical distributions derived from finite (room) temperature, cannot include important contributions from conformers different from the absolute minimum. In other words, our MM analysis clearly suggests that (-)-Dolastatin 10 is a relatively rigid molecular system. iii) The 100 ps MD simulation of the molecule considered in gas phase gave in the long-time period variations in total energy not larger than ± 10 Kcal/mol with respect to the average value. This result confirms that the simulation has reached a nearly steady state and that, again, no large conformational freedom exists. iv) The global minimum has a pocket-like shape, with all the peptide groups pointing outside the folder frame. It is easy to recognize that such groups are those mostly involved in the hydration process, as it is confirmed by the MD calculations: the water molecules show a clear tendency to form clusters around the most polar groups. The hydration process, however, does not dramatically affect the peculiar conformational features. It must be noticed that inside the 'pocket' a relatively small area exists where some water molecules can penetrate, causing a sligtly less folded conformation in comparison with the gas phase (see Fig. 3) .

References

1. G.R. Pettit, S.B. Singh, F. Hogan, P. Lloyd-Williams, D.L. Herald, D.D. Burkett, P.J. Clewlow, *J. Am. Chem. Soc.* **111**, 5463, (1989).
2. Byosim Technologies Inc., San Diego, CA, U.S.A., Insight-Discover package.
3. Oxford Molecular Ltd (1991), Oxford, G.B., *Pimms* package.
4. P. Fantucci, F. Magugliani, E. Mattioli, A.M. Villa, L. Villa, *Atti 1° Convegno Nazionale di Informatica Chimica*, 74, Venezia, (1991).
5. P. Fantucci, E. Mattioli, A.M. Villa, L. Villa, *J. Comput.-Aided Mol. Design* **6**, 315, (1992).

In Vitro Solubilization of a Recombinant Elongation Factor Tu, Carrying the Asp138Asn Mutation

M.C. GAGLIANO, A. CROCCO, G. APA, and G. PARLATO.
Istituto di Biochimica Fisica e Patologia Molecolare e Cellulare
Facoltà di Medicina e Chirurgia, Università degli Studi di Reggio Calabria
88100 Catanzaro, Italy

1. Introduction

One of the problems with producing recombinant proteins in *E. coli* is the formation of intracellular insoluble aggregates, known as inclusion bodies [1]. The solution to this problem is essential when large amounts of biologically active recombinant proteins must be produced for pharmaceutical use as well as for biochemical or biophysical studies. A general procedure enabling to produce *in vivo* as soluble any recombinant protein, is still unknown. The production of recombinant proteins in soluble form was possible for the recombinant metalloenzymes of the superoxide dismutase, by supplementing the metal cations (Cu^{+2}, Zn^{+2}, Mn^{+2}) to culture medium [2,3]; for active recombinant subtilisin E by reducing the expression with a very low inducer concentration at low culture temperature of 23 °C [4], for recombinant human interferon by decreasing the growth medium temperature from 37 °C to 23 °-30 °C [5]; for other cases by secreting into extracellular medium or periplasmic space the foreign protein [6]. In the present work we have found the conditions to solubilize *in vitro* a *E. coli* recombinant elongation factor Tu, m.w. 43 KDa, modified via site-specific mutagenesis in the 138 position by substituting Asp with Asn (rmEF-Tu). According to X-ray crystallographic data, the mutated aminoacid participates in the interaction with the guanine base of GDP [7, 8]. In fact our initial aim was to dispose of a large amount of rmEF-Tu produced as *in vivo* soluble protein, in order to obtain crystals for X-ray studies. The failure in finding the conditions for the *in vivo* production of soluble rmEF-Tu, induced us to solubilize *in vitro* rmEF-Tu by chemical treatments, in order to insight the mechanism of the aggregate formation at molecular level.

2. Materials and Methods

The site-specific mutagenesis of *TufA*, one of the two genes enconding EF-Tu, has been already described (9). The mutated gene, transferred to the overexpression runaway vector pCP40 was under the control of the thermoinducibile λP_L promoter, regulated by the thermolabile pcI857 repressor. The host cell for the over-production was *E. coli* PM 455 rec⁻, a strain carrying only the *Tuf A* as active gene of EF-Tu.

211

N. Russo et al. (eds.), Properties and Chemistry of Biomolecular Systems, 211–217.
© 1994 *Kluwer Academic Publishers. Printed in the Netherlands.*

For the lysis, the cells, pelleted at 1000 g for 15 min, after the freezing and thawing processes were treated at 30 °C for 30 min with lysozyme (4 mg/ml) in 25 mM Tris-HCl, pH 7.8, 25 mM KCl, 5 mM $MgCl_2$, 2.5 mM 2-mercaptoethanol, 1 mM EDTA, 1mM phenylmethanesulfonyl fluoride (TMK buffer). For the optimal production of rmEF-Tu the cells were grown up to a density of 1.2 A_{550} units, at 40 °C, by adopting a value of 1:10 for the ratio between the volume of the growth medium and culture flask. After each chemical treatment, the sample with urea, deprived of the cyanate ions by a mixed bed of strong ionic exchangers (Dowex), was centrifuged at 64.000 g for 30 min. The surnatant as well as the pellet, resuspended in the same volume of buffer, were analyzed by sodium dodecyl sulphate polyacrylamide gel electrophoresis (SDS-PAGE), at pH 8, on 12.5 % polyacrylamide gels in the Mini Protean II (Bio-Rad) apparatus. The gels were loaded with 25 µl samples, containing 10 µl of surnatant solution or pellet suspension, 15 µl of 62.5 mM Tris-HCl pH 8.0, 70 mM 2-mercaptoethanol, 10% (v/v) glycerol, 2% SDS, 0.01% Bromophenol Blue. The gel was stained with 0.1% (g/l) Coomassie brilliant blue, dissolved in 50% water-methanol mixture. The antibiotics ampicillin (Serva) and kanamycine (Sigma), after sterilization by filtration through a 0.22 µ membrane, were added at final concentrations of 0.1 g/l and 0.05 g/l to the growth medium, autoclaved at 120 °C for 45 min and containing per liter: 10 g Nutrient Broth (Difco), 5 g Yeast extract (Difco), 8 g NaCl, 0.52 g Tris Base, 0.02 g Thymine. All chemicals were of analytical grade.

3. Results

We carried out experiments to produce *in vivo* soluble rmEF-Tu by lowering the temperature of the growth medium or by supplementing the growth medium either with GDP, the strong ligand of the wild-type EF-Tu with Kd= 10^{-9} M [10], or with XDP, a specific ligand of EF-Tu, harboring the mutation Asp 138 Asn [11], or with Mg^{+2}, involved in the formation of the binding site of the nucleotide [7, 8]. The unsuccessful results of these experiments induced us to solubilize rmEF-Tu *in vitro* by chemical treatments. After the cell lysis, the SDS-PAGE analysis revealed in the pellet (lane 3) the presence of rmEF-Tu as the most abundant protein (fig. 1).

After suspending the pellet in TMK buffer containing 1 M urea, by fast vortexing, the suspension was left at 30 °C for 30 min under gentle agitation before the centrifugation at 64.000 g for 30 min. In this case, rmEF-Tu was also substantially present in the pellet as shown by SDS-PAGE analysis (lane 6).

The same results were obtained after the lysate treatment as before but in presence of 2M urea. In figure 2, SDS-PAGE of triplicate samples of surnatant (lane A, B, C) and of suspended pellet (lane A', B', C') indicates that rmEF-Tu, together with lower molecular weight protein, is present in the insoluble fraction.

By treating with 4 M urea the pellet recovered after 2 M urea treatment, rmEF-Tu was present in the surnatant (fig. 3, lane 2), as practically purified protein and was soluble also after the removal of urea by overnight dialysis at 4 °C against a TMK buffer, brought at pH 10 by the addition of the NH_3 solution (fig. 4, lane 1).

The SDS-PAGE profiles were unmodified for rmEF-Tu aggregates suspended and electrophorized in absence of 2-mercaptoethanol (not shown).

rmEF-Tu, soluble at pH 10, was again a precipitate as it approached by dialysis the pH

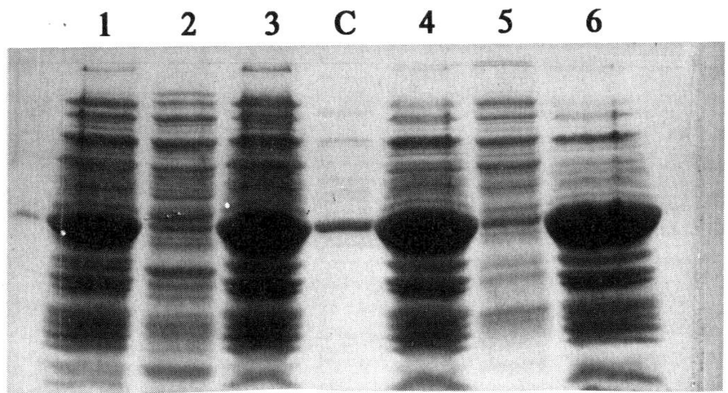

Figure 1. Analysis of solubility of rmEF-Tu by SDS-PAGE electrophoresis.
Lane 1: cell lysate. Lanes 2 and 3, surnatant and pellet respectively after centrifugation
of the cell lysate at 64.000 g, for 30 min. Lane 4: suspension obtained by treating with 1
M urea at 30 °C for 30 min the pellet of running in lane 3. Lane 5 and 6, surnatant and
pellet respectively obtained by centrifugating as before the above suspension. Lane C,
EF-Tu as protein size marker.

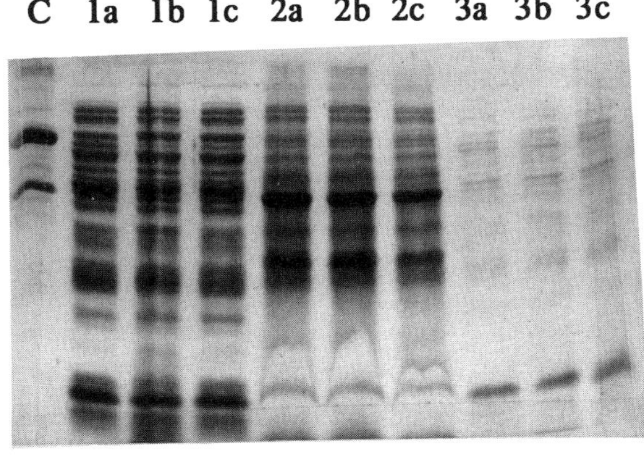

Figure 2. Analysis of solubility of rmEF-Tu by SDS-PAGE electrophoresis.
Lane C, EF-Tu and bovine serum albumin (upper) as protein size marker. Lanes 1a, 1b,
1c, triplicate samples obtained by treating for 30 min at 30 °C with 2 M urea the pellet of
the cell lysate, centrifuged at 64.000 g for 30 min. Lanes 2a, 2b, 2c and lanes 3a, 3b,
3c: the pellets(2) and surnatants (3) respectively after centrifugation as before of the
samples running in the lanes 1a, 1b, 1c.

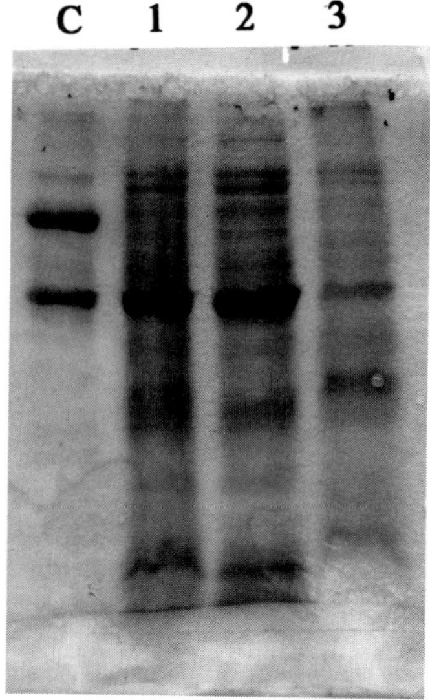

Figure 3. Analysis of solubility of rmEF-Tu by SDS-PAGE electrophoresis.
Lane 1, sample obtained by treating with 4M urea the proteins recovered as aggregates
after treatment with 2 M urea. Lane 2 and 3, surnatant and pellet respectively after
centrifugation of the sample running in Lane 1. Lane C, EF-Tu and bovine serum
albumin (upper) as protein size marker.

8, neither its aggregates were dissolved by the treatment with a TMK buffer at pH 10,
without urea.
It's worth while to note that the treatment with non-ionic surfactant as Triton X-100 was
unaffective to solubilize rmEF-Tu.

4. Discussion

The high expression of recombinant proteins in *E. coli* as aggregates is a well known
problem that has been rate limiting for the diffusion of DNA recombinant technology in
pharmaceutical field as well as in biochemical research, where the biological activity is a
crucial prerequisite. The enormous advantage to produce in bacterial strain proteins
normally present in small amount in living organisms has stimulated the scientists to
insight the molecular mechanism responsible for the aggregation.
Several factors favor the *in vivo* production of soluble recombinant proteins by
regulating the correct folding of the polypeptide chain [12-14]. Between these factors the
reduction of the growth rate or of the induction is the most common [5, 15]. Recent data

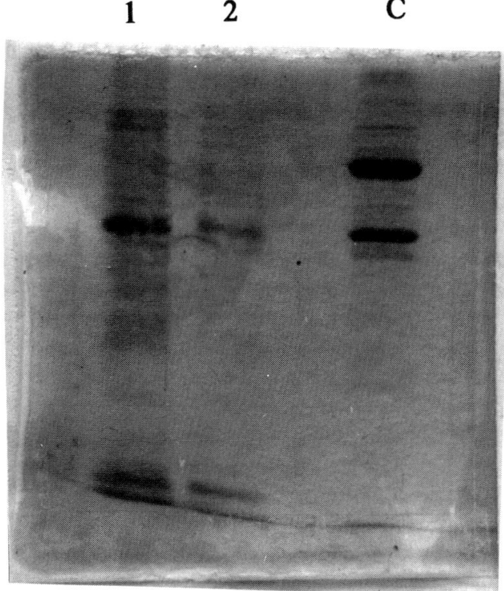

Figure 4. Analysis of solubility of rmEF-Tu by SDS-PAGE electrophoresis.
Lane 1, soluble proteins after the urea removal by overnight dialysis of the surnatant,
after centrifugation of the sample treated with 4M urea. The dialysis was against TMK
buffer, brought at pH 10 by NH3 addition. Lane 2, insoluble proteins recovered after
the dialysis. Lane C, EF-Tu and bovine serum albumin (upper) as protein size marker.

show that replacing a single aminoacid in the crucial site can affect the protein folding
[16-18]. The insolubility of rmEF-Tu may be due to the substitution of the Asp with
Asn residues in position 138 that, being located on the molecule surface, as indicated by
the crystallographic model [7, 8], could be critical for the proper folding. Also the
failure to produce *in vivo* soluble molecule of rmEF-Tu by lowering the temperature, in
order to reduce the folding rate, supports this possibility. Probably the mutated EF-Tu,
is carrying a rigid molecular modification, which hinders the proper folding. The 4 M
urea concentration useful for the rmEF-Tu solubility, is lower respect to the 6-8 M
concentration normally used to solubilize the recombinant proteins [6]. This could be
indicative that smaller and /or weaker inter/ intra-chain electrostatic interactions are
present in the rmEF-Tu aggregates respect to the other cases, where the urea
concentration used for the solubilization is higher. The absence of polymers indicated by
SDS-PAGE in absence of 2-mercaptoetanol excludes that the covalent disulphide bonds
are involved in the aggregation. The unaffectiveness of non-ionic surfactant to solubilize
rmEF-Tu supports that hydrophobic interactions are negligible for the aggregates
formation.
Perhaps the modified structure of rmEF-Tu do not fold correctly *in vivo,* because its
structure, modified by the Asp138Asn mutation, could not be recognized by the

molecular chaperones [19] or because the chaperones could not help the folding of the protein molecule, probably less flexibile.

The *in vitro* solubility at pH 10 after the removal of the denaturants confirms that the electrostatic interactions could be responsible for the formation of aggregates. In fact at pH 10 there are possible charge modifications, i.e. the positive charges are reduced as well as attractive forces. The irreversibility of the rmEF-Tu precipitation process, when the pH value increases from 7.8 to 10 in absence of urea, may indicate that cooperative molecular changes are involved in the aggregation process and these induce an uncorrect folding in which the strong noncovalent bonds between polypeptide side-chains are buried.

The availability of rmEF-Tu soluble in aqueous solutions will give us the opportunity to insight by spectroscopic and/or biochemical approach the intra-intermolecular interactions involved in the formation of the aggregates.

Acknowledgement

The work has been supported by a grant from CNR, Roma, Progetto Finalizzato, Biotecnologie e Biostrumentazioni.

References

1. R.G. Schoner, L.F. Ellis, and B.E. Schoner, "Isolation and purification of protein granules from *Escherichia coli* cells overproducing bovine growth hormone", *Bio/Tecnology* **3**, 151-154, (1985).
2. J.R. Hartman, T. Geller, Z. Yavin, D. Bartfeld, D. Kanuer, H. Aviv, and M. Gorecki, "High-level expression of enzymatically active human CuZn superoxide dismutase, in *Escherichia coli* " *Proc. Natl. Acad. Sci USA* **83**, 7142-7146, (1986).
3. Y. Beck, D. Bartfeld, Z. Yavin, A. Levanon, M. Gorecki, and J.R. Hartman, "Efficient production of active human manganese superoxide dismutase, *Escherichia coli* " *Bio/Tecnology* **6**, 930-935, (1988).
4. H. Takagi, Y. Morinaga, M. Tsuchiya, H. Ikjemura, and M. Inouye, "Control of folding of proteins secreted by a high expression secretion vector, pIN-III-omp A: 16-fold increase in production of active subtilisin E, in *Escherichia coli*'" *Bio/Tecnology* **6**, 948-950, (1988).
5. H.C. Schein, and M.H.M. Noteborn, "Formation of soluble recombinant proteins in *Escherichia coli* is favored by lower growth temperature", *Bio/Tecnology* **6**, 291-294, (1988).
6. F.A.O. Marston, "The purification of eukariotic polypeptides expressed in *Escherichia coli,* in DNA Cloning. A practical approach", Glover, D.M., ed. 59-88, (1987), <u>Practical Approach Series</u> **3**, IRL Press, Oxford.
7. F. Jurnak, "Structure of the GDP domain of EF-Tu and location of the aminoacids homologous to *ras* oncogene proteins", *Science* **230**, 32-36, (1985).

8. T.F.M. La Cour, J. Nyborg, S. Thirup, and B.F.C. Clark, "Structural details of the binding of guanosine diphosphate to elongation factor Tu from *E. coli* as studied by X-ray crystallography", *EMBO J.* **4,** 2385-2388, (1985).

9. P.H. Anborgh, R.H. Cool, E. Jacquet, M. Jensen, G. Parlato, and A. Parmeggiani, "Structure-function relationships of the GTP-binding domain of elongation factor Tu, in the guanine nucleotide binding. Common Structural and Functional Properties L. Bosch, B. Kraal, B., and A. Parmeggiani, eds., NATO-ASI Series, **165**, 67-75 Plenum Press, New York (1989).

10. O. Fasano, W. Bruns, J.B. Crechet, G. Sander, and A. Parmeggiani, "Modifications of elongation factor Tu. Guanine nucleotide interaction by kirromycin: a comparison with the effect of aminoacyl tRNA and elongation factor Ts", *Eur. J. Biochem.* **89**, 557-565, (1978).

11. Y.W. Hwang, and D.L. Miller, "A mutation that alters the nucleotide specificity of elongation factor Tu, a GTP regulators protein", *J. Biol. Chem.* **262**, 13081-13085, (1987).

12. Y.S.E. Cheng, "Increased cell buoyant densities of protein overproducing *Escherichia coli* cells", *Biochem. Biophys. Res. Comm.* **111**, 104-111, (1983).

13. G. Georgiou, J.N. Telford, M.L. Shuler, and D.B. Wilson, "Localization of inclusion bodies in *Escherichia coli* over producing β-lactamase or alkaline phosphatase", *Appl. Environ. Microbiol.* **52**, 1157-1161, (1986).

14. R.A. Hart, U. Rinas, and J.E. Bayley, "Protein composition of *Vitreoscilla* hemoglobin inclusion bodies produced in *Escherichia coli* ", *J. Biol. Chem.* **265**, 12728-12733, (1989).

15. E. Kopetzki, G. Schumacher, and P. Buckel, "Control of formation of active soluble or inactive insoluble bakers yeast α-glucosidase PI in *Escherichia coli* by induction and growth condition", *Mol. Gen. Genet.* **216**, 149-155, (1989).

16. R. Wetzel, L.J. Perry, and C. Veilleux, "Mutations in human interferon gamma affecting inclusion body formation identified by a general immunochemical screen", *Bio/Technology* **9**, 731-737, (1991).

17. U. Rinas, L.B. Tsai, D. Lyons, G.M. Fox, G. Stearns, J. Fieschko, D. Fenton, and J.E. Bailey, "Cysteine to serine substitutions in basic fibroblast growth factor: effect on inclusion body formation and proteolytic susceptibility during in vitro refolding", *Bio/Tecnology* **10**, 435-440, (1992).

18. A.K. Dill, and D. Shortle, "Denatured States of proteins", *Ann. Rev. Biochem.* **60**, 795-825, (1991).

19. R.J. Ellis, and S.M. Van der Vies, *Molecular Chaperones* **60**, 321-347, (1991).

Search for Preparing Antibiotics Active Against Resistant Strains of Bacteria and Wider Spectrum of Activity

M.P. GEORGIADIS
*Chemical Laboratories, Agricultural University of Athens, Iera odos 75 11855 Athens
Geece*

Introduction

The use and overuse of antibiotics resulted in the development of resistant strains bacteria. The two main categories of antibiotics in clinical use to day are, β-lactams and aminoglycosides. Since these are not active against the resistant strains of bacteria, there is an urgent need of discovering new compounds of better antibiotics active against the resistant strains of bacteria.
It is believed that antibiotics such as aminoglycosides develop their antibiotic activity through their binding to the ribosomes thus inhibiting the protein synthesis of bacteria. Modifying enzymes of resistant strain of bacteria are considered to be responsible for preventing the aminoglycoside either from entering into the cell or from binding to the ribosome. Considering that both procedures consist essentially of aminoglycoside binding to several proteins (transport or ribosomal), we can say that generally speaking, these enzymatic modifications aim at preventing the aminoglycoside from binding to proteins.
Our approach for making better antibiotics has been the following:

A)   Preparation of modified or semi-synthetic antibiotics invulnerable by the modifying enzymes of the resistant strain of bacteria.
B)   Preparation of "hybrid" or mixed structure antibiotics [2], with wider spectrum of activity, that is, β-lactams x aminoglycosides and quinolones x aminoglycosides.

Designing the above class of semisynthetic antibiotics A and B we have aimed, not only at more active antibacterials, but also at less toxic ones [3].

1.1. APPROACH A

I. Whereas, resistance to antibiotics such as, β-lactams is due to the presence of one type of enzymatic modification (β-lactams), resistance to the aminocyclitols has been shown to involve any of several different enzymatic modifications that include O-phosphorylation, O-adenylation or N-acetylation. To date, about 12 different enzymatic modifications have been characterized in clinical isolates of Gram negative and Gram positive bacteria [1].

N. Russo et al. (eds.), Properties and Chemistry of Biomolecular Systems, 219–232.
© 1994 *Kluwer Academic Publishers. Printed in the Netherlands.*

The structural modifications occur at several of the hydroxy- and amino- groups as exemplified in the case of Neomycin B (Figure 1).

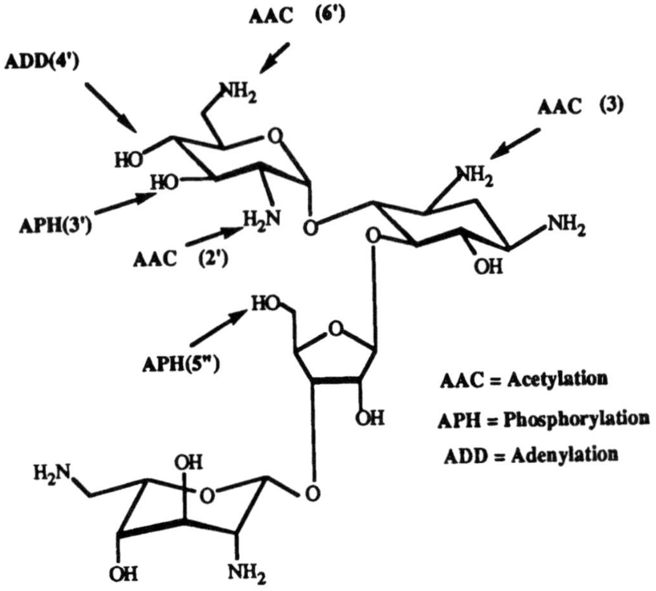

Figure 1. Enzymatic modification of neomycin B by resistant strains.

Inversion of chirality on the carbon whose substituent is attacked by modifying enzymes, may produce compounds invulnerable by the modifying enzymes. More specifically inversion of chirality at the (C-5') carbon of the neomycin B (ring A, scheme I) may produce indeed an invulnerable molecule by the modifying enzymes. This modification may be an appropriate one since neomycin B is about ten times more active than neomycin C which bears the (C-5') epimeric D-glucopyranosyl structure (Figure 2). Although all four ring of neomycin are essential for its activity, it is a fact that the L-do-configuration of the D ring enhances the antibacterial activity of neomycin B. The inversion of chirality on one carbon out of the twenty three carbons of the neomycin B is not an easy task, that is why we have focussed our attention towards neamine (see Scheme I) which is a biosynthetic building block of neomycin B as well as of other aminocyclitol antibiotics.

Studies of the biosynthesis of aminocyclitol antibiotics have led to the development of the mutasynthetic technique for the preparation of related antibiotics (Neamine is practically apart or common moiety of most aminocyclitol antibiotics) [4].

The term mutasynthesis refers to a technique for the synthesis of new antibiotics using specially selected mutants of strains of microorganisms which normally produce antibiotics. The mutants used are selected for their inability to produce their usual

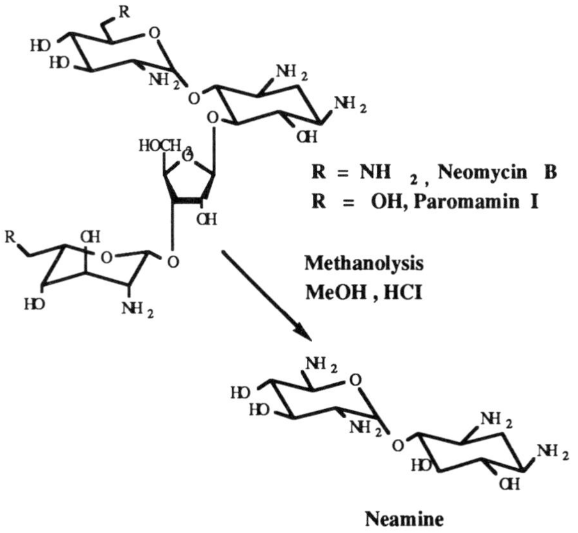

Scheme I

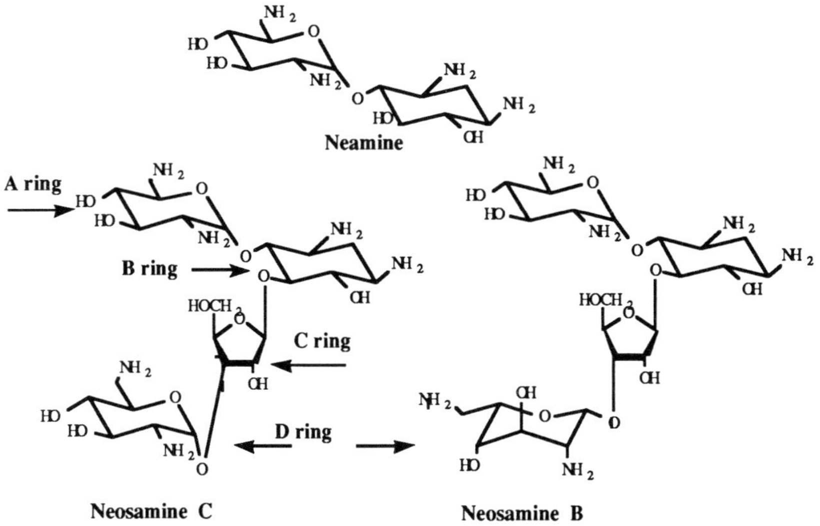

Figure 2. More modifications of Neomycin B.

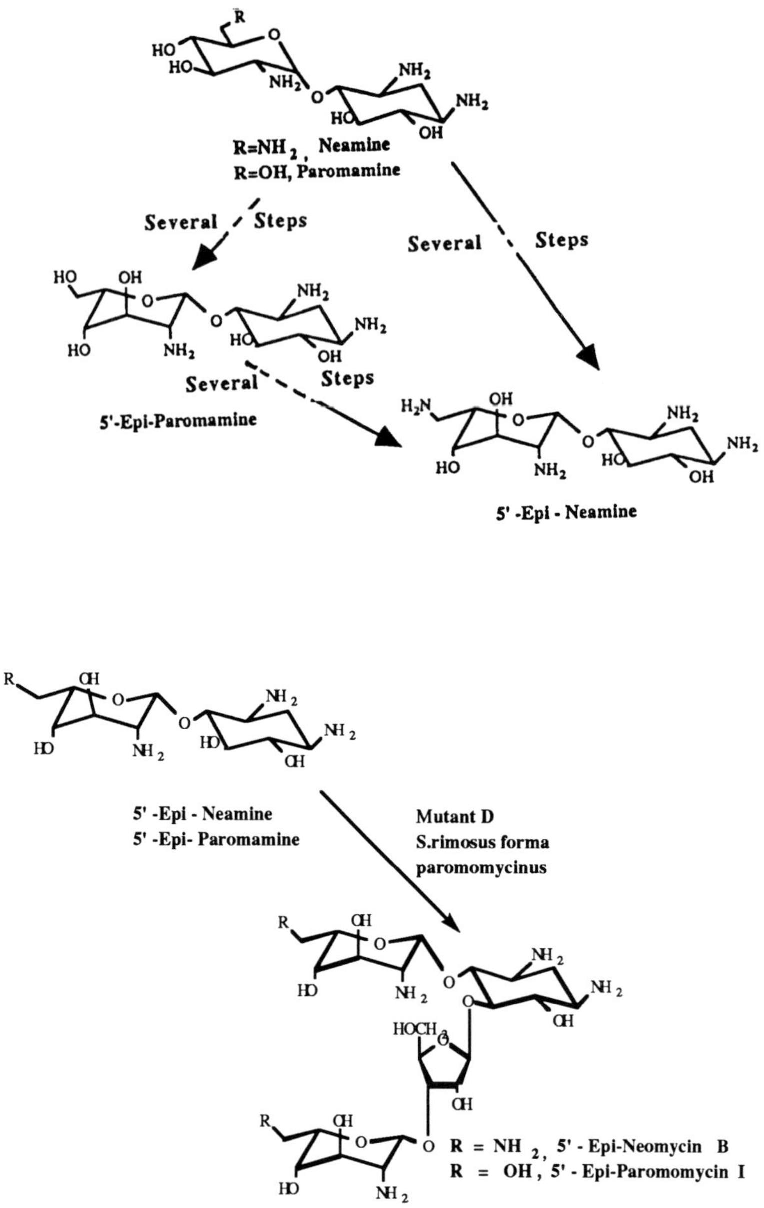

Scheme II : Targeted mutasynthons

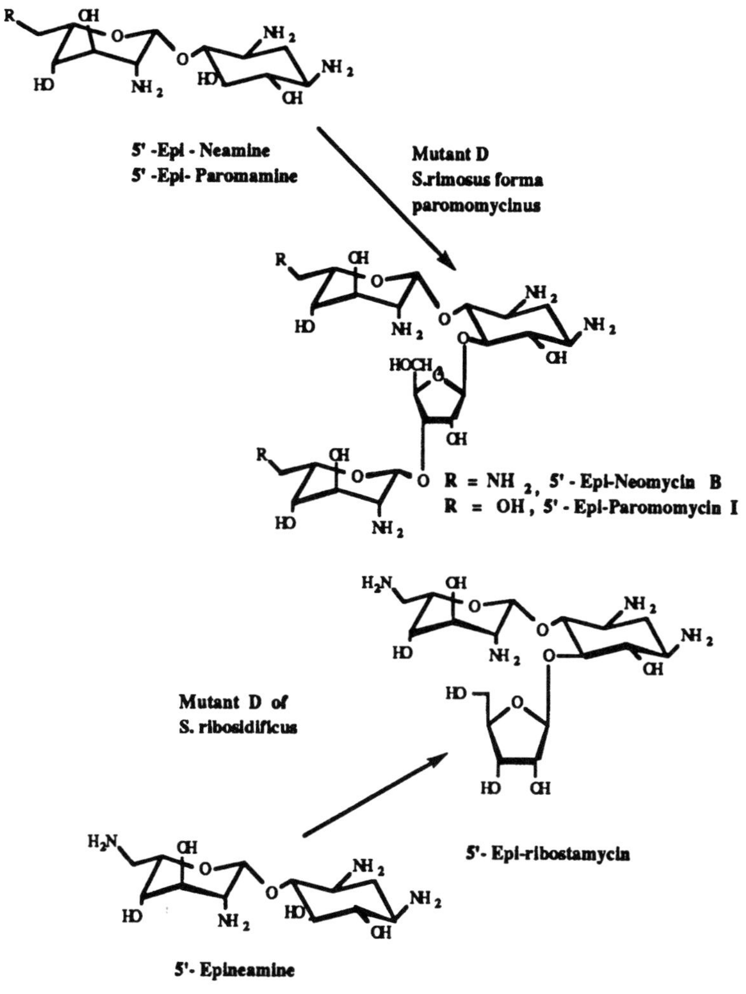

5' -Epi - Neamine
5' -Epi- Paromamine

Mutant D
S.rimosus forma paromomycinus

R = NH$_2$, 5' - Epi-Neomycin B
R = OH, 5' - Epi-Paromomycin I

Mutant D of S. ribosidificus

5'- Epi-ribostamycin

5'- Epineamine

Scheme III

antibiotic action unless one of the component parts (building block) of the usual antibiotic is added to the fermentation medium. Such mutants, which are clearly deficient in their ability to synthesize that component, have been refferred to as idiotrophs. Maintaining with the term mutasynthesis [4] for the procedure, the related compounds supplementing the medium are referred to as mutasynthons, and the new antibiotic as a mutasynthetic one (Scheme II).

Conversion of neamine to C-5'epi-neamine may produce an excellent synthon for preparing C-5'epi-neomycin B, C-5'epi paromamine, C-5'epi ribostamycin as well as

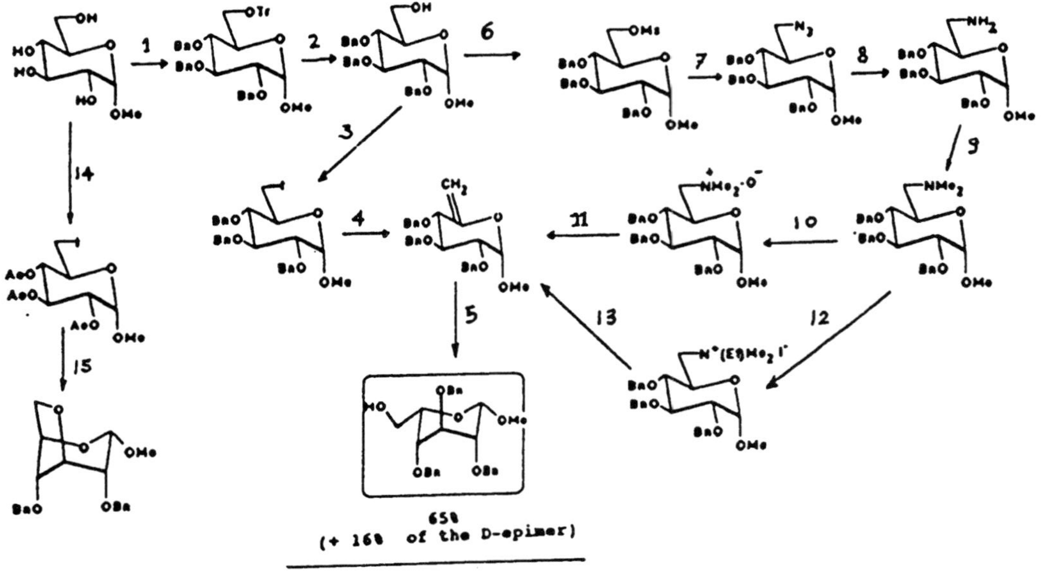

Scheme IV. Model works for C-5' epimerization of paromamine and neamine [5].
1. a) TrCl / Py b) BnBr / NaOH (powder) / Bu_4 $NHSO_4$ / C_6 H_6 , reflux.
2. HCOOH / ether 3. Pf_3 / I_2 / imidazole / toluene, 70 °C 4. NaO (powder) / Bu_4 $NHSO_4$ / C_6 H_6 , reflux 5. a) B_2 H_6 b) H_2 O_2 / OH^- 6. MaCl / Py 7. NaN_3 / DMF 8. H_2 :Pd, AcOEt 9. (HCHO)n / MeOH $NaCNBH_3$ 10. Davis' reagent 11. pyrolysis (150 °C) 12. EtI /EtOH reflux 13. Ag_2 O/ MeOH / H_2 O, 60°C 14. a) Pf_3 / I2: imidazole toluene, 70°C b) Ac_2O / Py 15. BnBr / NaOH (powder) / Bu_4NHSO_4 C_6H_6,reflux.

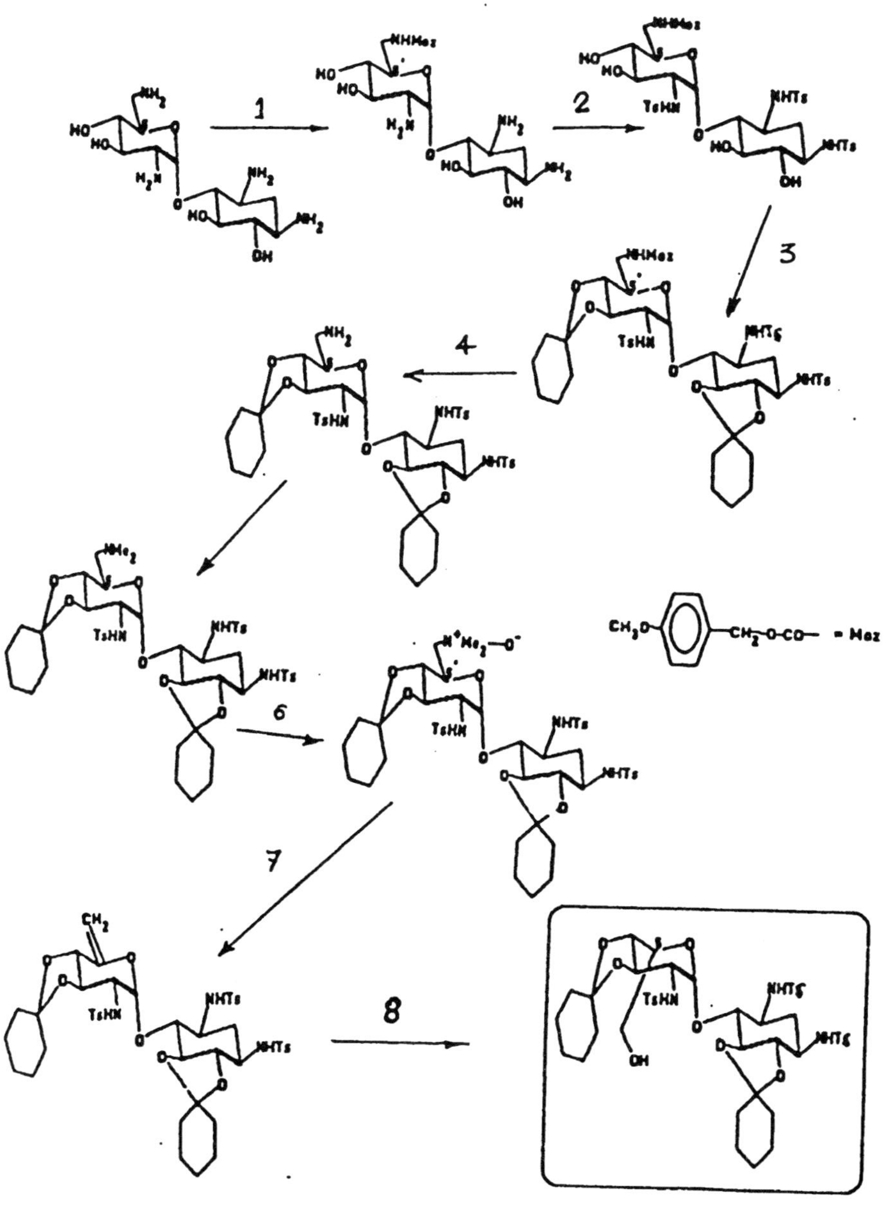

Scheme V. 1 The (C-5')-epimerization of neamine [5b] (1. Moz / dioxan / H_2O 2. TsCl / dioxan / H_2O 3. 1,1-dimethoxycyclohexane / TsOH / DMF , 70 °C 4. t-BuO⁻K⁺ / H_2O / MeOH 5. $(HCHO)_n$ / MeOH / $NaCNBH_3$ 6. Davis' reagent 7. pyrolysis , 150 °C 8. a) B_2H_6 b) H_2O_2 / OH⁻)

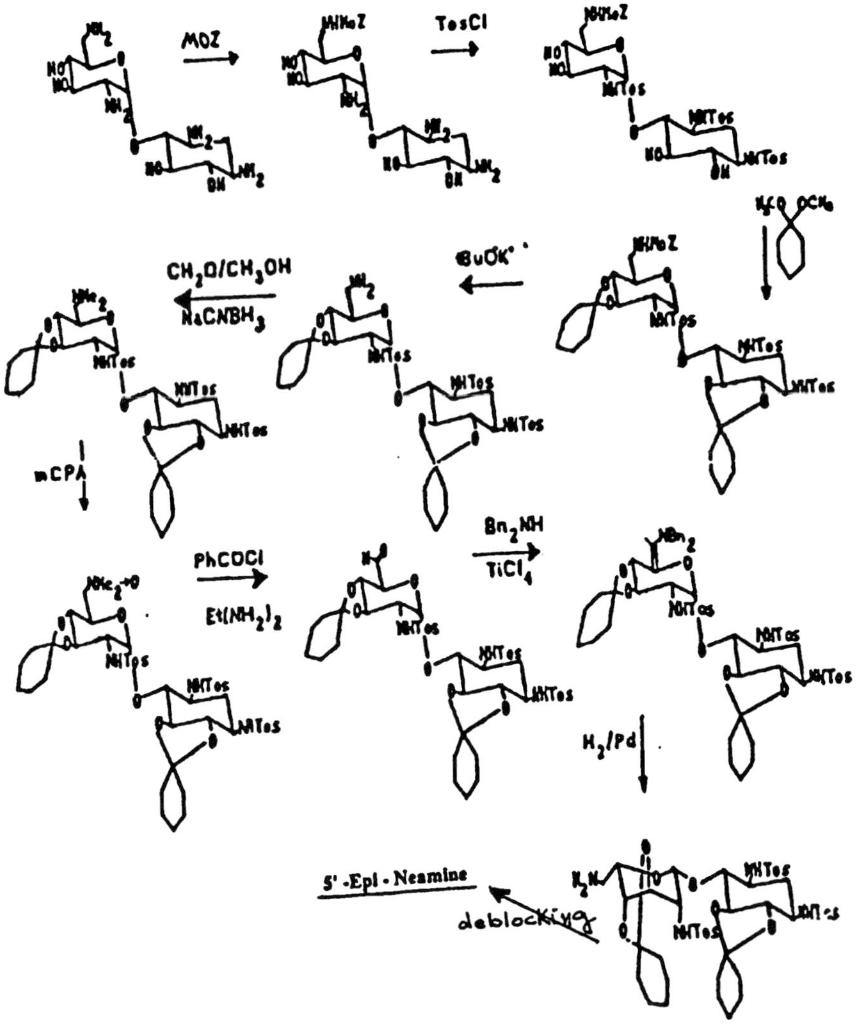

Scheme VI

their aminocyclitol antibiotics by mutational biosynthesis or even by chemical synthesis (Scheme III). Prior to the synthesis of our targeted mutasynthons we have used simple molecules as models for gaining chemical experience. Model works with monosaccharide analoques for the preparation of C-5'epi-neamine or C-5'epi-paromanime are depicted below (Scheme III). Hydroboration of methyl αD-xylo-exen-5-pyranoside gave 16% (scheme IV) of the undesired methyl αD-glycopyranoside since the hydride attack was possible from either α and , β site. On the other hand, when methyl-L-idopyranoside was converted to its 6-amino-6-deoxy-derivative an unusual epimerization at C-1 occurred which was attributed to the conformation adopted by the L-idopyranosyl structure in solution. The above two undesired complications were not observed in analogous procedures that is hydroboration of the C' 5'-6'- unsaturated key intermediate received either from paromamine or neamine, because its stereochemistry and rigidity favors the formation of C-5'epi-paromamine. However the latter compound needs further work in order to be converted to C-5' epi-neamine targeted synthon (see scheme V)[5b]. Thus conversion of the 6'-aminogroup (6'-NH_2) of neamine, a readily available inexpensive starting material, to an aldehydic group, and subsequent formation to enamine followed by hydrogenation-hydrolysis yielded the desired C-5' epi-neamine in fewer steps and much higher yield (see scheme VI) [5c].

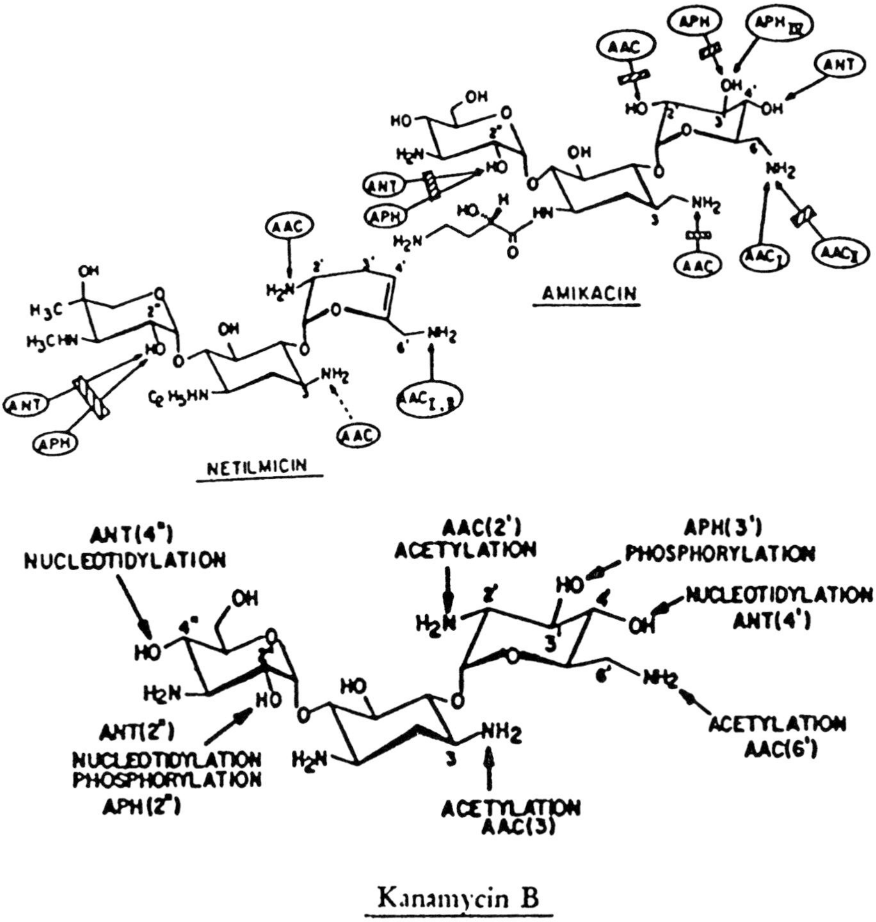

Figure 3. Aminoglycoside-modifying enzymes.

II. Chemical derivatization of the functionalities of aminoglycosidic antibiotics which are attacked by the modifying enzymes, will produce antibiotics invulnerable by these enzymes on the same functional group of these molecules. If such a derivatization could

produce at the same time a molecule with better or equal antibacterial activity than the original one, then a better semisynthetic antibiotic will be produced.
Searching, on the above approach, for a simple way to prepare antibiotics against the resistant strains of bacteria, and, on the other hand, having in mind that the clinical use of amino-glycosidic antibiotics is limited because of their oto-and nephrotoxicity, we have designed and prepared aminoacid derivatives of some commonly used aminocyclitol antibiotics i.e. Netilmicin, Kanamycin. The attack on the molecules of these antibiotics by the modifying enzymes is shown on Figure 3.
Oto-and nephrotoxicity of aminoglycosidic antibiotics are related to the binding of these amino pseudopolysaccharides (in a protonated or neutral form) to the charged phospholipids and subsequent inhibition of the lysosomal phospholipases [6]. Preliminary work in our laboratories has shown that the 6' aminoacid derivatives of neamine were active [7]. This stimulated us to extend our work to the synthesis of amino acid and peptide derivatives of kanamycin (1) and metilimicin (2). 6'-N or 1-N derivatives of 1 and 6'-N derivatives of 2 with common amino acids and peptide (L-Ala-OH, D-Ala-OH, Gly-OH, L-Leu-OH, L-Asp-OH, L-Ala-L-Ala-OH), were prepared using the active ester method for coupling. The selective formation of the amide bond at the 6'N position of aminoglycosides was performed in the presence of Cu^{2+} ions in one step [8]. The preferential complexation of vicinal hydroxy amino groups by Cu^{2+}, aimed us to synthesize the 1-N derivatives after selective protection of 6'-amino group of kanamycin A by tert-butoxycarbonyl group. The antibacterial properties are under investigation.

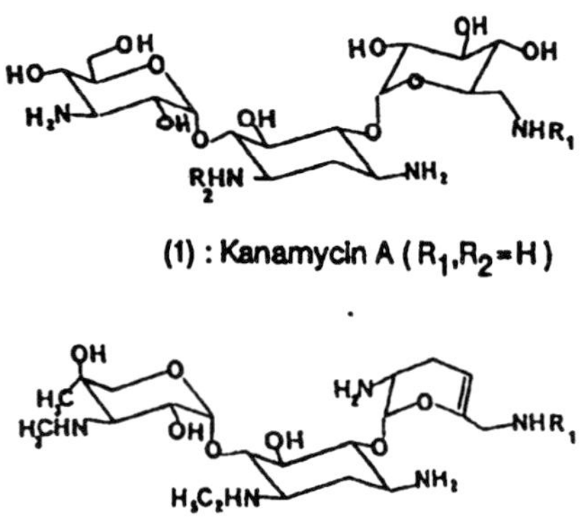

(1) : Kanamycin A (R_1,R_2=H)

1a : R_1 = Amino acid or peptide residue L-Ala-OH,D-Ala-OH,Gly-OH, L-Leu-OH,L-Asp-OH, L-Ala-L-Ala-OH)
R_2 = H

1b : R_1 = H
R_2 = Amino acid or peptide residue (L-Ala-OH,D-Ala-OH,Gly-OH, L-Leu-OH,L-Asp-OH, L-Ala-L-Ala-OH)

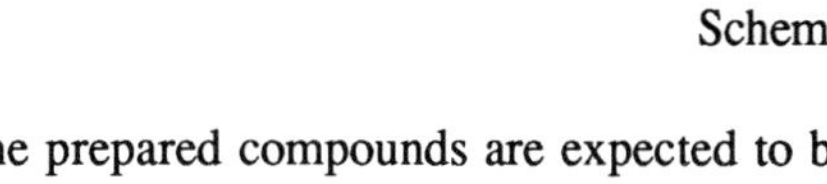

(2) : Netilmicin (R_1=H)

2a : R_1 = Amino acid or peptide residue (L-Ala-OH,D-Ala-OH,Gly-OH, L-Leu-OH,L-Asp-OH, L-Ala-L-Ala-OH)

Schema VII

The prepared compounds are expected to be not only invulnerable by the modifying enzymes, but also less toxic, since their structure and conformation [10] will prevent their interaction to the phospholipids of the proximal tubular cells in the kidney etc [11], and since the high toxicity of the aminoglycosidic antibiotics is the main concern in their clinical use. The prepared compounds, which are currently under biological evaluation, may be valuable against microbial infections.

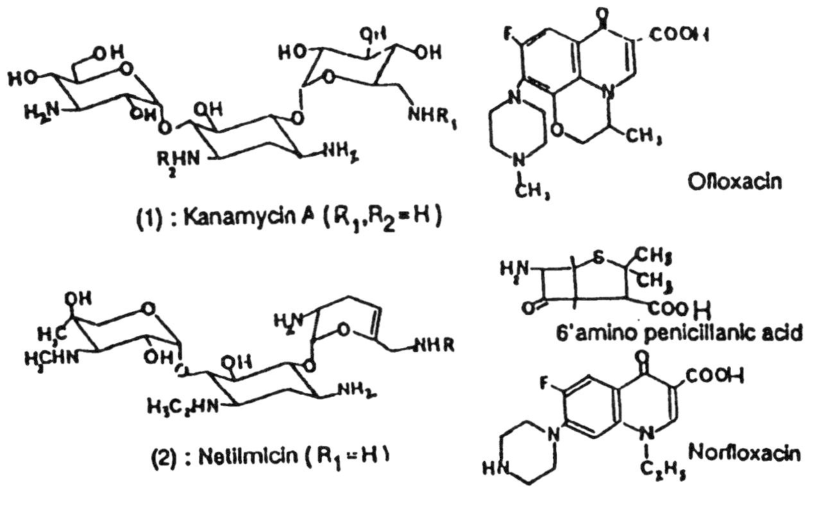

(1) : Kanamycin A (R_1, R_2 = H)

Ofloxacin

6'amino penicillanic acid

(2) : Netilmicin (R_1 = H)

Norfloxacin

1a: R_1 = Fluoroquinolones (Ofloxacin, Norfloxacin) or β-lactam (6'-amino penicillanic acid) residue
    R_2 = H
2a: R_1 = Fluoroquinolones (Ofloxacin, Norfloxacin) or β-lactam (6'-amino penicillanic acid) residue

Scheme VIII

1.2. APPROACH B

The knowledge assembled on the mechanisms of uptake and distribution of drugs in cells and the mechanisms of cellular toxicities has prompted us to synthesize aminoglycosidic conjugates with ,β-lactams and quinolones with unusual pharmakokinetic properties [12]. These compounds are expected to have combined properties (and synergism) of the two major classes of antibiotics and thus wider spectrum of antibacterial activity.

We have prepared two kinds of conjugates:

a) We have incorporated Kanamycin A (1) and netilmicin (2) as the aminoglycoside component and ofloxacin, norfloxacin or 6-amino-penicillanic acid as the quinolone or ,β-lactame component. Benzylocarbonyulamino-penicillanic acid and fluoroquinolones were coupled with 6'-amino group of aminoglycosides by an amide bond. The selective coupling was performed after treatment of aminoglycosides with DMSO solution of Cu^{2+}. The antibacterial properties are under investigation.

b) We have activated the C-6' or C-6" hydroxy or C-6" amino group of kanamycin A by converting them to activated carbonates with phosgene (or we have converted the

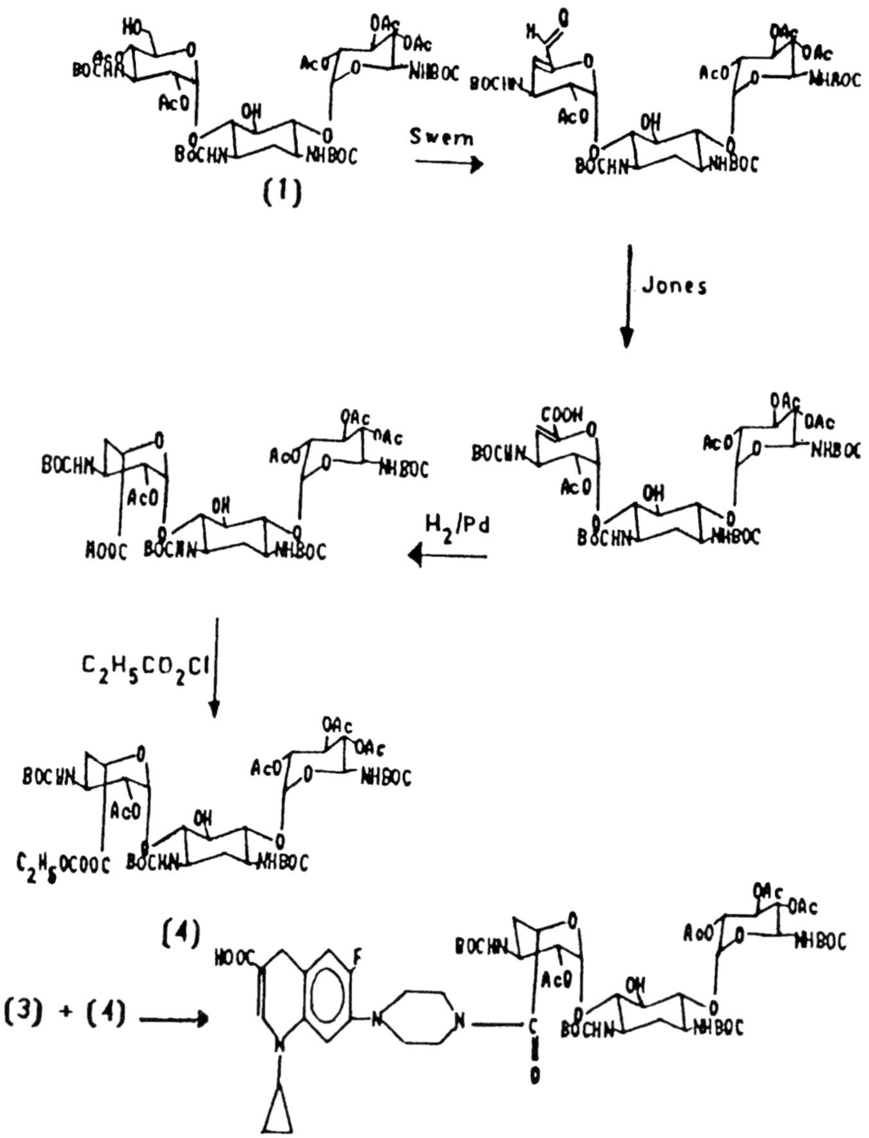

Scheme IX

same group to carboxylate functions) before coupling them with, β-lactams or quinolones as depicted below.

Acknowledgement

I thank my collaborators and graduate fellows (chemists) at the Agricultural University of Athens for making this presentation possible, namely : Ioannis Grapsas, Ioannis Ierapetritis, Violetta Constantinou-Kokotou, Stamatia Kotretsou.
All of us thank the National Drug Organization (EO$^{\circ}$) for the main financial support and its president till 1989 (Gerassimos Kavadias) for encouraging mission oriented research. We also thank the Ministry of Industry Section of Research and Technological Development for a supplementary grant. We are grateful to the following Pharmaceutical Companies for donating chemicals or antibiotics:
GAP Pharmaceutical Co Athens Greece (chemicals)
Schering Pharmaceutical Corporatim Bloomfield N.J.(netilmicin)
UpJohn Pharmaceutical Co Kalamazoo Michigan (neomycin)
HELP Pharmaceutical Co, Ioannina-Greece (kanamycin)
Finally we thank Professor Kenneth L.Rinehart Jr of the University of Illinois for inspiring the initiation of our work on aminocyclitol antibiotics.

References

1. a) P.C. Tai , B.J. Wallace, B.D. Davies, *Proc. Natl. Acad. Sc.* **75**, 275, U.S.A. (1978).
    b) J. Davies *Aminocyclitol Antibiotics* Chapter **8**, K.L.Jr Rinehart , T. Suami , Editors, ACS Symposium Series No 125, (1980).
2. a) G. Beskid , H.A. Albrecht , V. Fallat , D.D. Keith , E.R. Lipschitz , D.H.Mc Garry, P. Rossman and J. Siebelist *30th ICAAC* **403**, (1990).
    b) J.L. Roberts, G. Bescid , J. Borgese, J.G. Christensen , N.H. Georgopapadakou, D.D. Keith , D.L. Pruess , C.C. Wei , and R. Yang, *30th ICAAC* , **406**, (1990).
    c) H.A. Albrecht , G. Beskid, J.G. Christenson , N.H. Georgopapadakou, D.D. Keith, F.M. Konzelmann, D.L. Pruess, P.L. Rossman and C.C. Wei, *J. Med. Chem.* **34**, 2857-2864, (1991).
3. M.P. Mingeot-Leclercq, A. VanSchepdael, R. Brasseur, R. Busson, H.J. Vanderhoeghe, J.P. Claes and P.M. Tulkens, *J. Med Chem.,* **34**, 1476-1482; ibid 1483-1492 and references therein, (1991).
4. K.L.Jr Rinehart , *Aminocyclitols Antibiotics* Chapter 19, K.L.Jr. Rinehart, T. Suami, Editors ACS Symposium Series No 125, 1980, and references therein ; See also K.L.Jr Rinehart *Pure and Appl.Chem.,***49**, 1361-1384, (1977).
5. a) I. Grapsas, *Ph.Thesis Agricultural University of Athens,* General Dept. of Physical Science, Athens, Greece, (1991).
    b) I. Grapsasand, I. Ierapetritis, Internal Report Agricultural University of Athens, and Ref. 5a and 5e, (1991).
    c) I. Ierapetritis, *Ph. Thesis Agricultural University of Athens* submitted, General Dept. of Physical Science, Athens,Greece, (1992).
6. M.P. Mingeot-Leclercq, J. Piret , R. Brasseur , and P.M. Tulkens *Biochem. Pharmacology,* **40**, 489-497, (1990) ; *ibid* **40**, 499-506, - see also ref. 3 and references therein, (1990).

7. M.P. Georgiadisand, V.N. Constantinou-Kokotou, G.C. Kokotos, *J. Carbohydrate Chem.* 10, 739, (1991).

8. a) T.L. Nagabhushan, A.B. Cooper, W.N. Twrner, H. Tsai, S.Mc Combie, A.K. Maltams, D. Rane, J.J. Wright, T. Recehert, D.L. Boxler, J.V. Weinstein, *J. Am. Chem. Soc.* **5254**, (1978).

b) S. Hanessian, G. Patil, *Tetrahedron Let.*, **1035**, (1978).

9. a) St. Kotretsou *Ph. Thesis Agricultural University of Athens* submitted, General Dept. of Physical Science, Athens,Greece, (1992).

b) V. Constantinou-Kokotou, Personal Communication *See also Ref.7.*

10. *See Ref.3*

11. a) P.S. Lietman <u>Principles and Practise of Infections Deseases</u>, G.L. Mandell, R.G. Douglas, J.E. Bennet (Eds) : John Willey & Sons : , pp 192-206, New York, (1985).

b) O. Alftan, O.V. Renkonen, A.Sivoren, *Acta Pathol. Microbiol. Scand.*, **241 B**, Suppl.92, (1973).

c) J. Fabre, M. Rudhardt, P. Blanchard, C. Regamey, P. Chauvin, *Kidney Int.*, **10**, 444, (1976).

d) R.A. Parker, W.H.Bennet, G.A. Porter, <u>Aminoglycosides : Microbiology Clinical Use and Toxicology</u>, A. Whelton, H.C. Neu, Eds Marcel Dekker Inc., pp 235-267, New York; Appel G.B. *ibid* pp 269-282, (1982).

The Mutagenic Heterocyclic Amines in Cooked Foods

S. GRIVAS
Department of Chemistry, Swedish University of Agricultural Sciences,
P.O. Box 7015, S-750 07 Uppsala, Sweden

1. Introduction

A careful and objective review of the current data base shows that most of the main cancers in main throughout the world do not stem from chemical contaminants in the environment, but relate to personal life-style. Personal life-style covers such habits as drinking coffee, smoking cigarettes, drinking alcoholic beverages, consuming a diet with a high fat and/or protein content, etc. (Sugimura 1988; Weisberger 1985). Also, many studies indicate a rise in colon cancer incidence with increasing westernization of dietary habits, including increased consumption of meat.

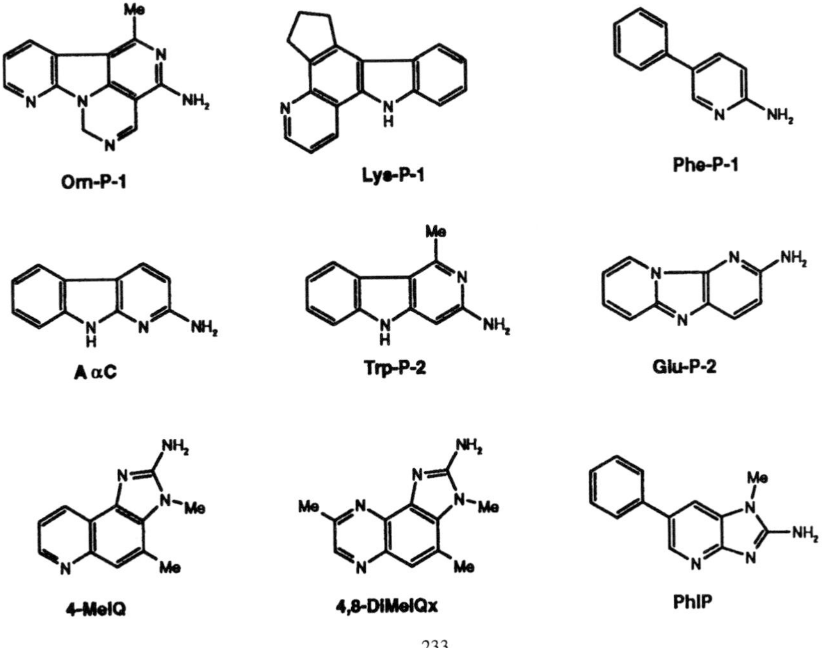

N. Russo et al. (eds.), Properties and Chemistry of Biomolecular Systems, 233–242.
© 1994 *Kluwer Academic Publishers. Printed in the Netherlands.*

The pattern of cancer incidence in migrants often change from those prevalent in their native countries to approximate those of the host countries (Vuolo and Schussler, 1985). Thus, the genesis of mutagenic activity during cooking of food is now well documented and interesting findings continue to be made. More than 25 years have elapsed since traces of strong carcinogens like benzo[a]pyrene and other polyaromatic hydrocarbons (PAH) were identified in grilled meats (Lijinsky and Shubik, 1964).

Interest in the heterocyclic amine carcinogens (HA) began in 1977-78 when Sugimura et al. followed by Commoner et al. reported that the mutagenicity observed in fried or grilled fish and meat was higher than could be accounted for by the non-polar PAH (Sugimura et al. 1977; Commoner et al. 1978).

Mutagenicity was observed after pyrolysis (500-700 °C) of proteins or their constituents and by cooking meat under usual household conditions. For instance, Orn-P-1, Lys-P-1, Phe-P-1, Trp-P-2 and Glu-P-2 were identified as pyrolysis products of ornithine, lysine, phenylalanine, tryptophan and glutamic acid. AαC was isolated from the pyrolysis products of soya bean globulin. 4-MeIQ, 4,8-DiMeIQx and PhIP belong to that group of HAs isolated from fried or grilled meat and fish. The amounts of these food mutagens range from less than 1 to ca. ng/g of meat (Felton and Knize, 1991; Övervik and Gustafsson, 1990). More than twenty HAs have been identified, new ones are constantly found and there are numerous unknown (e.g., Kurosaka, et al. 1992).

2. Formation of the HAs

It has been found that boiling, stewing, roasting or microwaving animal protein-rich foods results to much lower or no mutagenic activity, in marked contrast to grilling and frying. However, proteinaceous foods like beans and tofu produce almost no mutagenicity. Indeed, creatine and creatinine have been found to partecipate in mutagen forming reactions. Simple model reaction systems have been devised in order to mimic the reactions that might occur while broiling meat products. For instance, creatinine, fructose and alanine was heated in diethylene glycol containing 14% water for 2 h at ca. 128 °C. The food mutagen 4,8-DiMeIQx was one of the products formed (Grivas et al., 1985).

This result reinforced the hypothesis that Maillard reaction products from sugars and amino acids (Ledl and Schleicher, 1990) together with creatinine might be the precursors of these potent mutagens as shown (Grivas et al., 1986).
For references on the binding of HAs to fibres and the effect of factors like the presence of inhibitors, water or fat content, and cooking time and temperature on the formation of HAs see Lee et al., 1992; Vikse et al., 1992; Wakabayashi et al., 1992; Felton and Knize, 1991; Barrington et al., 1990; Övervik and Gustaffson 1990, and references therein.

3. Detection, isolation and identification of the HAs

Suitable analysis methods for detection and isolation of the HAs formed during cooking are essential for their identification and for estimation of their risk to humans. The methods employed should permit reproducible quantification of the HAs if the real risk to humans is to be assessed. The quantities of mutagens isolated, even from multigram samples of cooked foods, are generally very small. Accordingly the detection and efficient extraction of these compounds from the complex mixtures are crucial analytical steps if the problem of the needle in the haystack is to be solved and their structures determined. The speed and sensitivity of the Ames' test (Maron and Ames, 1983) has allowed the use of the HAs' high mutagenicity to follow their extraction from food matrices. The isolation of the mutagens has usually been accomplished by their extraction (acetone or methanol) from the homogenized cooked food, acid-base partition, column chromatography (Amberlite XAD-2, Sephadex LH-20, silica gel, cation exchange etc.), C_{18} reverse-phase Sep-Pak treatment, "blue cotton" treatment (trisulfocopper phtalocyanine bound to cotton) (Hayatsu, 1992) and repetitive HPLC steps on reverse- and normal-phase columns (Grivas et al., 1986). A recently developed method involves solid-phase extraction on diatomaceous earth followed by purification on propylsulfonic acid silica gel (Gross and Grüter, 1992). Chromatographic coelution and comparison of retention times with synthetic internal standards of labelled HAs is the method of choice for the identification of isolated HAs. Gas chromatography-mass spectrometry has been employed for detection and quantification of some HAs (Murray et al., 1988; Turesky et al., 1988c). However, derivatization is often necessary for the GC separation. HPLC analysis of the amines is thus more popular but of lower sensitivity. The method of detection most commonly used is UV absorption, fluorescence and mass spectrometry (Yamaizumi et al., 1986). Electrochemical detection of the HAs has also been used since it offers good selectivity obtainable by adjusting the operating redox potential to the compounds of interest, and low detection limits (Grivas and Nyhammar, 1985). Accelerator mass spectrometry has recently been employed with success (Felton et al., 1990).

4. Synthesis of the HAs

The amounts of the toxic amines formed during cooking or isolated from model reaction systems are very small in order to allow unequivocal structure determination. Therefore, efficient and unambiguous synthetic methods for preparing the compounds in labelled and isomerically pure form are required for comparison purposes, internal standards, in metabolism studies and structure-biological activity relationship studies. The compounds are also required in multigram quantities for long-term animal feeding studies. Kasai et al. (1981) were the first to isolate 8-MeIQX from fried beef. Synthetic material was required in order to confirm its structure; condensation of methylglyoxal (MeCOCHO) with triamine **1** afforded the two isomeric quinoxalines **2** and **3**. Removal of **2** by HPCL followed by tosylation and nitration of **3** gave quinoxaline **4**. Detosylation, methylation and reduction of **4** afforded **5** which on treatment with cyanogen bromide yielded 8-MeIQx.

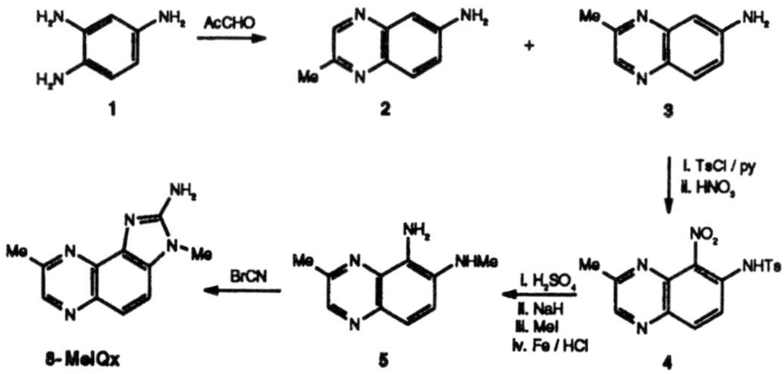

Since then a number of different methodologies have been reported (Grivas and Olsson, 1985; Grivas, 1986; Knapp et al., 1989; Bierer et al., 1992). However, the shortest and most efficient published methodology to obtain 8-MeIQx is via the readily available benzoselenadiazole **7** with allows the regiocontrolled conversion into suitable quinoxalines (e.g. **11**; Grivas et al., 1992; Grivas and Tioan, 1992).

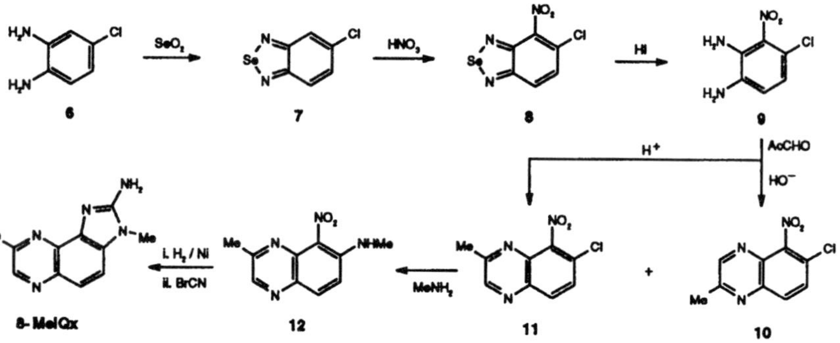

Thus, nineteen compounds in the IQx series could be easily synthesized via suitably substituted benzoselenadiazoles (Grivas et al., 1993; Tian et al., 1993; Tian and Grivas, 1992) and tested for mutagenicity (Vikse et al., 1993; Hatch et al., 1992). A number of closely related mutagens have been synthesized. For instance, the presence of the 2-amino and 3-methyl group in 8-MeIQx were shown to influence its mutagenic activity (Grivas and Jagerstad, 1984; Jägerstad and Grivas, 1985). The 2-nitro analogue of 8-MeIQx has been synthesized by diazotization and subsequent nitrite treatment (Grivas, 1988) and was used as starting material for the synthesis of the mutagenic 2-hydroxyamino 8-MeIQx which was involved in ^{32}P-postlabelling studies by Yamashita et al. (1988). Recently, the three isomers of IQ (a potent hepatocarcinogen in nonhuman primates) shown below were synthesized (Ronne and Grivas, 1993) and found to have much lower mutagenicity (Vikse et al., submitted).

5. Metabolism of the HAs and formation of adducts

The HAs are indirect bacterial mutagens and therefore require metabolic activation in order to exhibit mutagenic activity. They also induce DNA damage in mammalian cells. It has been demonstrated in a number of species including rodents, monkey and man that liver is the most active tissue in metabolising the HAs to mutagenic agents. The main initial activation steps are ring or side chain hydroxylation and oxidation of the exocyclic amino group to the hydroxylamine by cytochrome P-450. The major part of the hydroxylated forms of the HAs is conjugated to glucuronic acid or sulfate and thus excreted as detoxified metabolites (urine, bile faeces). It is only a minor part of the N-OH derivatives of the HAs that are further activated and thus exert their genotoxicity (Aeschbacher and Turesky, 1991).

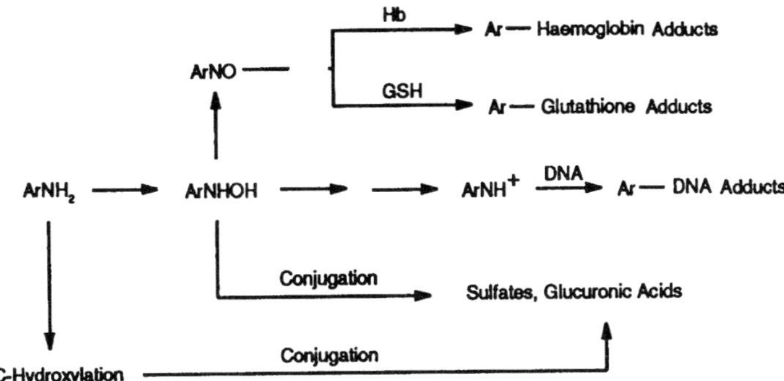

A number of sulfates, glucuronides and adducts with DNA deoxyguanosine, glutathione and haemoglobin have been isolated and identified (Alexander et al., 1991; Frandsen et al., 1992; Lin et al., 1992; Nagaoka et al., 1992; Umemoto et al;, 1988a,b and 1992, and references therein). Some of them are shown overleaf.

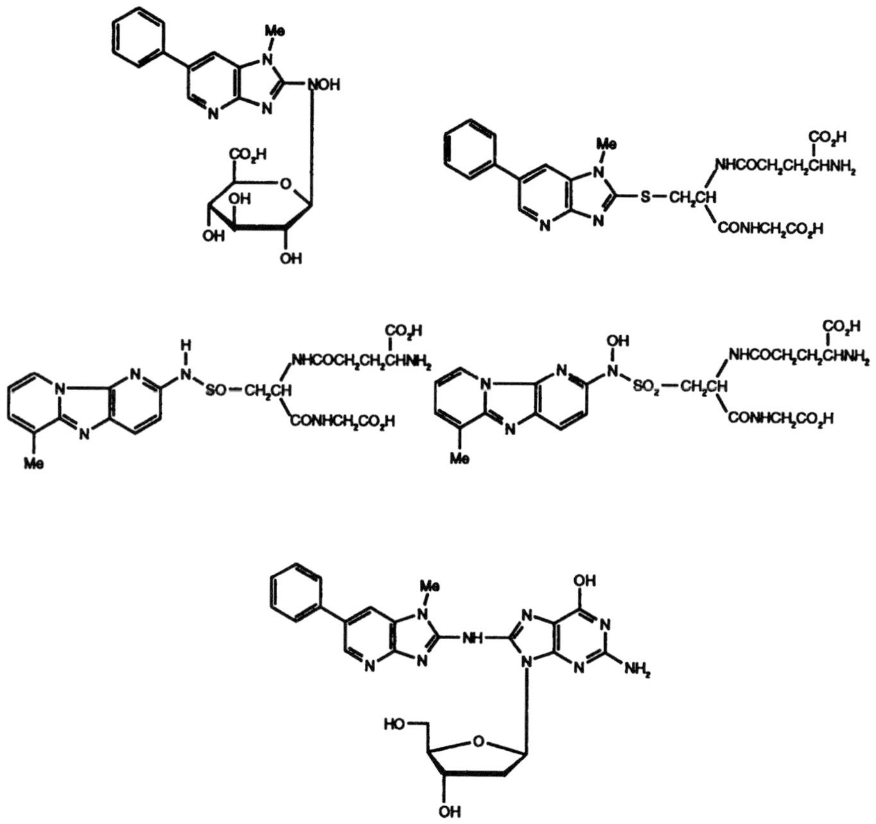

6. Animal studies, epidemiology and risk evaluation

The heterocyclic amines have been shown to be potent liver carcinogens and potential human carcinogens. Oral administration to mice and rats have caused tumours in the liver, stomach, intestines, mammary gland and Zymbal gland (Wakabayashi et al; 1992). Experiments by Adamson et al. at the National Institutes of Health, Bethesda, with nonhuman primates (macaque monkeys) have shown that IQ is a potent liver carcinogen and thus a potential carcinogen for humans. Light and microscopy studies of the myocardium of IQ treated macaques have revealed myocyte necrosis, hypertrophy and other abnormalities. Similar studies with 8-MeIQx and PhIP are in progress. The results from animal experiments cannot be extrapolated to humans easily. There are certainly differences in the absorption and metabolism of the HAs between animals and humans. The diet we consume contains more than twenty HAs and the combined effect could be very different than that of the individual amine. Further, the doses given to animals has been much greater than the average content of our diet. Moreover, our diet contains anticarcinogens, fibres and other inhibitors. See Steineck et al., 1990, Dooley et al., 1992, and references therein. Studies on the relation between diet or its components and diseases like cancer gains attention constantly. Three population based case control studies in Sweden have shown a connection between broiled foods and pancreatic, colorectal and urothelial cancers (Gerhardson de Verdier et al., 1991). These might be related to the HAs formed during cooking. For further details see the reference list compiled by Hatch et al., (1988) and the papers by de Meester, 1989; Övervik and Gustafsson, 1990; Felton and Knize, 1991, and Wakabayashi et al., 1992.

Acknowledgements

Parts of the work presented were financed by the Swedish Council for Forestry and Agricultural Research, and the Foundation for Promotion of Cancer Research, Japan.

References

H.-U. Aeschabacher, and R.J. Turesky, "Mammalian cell mutagenicity and metabolism of heterocyclic aromatic amines" *Mutation. Res.* **259**, 235-250, (1991).

J. Alexander, H. Wallin, O.J. Rossland, K. Solberg, J. Holme, G. Becher, R. Andersson & S. Grivas, "Rat liver forms a glutathione conjugate and a semistable transportable glucuronide conjugate of 2-hydroxamino-1-methyl-6-phenylimidazo [4,5-b]pyridine (2-hydroxamino-PhIP)", *Carcinogenesis* **12**, 2239-2245, (1991).

P.J. Barrington, R.S.U. Baker, A.S. Truswell, A.M. Bonin, A.J. Ryan and A.P. Paulin, "Mutagenic of basic fractions derived from lamb and beef cooked by common household methods", *Food Chem. Toxic.* **28**, 141-146, (1990).

D.E. Bierer, J.F. O'Connell, J.R. Parquette, C.M. Thompson and H. Rapoport, "Regiospesific synthesis of the aminoimidazoquinoxaline (IQX) mutagens from cooked foods", *J. Org. Chem.* **57**, 1390-1405, (1992).

B. Commoner, A.J. Vihayathil, P. Dolara, S. Nair, P. Madyastha and G.C. Cuca, "Formation of mutagens in beef and beef extract during cooking" *Science* **201**, 913-916, (1978).

C. de Meester, "Bacterial mutagenicity of heterocyclic amines found in heat-processes food" *Mutation Res.* **221**, 235-262, (1989).

K.L. Dooley, L.S. von Tungeln, T. Bucci, P.P. Fu and F.F. Kadlubar, "Comparative carcinogenicity of 4-aminobiphenyl and the food pyrolysates, Glu-P-1, IQ, PhIP, and 8-MeIQx in the neonatal B6C3F1 male mice", *Cancer Lett.* **62**, 205-209, (1992).

J.S. Felton, K.W. Turteltaub, J.S. Vogel, R. Balhorn, B.L. Gledhill, J.R. Southon, M.V. Caffee, R.C. Finkel, D.E. Nelson, I.D. Proctor and J.C. Davis, "Accelerator mass spectrometry in the biomedical sciences: applications in low exposure biomedical and environmental dosimetry" *Nuclear Instr. Meth. Physics Res.* **B52**, 517-523, (1990).

J.S. Felton, and M.G. Knize, "Occurrence, identification, and bacterial mutagenicity of heterocyclic amines in cooked food" *Mutation Res.* **259**, 205-207, (1991).

H. Frandsen, S. Grivas, R. Andersson, L. Dragsted & J. Larsen, "Reaction of the N^2-acetoxy derivative of 2-amino-1-methyl-6-phenylimidazo[4,5-b]pyridine (PhIP) with 2'-deoxyguanosine and DNA. Synthesis and identification of N^2-(2'-deoxyguanosin-8-yl)-PhIP" *Carcinogenesis* **13**, 629-635, (1992).

M.Gerhardsson de Verdier, U. Hagman, R. Peters, G. Steineck and E. Overvik, "Meat, cooking methods and colorectal cancer: a case referent study in Stockholm", *Int. J. Cancer* **49**, 520-525, (1991).

S. Grivas & M. Jägerstad, "Mutagenicity of some synthetic quinolines and quinoxalines related to IQ, MeIQ or 8-MeIQx in Ames test." *Mutation Res.* **137**, 29-32, (1984).

S. Grivas and K. Olsson, "An improved synthesis of 3,8-dimethyl-3H-imidazo [4,5-f] quinoxaline-2-amine ("MeIQx") and its 2-^{14}C-labelled analogue" *Acta Chem. Scand., Ser. B* **39**, 31-34, (1985).

S. Grivas, & T. Nyhammar, "Analysis of mutagenic imidazo[4,5-f]quinolines and -quinoxalines (IQ compounds) Comparison of electro-chemical and ultraviolet detection" *Mutation Res.* **142**, 5-8, (1985).

S. Grivas, T. Nyhammar, K. Olsson & M. Jägerstad, "Formation of a new mutagenic DiMeIQx compound in a model system by heating creatinine, alanine and fructose" *Mutation Res.* **151**, 177-183, (1985).

S. Grivas, "Efficient synthesis of mutagenic imidazo[4,5-f]quin-oxalin-2-amines via readily accessible 2,1,3-benzoselenadiazoles", *Acta Chem. Scand., Ser. B.* **40**, 404-406, (1986).

S. Grivas, T. Nyhammar, K. Olsson and M. Jägerstad, "Isolation and identification of the food mutagens IQ and MeIQx from a heated model system of creatinine, glycine and fructose" *Food Chem.* **20**, 127-136, (1986).

S. Grivas, "Synthesis of 3,8-dimethyl-2-nitro-3H-imidazo[4,5-f]quinoxaline, the 2-nitro analogue of the food carcinogen MeIQx" *J. Chem. Res. (S)* **84**, (1988).

S. Grivas & W. Tian, "Convenient synthesis of 4-methoxy-3-nitro-1,2-benzenediamine and its conversion into 6-methoxy-5-nitro-quinoxalines" *Acta Chem. Scand.* **46**, 1109-1113, (1992).

S. Grivas, W. Tian and R. Andersson "A method for the regioselective synthesis of 2-methyl-5-nitro- and 2-methyl-8-nitro-quinoxalines, and the application of proton-coupled ^{13}C NMR to assign their structure" *J. Chem. Res. (S)* 328-329, *(M)* 2701-2712, (1992).

S. Grivas W. Tian, E. Ronne, S. Lindstrom and K. Olsson, "Synthesis of mutagenic methyl- and phenyl-substituted 2-amino-3H-imidazo[4,5-f]quinoxalines via 2,1,3-benzoselenadiazoles", *Acta Chem. Scand.*, in press.

G.A. Gross, and A. Grüter, "Quantitation of mutagenic/carcinogenic heterocyclic aromatic amines in food products" *J. Chrom.* **592**, 271-278, (1992).

F. Hatch, M. Knize, S. Healy T. Slezak and J. Felton, "Cooked-food mutagen reference list and index" *Environ. Mol. Mutagen.* **12**, *(Suppl. 14)*, 1-85, (1988).

F.T. Hatch, M.G. Knize, D.H. Moore II and J.S. Felton, "Quantitative correlation of mutagenic and carcinogenic potencies for heterocyclic amines from cooked foods and additional aromatic amines" *Mutation Res.* **271**, 269-287, (1992).

H. Hayatsu, "Cellulose bearing covalently linked copper phthalocyanine trisulfonate as an adsorbent selective for polycyclic compounds and its use in studies of environmental mutagens and carcinogens" *J. Chrom.* **497**, 37-56, (1992).

M. Jägerstad, M. & S. Grivas, "The synthesis and mutagenicity of the 3-ethyl analogues of the potent mutagens IQ, MeIQ, MeIQx and its 3,7-dimethyl isomer, *Mutation Res.* **144**, 131-136, (1985).

H. Kasai, T. Shiomi, T. Sugimura and S. Nishimura, "Synthesis of 2-amino-3,8-dimethylimidazo[4,5-f]quinoxaline (MeIQx), a potent mutagen isolated from fried beef" *Chem. Lett.*, 675-678, (1981).

S. Knapp, J. Ziv and J.D. Rosen, "Synthesis of the food mutagens MeIQx and 4,8-DiMeIQx by copper(I) promoted quinoxaline formation" *Tetrahedron* **45**, 1293-1298, (1989).

R. Kurosaka, K. Wakabayashi, H. Ushiyama, H. Nukaya, N. Arakawa, T. Sugimura, and M. Nagao, "Detection of 2-amino-1-methyl-6-(4-hydroxyphenyl) imidazo[4,5-b]pyridine in broiled beef" *Jpn. J. Cancer Res.* **83**, 919-922, (1992).

F. Ledl and E. Schleicher, "New aspects of the Maillard reaction in foods and in the human body" *Angew. Chem. Int. Ed. Engl.* **29**, 565-594, (1990).

H. Lee, C-Y. Jiaan and S-J. Tsai, "Flavone inhibits mutagen formation during heating in a glycine/creatine/glucose model system" *Food Chem.* **45**, 235-238, (1992).

W. Lijinsky, and P. Shubik, "Benzo[a]pyrene and other polynuclear hydrocarbons in charcoal-broiled meat" *Science* **145**, 53-55, (1964).

D. Lin, K.R. Kaderlik, R.J. Turesky, D.W. Miller, J.O. Lay and F.F. Kadlubar, "Identification of N-(deoxyguanosin-8-yl) -2-amino-1-methyl-6-phenylimidazo(4,5-b]pyridine as the major adduct formed by the food-borne carcinogen, 2-amino-1-methyl-6-phenylimidazo[4,5-b]pyridine, with DNA" *Chem. Res. Toxicol.* **5**, 691-697, (1992).

D. M. Maron and B.N. Ames, "Revised methods for the Salmonella mutagenicity test" *Mutation Res.* **113**, 173-215, (1983).

S. Murray, N.J. Gooderham, A.R. Boobis and D.S. Davies, "Measurement of MeIQx and DiMeIQx in fried beef by capillary column gas chromatography electron capture negative ion chemical ionisation mass spectrometry" *Carcinogenesis* **9**, 321-325, (1988).

H. Nagaoka, K. Wakabayashi S.-B. Kim, I.-S. Kim, Y. Tanaka, M. Ochiai, A. Tada, H. Nukaya, T. Sugimura and M. Nagao, "Adduct formation at C-8 of guanine on in vitro reaction of the ultimate form of 2-amino-1-methyl-6-phenylimidazo[4,5-b]pyridine with 2'-deoxyguanosine and its phosphate esters"*JPN. J. Cancer Res.* **83**, 1025-1029, (1992).

E. Övervik, and J.A. Gustafsson, "Cooked-food mutagens: current knowledge of formation and biological significance", *Mutagenesis* **5**, 437-446, (1990).

E. Ronne and S. Grivas, "The synthesis of three isomers of the food mutagen IQ", *Heterocycles* **36**, 101-105, (1993).

G. Steineck, U. Hagman, M. Gerhardsson and S.E. Norell, "Vitamin A supplements, fried foods, fat and urothelial cancer. Acase-referent study in Stockholm in 1985-87", *Int. J. Cancer* **45**, 1006-1011, (1990).

T. Sugimura, T. Kawachi, M. Nagao,, T. Yahagi, Y. Seino, T. Okamoto, K. Shudo, T. Kosuge, K. Tsujik, K. Wakabayashi, Y. Jitaka and A. Itai, "Mutagenic principle(s) in tryptophan and phenylalanine pyrolysis products" *Proc. Japan. Acad.* **53**, 58-61, (1977).

T. Sugimura, "New environmental carcinogens in daily life" *Trends Pharmacol. Sci.* **9**, 205-209, (1988).

W. Tian, & S. Grivas, "Synthesis of 6-halo-5-nitroquinoxalines" *J. Heterocycl. Chem.* **29**, 1305-1308, (1992).

W. Tian, S. Grivas & K. Olsson, "Nitration of 5-fluoro-2,1,3-benzoselenadiazoles, and the synthesis of 4-fluoro-3-nitro-, 4-fluoro-6-nitro-, 5-fluoro-3-nitro-1,2-benzenediamines and 3,4-diamino-2-nitrophenols by subsequent deselenation, *J. Chem. Soc.*, Perkin Trans. 1, 257-261, (1993).

R.J. Tureski, H. Bur, T. Huynh-Ba, H.U. Aeschbacher and H. Milon, "Analysis of mutagenic heterocyclic amines in cooked beef products by high-performance liquid chromatography in combination with mass spectrometry" *Food Chem. Toxic.* **26**, 501-509, (1988)c.

A. Umemoto, Y. Monden, M. Tsuda, S. Grivas & T. Sugimura, "Oxidation of the 2-hydroxyamino derivative of 2-amino-6-methyldipyrido[1, 2-a: 3',2' -d]imidazole (Glu-P-1) to its 2-nitroso form, an ultimate form reacting with hemoglobin thiol groups", *Biochem. Biophys. Res. Commun.* **151**, 1326-1331, (1988)a.

A. Umemoto, S. Grivas, Z. Yamaizumi, S. Stato & T. Sugimura, "Non-enzymatic glutathione conjugation of 2-nitroso-6-methyldipyrido[1, 2-a: 3',2' -d]imidazole (NO-Glu-P-1) *in vitro*: N-hydroxy-sulfonamide, a new binding form of arylnitroso compounds and thiols", *Chem. Biol. Interact.* **68**, 57-69, (1988)b.

A. Umemoto, Y. Monden, S. Grivas, K. Yamashita & T. Sugimura, "Determination of human exposure to the dietary carcinogen 3-amino-1,4-dimethyl-5H-pyrido[4,3-b]indole (Trp-P-1) from hemoglobin adduct: the relationship to DNA adducts", *Carcinogenesis* **13**, 1025-1030, (1992).

R. Vikse, B. Bålsrud Mjelva and L. Klungsøyr, "Reversible binding of the cooked mutagen MeIQx to lignin-enriched preparations from wheat bran" *Food Chem. Toxic.* **30**, 239-246, (1992).

R. Vikse, A. Knapstad, L. Klungsøyr and S. Grivas, "Mutagenic activity of the methyl and phenyl derivatives of the food mutagen 2-amino-3-methylimidazo[4,5-f]quinoxaline (IQx) in the Ames test", *Mutation Res.* **298**, 207-214, (1993).

R. Vikse, L. Klungsøyr and S. Grivas, "Mutagenic activity of three synthetic isomers of the food carcinogen IQ in the Ames test", Submitted.

L. Vuolo, and G. Schuessler, "Putative mutagens and carcinogens in foods. VI. Protein pyrolysate products", *Environ. Mutagen.* **7**, 577-598, (1985).

K. Wakabayashi, M. Nagao, H. Esumi and T. Sugimura "Food-derived mutagens and carcinogens" *Cancer Res.* **52**, 2092-2098, (1992).

J. Weisburger, "Nutrition and colon cancer" *Clin. Nutr.* **4**, 131-137, (1985).

Z. Yamaizumi, H. Kasai, S. Nishimura, C. Edmonds, J. McCloskey "Stable isotope dilution quantification of mutagens in cooked foods by combined liquid chromatography-thermospray mass spectrometry", *Mutation Res.* **173**, 1-7, (1986).

K. Yamashita, A. Umemoto, S. Grivas, S. Kato & T. Sugimura, "*In vitro* reaction of hydroxyamino derivatives of MeIQx, Glu-P-1 and Trp-P-1 with DNA: ^{32}P-postlabelling analysis ofDNA adducts formed *in vivo* by the parent amines and *in vitro* by their hydroxyamino derivatives" *Mutagenesis* **3**, 515-520, (1988).

On the Possible Structure-Activity Relationships in the Food Mutagen Imidazoquinoxaline (IQx) Series

S. GRIVAS[a], N. RUSSO[b] and M. TOSCANO[c]

a) *Swedish University of Agricultural Sciences, Department of Chemistry, Box 7015, S-750 07 Uppsala, Sweden;* b) *Dipartimento di Chimica, Universita' della Calabria, I-87030 Arcavacata di Rende (CS), Italy ;* c) *Facolta' di Farmacia, Universita' della Calabria, I-87030 Arcavacata di Rende (CS), Italy*

1. Introduction

The relationship between diet and cancer is well known. The most consistent correlations between diet and human colon cancer are associated with the " Western life style" which includes a high intake of dietary fat and meat [1-3]. Some of the IQx food mutagens shown in the Figure are found together with several other heteroaromatic amines in cooked proteinaceous foods [4,5]. The fifteen methyl derivatives of IQx were recently synthesized [6] and Ames tested [7], thus complementing previous studies [8-10] on these structure-activity relationships. In this paper we report our preliminary molecular mechanics calculations and the possible structure-activity relationships of the IQx food mutagens.

2. Computational details

Full geometry optimization has been performed using Molecular Mechanics method [11] employing the MMX force field within the MODEL code [12]. The total energy results from the following summation of contributions:

$$E_{total}= E_{stretch} + E_{bend} + E_{torsion} + E_{VDW} + E_{elec} \quad \text{with:}$$

$$E_{stretch}= K_{str} \, \Sigma \, [l_{A,B}{}^2 - (C_{radA} + C_{radB})^2]^2$$

$$E_{bend}= K_{bnd} \, \Sigma[l_{A,C}{}^2 - (C_{radA} + C_{radB})^2 + (C_{radB} + C_{radC})^2 - 2(C_{radA} + C_{radB})^2 (C_{radB} + C_{radC}) \cos Q_{ABC}]^2$$

$$E_{torsion}= \Sigma \, 0.5 \, V_2 \, (1-\cos(2w) + 0.5 \, V_3 \, (1-\cos(3w))$$

$$E_{VDW}= \Sigma \, e[R_0{}^{12}/ R^{12} - 2(R_0{}^6/ R^6)] \quad \text{and} \quad E_{elec}= \sum_{i>j} Q_i \, Q_j/R_{ij}{}^2$$

243

N. Russo et al. (eds.), Properties and Chemistry of Biomolecular Systems, 243–247.

© 1994 *Kluwer Academic Publishers. Printed in the Netherlands.*

where E_{total}, $E_{stretch}$, E_{bend}, $E_{torsion}$, E_{VDW} and E_{elec} are the total, stretching, bending, torsional, van der Waals and electrostatic energies respectively; K_{str}, K_{bnd} are constants; $l_{A,B}$ is the A-B distance; $Q_{ABC}l$ is the optimal bond angle for ABC; C_{radA} is the covalent radius of atom N; w is the ABCD torsional angle; V_N is the N-fold torsional potential barrier; R_0 and R are the sum of the VDW radii of atoms A, E and the A, E distance respectively; Q_N represents the charge on atom N and R_{ij} the i-j distance.

3. Results and Discussion

With the aim to investigate the possible relationships between the structural modifications depending on the number and the position of the methyl groups and the biological activity, in this first report we present the study on 2-amino-3-methylimidazo[4,5-*f*]quinoxaline (IQx) and its mono-, di-, tri- and tetramethyl derivatives (see Figure and Table 1).

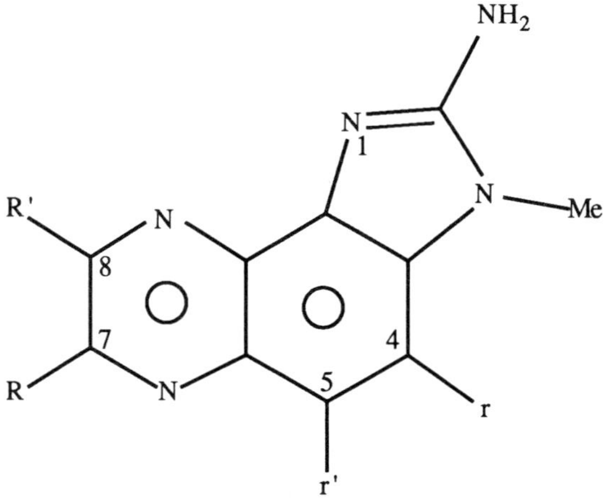

Figure. Structural formula and labelling of IQx and its mono-, di-, tri- and tetramethyl derivatives.

In the first step of the work we have performed a full geometry optimization of all shown compounds. As structural parameters we have considered the total surface area (Tsurf), the polar surface area (PSA), the nonpolar unsaturated area (NUSA), and the maximum length (Mleng) and width (Mwid) that characterize the shape of the systems in the x and y directions. These parameters have been correlated with the biological activity of the IQx compounds in the Ames test [7].

The results are collected in the Table. The compounds that show high biological activity (i.e. 4-MeIQx, 5-MeIQx, 7-MeIQx, 4,7-DiMeIQx and 7,8-DiMeIQx) have a NUSA value that falls in the range of 41-44 $Å^2$, while those with very low activity (4,5,7-TriMeIQx, 4,5,8-TriMeIQx, 5,7,8-TriMeIQx and 4,5,7,8-TetraMeIQx) have a low NUSA value (33-35 $Å^2$). The latter compounds have also the largest dimensions (about 8.5 and 9.9 Å as maximum length and width, respectively).

Table 1 The studied IQx derivatives (see Figure)

r	r'	R	R'	Name
H	H	H	H	IQx
Me	H	H	H	4-MeIQx
H	Me	H	H	5-MeIQx
H	H	Me	H	7-MeIQx
H	H	H	Me	8-MeIQx
Me	Me	H	H	4,5-DiMeIQx
Me	H	Me	H	4,7-DiMeIQx
Me	H	H	Me	4,8-DiMeIQx
H	Me	Me	H	5,7-DiMeIQx
H	Me	H	Me	5,8-DiMeIQx
H	H	Me	Me	7,8-DiMeIQx
Me	Me	Me	H	4,5,7-TriMeIQx
Me	Me	H	Me	4,5,8-TriMeIQx
Me	H	Me	Me	4,7,8-TriMeIQx
H	Me	Me	Me	5,7,8-TriMeIQx
Me	Me	Me	Me	4,5,7,8-TetraMeIQx

Table 2. Structural parameters (NUSA= nonpolar unsaturated area; PSA= Polar surface area; TSA= total surface area; Mleng= maximun length; Mwid= maximum width) and biological activities (BA) of IQx and its derivatives. Distances are in Å, surfaces in Å^2 and biological activities in revertants/mmol.

Compound	NUSA	PSA	TSA	Mleng	Mwid	BA
IQx	46.3	50.0	173	7.52	8.71	106
4-MeIQx	44.0	53.3	193	7.52	8.68	859
5-MeIQx	43.3	48.4	191	8.49	8.69	584
7-MeIQx	43.1	49.4	195	7.52	8.71	828
8-MeIQx	42.8	51.0	196	7.52	9.99	106
4,5-DiMeIQx	42.5	49.1	206	8.46	8.64	245
4,7-DiMeIQx	39.5	50.3	212	7.52	8.87	880
4,8-DiMeIQx	39.8	50.3	213	7.52	9.96	431
5,7-DiMeIQx	39.5	47.7	214	8.48	8.70	266
5,8-DiMeIQx	40.0	49.3	214	8.48	9.98	104
7,8-DiMeIQx	41.8	48.7	215	7.52	9.99	622
4,5,7-TriMeIQx	33.1	49.0	220	8.59	8.94	71
4,5,8-TriMeIQx	35.9	50.7	229	8.59	9.92	74
4,7,8-TriMeIQx	34.8	48.2	226	7.52	9.95	281
5,7,8-TriMeIQx	37.3	46.6	234	8.48	9.96	16
4,5,7,8-TetraMeIQx	32.8	49.8	245	8.58	9.92	15

The lack of any direct relationship for the rest of the values means that these data cannot be univocally correlated to the measured mutagenic activity. However taking into account that the biological activities are related to the final step of the mechanism (i.e. the reaction with the DNA substrate) we can understand the lack of a clear relationship between the geometrical parameters and the activity. Before binding with DNA, the IQx and the studied compounds are involved in other biological and chemical mechanisms such as oxidation and interactions with membranes. Our data can be better correlated with the membrane interaction step. In order to bypass this obstacle the shape and the dimensions of the molecules play a significant role. Systems that are too large are not able to diffuse into the cell. The very low biological activity of 4,5,7-TriMeIQx, 4,5,8-TriMeIQx, 5,7,8-TriMeIQx and 4,5,7,8-TetraMeIQx, can be explained by their large dimensions and their more spherical shape. These results are also confirmed by recent data on similar compounds in which a methyl group is replaced by a phenyl ring. These compounds were found to have very low mutagenic activity [7].

Further work including the determination of other chemical-physical parameters (i.e. charge distribution, electrostatic potentials, etc.) and the study of the interaction with the guanine DNA bases is in progress and can increase the possibility to find a more reliable and exhaustive relationship between structure and mutagenic activity for the IQx mutagen and its derivatives.

4. Conclusions

Molecular mechanic computations have been performed on the potent food mutagen 2-amino-3-methylimidazo[4,5-*f*]quinoxaline (IQx) and its mono-, di-, tri- and tetramethyl derivatives with the aim to correlate some structural parameters with the wide range of mutagenic activity in the Ames test determined for these compounds. On the basis of our results we can postulate that the compounds with large dimensions have more difficulty to penetrate the membrane and consequently to diffuse into the cell. Our determination of geometrical parameters and other structural data can be only regarded as the starting point of a more systematic study on the structure-activity relationships for these compounds that require to take into account the oxidation processes and the binding with DNA bases other than the interaction with the membrane cell. Work is in progress in these directions.

Acknowledgements

The financial support of the Swedish Council for Forestry and Agricultural Research and Ministero Universita' e Ricerca Scientifica e Tecnologica (MURST 60%) is gratefully acknowledged.

References

1. R. Doll and R. Peto, " Epidemiology and the prevention of cancer: some recent developments" *J. Cancer Res. Clin. Oncol.* **114**, 447-458, (1988), and references therein.

2. T. Byers, " Diet and cancer, any progress in the interim?" *Cancer* **62**, 1713-1724, (1988).

3. W. Willet, " The search for the causes of breast and colon cancer" *Nature* **338**, 389-394, (1989).

4. E. Overvik and J-Å. Gustafsson, , " Cooked-food mutagens: current knowledge of formation and biological significance" *Mutagenesis* **5**, 437-446, (1990) and references thereim.

5. J.S. Felton and M.G. Knize, "Occurrence, identification and bacterial mutagenicity of heterocyclic amines in cooked food" *Mutation Res.* **259**, 205-217, (1991) and references therein.

6. S. Grivas, W. Tian, E. Ronne, S. Lindstrom and K. Olsson, " Synthesis of mutagenic methyl- and phenyl- substituted 2-amino-3H-imidazo[4,5-*f*]quinoxalines via 2,1,3-benzoselenadiazoles" *Acta Chem. Scand.*, in press.

7. R.Viske, A. Knapstad, L. Klungsoyr and S. Grivas, " Mutagenic activity of the methyl and phenyl derivatives of the food mutagen 2-amino-3-methylimidazo[4,5-*f*]quinoxaline (IQx) in the Ames test" *Mutation Res.* **298**, 207-214, (1993).

8. F.T. Hatch, M.G. Knize, and J.S. Felton, " Quantitative structure-activityt relationships of heterocyclic amine mutagens formed during the cooking of food" *Envir. Molecular Mutag.* **17**, 4-19, (1991).

9. F.T. Hatch, M.G. Knize, D.H. Moore II and J.S. Felton, " Quantitative correlation of mutagenic and carcinogenic potencies for heterocyclic amines from cooked foods and additional aromatic amines" *Mutation Res.* **271**, 269-287, (1992).

10. G. Sabbioni and D. Wild, " Quantitative structure-activity relationships of mutagenic aromatic and heteroaromatic azides and amines" *Carcinogenesis* **13** , 709-713, (1992).

11. N.L. Allinger and U. Burker, <u>Molecular Mechanics</u>, ACS Monography n. 177, Washington, 1982.

12. Program MODEL, Serena Software, Bloomington, USA.

Elongation Factor Tu, a Model Protein for Studying Structural-Functional Relationships and Regulatory Mechanisms

K. HARMARK[1]*, P.H. ANBORGH[1], A. WEIJLAND[1], J. JONAK[2], G. PARLATO[3], and A. PARMEGGIANI[1]
[1]*Structure Diverse d'Interventions n° 61840, Centre National de la Recherche Scientifique, Laboratoire de Biochimie, Ecole Polytechnique F-91128 Palaiseau Cedex, France*
[2]*Institute of Molecular Genetics, Czechoslovak Academy of Sciences CS-16637 Prague, Czechoslovakia*
[3]*Istituto di Biochimica Fisica e Patologia Molecolare e Cellulare, Facoltà di Medicina e Chirurgia, università degli Studi di Reggio Calabria I-88100 Catanzaro, Italy*

1. General Properties

In *Escherichia coli* elongation factor Tu (EF-Tu) is a monomeric protein of 393 amino acid residues encoded by two unlinked, almost identical genes, *tufA* and *tufB*, situated at 72 and 88 min on the genetic map (for reviews, see Bosch et al. 1983; Parmeggiani and Swart, 1985; Weijland et al., 1992). The products of these two genes are identical but for the C-terminus (glycine in EF-TuA, 43195 Da, and serine in EF-TuB, 43225 Da) and constitute the most abundant protein in the *E. coli* cell (5-10 % of the total cell proteins). EF-Tu, an essential component of protein biosynthesis, is an ubiquitous protein found in all bacterial organisms with counterparts in the eukaryotic and vegetal systems. EF-Tu is, together with protein Ha-*ras* p21, the best characterized GTP-binding protein. Besides being the object of extensive functional investigations- that in the case of EF-Tu span more than two decades - EF-Tu and p21 are the only two GTP binding proteins with known threedimensional (3-D) structure. The X-ray analysis of the EF-Tu•GDP complex (Kjeldgaard and Nyborg, 1992) has shown the existence of three domains [N-terminal- (N), middle (M) and C-terminal (C) domain, Figure 1]. The N-terminal domain displays a typical α/β structure and contains the nucleotide binding site that is defined by four loops (residues 18-23, 82-83, 136-142 and 173, respectively). The first three loops are part of the three consensus sequence elements (G18HVDHGK24; D80CPG83; and N135KCD138, respectively) that characterize the class of the GTP-binding proteins and contain amino acid residues interacting with the guanine nucleotide; the first two elements with the phosphoryl groups and the third one with the purine ring. The M- and C-domain do not contain any α-helices; functionally they mediate the allosteric mechanisms, that regulate the EF-Tu activities and the interaction with ligands. In protein synthesis, EF-Tu acts as the carrier of aminoacyl-tRNA (aa-tRNA) to the messenger (m) RNA-programmed ribosome. As for all GTP-binding proteins, the GTP-bound EF-Tu is the "on" state that allows a

N. Russo et al. (eds.), Properties and Chemistry of Biomolecular Systems, 249–266.
© 1994 *Kluwer Academic Publishers. Printed in the Netherlands.*

productive interaction with aa-tRNA and ribosomes, whereas its complex with GDP represents the "off" state that terminates these interactions.

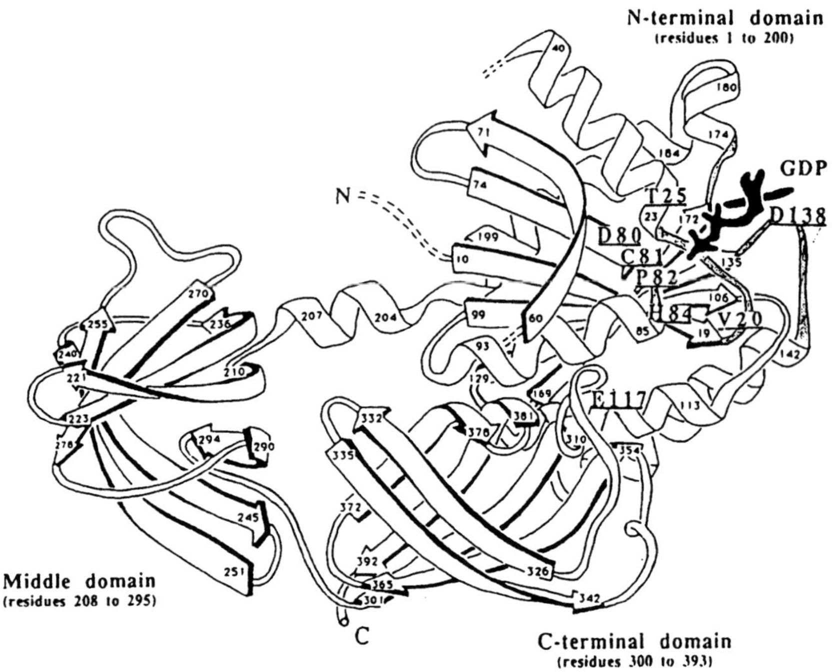

Figure 1. 3-D model of a nicked EF-Tu•GDP complex lacking residues 45-58, developed from X-ray diffraction analysis. The four loops defining the GDP binding pocket are shaded. The point substitutions mentioned in this article are indicated. This illustration is derived from Clark et al. (1990) and Kjeldgaard and Nyborg (1992).

The elongation cycle in protein biosynthesis is initiated by the regeneration of the active EF-Tu•GTP complex from EF-Tu•GDP, an exchange reaction that needs the intervention of elongation factor Ts to accelerate the dissociation of the tight EF-Tu•GDP complex. Indeed, the affinity of EF-Tu for GDP is 100-fold higher than that for GTP and EF-Tu•GDP displays a much slower dissociation rate. The regenerated EF-Tu•GTP is strongly stabilized by the bindng of aa-tRNA. The complex between EF-Tu•GTP and aa-tRNA interacts with the A-site of the mRNA-programmed ribosome carrying peptidyl (p) -tRNA in the P-site. The codon-anticodon interaction triggers a burst-like hydrolysis of the EF-Tu-bound GTP, that causes the release of EF-Tu in complex with GDP, since the latter nucleotide dramatically decreases the affinity of EF-Tu for aa-tRNA and the ribosome. The dissociation of EF-Tu•GDP from the ribosome precedes the formation of a new peptide bond between the N-terminus of aa-tRNA situated in the ribosomal A-site and the C-terminus of peptidyl-tRNA bound to the P-site. The subsequent translocation of peptidyl-tRNA from the A-site to the P-site, catalyzed by elongation factor G, leads to a vacant A-site and the displacement of tRNA$_{OH}$ to the ribosomal exit (E-) site; a new elongation cycle can now start.

The canonical scheme of the EF-Tu cycle is resumes by the following equations:

1)   EF-Tu•GDP + GTP + EF-Ts $\leftrightarrow$ EF-Tu•GTP + GDP + EF- Ts

2)   EF-Tu•GTP + aa-tRNA $\rightarrow$ EF-Tu•GTP•aa-tRNA

3)   EF-Tu•GTP•aa-tRNA + mRNA•ribosome•p-tRNA$_{n}$(P-site)•tRNA$_{OH}$(E-site) $\leftrightarrow$
     EF-Tu•GTP•aa-tRNA$_{(A-site)}$•mRNA•ribosome•p-tRNA$_{n}$(P-site) + tRNA$_{OH}$

4)   EF-Tu•GTP•aa-tRNA$_{(A-site)}$•mRNA•ribosome•p-tRNA$_{n}$(P-site) $\rightarrow$ p-tRNA$_{n+1}$(A-site) mRNA•ribosome•tRNA$_{OH}$(P-site) + EF-Tu•GDP +P$_i$

4A) p-tRNA$_{n+1}$ (A-site)mRNA • ribosome • tRNA$_{OH}$(P-site) (translocation) $\rightarrow$
     mRNA •ribosome •p-tRNA$_{n+1}$(P-site) • tRNA$_{OH}$(E-site)

Concerning the partecipation of two EF-Tu•GTP molecules in this cycle and the existence of a pentameric complex, (EF-Tu•GTP)$_2$•aa-tRNA, see 2.2.3. The rate of polypeptide chain elongation in the cell is 10 to 20 aa•$^{s-1}$ and the complete EF-Tu cycle in the cell takes approximately 50 ms (Bilgin et al., 1988). Similar values can also be obtained in *in vitro* systems by using ionic conditions and concentrations of biological components mimicking as far as possible the *in vivo* conditions (Jelenc and Kurland, 1982). As all GTP-binding proteins, EF-Tu displays in the absence of ribosomes a very low but significant intrinsic GTPase activity whose rate can be enhanced by the mRNA-programmed ribosomes by five orders of magnitude. Simple codon-anticodon interaction cannot guarantee the 10^{-4} to 10^{-3} overall missense error frequency as observed for protein biosynthesis *in vivo*. To explain this translaion fidelity, the choice of the proper aa-tRNA has been proposed to take place via an initial selection step followed by a proof-reading process, in which the EF-Tu-bound GTP is hydrolyzed. In this model, the accuracy is amplified by the interactions between EF-Tu•GTP, aa-tRNA and the mRNA-.

1.1. ANTIBIOTICS ACTING SPECIFICALLY ON EF-Tu

The importance of EF-Tu in the cell is highlighted by the existence of several antibiotics specifically inhibiting its activity in protein synthesis. The firts identified was kirromycin (Wolf et al., 1974), followed by pulvomycin (Wolf et al., 1978) and recently by GE2270 A (Anborg and Parmeggiani, 1991). A common property of these strcturally unrelated antibiotics (Figure 2) is their tight binding to EF-Tu and their ability to retard the dissociation of the EF-Tu•CTP complex. The main effect of kirromycin is to block the release of EF-Tu from the mRNA-programmed ribosome, since in its presence EF-Tu•GDP displays a much higher affinity for aa-tRNA•mRNA•ribosome. The subsequent pausing of EF-Tu•GDP•kirromycin on the ribosome stops the elongation cycle. Kirromycin (for refs Parmeggiani and Swart, 1985), stimulates considerably the very low intrinsic GTPase activity but to a moch lower extent than the mRNA-programmed ribosomes (two vs five orders of magnitude). The effect of kirromycin allowed the identification of the GTP hydrolysis as an intrinsic activity of EF-Tu. Pulvomycin and GE2270 A both hinder the formation of the complex between EF-Tu•GTP and aa-tRNA, each of the two antibiotics showing however a number of selective differences.

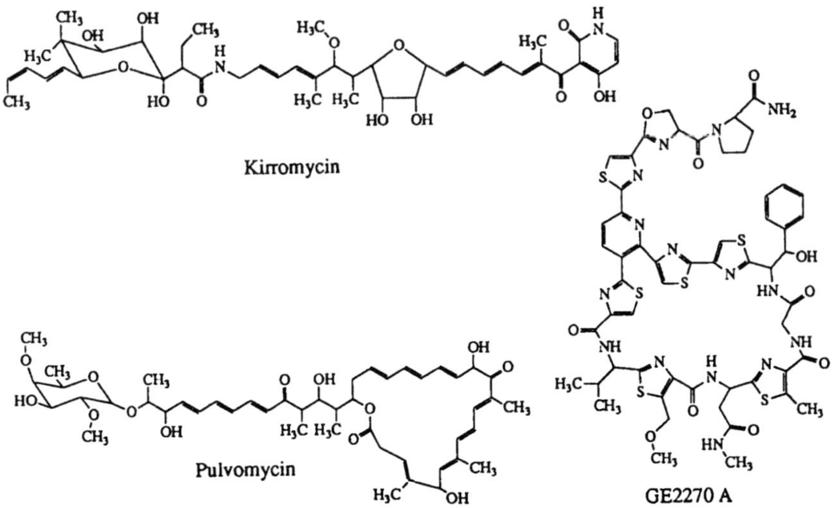

Figure 2. Structures of kirromycin, pulvomycin and GE2270 A.

2. The Use of Site-directed Mutagenesis to study the Functions of EF-Tu

Our approach to study the structure-function relationships in EF-Tu consisted in modifying the *tuf*A gene or regions of this gene, cloned on suitable plasmids, using synthetic oligodeoxynucleotides and expressing the modified *tuf* gene in suitable host cells (for refs: Parmeggiani et al., 1987; Anborgh et al., 1991). The N-terminal domain of EF-Tu, which according to the X-ray crystallography contains the GDP binding site, was isolated by deleting the M-and C- domain and named G domain (Parmeggiani et al., 1987). Engineered EF-Tu deleted of the M-domain (EF-TuΔM-domain) or the C-domain (EF-TuΔC-domain) were also recently constructed. Point substitutions of residues, that the crystallographic analysis or the homology with other guanine nucleotide binding proteins suggested to be important for the various activities, were introduced into EF-Tu or its G domain.

2.1. FUNCTIONAL ROLE OF THE EF-Tu DOMAINS

The functional characterization of a truncated EF-Tu molecule deleted of the M-and C-domain, consisting of the first 202 N-terminal amino acid residues (Parmeggiani et al., 1987), showed that the binding site for GDP and GTP and the intrinsic GTPase of EF-Tu are active independently of the presence of the other two domains. However, these two domains are required for the marked differential affinity that the intact molecule displays towards GDP and GTP. Indeed, the affinity of the G domain for GDP is not much stronger that for GTP [$K_{d(GDP)}$ = 2-3 µM vs $K_{d(GTP)}$ = 3-5 µM, at 0 °C], whereas the intact EF-Tu binds GDP two orders of magnitude tighter than GTP (K_d = 1nM and 0.6 µM, respectively, at 0 °C). Accordingly, this decrease in the affinity for GDP of the G domain induces a more favorable thermodynamic situation, allowing a higher GTPase turnover rate than that with the intact molecule, in which the much stronger affinity for GDP inhibits the regeneration of the active EF-Tu•GTP complex from EF-Tu•GDP (Fasano et al., 1982). The study of the function of the G domain has shown that the M-domain and the C-domain are determinant for a productive interaction between the N-terminal domain and EF-Tu ligands, such as ribosomes, aa-tRNA, EF-Ts and antibiotics. Only the 70 S ribosome could enhance the GTPase activity of the G domain, but only to a very small extent and not with all ribosomal preparations. That the G domain is capable of a weak interaction with the ribosome was nonetheless later on confirmed by the observation that both EF-Tu and G domain give the same footprint of 23S rRNA, resulting in the protection of the nucleotide A2665 in the α-sarcin loop (J. M. Robertson, A. Parmeggiani and H.F. Noller, unpublished results as quoted in Noller, 1991). Monovalent cations can strongly enhance the intrinsic GTPase of EF-Tu in a specific fashion that depends on their concentration and nature. The G domain-dependent GTPase activity was found to be influenced by increasing the concentrations of monovalent and divalent cations in a way resembling that of the intact molecule in the presence or absence of kirromycin; specific differences in the effects of the single cations were however observed (Harmark et al., 1990, Fasano et al., 1982; Ivell et al., 1981). Divalent cations are required for the GTPase activity for both EF-Tu and G domain (Harmark et al., 1992; Mistou et al., 1992). These results indicated that the deletion of the M- and C-domain did not modify essentially the nucleotide binding pocket of the N-terminal domain. This conclusion was reinforced by nuclear magnetic resonance studies, indicating that the GTP binding site of the G domain conserves an ordered structure (Lowry et al., 1992). In particular, the amide of lysine 24, a crucial amino acid residue interacting with the substrate phosphoryl groups, showed the same proton shift and resonance as found for the corresponding residue (Lys 16) in ras p21 (for the similarity of the 3-D structures of EF-Tu and ras p21, see below). Preliminary results from our laboratory (P.H. Anborgh, unpublished results) have shown that the removal of either the M-domain or the C-domain alone is sufficient to abolish almost entirely the ability of EF-Tu to distinguish between GTP and GDP, bringing about an effect similar to the deletion of both domains. On the other hand, the action of EF-Tu effectors, such as kirromycin and aa-tRNA, is specifically influenced by the presence of either one of these two domains. By determining the electrophoretic mobility, Peter et al. (1990) concluded that the combination of the M- and C-domain is able to interact with EF-Ts. On the other hand, Hwang et al. (1992) investigating the effect of point substitutions in EF-Tu, concluded that the C-terminal part of the N-domain of EF-Tu should contain an interaction site with EF-Ts. The ability of the G domain to bind GDP and GTP and to support the GTPase activity suggested its use as a suitable tool to study the structure-function relationships of these basic activities of EF-Tu, avoiding the constraints imposed by the M- and C-domain.

2.2. SUBSTITUTION OF SINGLE RESIDUES

So far only the 3-D structure of the GDP complex of EF-Tu from *E. coli* has been elucidated at middle resolution (2.6 Å). Moreover, the known model of EF-Tu•GDP (Figure 1) draws the 3-D structure of a nicked molecule, since in order to obtain crystals suitable for X-ray-diffraction studies EF-Tu was subjected to partial proteolysis (digestion of residues 45-58). To by-pass these constraints we had to take into account the situation of the *ras* p21 protein, whose 3-D structure is known at a higher resolution with respect to both the GDP- (at 2.2 Å, Privé et al., 1992) and the GTP-bound (1.4 Å, Wittinghofer and Pai, 1991) forms. This was made possible by the remarkable observation that the secondary elements of the N-terminal domain of EF-Tu and ras p21 show the same topology, despite only 17% identity in terms of primary structure (Wittinghofer and Pai, 1991). It should also be mentioned that soon also the 3-D model of the GTP complex of EF-Tu from *Thermus thermophilus* will be available at high resolution (1.9 Å, Reshetnikova et al., 1991).

2.2.1. *Mutations in the First Consensus Sequence Element: EF-TuV20G, EF-TuT25S and EF-TuT25A*

To examine the structural and functional role of the consensus motifs involed in the binding of the substrate we decided to carry out the substitution of valine 20, that is a variable residue within the first consensus element (Jacquet & Parmeggiani, 1988, 1989). Position 20 in EF-Tu corresponds to position 12 in *ras* p21, which is occupied by glycine instead of valine. In p21, the substitution of glycine 12 by any amino acid residue, except proline, strongly inhibits the GTPase activity and induces the power to transform the normal mammalian cell into an oncogenic one, by increasing the life-span of the active p21•GTP complex. Despite the different residues involved, the substitution Val20 $\rightarrow$ Gly in EF-Tu was also found to inhibit the GTPase activity, as in p21 (Figure 3). The association and dissociation rates of EF-Tu•GDP were strongly increased, while the dissociation rate of EF-Tu•GTP were decreased, a situation that favors higher levels of the GTP-bound form. The low GTPase activity of the more stable EF-TuV20G•GTP complex was slightly sensitive to the stimulation by the ribosome. It is important to mention that in p21 the GTPase activating protein (GAP) is also unable to enhance the GTPase of p21G12V. The similarity of these effects in EF-Tu and p21 concerning the substitution of a corresponding position with different amino acid residues shows that corresponding positions in GTP-binding proteins, even when nonconserved, can mediate the same function (Gümüsel et al., 1990). In these proteins the 3-D position of a residue can be more important than the nature of the amino acid itself. As a consequence, function-structure relationships derived from one GTP-binding protein can be applied to members of other families of this kind of proteins.

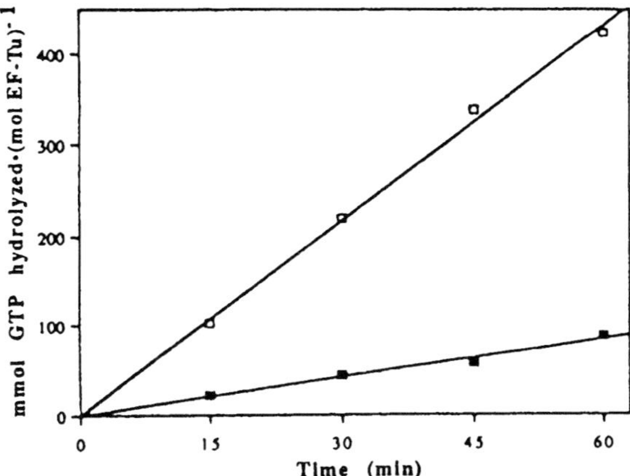

Figure 3. Kinetics of the intrinsic GTPase activity of EF-TuV20G (■) and wild-type EF-Tu (◻). (From Jacquet and Parmeggiani, 1988, modified).

Table I. Influence of substitution Val20 → Gly on the interaction with GTP and GDP. (From Jacquet and Parmeggiani, 1988, modified).

	GDP		GTP	
	EF-TuV20G	(EF-Tu wt)	EF-TuV20G	(EF-Tu wt)
Association rate constant $[10^{-4}k_{+1}(M^{-1}s^{-1})]$	260	(26)	0.7	(1)
Dissociation rate constant $[10^{4}k_{-1}(s^{-1})]$	53	(2.3)	11.8	(59)
Dissociation half-life (s)	130	(3100)	585	(117)
Dissociation constant (K_d, nM)	2	(0.9)	170	(590)

Another interesting observation was that the activity of EF-TuV20G in polypeptide synthesis was largely indipendent of the presence of EF-Ts, a ligand that in the case of wild-type EF-Tu is essential for a fast rate of elongation. This effect was due to the intrinsically increased dissociation rate of EF-Tu•GDP (see above) that "per se " sustained an efficient GDP to GTP exchange rate on EF-Tu in the absence of EF-Ts. We also carried out this substitution in the G domain (M.Y. Mistou and E. Jacquet, unpublished results). The modifications of the GDP and GTP binding and of the GTPase activity were of the same type as those obtained with the intact molecule. Threonine 25 is a residue at the edge of the first consensus motif, that interacts with the magnesium molecule coordinated to the guanine nucleotide. A preliminary characterization of EF-TuT25S or T25A has shown that both these mutants are able to support poly(Phe) synthesis, though at a reduced rate (K. Harmark, K. Tasiouka and A. Weijland, unpublished results). Their activity is sensitive to the action of kirromycin.

2.2.2. *Mutation in the Second Consensus Sequence Element*

From the analysis of the primary structure, the 3-D situation and homology with other enzymes, such as DNAse 1 (see below), this motif was expected to be involved in the catalytic mechanism of the GTP hydrolysis. Indeed, the functional effects of substitutions supported this possibility. The introduced point mutations will be described in the following order: Pro82 $\rightarrow$ Thr, His84 $\rightarrow$ Gly, Asp80 $\rightarrow$ Asn and Cys81 $\rightarrow$ Gly. Proline 82 is a variable residue of this consensus motif. Its substitution was suggested to us by the observation that mutation of the residue situated in the corresponding position of ras p21 (Ala59 $\rightarrow$ Thr) was associated with inhibition of the GTPase activity and appearance of autophosphorylating activity (Shih et al., 1982). A similar effect was also found for EF-TuP82T (Cool et al., 1991). The analysis of the amino acid and peptide composition of the ^{32}P-labeled G domainP82T and the Edmann degradation of the tryptically digested peptide containing the covalently bound ^{32}P demonstrated that the phosphorylated residue was threonine 82. It was shown kinetically that the cleaved γ-phosphate of GTP is stoichiometrically transferred to this residue. As shown in Figure 4, the amount of ^{32}P-G domainP82T is linearly dependent on the protein concentration.

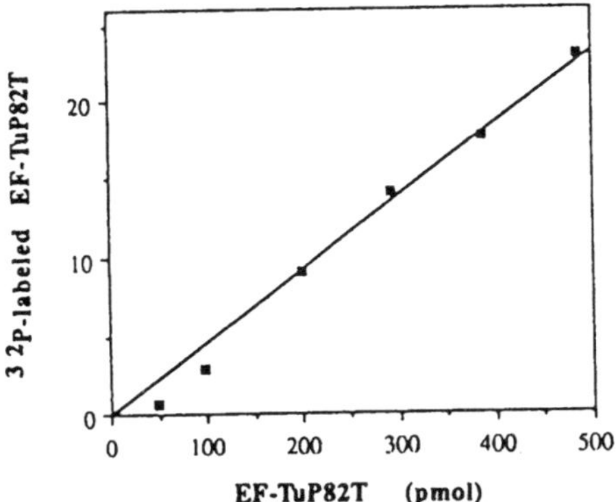

Figure 4. The autophosphorylation reaction is a first order rate reaction. (From Cool et al., 1991 modified).

The apparent first order rate of the autophosphorylation points to an intramolecular mechanism (John et al., 1988). The equilibrium and the dynamics of the binding of GDP and GTP were slightly affected by the mutation. These results show that threonine 82 is in close proximity to the γ-phosphate of GTP, as also found for *ras* p21, further confirming the similarity of the 3-D situation in these two proteins and the validity of the available 3-D model of EF-Tu. Already several years ago Eccleston and Webb (1982), using chiral analogs of GTP in the presence of kirromycin, could show that the stereochemical course of the EF-Tu-dependent GTP hydrolysis in the presence or absence of ribosome the γ-phosphate cleavage of GTP takes place in a single step, in line transfer of the γ-phosphate to a water molecule, with inversion of the configuration at the transferred phosphorus. This suggested that the cleavage of the γ-phosphate of GTP follows the nucleophilic attack of a water molecule. From the 3-D model of EF-Tu and analogy with DNAse 1 (Suck and Oefener, 1986), Swart proposed, already in 1987, a mechanism of general acid-base catalysis, whose structural background involved the transfer of a proton from a water molecule via the side chains of histidine 84 to glutamate 117 that in the 3-D model of EF-Tu lie in hydrogen bond distance. This mechanism of general acid-base catalysis was tested by substituting glutamate 117 with glutamine, but the experimental evidence did not support the involvement of glutamate 117 in a proton transfer; the GTPase of EF-TuE117Q was nearly as strong as that of wild-type EF-Tu (Harmark et al., 1990). To test the possibility of a basic catalysis supported by histidine 84, we replaced this residue with glycine. Histidine 84 is situated at the edge of the second consensus sequence element and is conserved in all elongation and initiation factors. In the EF-Tu•GDP 3-D model, the side-chain of histidine 84 points away from the phosphoryl groups and its amino group lies 11.4 Å from the next oxygen of the β-phosphoryl group (J. Nyborg, personal communication). Although this orientation does not allow neither a direct nor an indirect interaction between this side-chain and the γ-phosphate, the possibility of a marked conformation change during the transition state remained open. G domainH84G revealed slight, if any, alterations of nucleotide binding, but its GTPase activity was strongly reduced (approx. 95 %) as compared to wild-type EF-Tu (Figure 5). The change of the free energy of activation inferred the disruption of an uncharged hydrogen bond (Table II), suggesting that an interaction with a water molecule was involved. This possibility was supported by the 3-D situation of glutamine 61, the corresponding residue in the p21•GppNp model (Pai et al., 1990; Krengel et al., 1990), but contradicted by the presence in the mutated EF-Tu of a significant residual GTPase activity that showed the same characteristics (cationic and pH-dependence) as the GTPase activity of the wild-type G domain (Cool and Parmeggiani, 1991). As an alternative mechanism, the side-chain of histidine 84 could exert a stabilizing effect on the transition state possibly via a water molecule interacting with an oxygen of the γ-phosphate. We feel that the second mechanism, in which the stabilization of the transition-state configuration of the substrate by the enzyme lowers the energy barrier of the hydrolysis via the binding energies of the interacting residues, thus facilitating the γ-phosphate cleavage, is more probable. This mechanism removes the need for a strong nucleophile, such as an activated water molecule, and was shown to represent a major catalytic factor for the hydrolysis of ATP by aminoacyl-tRNA synthetase (Fersht et al., 1988). Other authors have recently reached the same conclusions by investigating the intrinsic GTPase activity of protein *ras* p21 (Privé et al., 1992; Langen et al., 1992). To deepen our knowledge on the mechanisms supporting the GTPase of EF-Tu, we also substituted Asp80 (with Asn), a residue involved in the binding of the magnesium ioncoordinated with the substrate. Asp80, the first residue of the second consensus sequence element, is conserved in all GTPases. The 3-D model of EF-Tu•GDP and the p21 model strongly suggests that Asp80 is a member of a ring (Asp80-H_2O-Mg^{2+} -Thr25-Asp80) chelating the magnesium ion coordinated with the guanine nucleotide. Our results show that Asp80 plays a crucial role in the overall stabilization of the G domain, in the coordination of the Mg^{2+}•GTP complex and in the catalytic reaction.

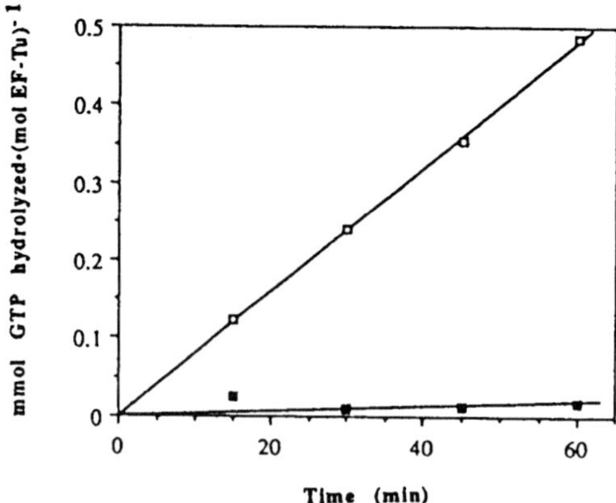

Figure 5. GTPase activity of G domain H84G (■) and wild-type G domain (▢). (From R.H. Cool and A. Parmeggiani, 1991, modified).

Table II. Properties of the GTPase activity of G domainH84G as compared to that of wild-type G domain (From Cool and Parmeggiani, 1991, modified).

	wild-type G domain		G domain H84G
	50mM NH4Cl	2M KCl	2M KCl
$K_{cat}(s^{-1} \cdot 10^3)$	0.125	1.4	0.078
$K_m(\mu M)$	8.0	7.8	7.2
$\Delta(\Delta G(kJ \cdot mol^{-1}))$			+ 7.1

Its substitution destabilizes the activities of the G domain and the binding of GTP, and most dramatically it strongly enhances the intrinsic GTPase of the G domain by almost two orders of magnitude. The marked instability of the GTPase of the G domain D80N was overcome by the presence of 30-40 % glycerol that induced a linear multiple-round GTPase activity (Figure 6). That glycerol did not influence directly the catalysis, since the initial GTPase in the absence of glycerol corresponded closely to the turnover linear activity obtained in its presence. Furthermore, under the same conditions the wild-type G domain was slightly stimulated (Harmark et al., 1992). Interestingly, the binding affinity for GDP of the G domain D80N and G domain wt, were approximately the same, whereas the affinity for GTP of the G domain D80N was almost 50 times weaker. Moreover, addition of EDTA mimicked the effect of the mutation: the affinity for GTP of wild-type G domain strongly decreased and became similar to that of the mutated complex that was virtually unaffected by EDTA. The GDP complexes of both mutated and wild-type G domain were slightly affected by EDTA, showing that this effect of magnesium was directed towards EF-Tu•GTP. These observations suggest that the γ-phosphate cleavage is associated with a relieve of the constraints imposed by asparagine80 on the magnesium•GTP coordination. It is known that metal ions can favor the phosphate transfer by direct and indirect mechanism (Cooperman, 1982).

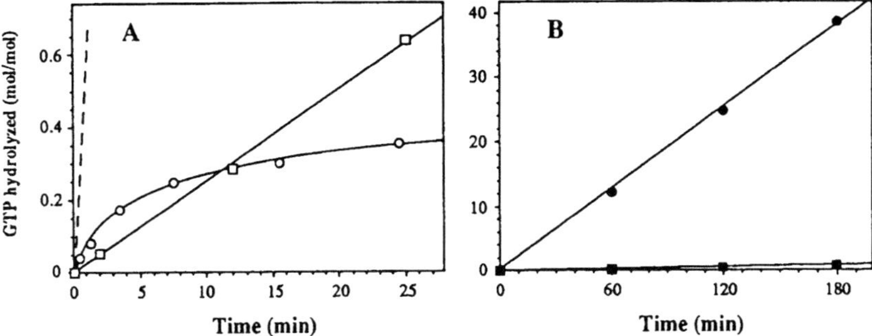

Figure 6. Kinetics of the GTPase activity of G domain D80N (O, ●) and wild-type G domain (□,■) in the absence of glycerol (A) and in the presence of 40% glycerol (B). The dashed line in A: initial velocity. (From Harmark et al., 1992, modified).

They can act via a direct mechanism, implying a nucleophilic attack by a Mg^{2+} -bound hydroxide in an intramolecular fashion, leading to a penta-coordinate transition state of the γ-phosphate, as taking place in other enzymatic systems (Sigel et al., 1984). It is suggestive to propose that a mechanism involving the substrate-coordinated magnesium could take place in the burst-like GTP hydrolysis, catalyzed by EF-Tu in the presence of mRNA-programmed-ribosomes, but at present the structural modifications involved in this hypothetical model would be too speculative. Mutagenesis of cysteine 81, the only residue of the second consensus motif, that we have mutated in the intact EF-Tu molecule, confirmed the indications of the 3-D model that this residue is important for stabilizing the architecture of the guanine nucleotide binding region. cysteine 81 contributes to the property of EF-Tu to have a stronger affinity for GDP than for GTP, a characteristic of all prokaryotic EF-Tu's. A decreased preference for GDP was confirmed by the observation that the thermal stability of both GTP- and GDP -bound forms of EF-TuC81G was virtually the same (Figure 7). This substitution induces a pattern similar to that found in the eukaryotic counterparts of EF-Tu; EF-1α, that has an alanine in the corresponding position, displays essentially the same affinity for GDP and GTP, the rates of the thermal inactivation of the relative complexes being the same (Nagata et al., 1976). Therefore, the nature of the amino acid residues (alanine in EF-1α) and glycine(in EF-Tu) does not appear to influence this property. The GTPase activity of EF-TuC81G was several times lower than that of the wild-type species.

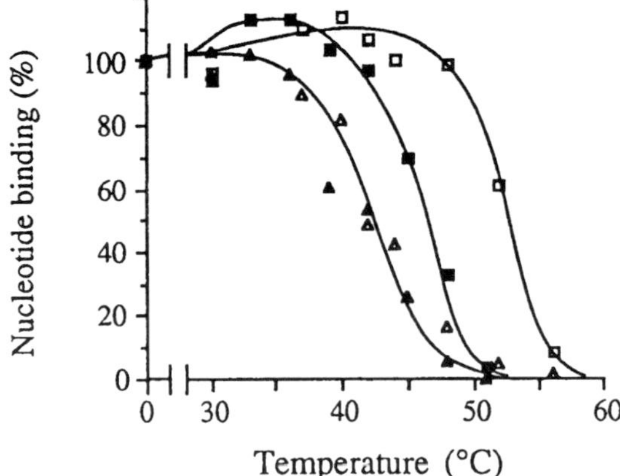

Figure 7. Temperature-induced denaturation of EF-TuC81G (Δ, $\blacktriangle$) and wild-type EF-Tu ($\square$, $\blacksquare$) in the presence of GDP (open symbols) or GTP (closed symbols). (From Anborgh et al., 1992, modified).

Cysteine 81 is known to be a specific terget of N-tasyl-L-phenylalanyl chloromethane (TPCK) in all prokaryotic elongation factors tested so far (for references, see Anborh et al., 1992). The modification of cysteine 81 by TPCK blocks the binding of aa-tRNA to EF-Tu (Jonàk et al., 1982; Sedlàcek et al., 1974), their affinity being decreased 1000 times (Ott et al., 1990). On the basis of competition experiments between TPCK and aa-tRNA and its 3' terminal analogs it was proposed that the binding site for the aminoacylated 3' end of aa-tRNA is localized around cysteine 81 (Jonàk et al., 1980). The direct involvement of this residue in the binding process wih aa-tRNA remains an open question, because the substitution of cysteine 81 with glycine only induced a partial inhibition of the interaction with aa-tRNA (Anborgh et al., 1992, and unpublished results). Thus, cysteine 81 is not essential for the binding of aa-tRNA to EF-Tu. Concerning the substitution of the conserved glycine 83 with alanine, Hwang et al., 1987 observed that the formation of a stable ternary complex was inhibited. Further experiments are required for a more detailed characterization of EF-TuG83A.

2.2.3. Mutation in the Third Consensus Sequence Element

The possibility of disposing of an EF-Tu with an altered substrate specificity, is of crucial importance to distinguish the GTPase of EF-Tu in the overall elongation process, with respect to that of EF-G, which is also a GTPase and catalyzes the translocation of the peptidyl-tRNA chain from the ribosomal A-site to the P-site after the release of EF-Tu•GDP and peptide bond formation. In 1987, Hwang and Miller, using a maxi-cell system, described that EF-TuD138N was able to bind XTP instead of GTP in the presence of aa-tRNA and ribosomes. This result was predicted by the 3-Dmodel of EF-Tu in which the exocyclic amino group of the base forms a hydrogen bond with the side-chain carboxyl group of Asp138. To avoid the disadvantages of the maxi-cell system, which only yields a low quantity of cells and consequently of mutated EF-Tu that is moreover contaminated with EF-Tu wt, the substitution Asp138 $\rightarrow$ Asn was introduced into the *tuf*A gene cloned on pEMBL19 and the mutated gene expressed from pCP40 under control of λP_L promotor (Parlato and Parmeggiani, 1988). After transfer of the mutated gene into a pTTQ18 vector, under control of *tac* promotor, EF-TuD138N was produced in *E. coli* and purified to homogeneity, free the chromosome-borne kirromycin-resistant EF-Tu (A. Weijland and A. Parmeggiani, in preparation). EF-TuD138N dispayed the same activity in poly(U)-directed poly(Phe) synthesis as wild-type EF-Tu, but its activity was dependent on XTP. These properties suggested the use of EF-TuD18N to analyze the EF-Tu-mediated energy consumption during the elongation cycle. The obtained results showed that two nucleotide triphosphates were hydrolyzed per each phenylalanine incorporated into the growing polypeptide chain. This was in agreement with the kinetic studies of Ehrenberg and Kurland (1990), suggesting that two GTP molecules from two EF-Tu•GTP complexes were hydrolyzed for each peptide bond synthesized. These authors proposed the formation of a pentameric complex consisting of two EF-Tu•GTP bound to a molecule of aa-tRNA. In the past, genetic (Vijgenboom and Bosch, 1989) and biochemical evidence (Anborgh et al., 1991) had suggested intermolecular interactions, but only in the case of two mutated EF-Tu species. The existence of this pentameric complex is however controversial; experiments of other authors (Bensch et al., 1991) supported the canonical 1:1:1 stoichiometric composition of the complex between EF-Tu•GTP and aa-tRNA. Whatever the composition of the complex between EF-Tu•GTP and aa-tRNA may be, the fact that with our system a 2 to 1 stoichiometry between the EF-Tu-dependent XTP hydrolysis and phenylalanine incorporation was found in diverse experimental conditions, suggests that both nucleotide triphosphates participate in the basic mechanism of elongation. Why should two γ-phosphates, very probably part of two EF-Tu•GTP complexes, be split per each cognate aa-tRNA properly positioned on the ribosome? Preliminary results suggest that this stoichiometry may be correlated to the mechanism selecting the proper amino acid residues to be incorporated into the polypeptide chain. A proof-reading mechanism with two molecules of GTP hydrolyzed sequentially would have the ability to increase the accuracy of the system, according to Hopfield (1974) and Ninio (1975). Additional experiments, also involving a reexamination of the EF-Tu•GTP interaction with aa-tRNA, possibly utilizing biochemical and genetic techniques in a coordinated fashion, will be required to solve these questions in a less hypothetical way.

3. Conclusions

The results obtained by modifying the EF-Tu molucule emphasize the potential of site-directed mutagenesis for the investigation of the structure-function relationships of an enzyme. It is nevertheless evident that, particularly, in the substitution of single residues, the interpretation of the results obtained with this method are subjected to severe limitations and cannot always be directly related to the structure of the replacing amino acid residue. Proteins are indeed very flexible structures and modification of a residue in most cases induces long-range effects that are difficult to interpret. A typical example in EF-Tu is found by the substitution of aspartate 80 in the G domain, which, besides inducing functional modifications that can be attributed to a direct effect on the γ-phosphate of GTP, also greatly modifies the overall stability of the G-domain by long-range actions. Despite these limitations, site-directed mutagenesis remains a crucial method for defining the structure-function relationships. This is particularly true for defining the function of the domains of a protein and tracing the transmission pathway of signals induced by low or high molecular ligands. The results achieved utilizing site-directed mutagenesis show the huge progress that one has recently achieved by coordinating the diverse techniques of modern biology in the investigation the properties of a protein. Concerning EF-Tu, the last decade has been crucial for revealing key-aspects of the mechanism of its basic activities and for initiating the exploration of the interactions with its specific ligands. It is conceivable that in the next few years the progress of our knowledge of its 3-D structure will lead to even more rewarding results that may be of great utility for interpreting the mechanisms involving other components of the diverse families of GTP-binding proteins.

Acknowledgement

This work was carried out in the framework of grant BAP-0066-F (CD) from the Biotechnology Action Programme of the Commission of the European Community and was supported by the "Association pour la Recherche contre le Cancer" and the "Institut National de la Santé et de la Recherche Medicale". K.H. was recipient of a fellowship from the Carlsberg Foundation.

Footnotes

* Present address: Russell Grimvade School of Biochemistry, University of Melbourne, Parkville, Victoria 3053, Australia.

References

P.H. Anborgh, and A. Parmeggiani, "New antibiotic that acts specifically on the GTP-bound form of elongation factor Tu" *EMBO J.* **10**, 779-774, (1991).

P.H. Anborgh, G.W.M. Swart, and A. Parmeggiani, "Kirromycin-induced modifications facilitate the separation of EF-Tu species and reveal intermolecular interactions" *FEBS Lett.* **292**, 232-236, (1991).

P.H. Anborgh, A. Parmeggiani, and J. Jonàk, "Site-directed mutagenesis of elongation factor Tu. The functional and structural role of residue Cys81" *Eur. J. Biochem.* **208**, 251-257, (1992).

K. Bensch, U. Pieper, G. Ott, N. Schirmer, M. Sprinzl, and A. Pingoud, "How many EF-Tu molecules participate in aminoacyl-tRNA binding?" *Biochimie* **73**, 1045-1050, (1991).

N. Bilgin, L.A. Kirsebam, M. Ehrenberg, and C.G. Kurland, "Mutations in ribosomal proteins L7/L12 perturb EF-G and EF-Tu functions" *Biochimie* **70**, 611-618, (1988).

K. Boon, E. Wijgenboom, L.V. Madesen, A. Talens, B. Kraal, and L. Bosch, "Isolation and functional analysis of His-tagged elongation factor Tu" *Eur J. Biochem.* **210**, 177-183, (1992).

L. Bosch, B. Kraal, P.H. Van der Meide, F.J. Duisterwinkel, and J.M. Van Noort, "The elongation factor EF-Tu and its encoding genes" *Progr. Nucl. Acides. Mol. Biol.* **30**, 91-126, (1983).

B.F.C. Clark, M. Kjeldaard, T.F.M. la Cour, S. Thirup, and J. Nyborg, "Structural determination of the functional sites of *E. coli* elongation factor Tu" *Biochim. Biophys. Acta* **1050**, 203-208, (1990).

R.H. Cool, and A. Parmeggiani, "Substitution of histidine 84 and the GTPase mechanism of elongation factor Tu" *Biochemistry* **30**, 362-366, (1991).

R.H. Cool, M. Jensen , J. Jonàk, B.F.C. Clark, and A. Parmeggiani, "Substitution of pro82 by threonine induces autophosphorylating activity in GTP-binding domain of elongation factor Tu" *J. Biol. Chem.* **256**, 6744-6749, (1990).

B.S. Cooperman, "The mechanism of action yeast inorganic pyrophosphatase" *Methods in Enzymology* **87**, 526-548, (1982).

J.F. Eccleston, and M.R. Webb, "Characterization of the GTPase reaction of elongation factor EF-Tu. Determination of the stereochemical course in the presence of antibiotic X5108" *J. Biol. Chem.* **257**, 5046-5049, (1982).

M. Ehrenberg, A-M Rojas, J. Weiser, and C.G. Kurland, "How many EF-Tu molecules partecipate in aminoacyl-tRNA binding and peptide bond formation in Escherichia coli translation" *J. Mol. Biol.* **211**, 739-749, (1990).

O. Fasano, E. De Vendittis, and A. Parmeggiani, "Hydrolysis of GTP by elongation factor Tu can be induced by monovalent cations in the absence of other effectors" *J. Biol. Chem.* **257**, 3145-3150, (1982).

A.R. Fersht, J.W. Knill-Jones, H. Bedouelle, and G. Winter, "Reconstruction by site-directed mutagenesis of the transition state for the activation of tyrosine by tyrosyl-tRNA synthetase: a mobile loop envelopes the transition state in a induced-fit mechanism" *Biochemistry* **27**, 1581-1587, (1988).

F. Gümüsel, R.H. Cool, A. Weijland, P.H. Anborgh, and A. Parmeggiani, "Mutagenesis of the NH_2-terminal domain of elongation factor Tu" *Biochim. Biophys; Acta* **1050**, 215-221, (1990).

K. Harmark, R.H. Cool, B.F.C. Clark, and A. Parmeggiani, "The functional and structural roles of residues Gln114 and Glu117 in elongation factor Tu" *Eur. J. Biochem.* **194**, 731-737, (1990).

K. Harmark, P.H. Anborgh, M. Merola, B.F.C. Clark, and A. Parmeggiani, "Substitution of Asp80, a residue involved in the coordination of magnesium, weakens the GTP binding and strongly enhances the GTPase of the G domain of elongation factor Tu" *Biochemistry* **31**, 7367-7372, (1992).

J.J. Hopfield, "Kinetic proofreading: a new mechanism for reducing errors in biosynthetic processes requiring high specificity" *Proc. Natl. Acad. Sci. U.S.A.* **71**, 4135-4139, (1974).

Y. -W. Hwang, and D. Miller, "A mutation that alters the nucleotide specificity of elongation factor Tu, a GTP regulatory protein" *J. Biol. Chem.* **262**, 13081-13085, (1987).

Y.-W. Hwang, F. Jurnak, and D.L. Miller, "A mutation that hinders the GTP induced aminoacyl-tRNA binding of elongation factor Tu". In: <u>The guanine nucleotide binding proteins. Common structural and functional properties</u> (L. Boscch, B. Kraal, and A. Parmeggiani, eds) NATO-ASI series, vol. **165**, Plenum Press, New York, p. 77-85, (1988).

Y.-W. Hawng, M. Carter, and D.L. Miller, "The identification of a domain in *Escherichia coli* elongation factor Tu that interacts with elongation factor Ts" *J. Biol. Chem.* **267**, 22198-22205, (1992).

R. Ivell, G. Sander, and A. Parmeggiani, "Modulation by monovalent and divalent cations of the guanosine-5'-triphosphatase activity dependent on elongation factor Tu" *Biochemistry* **20**, 1355-1361, (1985).

E. Jacquet and A. Parmeggiani, "Function-structure relationships in the GTP-binding domain of EF-Tu: mutation of Val20, the residue homologous to position 12 in p21" *Embo J.* **7**, 2861-2867 (1988).

E. Jacquet and A. Parmeggiani, "Substitution of Val20 by Gly in elongation factor Tu. Effects on the interaction with elongation factor Ts, aa-tRNA and ribosomes" *Eur. J. Biochem.* **185**, 247-255, (1989).

P.C. Jelenc, and C.G. Kurland, "Nucleoside triphosphate regeneration decreases the frequency of translation errors" *Proc. Natl. Acad. Sci. U.S.A.* **76**, 3174-3178, (1982).

J. John M. Frech and A. Wittinghofer, "Biochemical properties of Ha-*ras* encoded p21 mutants and mechanism of autophosphorylation reaction" *J. Biol. Chem.* **263**, 11792-11799, (1988).

J. Jonàk, J. Smrt, A. Holy, and J. Rychlìk , "Interaction of *Escherichia coli* EF-Tu•GTP and EF-Tu•GDP with analogues of the 3' terminus of aminoacyl-tRNA" *Eur. J. Biochem.* **105**, 315-320, (1980).

J. Jonàk, T.E. Petersen, B.F.C. Clark and I. Rychlik, "N-Tosyl-L-phenylalanylchloromethane reacts with cysteine 81 in the molecule of elongation factor Tu from *Escherichia coli*" *FEBS Lett.* **150**, 485-488, (1982).

M. Kjeldgaard, and J. Nyborg, "Refined structure of elongation factor EF-Tu from *Escherichia coli*" *J. Mol. Biol.* **223**, 721-742, (1982).

U. Krengel, I. Schlichting, A. Scherer, R. Shumann, M. Frech, J. John, W. Kabsch, E.F. Pai, and A. Wittinghofer, "Three-dimensional structures of H-*ras* p21 mutants: molecular basis for their inability to function as signal switch molecules" *Cell* **62**, 539-548, (1990).

R. Langen, T. Schweins, and A. Warshel, "On the mechanism of guanosine triphosphate hydrolysis in *ras* p21 proteins" *Biochemistry* **31**, 8691-8696, (1992).

D.F. Lowry, R.H. Cool, A.G. Redfield, A. Parmeggiani, "NMR study of the phosphate-binding elements of *Escherichia coli* elongation factor Tu catalytic domain" *Biochemistry* **30**, 10872-10877, (1991).

M.Y. Mistou, R.H. Cool, and A. Parmeggiani, "Effects of ions on the intrinsic activities of the c-H-*ras* protein p21. A comparison with elongation factor Tu" *Eur. J. Biochem.* **204**, 179-185, (1992).

S. Nagata, K. Iwasaki, and Y. Kaziro, "Interaction of the low molecular weight form of elongation factor 1 with guanine nucleotides and aminoacyl-tRNA" *Arch. Biochem. Biophys.* **172**, 168-177, (1976).

J. Ninio, "Kinetic amplification of enzyme discrimination" *Biochimie* **57**, 587-595 (1975).

H.F. Noller, "Ribosomal RNA and translation" *Annu. Rev. Biochem.* **60**, 191-227, (1991).

G. Ott, J. Jonàk, J.P. Abraham, and M. Sprinzl, "The influence of different modifications of elongation factor Tu from *Thermus thermophilus*" *Nucleic Acids Res.* **18**, 437-441, (1990).

E.F. Pai, U. Krengel, G.A. Petsko, R.S. Goody W. Kabsch, and A. Wittinghofer, "Refined crystal structure of the triphosphate conformation of H-*ras* p21 at 1.35 Å resolution: implications for the mechanism of GTP hydrolysis" *Embo J.* **9**, 2351-2359, (1990).

G. Parlato, and A. Parmeggiani, "Effect of mutation Asp138 $\rightarrow$ Asn, obtained via site-directed mutagenesis on the binding of elongation factor tu with GDP" *Italian J. Biochem.* **37**, 353A-355A (1988).

A. Parmeggiani, G.W.M. Swart, K.K. Mortensen, M. Jensen, B.F.C. Clark, L. Dente, and R. Cortese, "Properties of a genetically engineered G domain of elongation factor Tu" *Proc. Natl. Acad. Sci. U.S.A.* **84**, 3141-3145, (1987).

A. Parmeggiani, and G.W.M. Swart, "Mechanism of Action of kirromycin-like antibiotics" *Ann. Rev. Microbiol.* **39**, 557-577, (1985).

M.E. Peter, C.O.A. Reiser, N.K. Shirmer, T. Kiefhaber G. Ott, N.W. Grillenbeck, and M. Sprinzl, "Interaction of the isolated domain II/III of *Thermus thermophilus* elongation factor Tu with the nucleotide exchange factor EF-Ts" *Nucleic Acids Res.* **18**, 6889-6893, (1990).

G.G. Privé, M.V. Milburn, L. Tong, A.M. De Vos, Z. Yamaizumi, S. Nishimura, and S.-H. Kim, "X-ray crystal structures of transforming p21 *ras* mutants suggest a transition-state stabilization mechenism for GTP hydrolysis" *Proc. Natl. Acad. Sci. U.S.A.* **89**, 3649-3653, (1992).

L.S. Reshetnikova, C.O.A. Reiser, N.K. Schirmer, H. Berchtold, R. Storm, R. Hilgenfeld, and M. Sprinzl "Crystals of intact elongation factor Tu from *Thermus thermophilus* diffracting to high resolution" *J. Mol. Biol.* **221**, 375-377, (1991).

J. Sedlàcek, I. Rychlìk, and J. Jonàk, "The role of aminoacyl-tRNA binding site of the factor EF-T in uncoupled GTPase reaction" *Biochim. Biophys. Acta* **349**, 78-83, (1974).

T.Y. Shih, P.E. Stokes, G.W. Smythers, R. Dhar, and S. Orozlan, "Characterization of the phosphorylation sites and the surrounding amino acid sequences of the p21 transforming proteins coded for by the harvey and kirsten strains of murine sarcoma viruses" *J. Biol. Chem.* **257**, 11767-11773, (1982).

H. Sigel, F. Hofstetter, R.B. Martin, R.M. Milburn, V. Scheller-Krattiger, and K. Sheller, "General considerations on transphosphorylations: mechanism of the metal ion facilitates dephosphorylation of nucleoside 5'-triphosphates, including promotion of ATP dephosphorylation by addition of adenosine 5'-monophosphate" *J. Am. Chem. Soc.* **106**, 7935-7946, (1984).

D. Suck, and C. Oeffner, "Structure of DNase I at 2.0 Å resolution suggests a mechanism for binding to and cutting DNA" *Nature* **321**, 620-625, (1986).

G.W.M. Swart, Ph. D. Thesis, Lieden University (1987).

E. Vijgenboom, and L. Bosch, "Translational frameshits induced by mutant species of the -polypeptide chain elongation factor Tu of *Escherichia coli*" *J. Biol. Chem.* **264**, 13012-13017, (1989).

A. Weijland, K. Harmark, R.H. Cool, P.H. Anborgh, and A. Parmeggiani, "Elongation factor Tu: a molecular switch in protein biosynthesis" *Mol. Microbiol.* **6**, 683-688, (1992).

A. Wittinghofer, and E.F. Pai, "The structure of Ras protein: a model for a universal molecular switch" *Trends Biochem. Sci.* **16**, 382-387, (1991).

H. Wolf G. Chinal, and A. Parmeggiani, "Kirromycin, an inhibitor of protein biosynthesis, that acts on elongation factor Tu" *Proc. Natl. Acad. Sci. U.S.A.* **71**, 4910-4914, (1974).

H. Wolf, D. Assmann, and E. Fischer, "Pulvomycin, an inhibitor of protein biosynthesis preventing ternary complex formation between elongation factor Tu, GTP and aminoacyl-tRNA" *Proc. Natl. Acad. Sci. U.S.A.* **75**, 5324-5328, (1978).

Hybrid Antibiotics: Aminoglycoside 3-Quinolone or β-Lactam Amides

S.I. KOTRETSOU, V. CONSTANTINOU-KOKOTOU, and M. P. GEORGIADIS
Chemical Laboratories, Agricultural University of Athens, Iera Odos 75, Athens 11855, Greece

1. Introduction

A wide variety of derivatives of aminoglycosides have been prepared over many years, in the effort to improve the antibacterial activity and decrease their toxicity [1].
The knowledge assembled on the mechanisms of uptake and distribution of drugs in cells, and on the mechanisms of cellular toxicities [2], has prompted us to synthesize new bifunctional aminoglycosides with unusual cellular pharmacokinetic properties. These compounds were based upon the concept of hybrid molecules, which combine the properties of two major classes of antibacterials [3,4,5].
Aminoglycosides are active against more aerobic *Gram*-negative organisms, staphylococci and *Gram*-positive bacilli. Their combination with penicillins is in clinical use and show synergistic effects, especially against *Pseudomonas aeruginosa*. However, high concentrations of penicillins inactivate the aminoglycosides [6]. On the other hand, fluoroquinolones have potent activity against a broad spectrum of *Gram*-positive and *Gram*-negative aerobic and anaerobic organisms, as well as mycobacteria, even though they display low water solubility [7]. The latter is in relation to the zwitterionic character of quinolones having a basic substituent at the 7- position [8].
In an effort to overcome these problems, we have incorporated kanamycin A (1) and netilmicin (2) as the aminoglycoside component and ofloxacin, norfloxacin or 6-amino-penicillanic acid as the quinolone or β-lactam component, in one molecule. Such molecule may be used as a pro-drug derivative, which liberates two different compartments with their antibacterial activity intact.
The term "dual action" aminoglycoside antibiotics can also been used to describe these hybrids. This mechanism gives the opportunity to expand the antibacterial spectrum including organisms which are resistant to aminoglycosides. It may also be possible to design aminoglycosides with significant advantages over those currently in use.

2. Experimental section

The dual action aminoglycosides of the amide type were synthesized according to Scheme, using Kanamycin A or Netilmicin as the aminoglycoside compartment.

N. Russo et al. (eds.), Properties and Chemistry of Biomolecular Systems, 267–270.
© 1994 *Kluwer Academic Publishers. Printed in the Netherlands.*

Solution of 1 or 2, free base, in DMSO was treated with $CuCl_2$ and stirred at room temperature for 4 h. The preferential complexation of vicinal hydroxy and amino groups by Cu^{+2} permitted the reaction of the carboxy group of fluoroquinolones or benzylocarbonylamino-penicillanic acid with the free 6'-amino group of aminoglycosides, using conventional methods of peptide chemistry (Table I).

The bifunctional amides 1a-c and 2a-c were generally purified by ion-exchange column chromatography using 1,4-dioxan-water-conc. NH_3 (2:1:0.4 %) as eluent. The 6'-N acyl derivatives were obtained in one step, in moderate yield.

Table I. Acylation Method

1a	Mixed anhydride
1b	N-Hydroxysuccinimide
1c	Mixed anhydride
2a	Mixed anhydride
2b	N-Hydroxysuccinimide
2c	Mixed anhydride

3. Results and Discussion

The applied methodology for the synthesis of the above mentioned compounds is quite general and can be used with any other aminoglycoside which has a primary amino group in its structure.

The products were confirmed by FAB-mass (Table II), ^{13}C-NMR spectroscopy. Specially for the hybrid aminoglycoside-6'-amino-penicillanic acid the high frequency stretching absorption of the β-lactam CO group at 1795 cm^{-1} in the IR spectrum is very characteristic (Table III).

Table II. FAB-mass spectra (m/z)

1a	829 $[M+2H^+]$, 857 $[M+2H^++2Na]$, 812 $[M-CH_3]$, 677, 667
1b	817 $[M+2H^+]$, 840 $[M+2H^++Na]$, 785 $[M-2CH_3]$, 683, 655
1c	787 $[M+2H^+]$, 766 [M-F], 635, 435
2a	819 $[M+H^+]$, 800 $[M+H^+-F]$, 744, 660
2b	807 $[M+H^+]$, 671 [M-135], 585, 514, 428
2c	777 $[M+H^+]$, 759 $[M+2H^+-F]$, 645, 398

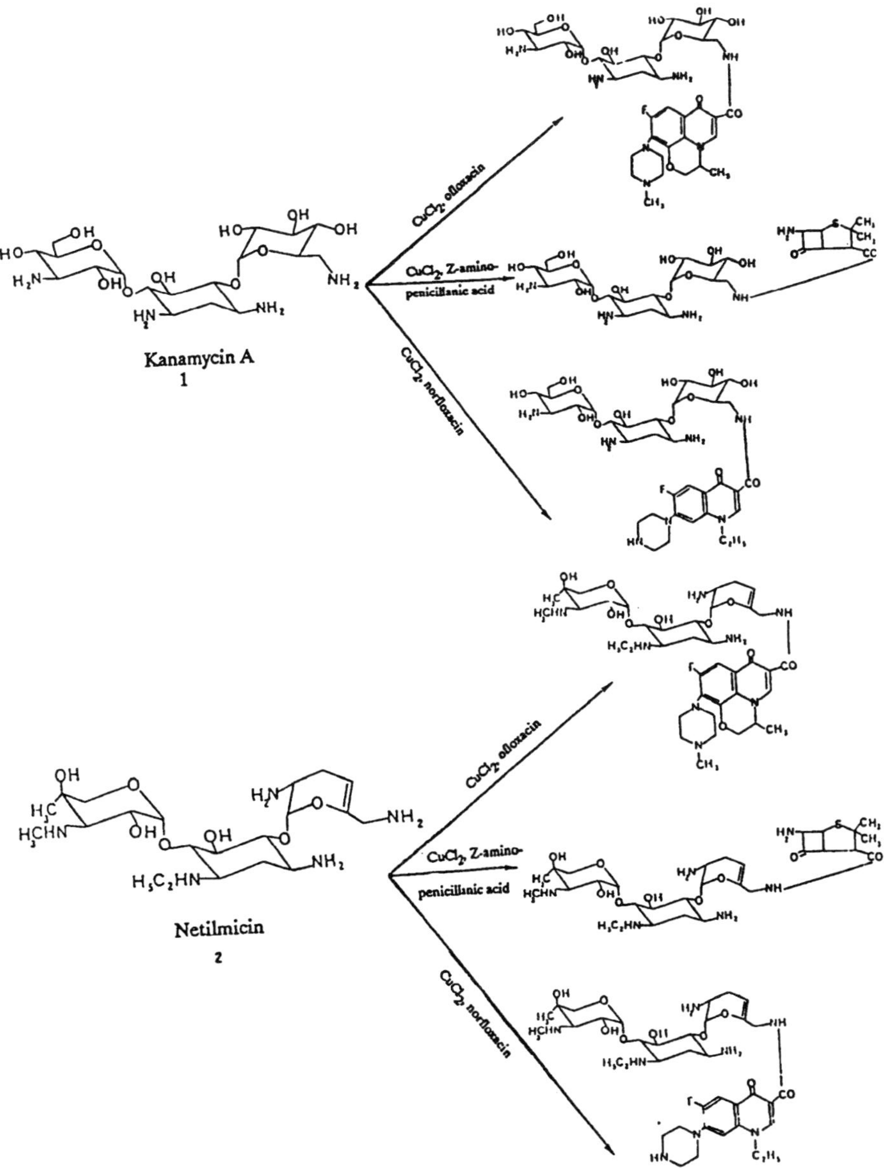

Scheme

Table III. Some of the strong IR bands of the compounds

	OH, NH$_2$	NHCO	OCONH	CO	C=C	C=O(β-lactam)
1a	3600-3200	1680	----	1660	1620	----
1b	3600-3100	1670	1700	----	----	1785
1c	3600-3200	1675	----	1660	1620	----
2a	3600-3200	1670	----	1655	1615	----
2b	3600-3200	1670	1700	----	----	1785
2c	3600-3200	1680	----	1660	1620	----

The antibacterial tests are under investigation.

References

1. K.,Jr. Rinehart , T. Suami , <u>Aminocyclitol Antibiotics,</u> ACS Symposium Series 125, Am. Chem. S.: Washington, D.C., (1980).
2. P.S. Lietman in <u>Principles and Practice of Infectious Diseases,</u> G.L. Mandell , R.G. Douglas, J.E. Bennet, Ed., John Wiley & Sons: New York, pp 192-206, (1985).
3. H.A. Albercht, G. Beskid, K-K. Chan, J. Christenson, R. Cleeland, K. Deitcher, N. Georgopapadakou, D. Keith, D. Pruess, J. Sepinwall , Specian A. Jr., Then R.,Weigele M., West K., Yang R., *J. Med. Chem.* **33**, 77, (1990).
4. H. A. Albercht, G. Beskid, J. Christenson, W. Durkin, V. Fallat, N. Georgopapadakou, D. Keith, F. M. Konzelmann, E. R. Lipschitz, D.H. McGarry, J. Siebelist , C.C. Wei, M. Weigele, R. Yang, *J. Med. Chem.* **34**, 669, (1991).
5. H.A. Albercht, G. Beskid, J. Christenson, N. Georgopapadakou, D. Keith, F.M. Konzelmann, D. L. Pruess, P. L. Rossman, C.C. Wei, *J. Med. Chem.* **34**, 2857, (1991).
6. A.E. Brown, O. Quesada, D. Armstrong, *Antimicrob. Agents Chemother.* **21**, 592, (1982).
7. D. Bouzard, <u>Recent Progress in the Chemical Synthesis of Antibiotics,</u> Springer-Verlag: Heidelberg, (1990).
8. J.P. Sanchez, J.M. Domagala, C.L. Heifetz, S.R. Priebe, J.A. Sesnie, A.K. Trehan, *J. Med. Chem.* **35**, 1764, (1992).

Complexes of Dirhodium Trifluoroacetate with Nucleosides and Nucleotides

A. KOUTSODIMOU[1], C.I. STASSINOPOULOU[2] AND N. KATSAROS[1]
Inst. of Physical Chemistry[1] and Biology[2], NCSR "Demokritos", 153 10 Ag. Paraskevi Attikis, Greece

1. Introduction

The effect of transition metals on nucleotides, nucleic acids and their constituent bases has implications relating to heavy metal toxicity. The efficacy of transition metal antitumor agents and the use of metals as labels in structural studies by x-ray or electron microscopy techniques is well known. Rhodium carboxylates are amongst the most promising of the second generation platinum metal anticancer compounds [1].

They appear to play an important role in the inhibition of DNA (possibly by inhibition of enzymes essential for its synthesis) and protein synthesis. The inhibition of DNA synthesis has also shown minimal inhibition of RNA synthesis.

The inhibition of DNA synthesis in vitro follows an order of methoxyacetate < acetate < propionate < butyrate [2]. Investigations into the chemical properties and biological effects of the rhodium(II) carboxylates, have shown that the species containing unprotonated amino groups, including adenine nucleotides, single stranded DNA, RNase A, bovineserum albumin and certain amino acids (notably histidine and methionine) bind tightly but reversibly in the axial position, whereas sulphydryl compounds (cystein for example) bind irreversibly with rhodium and liberate carboxylate ions upon reaction.

Relatively few complexes of dirhodium acetate have been prepared with various types of ligands and even less with nucleosides and nucleotides [3,4]. No complexes have been prepared so far with nucleic acid derivatives and dirhodium trifluoroacetate, $Rh_2(CF_3CO_2)$ (RT).

In order to understand the antitumor properties of rhodium carboxylates and to obtain complexes other than adenine derivatives, we used dirhodium trifluoroacetate instead of the acetate derivative. The highly electronegative CF_3 substituents enhance the Lewis

acidity of the Rh atoms so as to allow axial interactions with relatively weak donor molecules. We report here, the synthesis of RT complexes with cytidine (Cyd) and the disodium salts of cytidine-5'-monophosphate (CMP) and guanosine-5'-monophosphate (GMP), the nature of the isolated adducts and the bonding sites of the metal with the bases.

271

N. Russo et al. (eds.), Properties and Chemistry of Biomolecular Systems, 271–275.

© 1994 *Kluwer Academic Publishers. Printed in the Netherlands.*

2. Experimental section

Preparation of $Rh_2(CF_3CO_2)_4$ complexes with: a) Cyd: An aqueous solution of cytidine was mixed with a methanolic solution of dirhodium trifluoroacetate to a ratio of 5:1. The mixture was allowed to evaporate to dryness, the product was washed repeatedly with acetone, diluted in methanol, filtered and dried. b) CMP: The nucleotide was added to a methanolic solution of RT in a ratio of 2:1. The solution was stirred overnight and upon cooling it gave a blue-green precipitate. It was washed with methanol and acetone and was dried under vacuum over P_2O_5. c) GMP: To an aqueous solution of guanosine-5'-monophosphate was added an aqueous solution of RT to a ratio 1:1. The mixture was stirred for four days and a green precipitate was obtained which was washed with hot water, acetone and diethyl ether. The product was finally dried in vacuum.

The aforesaid complexes were characterized by elemental analysis, magnetic susceptibility measurements, UV-Vis, IR and NMR spectroscopies. Solubility restrictions permitted solution NMR studies only to the product formed by the reaction of RT with cytidine.

3. Results and discussion

Reaction of dirhodium trifluoroacetate with Cyd yields $Rh_2(CF_3CO_2)_2(Cyd)_4 \cdot 3H_2O$ (RCD), with CMP gives $Rh_2(CF_3CO_2)_2(CMP)_4 \cdot 18H_2O$ (RCM) and with GMP yields $Rh_2(CF_3CO_2)_2(GMP)_2 \cdot 10H_2O$ (RGM). All complexes were diamagnetic. Their elemental analysis and physical properties are given in Table 1.

Table 1. Analytical data and physical properties of the complexes

Compound		C%	H%	N%	F%	Colour	Decomposition point (°C)
RCD	Found	32.34	4.12	10.79	8.71	blue	200
	Calcd	32.94	4.01	11.52	7.82		
RCM		22.04	3.43	6.98	4.15	blue-green	215
		21.59	3.83	7.55	5.12		
RGM		20.56	2.94	10.13	7.01	green	250
		20.21	3.11	9.82	7.99		

Rhodium carboxylates usually form adducts with molecules containing donor atoms with bond formation occurring at the two axial positions. Few Rh_2^{+4} complexes containing mixed ligands have been isolated and structurally characterized [5,6]. The visible spectra of the adducts are very sensitive to the nature of the donor atom [7]. Those adducts in which an oxygen atom of the ligand is bound to rhodium are blue or green; those involving nitrogen are pink, rose-red or violet, while those containing sulfur or phosphorus range from burgundy to orange. As it is shown in Table 1 the colour of all three complexes was green or blue indicating that oxygen is involved in the bonding of Rh(II) with the ligands in the solid state.

The oxygen involvement in the bonding of the metals with the bases is also confirmed by the visible spectra of the complexes in water. The electronic spectrum of rhodium trifluoroacetate in water consists of two bands in the visible region, band I (574nm), and band II (448nm) with two stronger bands, III and IV, in the UV (ca. 250nm (shoulder) and 219nm respectively). Upon complexation band I is not affected very much impliciting interaction of the Rh_2^{+4} core with an oxygen atom [3,7].

The IR spectra of the complexes are given in Table 2.

Table 2. Some of the strong IR bands of the compounds

Cyd	RCD	Assign.	CMP	RCM	Assign.	GMP	RGM	Assign.
1655	1670	νC=O	1659	1660	νC=O	1703	1650	νC=O
1610	1585	νC=N	1652	1580	δNH_2+νC=N +νc=C	1638		δNH_2
1535	1525	ring	1110	1115	vasPO_3+νC-O sugar	1085	1062	νPO_3deg
1503	1500	vibrat	1080	1060				
1380			980	972	νPO_3sym	982	970	νPO_3sym
1295	1290	νC-N_{ext}	820	805	νP-O	820	822	νP-O
985	995	OH$_{def.}$					800	
790	775							

From the IR spectra the following conclusions can be drawn:

a) For RCD: Small shift of the carbonyl stretching band and decrease in the intensity of the νC=N band means probable interaction with the exocyclic oxygen of the pyrimidine ring.

b) For RCM: Broadening of the νC=O and shifting of the νC=N band at lower frequencies indicate possible binding to the carbonyl oxygen.

c) For RGM: Shifting (-40 cm^{-1}) of the band assigned to C6=O stretching, indicates coordination through C6=O. Shifting without splitting of the strong bands at 1085 and 982 cm^{-1} and splitting of the medium νP-O band at 820 cm^{-1} indicate an indirect phosphate-metal bonding via a hydrogen bonded water molecule [8].The ^{1}H and ^{13}C spectra of the cytidine complex reveal that Rh(II) is mainly bonded through the exocyclic oxygen of the nucleoside. The assignments are given in Tables 3, 4.

The ^{1}H-NMR spectra in Methanol-d$_4$ indicate two possible types of complexes: a, where $\Delta\delta$ of H(6) is +0.13 and b, where $\Delta\delta$ of H(6) is +0.28 in a ratio of a:b=2:1. At elevated temperature in DMSO-d6 it is possible that there are two different types of complexes. The main site of coordination at 80^0C is probably the N3 atom (pink colour of the solution, the chemical shift of H(5) +0.46 is greater than $\Delta\delta$ of H(6) +0.41) or perhaps the DMSO itself binds to the metal. The greater chemical shifting of H(6) than H(5) indicates C2=O as possible site of interaction in D_2O.

Table 3. ^{1}H-NMR spectra (ppm downfield from TMS or DSS)

Compound	H(6)	H(5)	Medium
Cyd	8.10d	5.95d	MeOD-d4
RCD	8.23d 8.23d	6.04d 6.23d	MeOD-d4
($\Delta\delta$)	(+0.13) (+0.28)	(+0.09) (+0.28)	
Cyd*	7.80d	5.80s	DMSO-d6
RCD*	7.90d 8.21m	5.85d 6.26m	DMSO-d6
	(+0.10) (+0.41)	(+0.05) (+0.46)	
Cyd	7.85d	6.05d	D_2O
RCD	8.02d	6.24d	D_2O
	(+0.27)	(+0.19)	

d=doublet, s=singlet, m=multiplet, * temperature at 80^{0}C

Binding at N(3), binding at O(2), or binding at both N(3) and O(2) will in general lead to withdrawal of electron density from about the same region of the cytosine ring. Downfield ^{1}H NMR shifts are expected in all cases. So with ^{1}H NMR alone it is not possible to distinguish these sites.

Table 4. ^{13}C-NMR chemical shifts (ppm downfield from TMS)

Compound	C(4)	C(2)	C(6)	C(5)	C(1')	C(4')	C(2')	C(3')	C(5')
Cyd*	166.4	156.4	142.4	94.9	90.1	85.0	74.8	70.3	61.6
RCD**	168.1	163.2	142.3	94.1	89.9	84.2	74.2	68.9	60.0
			139.8	96.4					

*in DMSO-d6 **in MeOD-d4

It is known that the ^{13}C chemical shifts of the heteroaromatic ring carbon are related to the electronic structure of the carbons. The characteristic change which signals the binding mode of the metal to cytidine, is the direction of shift of C(2) resonance. The complexation of this nucleoside with dirhodium trifluoroacetate leads to downfield shift (+6.8ppm) of the C(2) indicating that the main site of interaction is the carbonylic oxygen C2=O [9]. The appearence of two peaks corresponding to C(6) and C(5) implies the possible existence of two types of complexes in methanol-d$_4$, fact also observed in

the ^{1}H NMR spectrum.
It is thus concluded that the main site of interaction for dirhodium trifluoroacetate at room temperature is the exocyclic oxygen of the pyrimidine ring for cytidine and its nucleotide and the carbonyl oxygen for GMP.

References

1. E.B. Boyarand and S.D. Robinson "Rhodium (II) Carboxylates".*Coord. Chem. Rev.* **50**, 109-208, (1983).

2. A. Erck, L. Rainen, J. Whileyman, I.M. Chang, A.P. Kimball and J. Bear "Studies of Rhodium (II) Carboxylates as potential antitumor agents" *Proc. Soc. Exp. Biol. Med.* **145**, 1278-1283, (1974).

3. G. Pneumatikakis and N. Hadjiliadis "Interactions of Tetrakis-u-acetato-rhodium (II) with adenine nucleosides and nucleotides" *J. Chem. Soc. Dalton Trans.*, 596-599, (1979).

4. N. Farrell "Adenine and Adenosine derivatives of Rhodium acetate" *J. Inorg. Biochem.* **14**, 261-265, (1981).

5. P. Piraino, G. Brono, G. Tresoldi, S. Lo Schiavo and P. Zanello "Chemical oxidation of Binuclear Rhodium (I) Complexes with Silver Salts, X-ray Crystal structure, and Electrochemical Properties of the Rh_2^{+4} Mixed -Ligand Complex $Rh_2(Form)_2$- $(O_2CCF_3)_2(H_2O)_2 \cdot 5C_6H_6$ (Form= N.N-Di-p-tolylformamidinate Anion) *Inorg. Chem.* **26**, 91-96, (1987).

6. E.C. Morrison and D.A. Tocher "New ortho-Metallated Dirhodium (II) compounds" *Inorg. Chim. Acta* **157**, 139-140, (1989).

7. J. Kitchens and J.L. Bear "A study of some Rhodium (II) acetate adducts" *J. Inorg. Nucl. Chem.* **31**, 2415-2421, (1969).

8. H.A. Tajmir-Riahi and T. Theophanides "Platinum(II) and magnesium (II) nucleotide complexes. Synthesis, FT-IR spectra and structural properties" *Can. J. Chem.* **61**, 1813-1822, (1983)

9. L.G. Marzilli, B. de Castro, J.P. Caradonna, R.C. Stewart and C.P. van Vuuren "Nucleoside complexing. A Raman and ^{13}C NMR spectroscopic study of the binding of hard and soft metal species" *J. Am. Chem. Soc.* **102**, 916-924, (1980).

Synthesis of Amino Acid and Peptide Derivatives of Aminoglycosides Targeted Against Resistance of Bacteria

S.I. KOTRETSOU, V. CONSTANTINOU-KOKOTOU, and M.P. GEORGIADIS
Chemical Laboratories, Agricultural University of Athens, Iera Odos 75, Athens 11855, Greece

1. Introduction

Interest in aminoglycoside antibiotics, in spite of their long familiarity, remains high because of their clinical utility. Unfortunately, the latter is limited by oto- and nephrotoxicities and is related to the binding of these polycationic drugs to negatively charged phospholipids and to the subsequent inhibition of lysosomal phospholipases [1,4]. It is also well known that bacterial resistance is associated with specific chemical transformattions in the molecule This has resulted in the research for novel semisynthetic aminoglycoside derivatives that would not be capable of such enzymic modification. The discovery of the most prominent aminoglycoside antibiotics, natural or semisynthetic, are butirosin, amikacin, fortimycin, sporaricin and minosaminomycin, which contain amino acids linked by an amide bond, has emphasized the favorable effects of aminoacyl substituents on aminoglycosides [5].

In a continuation of our studies on the synthesis and pharmacology of aminoglycoside antibiotics [6], we present here the synthesis of 6'-*N* and 1-*N* amino acid and peptide derivatives of kanamycin A (1), as well as the synthesis of 6'-*N* derivatives of netilmicin (7), targeted against resistance of bacteria, utilizing transition-metal cations to selectively control the site of acylation.

2. Experimental section

Treatment of a mixture of 1 or 7, free base and $CuCl_2$ in DMSO gave after 1 h stirring at r.t. the suitable complex. *In situ* reaction of the free 6'-amino group with *N*-hydroxysuccinimide esters of *N*-protected amino acids and peptide gave, after purification using Amberlite CG-50 (NH_4^+), compounds 2 and 8. Thus 6'-*N* derivatives with L-Ala-OH, D-Ala-OH, Gly-OH, L-Leu-OH, L-Asp-OH, L-Ala-L-Ala-OH were obtained in one step. Removal of the protecting groups using HCl 4N in THF or TFA yielded 6'-*N* derivatives 3 and 9 (Scheme I, II).

The synthesis of 1-*N* derivatives of kanamycin was performed through the selective blocking of 6'- and 3'-amino group. Treatment of 1 with *t ert*-butyl-dicarbonate in the

277

N. Russo et al. (eds.), Properties and Chemistry of Biomolecular Systems, 277–280.
© 1994 *Kluwer Academic Publishers. Printed in the Netherlands.*

presence of Cu^{+2} gave after column purification 6',3-*N*,*N*-di-Boc-kanamycin A, which in turn was coupled with the above mentioned amino acids and peptide using the *N*-hydroxysuccinimide ester method. Deprotection of 5 yielded 1-N derivatives 6.

3. Results and Discussion

1 and 7 share a common pseudotrisaccharidic structure containing the diaminocyclitol 2-deoxystreptamine associated with two amino sugar moieties. The selective fomation of the amide bond at the 6'-*N* position of 1 and 7 based on the temporary protection of vicinal hydroxy and amino groups as copper chelates. Structural proof of the synthesized products was obtained from mass spectra (FAB) and ^{13}C-NMR analysis.

Scheme I

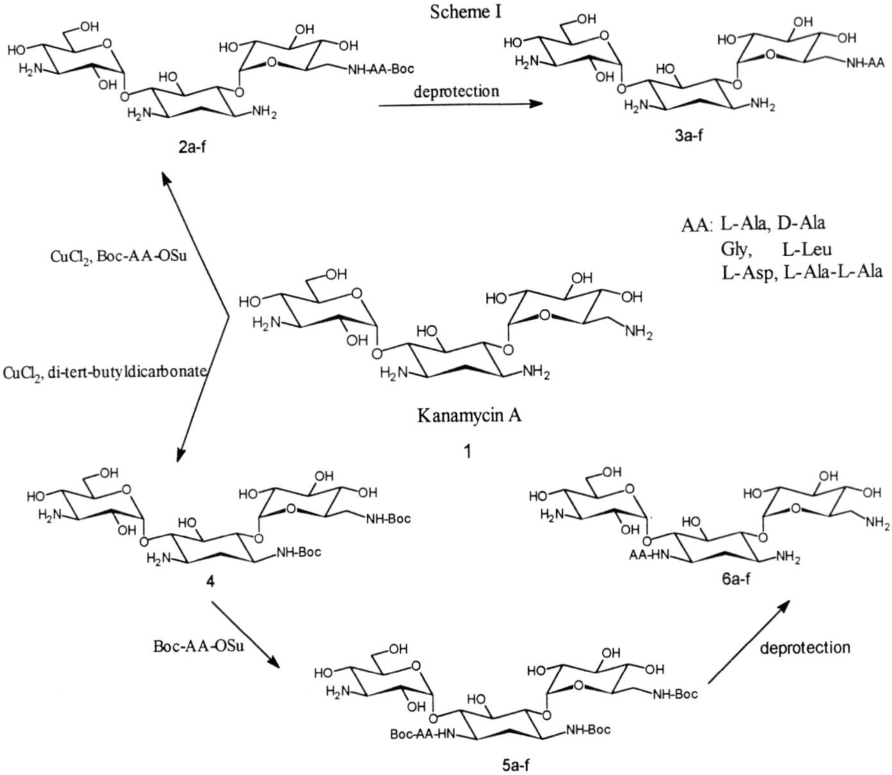

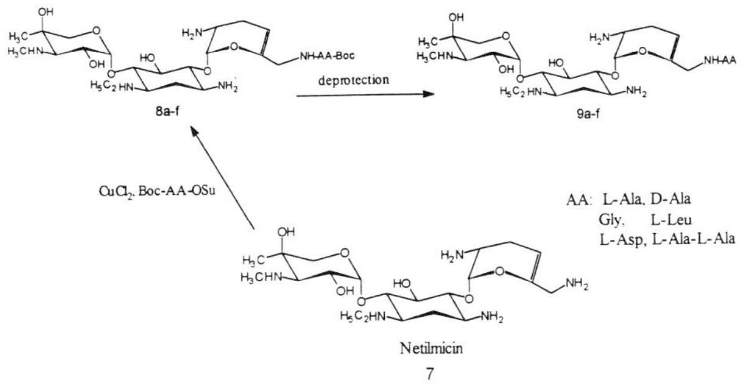

Scheme II

All the mass spectra showned a significant peak with m/z corresponding to the value calculated for the $[M+H^+]$ or $[M+23^+]$ ion of the free base (Table I).

In the ^{13}C-NMR spectra protoniation of amino groups causes upfield shift of the α- and β- carbon signals due to the shielding. However in the 6'-N derivatives, the protoniation shift for C-1' is about half of the ordinary value. On the other hand 6'-N acylation causes an upfield shift of the C-6' and C-5'.

Thus the upfield shifts of carbons C-2 and C-6 of 4 in comparison to those of 5 indicated that the amino acids or peptide were coupled with C-1 amino group. Similarly the presence of the amino acid at the C-6' position was indicated by the typical upfield shift [7] of the C-6' and C-5' of compound 2 in comparison with the free kanamycin A. Besides, the equal chemical shifts of C-4 and C-6 indicated the absence of substitution at C-1 or C-3 of compound 2.

The antibacterial tests are under investigation.

Table I. FAB-mass spectra

3a	556 $[M+H^+]$, 436,395,333
3c	542 $[M+H^+]$, 381, 348, 219
3d	598 $[M+H^+]$, 437, 275
3e	600 $[M+H^+]$, 439, 277
3f	627 $[M+H^+]$, 466, 448, 433
4	685 $[M+H^+]$, 506, 424, 262
6a	557 $[M+2H^+]$, 424, 396, 378
6c	542 $[M+H^+]$, 381, 363, 291
6f	627 $[M+H^+]$, 466, 448, 291
9a	547 $[M+H^+]$, 416, 388, 198
9c	533 $[M+H^+]$, 374, 274, 184
9d	589 $[M+H^+]$, 430, 412
9f	618 $[M+H^+]$, 458, 268

Table II. ^{13}C-NMR data (δ, ppm)

	4*	5a*	2a**	6a**	base H$^+$	1**
C-1	51.8	51.0	50.9	50.5	51.3	50.6
C-2	36.2	34.0	35.3	31.4	36.2	28.3
C-3	50.9	50.8	49.8	45.7	49.8	48.4
C-4	83.1	83.2	86.5	78.8	88.1	79.0
C-6	88.4	83.7	87.4	84.7	88.6	84.7
C-1'	101.9	102.0	100.0	101.4	100.3	96.3
C-2'	73.1	72.1	70.9	71.6	72.7	71.6
C-3'	74.2	73.3	72.4	72.7	73.7	73.0
C-4'	72.4	72.1	68.4	70.5	71.8	71.6
C-5'	72.0	72.0	72.0	69.4	73.7	69.5
C-6'	42.3	42.4	40.2	41.4	42.4	41.2
C-1"	101.3	98.9	99.9	95.8	100.8	101.2
C-2"	72.9	69.9	70.1	67.8	72.7	69.0
C-3"	58.9	57.3	55.0	54.6	55.0	55.8
C-4"	71.8	67.8	62.5	65.5	70.1	66.4
C-5"	74.3	74.3	69.5	72.8	72.9	73.0
C-6"	61.1	61.1	60.7	60.2	61.1	60.8

*in DMSO-d$_6$; **in D$_2$O

References

1. G. Kahlmeter, J. Dahlager, *J. Antimicrob. Chemother.* **13**, 9, (1984).
2. A. Van Schepdael, J. Delcourt, M. Mulier, R. Busson, L. Verbist, H.J. Vanderhaeghe, M.P. Mingeot-Leclercq, P.M. Tulkens, P.J. Claes, *J. Med. Chem.* **34**, 1468, (1991).
3. M.P. Mingeot-Leclercq, A. Van Schepdael, R. Brasseur, R. Busson, H.J. Vanderhaeghe, P.J. Claes, P.M. Tulkens, *J. Med. Chem.* **34**, 1476, (1991).
4. A. Van Schepdael, R. Busson, H.J. Vanderhaeghe, P.J. Claes, L. Verbist, M.P. Mingeot-Leclercq, R. Brasseue, P.M. Tulkens, *J. Med. Chem.* **34**, 1483, (1991).
5. T. Haskell, R. Rodebaugh, N. Plessas, D. Watson, R. Westland, *J. Carbohyd. Res.* **28**, 263, (1973).
6. M.P. Georgiadis, V. Constantinou-Kokotou, G. Kokotos, *J. Carbohydr. Chem.* **10**, 739-748, (1992).
7. K.F. Koch, G.A. Rhoades, E.W. Hagaman, E.Wenker, *J. Amer. Chem. Soc.* **96**, 3300-3305, (1974).

Fusion and Aggregation of Phospholipid Vesicles: Experimental Problem of The Distinction

C. LA ROSA and D. GRASSO
Dipartimento di Scienze Chimiche, Università di Catania
Viale A.Doria, 6-95125 Catania (ITALY)

1. Introduction

In recent years many experiments have been carried out on the fusion problem [1-4]. It is known that the fusion between lipid membranes is involved in many biological processes such as endocytosis, exocytosis, transport processes between cellular organelles, virus infection and so on [5]. The natural cell is a complex system, so a model system, like phospholipid vesicles, is used in this study, because of the biological reality. Membrane fusion results in communication between two aqueous compartments initially separated by the txo fusing membranes and involves two distinct steps: a) the close approach and adhesion of the membranes (aggregation) and b) the destabilization and merging of the lipid bilayers in the region of adhesion, with concomitant mixing of contents within the two acqueous compartments. Therefore, the fusion is subsequent to the aggregation process. Many papers in literature are available on this argument, and we have found that phosphatidylserine undergoes steps a) and b), while in phophatidylcoline the process is stopped at the first step. Quantitative tecniques for monitoring the membrane fusion kinetic are mainly based on fluorescence assays [6,7]. All of these methods make use of fluorescent probes which intermix after the fusion event, and the conseguent spectroscopic variations are recorded as a functon of time. In a recent paper some authors have pointed out that different fluorescent assays may lead to contrasting results [8]. Other experimental tecniques have been used. Amongst them, we recall NMR [9] light scattering [10] and turbidity spectroscopies [11]. Recently, we have developed a thermovomuletric (SD) technique which has been shown to be successfully used to measure various parameters, as well as the gel-liquid crystal transition temperature, thermal expansion coefficient, molar volumes in different phases [12], isothermal compressibility [13], lateral phase separation, together with is time-evolution [14] and kinetic fusion [15]. The evaluation of all these parameters can be useful in understanding the chemical-physical variables affecting the phenomenon of fusion. Besides, all these measurements are obtained by using the same apparatus and, often, the same samples. This allows the reduction of the experimental errors, loss of time and chemicals used. In this paper we carried out investigations on phosphatidylserine (fusion) and phophatidylcoline (aggregation) phospholipid vesicles, which have been already investigated by fluorimetric fusion assays as reported in literature, turbidimetric (TB), scanning and isothermal dilatometric (SD), and differential

N. Russo et al. (eds.), Properties and Chemistry of Biomolecular Systems, 281–289.

© 1994 *Kluwer Academic Publishers. Printed in the Netherlands.*

scanning calorimetric (DSC) measurements, in order to distinguish between fusion and aggregation phenomena.

2. Experimental

2.1. MATERIALS

Dipalmitoylphosphatidylcholine (DPPC), (FLUKA) and dipalmitoylphophatidylserine (DPPS) (SIGMA) were commercial products; the purity was greater than 99% as judged by bidimensional thin-layer chromatography in which a silica gel plate loaded with solution of the lipid in CHCl was developed first in $CHCl_3$-CH_3 OH-(7N)NH_4OH 60:30:5 (by volume) and successively in $CHCl_3$-CH_3OH-CH_3COOH-H_2O 12:60:8:2.5 (by volume). Sodium chloride were recrystallized from bidistilled water and tested for absence of calcium by atomic absorption.

2.2. PREPARATION OF LIPOSOMES

Aqueous dispersion of pure phospholipids and mixtures of DPPC, and DPPS were prepared by the following procedure. Pure lipids were dissolved in $CHCl_3$-H_3-OH 1:1 (by volume); the solvent was then removed at 30 °C on rotating evaporator by a nitrogen stream, followed by overnight high vacuum storage. On the thin film so obtained was added electrolitic water solution (NaCl 0.1 M) and heated at a higher temperature than that o the gel-liquid crystal phase transition, characteristic of each single phospholipid. The sample was vortexed twice for two minutes. This procedure was followed to make multilamellar vesicles (MLV). After each experiment the total phosphorus were determined spectrophotometrically [16]. Differential scanning calorimetry (DSC). A Mettler TC10A processor equipped with a DSC 20 measuring modulus was used for calorimetric analysis, after calibration in temperature and energy using indium, auric, capric and miristic acids as standards. The plotting range, as full scale deflection, was set to 1 mW. Entalphy change, was evaluated using the integration program (Bezout method) of IBM XT computer. The noise (RMS) was 0.006 mW; Thermograms was detected in heating and cooling modes, in the 30-80 °C range, with a precision of 0.1 °C.

About 5 mg of phospholipid (120 μl suspension) was used for experiment.

2.3. SCANNING AND ISOTHERMAL DILATOMETRY (SD,ID)

A Mettler TC 10A processor equipped with TMA40 thermomechanical modulus, previously calibrated for temperature and weight was used, in order to obtain measurements of light as function of temperature, at constant pressure. The accurate procedure for calibration and special sample-cell are reported in ref [13]. The sample's volume was about 250 μl (12 mg).

2.4. DENSITY MEASUREMENTS

The density was measured, mainly in order to obtain precise values for start volume in SD experiment. The densities were measured with an Anton Paar "vibrating tube" digital densitometer (DAM 602/60). Thermal stability of the liquid flowing through the jacket around the measuring cell was maintained within 0.001 °C by a Techne Tempette TE-8D temperature controller. The densitymeter was calibrated with bidistilled water, air free, every day. Reproducibility of the density measurements was better than 15 ppm.

2.5. TURBIDITY MEASUREMENT

The turbidity was measured using a Pelkin Elmer 330 spectrophotometer equipped with a thermostated cell. The absorbance was measured at 350 nm.

3. Results and discussion

Briefly, we recall the physical basis of the volumetric variation accompanying the fusion processes. When two small lipid vesicles fuse, the resulting volume is slightly smaller than the sum of the single volumes, because in small vesicles the packing of lipid molecules is strained owing to curvature effects. This strain leads the system to increase the curvature radius, allowing a better packing of the phospholipids. Consequently, in the hypothesis of constant thickness of the bilayer (this hypothesis is supported by x-ray measurements [17]), the total volume of the vesicles becomes smaller. By exploiting this property we can follow intervesicle fusion kinetics without introducing any foreign probes. In isothermal conditions, we observe that the relative variation v.s. time, decreases exponentially reaching an asymptotic value when the process is completed. This happens when most of the small vesicles are fused forming large aggregates; these do not give further volume variations in subsequent fusion events.

In previous papers [14,15] we have tested the validity of the association of volume variation with fusion process, through preliminary experiments at various vesicle concentrations. The aggregation process consists of the surface adhesion among vesicles and also of obtaining a vesicles cluster. In this process we do not observe any volume or heat change, but we detect one change in light scattering. In figure 1 we report a schematic drawing of aggregation and fusion processes.

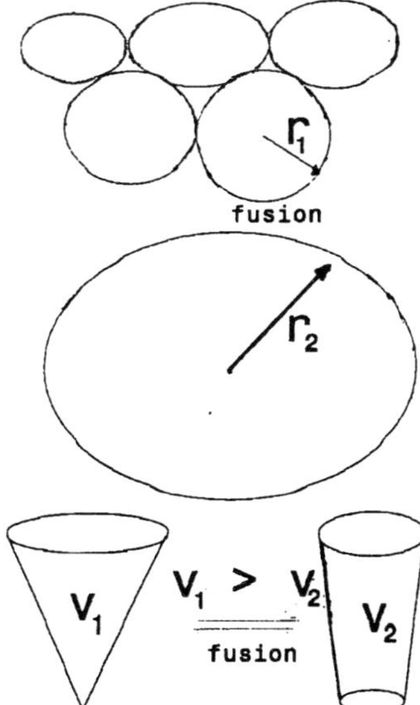

Figure1. Schematic drawing of fusion process and the related molar volume variation of phospholipid molecule.

In order to distinguish between fusion and adhesion we perform first DSC, SD and turbidity measurements on a) known fusogen system, such a DPPS MLV vesicles

[18,14]; b) a known aggregation system such as DPPC MLV vesicles [19]; c) known system which are neither aggregation nor fusion systems, such as DPPE MLV vesicles [20]; and d) on the above systems we also performed isothermal volumetric measurements. In this section we discuss the results of three distinct experimental investigations: (a) DSC and SD measurements; (b) turbidity measurements and (c) isothermal volumetric measurements.

(a) *Differential Scanning Calorimetry and Scanning Dilatometry measurements*. In Figure 2a and 2b we show both DSC and SD curve relative to DPPC in NaCl 0.1 M.

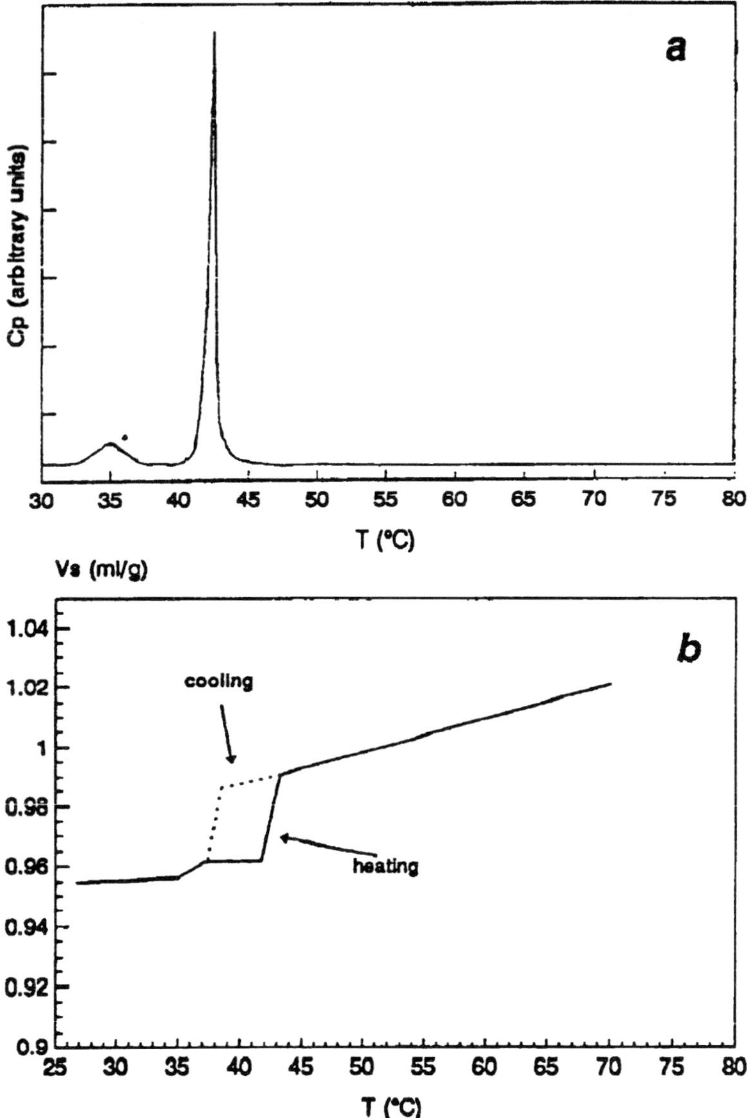

Figure 2. Differential scanning calorimetry (a) and scanning dilatometry (b) trace of DPPC vesicles in sodium chloride solution (0.1 M).

It is known [21] that DPPC MLV vesicles show three phases, L_β, called gel phase, P_β, called ripple phase and L_α called liquid crystal phase. DSC of this system show two peaks: a broad one with maximum of 35 °C (pre-transition) and a sharp one with a maximum of 42.2 °C (main-transition). Moreover, we observed that a lot of heating-cooling cycles are the same. from this we can only assume that the process is reversible. Figure 3a and 3b display both DSC and SD traces relative to DPPE in NaCl 0.1 M.

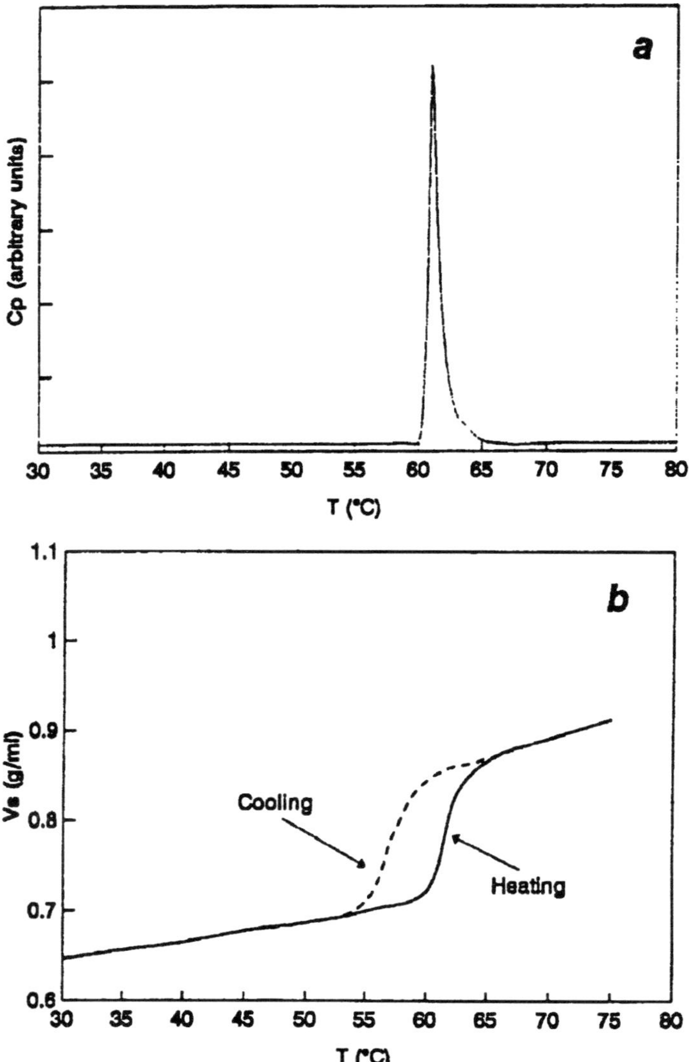

Figure 3. Differential scanning calorimetry (a) and Scanning dilatometry (b) trace of DPPE vesicles in sodium chloride solution (0.1 M).

These vesicles show two phases L_β and L_α, while the gel-liquid crystal transition occurs at 63 °C, and is reversible. In figure 4a and 4b DSC and SD traces from the DPPS system are shown, in NaCl 0.1 M.

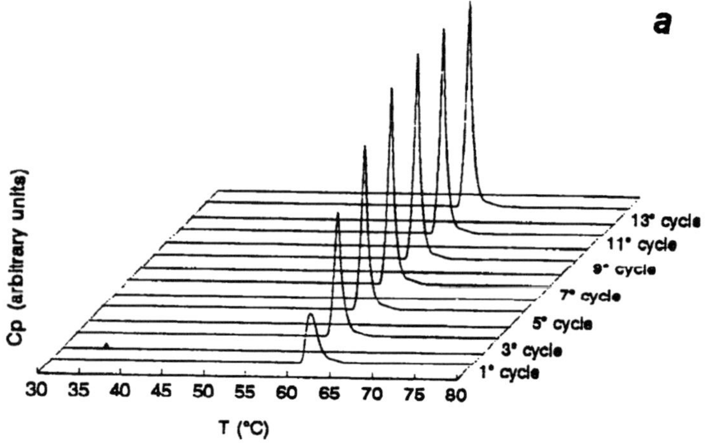

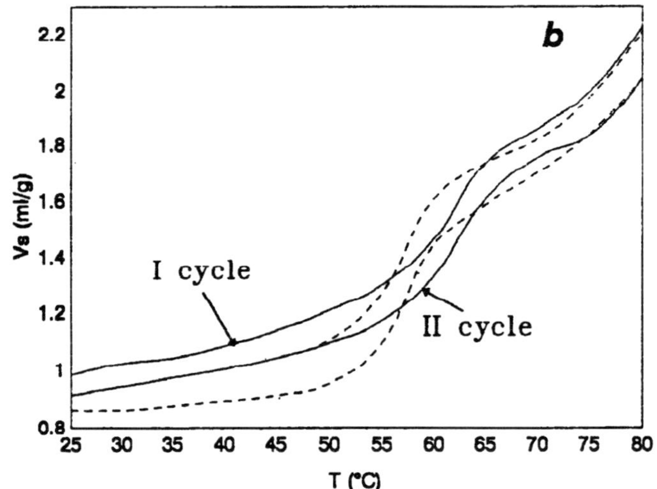

Figure 4. Differential scanning calorimetry (a) and scanning dilatometry (b) trace of DPPS vesicles in sodium chloride 0.1 M.

The heating profiles show the gel-liquid crystal phase transition at 62.5 °C. These vesicles are characterized by different behaviors from the heating-cooling cycles: from DSC traces it is possible to note in the first heating run, the existence of a broad peak due to the gel-liquid crystal transition. In the second heating run the peak becomes narrow and in subsequent cycles the peak shrinks, until it becomes stable. From SD registrations it is possible to note that the volume, after each thermal cycle, decreases. This irreversible phenomenon, observed by DSC and SD is due to the fusion process [21]. In fact, the first DSC broad peak is due to the low cooperativity of the system, which consequently increased at the fusion phenomenon, while the volume decrease is due to the surface area lowering of the phospholipid molecules (see above). In conclusion DSC and SD techniques are able to detect fusion phenomenon, giving as parameters half of the height of the peak and the volume change. They cannot be used for quantitative kinetic analysis. Moreover, they are not sensitive to the aggregation process. This was revealed by DPPE (these vesicles neither fused nor aggregated. Unfortunately, DSC and SD measurements were not used for quantitative kinetic analysis.

b) *Turbidimetric measurements*. Turbidimetric measurements were used in order to obtain quantitative kinetic data on fusion phenomenon [11]. In figure 5b we showed turbidity change versus time, for DPC, DPPE and DPPS vesicles.

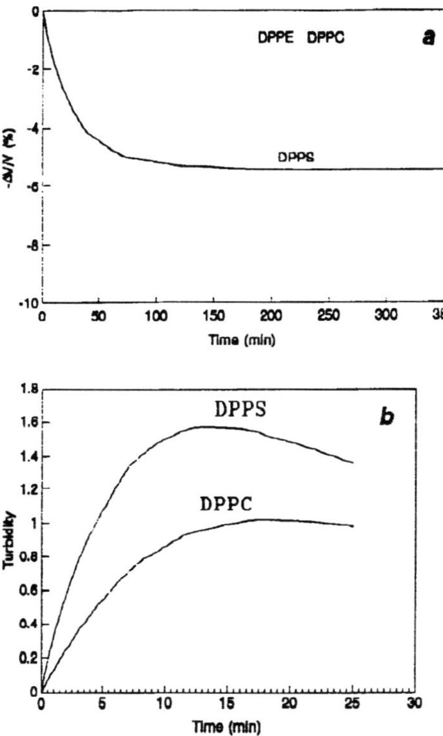

Figure 5. Isothermal volumetric measurements (b) as $\Delta V/V$ percentage versus time of DPPC, DPPE and DPPS vesicles in sodium chloride solution (0.1M). Turbidimetric measurements (a), as absorbance versus time of DPPC and DPPS vesicles in sodium chloride solution (0.1 M). In DPPE suspension did not observe any turbidity change. All measurements was carried out at 70 °C (liquid crystal state). In gel state did not observe any turbidity change.

The experimental conditions were the same as those for DSC and SD experiments. In the gel state we did not observe any turbidity change in all systems. In liquid crystal state we observed increasing turbidity in both the DPPS and DPPC systems. The turbidity increase was linked to the rising volume of one dark object. This phenomenon is ascribible to the fusion of two vesicles, to obtain one vesicle. In the aggregation case, a cluster of vesicles is obtained due to the contact of many vesicles. Therefore, in both cases we observed increasing turbidity. Hence it is impossible to distinguish between the two phenomena only from these experimental data. In fact, we observe turbidity change in both DPPC and DPPS systems, but not in DPPE. From figure 5b we noted that the turbidity at first increases, and after decreases. This apparently strange behavior is due to the cluster sedimentation, when a critical mass is reached.

c) *Isothermal Dilatometric measurements*. Since both fusion and aggregation occurs only in the liquid crystal state [14, 22] (see turbidity measurements) we performed isothermal volumetric measurements at 70 °C. In figure 5a we showed isothermal $\Delta V/V$ percentage versus time for DPPC, DPPE and DPPS phospholipid vesicles in saline solution (NaCl 0.1 M). In DPPC and DPPE vesicles, curves 1 and 2, we did not observe any volume variation, while, DPPS vesicles (curve 3) showed decreasing volume. In the gel state we did not observe any volume variation for any ystem which agreed with turbidity measurements. When an aggregation phenomenon occurs, no volume change is observed, because the molar volume of the phospholipid molecule does not change as a consequence of the collision between two vesicles, the result in only the adhesion between the vesicle surfaces.

Moreover, the density of the vesicle suspension is constant. In the case of a fusogen system, the molar volume of a single phospholipid molecule changes. After the collision between two vesicles the curvature radii, increases with a consequent decreasing of the phospholipid molar volume.

Macroscopically we observed a decreasing density. In fact, in the case of a fusogen system (see figure 4) we observed a decreasing volume, while in other systems the volume is constant. Comparing the isothermal volumetric and turbidimetric curves, the kinetic data are not in complete agreement. From turbidity measurements the fusion stops after 20 min.; on the contrary, from isothermal volumetric experiments, the fusion finishes after 150 min. In our opinion these very different results are due to the cluster sedimentation.

In conclusion, we can say that DSC and SD measurements are able to distinguish fusogen from non fusogen systems, but they cannot give quantitative kinetic data. The same techniques are not able to detect aggregation processes. Turbidity measurements are able to detect the two processes, but they are not able to distinguish between aggregation and fusion. Instead, the isothermal volumetric tecnique is able to detect the fusion processes quantitatively, but is not able to detect aggregation processes. In any case, in order to monitor the kinetics and to distinguish between fusion and aggregation phenomena, the combined use of the two tecniques is necessary. In our opinion the problem can be solved using isothermal volumetry (ID) and turbidity (TB) tecniques together. In this way very simple tecniques are used without molecular probes.

Acknowledgement

This work has been partially supported by the Italian MURST.

References

1. J.H. Prestegard and M.P. O'Brien, *Ann. Rev. Phys. Chem.* **38**, 383, (1987).
2. S. Ohki, <u>Membrane Fusion,</u> A. Sowers (Ed.), New York plenum, (1986).
3. A.J. Verkleij, *Biochim. Biophys. Acta, Rev. on Biomembranes* **43**, 779, (1984).
4. J. White, M. Kielian and A. Helenius, *Q. Rev. Biophys.* **16**, 151, (1983).

5. T. Stegman, D. Hoekstra, G. Scherphof and J. Wilschut, *Biochemistry* **24**, 3107, (1985).
6. N. Dtzgtnex and J. Bentz, <u>Spectroscopic Membranes Probes</u>, L.M. Loew ed. (CRC press), (1986).
7. J. Wilschut, N. Dtzgtnex, R. Flaley and D. Papahadjopoulos, *Biochemistry* **19**, 6011, (1980).
8. N. Dtzgtnex, T.M. Allen, J. Fedor and D. Pahadjopoulos, *Biochemistry* **26**, 8435, (1987).
9. H.L. Kantor and J.H. Prestegard, *Biochemistry* **14**, 3592, (1978).
10. J.H. Goll and G.B. Stock, *Biophys. J.* **19**, 265, (1977).
11. L. Stamatatos, R. Leventis, M. Zuckermann and J.R. Silvius, *Biochemistry* **27**, 3917, (1988).
12. A. Raudino, F. Zuccarello, C. La Rosa and G. Buemi, *J. Phys. Chem.* **94**, 4217, (1990).
13. C. La Rosa and D. Grasso, *Il Nuovo Cimento Sez. D* **12**, 1213, (1990).
14. D. Grasso, C. La Rosa, A. Raudino and F. Zuccarello, *Liq. Cryst.* **3**, 1699, (1988).
15. F. Zuccarello, A. Raudino, C. La Rosa and D. Grasso, *Il Nuovo Cimento Sez. D* **13**, 1101, (1991).
16. G.R. Bartlett, *J. Biol. Chem.* **234**, 466, (1959).
17 N.P. Franks, W.R. Lieb, <u>Liposome: From Physical Structure to Therapeutic Application</u>, C.G. Knight ed., Elsevier, Amsterdam, (1981).
18. D. Papahadjopoulos, W.J. Vail, K. Jacobson and G. Poste, *Biochim. Biophys. Acta* **394**, 483, (1975).
19. R.C. Aloia, <u>Membrane Fluidity in Biology</u> Vol **2**, R.C. Aloia (Ed.), Academic Press , (1983).
20. C. La Rosa, Ph.D.Thesis, (Catania University, Catania), (1988).
21. F. Szoka and D. Papahadjopoulos, <u>Liposome: From Physical Structure to Therapeutic Application</u>, c.g. Knight editor, (1981).
22. J. Bentz, N. Dtzgtnex and S. Nir., *Biochemistry* **24**, 1064, (1985).

A Molecular Dynamics Study of Tyr-DAla-Phe-Gly-NH2 and of its 18-crown-6-ether Complex. A Model μ-Opioid Peptide and its Interaction with a Simulated Receptor Site

F. LELJ* and P. GRIMALDI
Dipartimento di Chimica, Universita' degli Studi della Basilicata, Via N. Sauro 85, I-85100 Potenza, Italy

1. Introduction

Great effort has been put into the identification of the conformation of flexible opiod agonists with the aim of identifying the so called " biological active conformation" and/or "conformations" [1]. Particular attention has been reserved to opioid peptides due to their relevance. Among the studied peptides Tyr-DAla-Phe-Gly-NH$_2$ (hereafter : TdAPG) is the minimum N-terminal fragment of dermorphin that retains considerable μ activity and selectivity. A careful 500 MHz NMR study of its complex with 18-crown-6-ether (hereafter: 18C6) in CDCl$_3$ showed [2] that the peptide conformation is characterized by a single intramolecular H-bond involving the NH of Gly^4. Although this information alone is not sufficient to define the global conformation of the molecule, preliminary conformational energy computations [2] on the uncomplexed peptide and general considerations based on solid state studies, suggested a C$_{10}$ turn involving the carbonyl oxygen of Tyr^1, favouring a type II' β-turn as also found in the case of analogous peptides.

This study was of particular interest since it allowed a conformational characterization of the molecule in a situation which reproduces at least two features of the active site as inferred by many studies on rigid molecules, i.e. a hydrophobic envirionment and the anchoring of the cation moiety. Further preliminary measurements seem to suggest that the uncomplexed molecule have a lower flexibility than the complexed one. Owing to the possible biological implications of this increased flexibility and to the non complete experimental description of the molecule we tried to better characterize and analyze the conformational and dynamic behaviour of the molecule and its complex by theoretical methods in order to rationalize its behaviour.

2. Model

2.1. FORCE FIELD

Computations were performed using a molecular mechanics approach in rigid geometry (RGA) (i.e. assuming all valence angle and bond distance fixed at their equilibrium

N. Russo et al. (eds.), Properties and Chemistry of Biomolecular Systems, 291–299.
© 1994 *Kluwer Academic Publishers. Printed in the Netherlands.*

values) and flexible geometry approximation (FGA) (i.e. relaxing all the geometrical constraints: bond length, valence angle and dihedrals).
All atom AMBER force field [3] was used in all the computations. In the RGA case the internal energy was computed as:

$$E_{RGA} = \Sigma_i \, V_{\phi i} \, [\, 1 + \cos \, (n\phi_i - \gamma_i)] + \Sigma_{i>j} \, w \, (A_{ij} / r^6_{ij} - B_{ij} / r^{12}_{ij}$$
$$+ 332.2 \, q_i q_j / \varepsilon \, r_{ij}) + \Sigma_{i>j} \, (\, C_{ij}/r^{10}_{ij} - D_{ij}/r^{12}_{ij})$$

where the first summation is extended over the internal rotation angles which have been varied, the second over all the pairs of nonbonded atoms i and j and the third over the pair of hydrogen-bonding hydrogens and H-bond acceptor atoms. In the Coulomb term a constant value of 1. of the dielectric constant were used while the 1-4 non bond and electrostatic ones were weighted by a factor w = 0.5.
In the case of FGA the total internal energy is computed as:

$$E_{FGA} = \Sigma_{bond} K_{bond}(d_i - d_{0i})^2 + \Sigma_{angle} K_{angle}(\theta_i - \theta_{0i})^2 + \Sigma_{imp.dih.} K_{imp.dih.}$$
$$(\chi_i - \chi_{0i})^2 + E_{RGA}$$

where E_{RGA} is computed at the local structure.

2.2. MOLECULAR DYNAMICS COMPUTATIONS

All the computations were performed in vacuum using the same all atom force field used in the case of Molecular Mechanics energy minimization and with a cut off distance for the non-bond interaction of 5 nm. Motion equation were integrated with an integration step of 0.002 ps constraining all bond distances using the SHAKE algorithm [4]. The simulations were run after assigning initial random values to the cartesian component of the velocities of each chosen from a Gaussian distribution at a given temperature and by coupling the system to an external bath at the same temperature [5] and removing, after the first step, both the overall translation and rotation of the molecule as well. The final simulation was run using as time coupling constant of the temperature a value of 0.1 ps and the data collected every 10 integration steps.
Particular care was reserved to the thermalization and equilibration procedure on the grounds of our previous experience [6]. In fact, starting from the minimum energy conformation found after energy minimization in FGA and thermalizing the system directly to the external bath at the temperature of 300K, the molecule escaped from that minimum in few ps, reaching, after an equilibration time of 10-20 ps, a completely different conformation characterized by a higher potential energy. This behaviour was independent of the value of the time coupling constant with the external bath and of the quality of the minimized structure. In fact, in order to remove all the possible "hot spots", the starting structure was minimized in such a way as to obtain a value of the gradient norm lower than 0.0001 Kj/nm.
As far as thermalization concerns, short dynamics runs (1 ps) have been carried out at temperatures well below 300K. Starting from 100 K, the temperature was increased in steps of 25 K. In these short MD runs the allowed temperature deviation was 5K and the time constant for the temperature relaxation time 0.025 ps. After each step of 1 ps of

simulation it has been verified (from the torsion angles analysis) that the conformation remained the same as the lowest energy minimum one (with a tolerance of 15 on each torsion). Once the temperature of 300 K was obtained and the system equilibrated for 25 ps, a simulation of 35 ps was carried out resetting the time constant to 0.1 ps and increasing the temperature deviation tolerance to10 K.

2.3. SIMULATED ANNEALING

As a further procedure to search for low energy conformations we used a Molecular Dynamics Simulated Annealing procedure [7]. This approach has been shown as valuable as the Monte Carlo Simulated Annealing procedure in finding low energy conformations [9]. Starting from 1000 K the temperature was reduced every 3 ps by a constant factor of 0.9 if the fractional fluctuations of the internal potential energy did not show modifications. As soon as the fractional fluctuations showed an increase, the time of the simulation at a given temperature was increased to 6 ps until a new decrease of the fractional fluctuations occurred [10].

3. Results and discussion

3.1. 18C6 COMPLEXES

In order to study the interactions between the peptide and the 18-crown-6 it was necessary to obtain a correct description of the main aspects of the 18C6 behaviour and of its interaction with an ammonium cation.

As regards 18C6, starting coordinates were taken from the crystalline structure of the 18C6 complexed with KSCN [11]. The crown had a symmetry D_{3d}.

A careful study was undertaken in order to define the charge to be used in the case of the crown ether. In fact the reported studies were devoted to the conformational stability of the crown alone [12] or to the relative stability and geometries of alkaline metal complexes [13]. Besides in our study we were much more interested in a correct description of both the relative stability of crown conformation and of the complex structure when the cation has a more complex structure, i.e. a $R\text{-}NH_3^+$. To this aim the $[18C6 - CH_3NH_3]^+$ complex was chosen as benchmark because of its crystal structure it has been determined with a good degree of accuracy [14] allowing also the identification of the H atoms. After some trial computations with a coarsely grained grid of charge values, we have interpolated a set of charges which were able to reproduce the correct nitrogen arrangement: the so called "perching mode of binding". The obtained charge set was O=-0.3, C=0.03 and H=0.06.

As a further and independent check we also found that uncomplexed crown-ether was more stable in the C_i conformation than in the D_{3d} conformation (ΔE (D_{3d} - C_i)= 4.28 Kj/mole) and that the inverse was true in the case of the complexed crown-ether (ΔE (C_i - D_{3d})= 74.0 Kj/mole). These findings agree with the avalaible experimental data [14,15] for both the free and complexing 18C6, although some authors [13] maintain that the non-planarity of the cation respect to the crown-ether is mainly due to intermolecular crystalline interactions between positive adducts and negative counterions.

On the other hand the fact that also in the case of the crystalline structure of the 18C6-BF_3NH_3 complex [16] the nitrogen is out of plane does not match with their hypothesis. A further computation on this last complex by using electrostatic potential derived charges by ab initio calculation at 6-31G* [17] level for the BF_3-NH_3 adduct gives a distance, between the nitrogen atom and the mean plane of the oxygen atom of 18C6, of 0.86 Å. This is in agreement, within 16% deviation, with the experimental results of 1.02 Å [16]. Therefore we used the above mentioned charges in all the following computations without further refinement.

3.2. CONFORMATIONS OF FREE PEPTIDE AND 18C6-PEPTIDE ADDUCT

Regarding the study of the 18C6-TdAPG complex two ways were followed to explore the conformational space of the complex. In the former case we put the crown ether, assumed in a conformation with D_{3d} symmetry, in the x-y plane with the origin at the center of the ether and Nitrogen of the starting NH_3 moiety of the peptide along the z-axis. Then we optimized and progressively reduced the height of the peptide along the z-axis. In particular we began the optimizations keeping fixed the nitrogen of Tyr at 4 Å and then reducing the height by 0.5 Å until the nitrogen of Tyr was at 0.5 Å out of the plane of 18C6. We used this approach starting from different types of conformation both containing the same 1-4 β-turns [2] but with different values of the Phe χ_1 dihedral corresponding to 180°-60° and 60°. The second approach used was to perform a grid search in the torsional subspace of the peptide. In particular we retained the backbone (Gly^4/Tyr^1) 1-4 β-turn conformation and performed the research on ψ_1 χ_1, χ_3 and ϕ_4 dihedral angles in RGA. Then we selected between these conformations the ones that bind to the crown ether without a large steric hindrance.. The most stable obtained conformation loses the type II β bend and is characterized by the presence of three intramolecular hydrogen bonds: two of C_7 type (between the O of DAla and H of Gly and between the O of Phe and H of NH_2terminal group) $DAla^2$/Gly^4 and Phe^3/NH_2; one of C_{11} type (between H of DAla and O of Gly) Gly^4 /$DAla^2$. The most stable conformation obtained containing the 1-4 (Gly^4/Tyr^1) type II' β-turn conformations is less stable by 8.4 kJ/mol than the most stable conformation obtained by the docking procedure. In order to have a further check of the potential energy hypersurface of supramolecular complex we have submitted the whole system (Peptide+18C6) to a Simulated Annealing run starting from the most stable conformation obtained from the two above mentioned procedures.

Surprisingly the obtained conformation at the end of the simulated annealing, albeit does not contain any β-turn, shows the presence of two C_7 turn, which involves $DAla^2$/Gly^4 (1.919 Å) and Tyr^1/Phe^3 (1.999 Å) H-bond, and a C_{11} bend due to a Gly^4/$DGly^2$ (2.065Å) H-bond. Furthermore the obtained conformation (ψ_1=-162.1, ϕ_2=83.9, ψ_2=-59.0, ϕ_3=-68.8, ψ_3=86.9, ϕ_4=95.0, ψ_4=-142.0) (Figure 1) more stable than the starting conformation by 21 . kJ/mol, confirming that the simulated annealing procedure has a certain validity in finding very low energy conformations and suggesting that the conformations containing β-turns are not the most stable one when complexed to the 18C6 at least in the *in vacuo* simulations.

Moreover the TdAPG shows a H-bonds between Phe N-H moiety and the nearest oxygen atom of the 18C6 crown. Furthermore the Tyr^1 nitrogen atom remains within the 18C6 frame in a "perching mode" of complexation and its distance from the medium plane of the crown oxygen atoms is 0.69 Å.

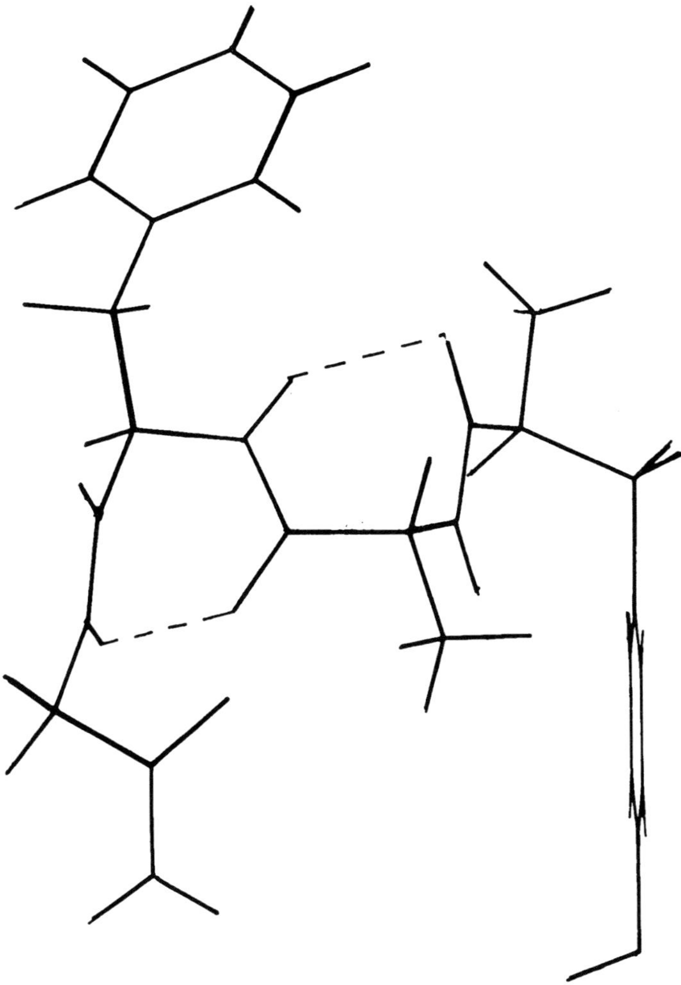

Figure 1. Plot of the minimum energy conformation of the complex [18C6 + peptide] obtained by simulated annealing procedure. The crown has been omitted in the picture.

The same simulated annealing procedure was applied in the case of the free TdAPG molecule (i.e. without the crown ether) starting from the most stable conformation containing the 1-4 type II' β-turn. The resulting structure more stable than the staring one by 5.3 Kj/mol is shown in Figure 2. Also in this case the molecule is characterized by Tyr^1/Phe^3 C_7 turn and one C_{11} , Phe^3/Tyr^1, and one C_{14}, Gly^4/Tyr^1.
These two rings (C_{11} and C_{14}) are due to the interactions of Phe^3 and Gly^4 carbonyl oxygen atoms with the hydrogen atoms of the NH_3^+ moiety of the Tyr^1. In this case,

due to the lack of the crown ether, the NH$_3^+$ moiety is able to interact with the last and second last carbonyl oxygen.

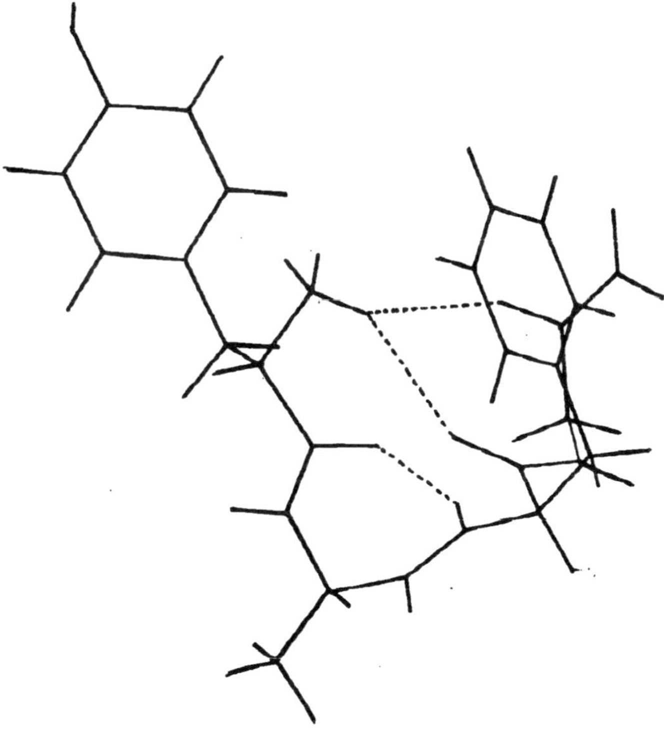

Figure 2. Plot of the minimum energy conformation of the free peptide obtained by simulated annealing procedure.

In order to evaluate in a quantitative way the internal mobility of the two molecules, i.e. free TdAPG and its complex we performed a Molecular Dynamics simulation for both the molecule and the complex. In both cases we started from the conformation obtained by the simulated annealing procedure. The simulation was run for 180 ps after 20 ps of thermalization at 300 K. The cartesian coordinates were collected and successively used in order to compare the history and averaged values of the relevant conformational parameteres (dihedrals and H-bond distances) for the two molecules. An inspection of the history of the backbone dihedrals showed a larger flexibility of the TdAPG+18C6 molecule with respect to the free TdAPG.

In order to compare the results, in a quantitative way we computed the RMSD of these parameters that are reported in Figure 3.

The reported results seem to indicate that also in a small time interval the isolated TdAPG molecule has a larger torsional freedom than the complexed molecule along the whole backbone. Besides the side chain χ_j1 of Tyr[1] shows opposite trend. This behaviour can be explained taking into account that the presence of the 18C6 crown lock the starting ammino moiety and reduces the freedom of rotation of the Tyr[1] side chain which can not have access to the [+180, +120., +60.] interval due to the presence of the crown as can be inferred by analyzing the distribution function of the χ_1 values. The increased freedom of the complexed TdAPG is understandable if we take into account that in the case of the free TdAPG the Tyr[1] NH_3^+ have the opportunity to interact with the Phe[3] and Gly[4] carbonyl oxygen.

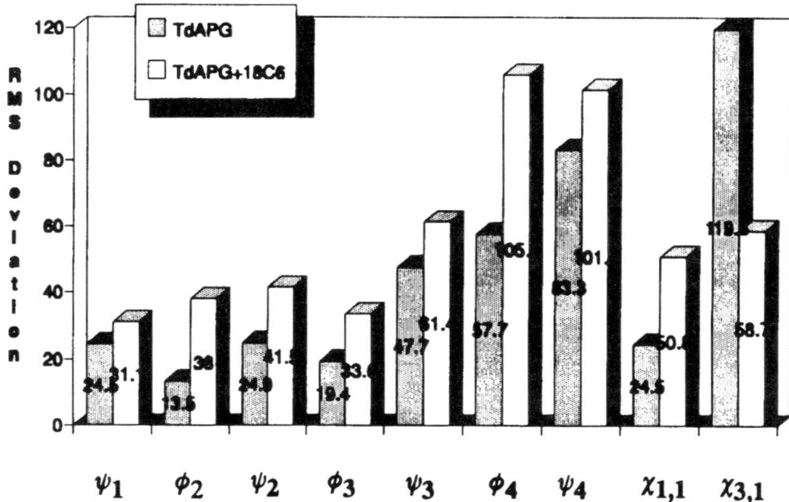

Figure 3. Comparison of the RMS Deviations of the backbone and selected side chain dihedrals for TdAPG and TdAPG+18C6.

4. Conclusions

MD analysis and energy minimization by standard techniques and MD annealing allows to give a detailed picture of the conformational behavior of this peptide with strong μ-opioid properties.

In particular MD simulations allows to point out that, in agreement with preliminary NMR results, the internal flexibility of the TdAPG molecule, when in a complex with the 18C6 ether, is larger than in the case of the free TdAPG.

The origin of the increased flexibility of the TdAPG when complexed with 18C6 can be explained by taking into account that the positive NH_3 group of Tyr[1] has a strong interaction with the terminal CO group in the case of the free molecule.

These results can have some biological implications. In fact if the molecule interacts with the receptor *via* its charged N-terminal group, it can gain a larger conformational flexibility which can be advantageous for its ability to adapt to the basic steric and electronic requirements of the receptor site.

Acknowledgements

We thank Prof. P.A.Temussi and Dr. T. Tancredi for helpful discussions and Prof. P.A
Kollman for providing us a copy of the AMBER 3.0 program.
We would like also to acknowledge also the CISIT of Universita' della Basilicata for a
generous gift of computer time. This work has been partially supported by a grant of
Italian "Ministero della Universita' e della Ricerca Scientifica e Tecnologica".

References

1. P.A. Temussi, D. Picone, M.A. Castiglione-Morelli, A. Motta and T. Tancredi,
   "bioactive Conformation of Linear Peptides in Solution: An Elusive Goal?", *Biopol.*
   28, 89-107, (1989).
2. M.A. Castiglione-Morelli, F. Lelj, A. Pastore, S. Salvadori, T. Tancredi, R.
   Tomatis, E. Trivellone and P.A. Temussi, "A 500 MHz Proton Nuclear Magnetic
   Resonance Study of μ-opioid Peptides in a Simulated Receptor Envirionment" *J.
   Med. Chem.* **30**, 2067-2073, (1987).
3. J. Weiner, P.A. Kollman, D.A. Case, U. Chandra Sing, C. Ghio, G. Alagona, S.
   Profeta Jr, and P. Weiner, "A New Force Field for Molecular Mechanical Simulation
   of Nucleic Acids and Proteins",*J.Am.Chem. Soc* **106**, 765-784, (1984).
4. W.F. van Gunsteren and H.J.C. Berends, "Algorithm for Molecular Dynamics and
   Constraint Dynamics", *Mol. Phys.* **34**, 1311-1327, (1977).
5. H.J.C. Berends, J.P.M. Postma, W.F. vanGunsteren, A. DiNola and J.R. Haak,
   "Molecular Dynamics with Coupling to an External Bath", *J.Chem.Phys.* **81**, 3684-
   3690, (1984).
6. P.L. Cristinziano, F. Lelj, P. Amodeo and V. Barone, "A Molecular Dinamics Study
   of Association in Solution. An NPT Simulation of Urea Dimer in Water",
   Chem. Phys. Lett. **140**, 401-405, (1987).
7. S. Kirkpatrick, C.D. Gelatt Jr. and M.P. Vecchi, "Optimization by Simulated
   Annealing", *Science* **220**, 4598-4606, (1983).
8. R. Carr and M. Parrinello, "Unified Approach for Molecular Dinamics and Density
   Functional Theory", *Phys. Rev. Lett.* **55**, 2471-2474, (1985).
9. S.R. Wilson, W. Cui, J. Moskowitz, K.E. Schimidt, "Application of Simulated
   Annealing to Chemical Conformational Analysis of Felxible Molecules",
   J. Comp. Chem. **12**, 342-349, (1991).
10. F. Lelj, P. Grimaldi and P.L. Cristinziano, "Conformational Energy
    Minimization by Simulated Annealing Using Molecular Dynamics: Some
    Improvement to the Monitoring Procedure", *Biopol.* **31**, 663-670, (1991).
11. J.D. Dunitz, M. Dobler, P. Seiler and R.P. Phizackerly, "Crystal Structure
    Analyses of 1,4,7,10,13,16-hexaoxa Cyclooctadecene and it Complexes with Alkali
    Thiocyanates", *Acta Cryst. B* **30**, 2733-2741, (1980).
12. M. Billetter, A.E. Howard, I.D. Kuntz and P.A. Kollman, "A New
    Technique to Calculate Low Energy Conformations of Cyclic Molecules Utilizing the
    ellipsoid Algorithm and Molecular Dynamics: Application to 18-crown-6",
    J. Am. Chem. Soc **110**, 8385-8391, (1988).

13. G. Wipff, P. Weiner and P.A. Kollman, "A Molecular Mechanics Study of 18-crown-6 and Its Alkali Complexes: an Analysis of Structural Flexibility, Ligand Specificity and the Macrocyclic Effect", *J. Am. Chem. Soc.* **104**, 3249-3258, (1982).

14. K.N. Trueblood, C.B. Knobler, D.S. Lawrence and R.V. Stevens, "Structures of 1:1 Complexes of 18-crown-6 with Hydrazinium Perchlorate, Hydroxylammonium Perchlorate and Methylammonium Perchlorate", *J. Am. Chem. Soc.* **104**, 1355-1362, (1982).

15. E. Meverick, P. Seiler, W.B. Schweizer and J.D. Dunitz, "1,4,7,10,13,16 Hexaoxycyclooctadecane: Crystal Structure at 100 K", *Acta Cryst. B* **36**, 615-620, (1980).

16. H.M. Colquhoum, G. Jones, J.M. Maud, J.F. Stoddart and J.D. Williams, "Crystal and Supramolecular Structure of Complexes of BF_3NH_3 and BH_3NH_3 with 18-crown-6", *J. Chem. Soc. Dalton Trans.*, 63-66, (1984).

17. W.J. Here, L. Radom, P.v.R. Schleyer and J.A. Pople, "Ab Initio Molecular Orbital Theory", John Wiley and Sons, New York, (1986).

Subunit Assembly in Bovine Seminal Ribonuclease

L. MAZZARELLA,[1,2] L. VITAGLIANO,[1] A. ZAGARI,[1] and S. CAPASSO,[1]
[1]University of Naples "Federico II", Department of Chemistry, Via Mezzocannone, 4.
I-80134 Naples, Italy.
[2]CE.IN.GE., Biotecnologie Avanzate, - 80131 Naples, Italy.

1. Introduction

The association of subunits in an oligomeric protein is a widespread phenomenon in the cellular organization,which promotes several advantages. A most important one is the regulation of catalytic activity through cooperative effects, mediated by specific interactions at subunit interface. Crystallographic studies of homologous proteins, which occur in different aggregation states, indicate that in most cases the globular monomer is stable on its own and is only marginally altered when it forms an oligomeric molecule. A notable exception is provided by some members of the ribonuclease family [1], which on aggregation show a substantial rearrangement of the tertiary structure of the monomer. The modification affects significantly the functionality of the active site, although the overall architecture of the site remains basically unchanged.

For many years, bovine pancreatic ribonuclease was the only member of the family to be well characterized and, in fact, physical chemical studies of this molecule have been important for the development of our basic knowledge on structural and functional features of protein systems [1]. In the last decade, a renewed interest in ribonuclease folding has arisen from the discovery of a series of extracellular proteins, homologous to pancreatic ribonuclease, which interfere with tumor cell growth [2]. Of particular interest is the dimeric ribonuclease isolated from bull seminal plasma [3] and extensively characterized by D'Alessio and his group[4].

Pancreatic ribonuclease catalyzes the degradation of RNA in a two step mechanism presented in Fig.1 . The substrate is a single stranded RNA, which must have a pyrimidine base on the 3' side of the phosphodiester bond to be hydrolyzed, whereas either a purine or a pyrimidine base can occur on the 5' side. In the first transphosphorolytic step, the 5' phosphoester bond to ribose is converted into a 2' bond to the preceding ribose moiety. The cyclic 2'-3' monophosphate is hydrolyzed to the final 3' phosphate nucleotide in the second step, which is the rate limiting step of the reaction. The two steps are belevied to be the reverse of each other, although no definite proof has been given: in the first step His12 activates the nucleophilic attack to the

301

N. Russo et al. (eds.), Properties and Chemistry of Biomolecular Systems, 301–312.
© 1994 Kluwer Academic Publishers. Printed in the Netherlands.

phosphorous atom and His119 protonates the leaving group, in the second step the two histidines interchange their role. Besides the two essential histidines in position 12 and 119, other residues such as Lys7, Lys41, Thr45 and Asp121 are also important for catalysis.

Pancreatic enzyme has two properties, which acquire new light in connection with seminal ribonuclease :

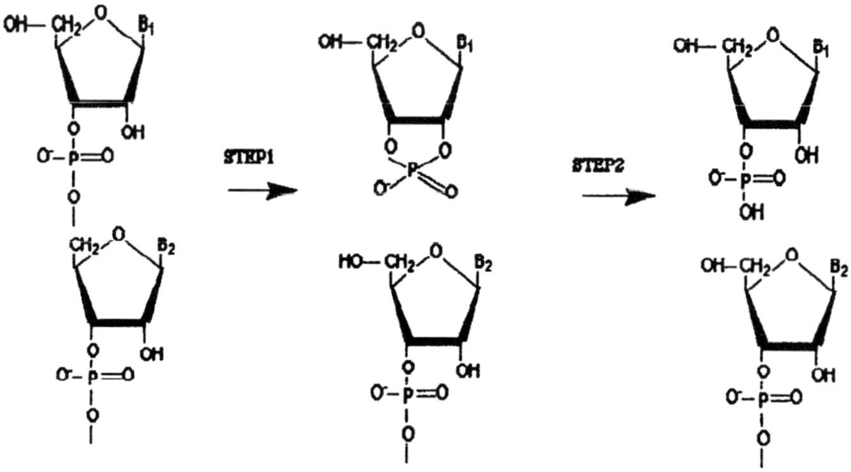

Figure 1. Scheme of the reaction mechanism of RNase A, showing the two-step degradation of RNA.

1) The peptide bond between Ala20 and Ser21 can be selectively cleaved by subtilisine. The proteolytic product, called RNase S, is a complex between the S-peptide (residues 1-20) and the S-protein (residues 21-124). The two polypeptide chains can be separated into completely inactive components and recombined to give a fully active enzyme [1]. The recombined molecule has a composite active site, formed by residues belonging to different polypeptide chains.

2) Under liophilization in presence of acetic acid, pancreatic ribonuclease forms a stable dimeric species in solution , which has two composite active sites. In this case, residues important for catalysis, such as His12 and His119, belong to different monomers [5]. The schematic model proposed at that time is shown in Fig.2. This was considered an interesting artifact which did not appear to possess biological relevance.

Seminal ribonuclease is a homodimeric enzyme, whose quaternary structure is stabilized by two consecutive disulfide bonds. Each chain formed by 124 residues is highly homologous to the pancreatic enzyme [1,4]. Most of the variations occur in the region 16 - 22, where a proline is found in position 19 of the seminal enzyme, and in the region 28 - 39. For the stabilization of the dimer, key substitutions are those of Cys31 and Cys32 for Lys and Ser, respectively, and of Leu28 for Asn. These residues occur at the subunit interface and the two half-cystines in position 31 and 32 form two disulfide bridges with residues 32 and 31, respectively, of the partner chain.

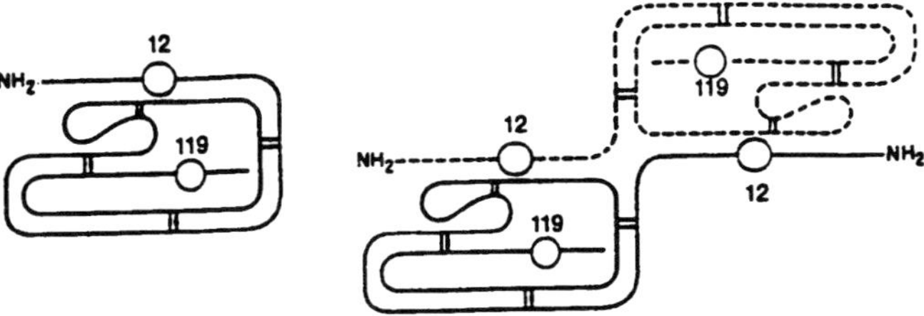

Figure 2 Scheme of RNase A dimerization (redrawn from Ref.5). His12 and 119 are part of the active site.

A few other variations are scattered along the chain but do not involve residues which are known to be important for the catalytic activity of the pancreatic molecule. Interchain disulfides can be selectively reduced and alkylated with iodoacetate [6]. Under these conditions, about one third of the carboxymethylated protein dissociates into stable and catalytically active monomers, while the remaining part is formed by dimeric species which can be dissociated by denaturing reagents only. These results indicate, that besides the interchain disulfides, non-covalent interactions are effective in stabilizing the dimeric structure of the seminal enzyme. Monomeric and dimeric species are not in equilibrium and can be separated by chromatographic methods.
Seminal ribonuclease has the same catalytic mechanism and specificity of the RNase A and follows classical Michaelis kinetics in the first step of the reaction. However, in the second step, non-hyperbolic kinetics is observed for the seminal enzyme, with negative cooperativity at low substrate concentrations and positive cooperativity at higher concentrations [7]. Furthermore, the binding of cytidine-3'-phosphate, the final product of the reaction, is biphasic, with negative cooperativity at low concentrations of the ligand and positive cooperativity at higher concentrations. The values of binding and kinetic constants, at very low concentrations of the ligand or substrates, are very similar to those of the monomeric derivative. On the other hand, the monomeric derivative of the seminal enzyme shows typical hyperbolic kinetics, whereas the artificial dimer of RNase A shows non-hyperbolic kinetics [7]. All these results suggest that pancreatic and seminal enzymes may be considered as the same enzyme in two different aggregation states; the native seminal enzyme being associated to the existence of cooperative phenomena.
In this paper we review the structural features of these two enzymes and present preliminary results on conformational energy calculations carried out for the peptide 16-22, which plays a fundamental role in subunit assembly of ribonucleases .

2. Structural studies

RNase A was one of the first enzymes studied by single crystal X-ray analysis, although the first available coordinates of a ribonuclease model were those of RNase S

[8]. Successively, several papers have been published, which describe high resolution refined models of RNase A and various complexes with inhibitors and substrate analogues, reviewed by A. Wloadawer [9]. More recently two independent structure analyses of the phosphate free enzyme at 1.26 [10] and 1.50 Å [11] resolution, respectively, have been performed.

A view of the molecule is shown in Fig. 3. It is characterized by the presence of an extensive β-structure and three helixes H1, H2 and H3 which encopasse residues 3-12, 24-32 and 50-60, respectively [9]. The structural model can be subdivided in a tail T (residues 1-15) and a body B (residues 23-124), linked by an irregularly folded heptapeptide, which will be denoted as the hinge peptide. The body is formed by an antiparallel β-sheet and the two helixes H2 and H3, each of them linked to the β-structure by a disulfide bond. Apart from the first and last two residues, the N-terminus tail has a helical conformation (helix H1) and interacts with the body B by means of several hydrogen bonds, which are important to properly fix the position of His12 side chain with respect to the phosphate in the active site.

The structure of RNase S is very similar to that of the intact enzyme, except for the region of the hinge peptide, which includes the bond selectively cleaved by subtilisine. This peptide is poorly defined in the electron density map and has the two additional end groups exposed to the solvent [8]. In particular the relative position of the two fragments is not affected by the break of the covalent linkage 20 -21.

For the seminal enzyme, the X-ray diffraction pattern, measured at room temperature, has an intrinsic lower resolution with respect to that of the pancreatic molecule. The dimer crystallizes from concentrated ammonium sulphate solution in the orthorombic space group $P22_12_1$ with four molecules per unit cell. Therefore the crystal symmetry

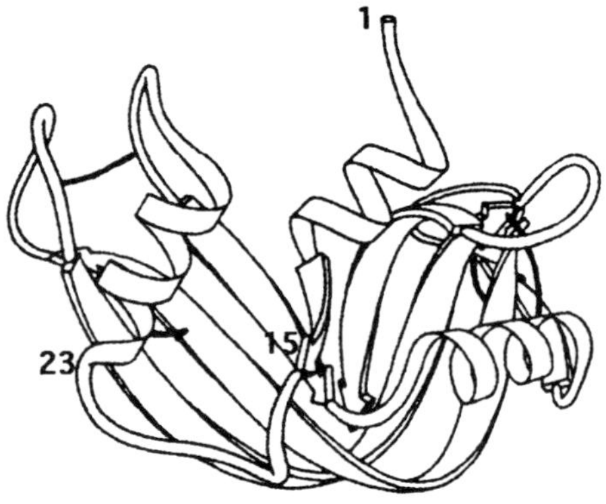

Figure 3. Model of RNase A. (Adapted from a figure prepared by J. Richardson, from Wlodawer,A. in <u>Biological Macromolecules and Assemblies</u>, Vol II, Nucleic Acids and Interactive Proteins (Jurnak & McPherson, Eds.) p.412, Wiley, New York, (1985). Copyright 1993 © by.. Reprinted by permission of J. Wiley & Sons, Inc.)

does not use the potential twofold symmetry of the molecule and the dimer constitutes the independent part of the unit cell [12].

The final model of BS-RNase [13], projected on a plane containing the approximate molecular twofold symmetry axis, is shown in Fig.4.

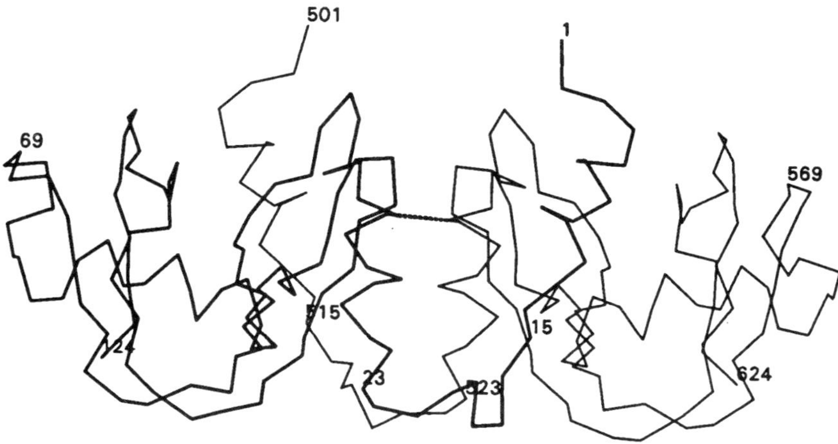

Figure 4. C$^\alpha$ carbon atoms drawing of BS-RNase. Light and heavy lines indicate the two subunits. Residue numbers of one subunit have been increased by 500.

The molecule closely satisfies the two-fold symmetry : the only regions which deviate significantly are the hinge peptide 16-22 and the external loop 65-72 (see also Fig. 5). With the exclusion of these two peptides, each half of the molecule is extremely similar to RNase A [14]. The N-terminal fragment T adopts an α-helical structure encompassing residues 3-13 and includes His12, whose side chain is of great importance for the enzyme activity. The rigidity of this peptide is enhanced by a salt bridge between Glu2 and Arg10 and by a network of hydrogen bonds involving two water molecules, which link one of the carboxylate oxygen of Glu2 to the carbonyl oxygen of Ala6. These structural details are perfectly preserved in both subunits, and in RNase A, in spite of the different crystallization medium used for the two enzymes. Helix H2 (residues 24-33) is also well preserved in the seminal enzyme, although in this structure, the helix is located at the subunit interface and it is preceded by one turn of a distorted 3/10 helix. All the β-regions are very similar in both enzymes.

Seminal ribonuclease has subunits linked by two consecutive disulfide bonds, which form a sixteen-membered cyclic structure across the molecular symmetry axis (Fig.6). This cycle, positioned at the end of the helixes H2 of the two subunits, severely restrains the relative positions of the two chains. The helix axes form an angle of about 40°, which allows favourable van der Waals interactions between the side chains of Leu28 of one chain with the same residue of the second chain. In RNase A, residue 28

Figure 5. Superposition of the two chains of the BS-RNase.

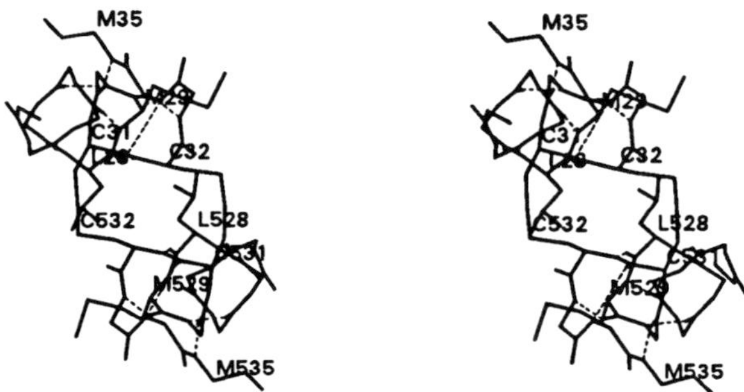

Figure 6. Stereoview of the cycle formed by the two consecutive interchain disulfide bonds. Torsion angles about the S-S bonds are close to +90°.

is Asn and is fully exposed to the solvent. Replacement of Asn by a Leu in the seminal enzyme, where it is buried at the subunit interface, certainly contributes to the stabilization of the dimer.

The most remarkable feature of the seminal enzyme is the way in which bodies and tails are connected in the dimer. Although the spatial relationships between T and B are practically indistinguishable from those found in pancreatic ribonuclease, the T and B moieties, contacting each other in the dimer, belong to different subunits.The similarity of the T/B interactions is also evidenced by the hydrogen bonding network observed in the crystal structure of the two enzymes. It includes the backbone NH...OC bond from Val47 to His12 and from Asp14 to Val47 and several backbone-side chain and side chain-side chain interactions. Therefore, the architecture of each active site at the T/B interface is similar to that of RNase A, although in BS-RNase it is formed by residues from different chains. The interchange of the tails between the two subunits requires a folding of the hinge peptide which is considerably different from that of the pancreatic molecule. It must be noted that the conformation of this peptide, besides being different from that of RNase A, also presents some differences in the two chains (Fig. 5).

The relevance of the crystallographic model of seminal ribonuclease in relation to the general understanding of subunit assembly in oligomeric proteins, together with the relatively poor quality of the electron density map in the region of the hinge peptide, led us to further improve the quality of experimental diffraction data and extend the resolution to the observed limit. A new data set, collected with an area detector on two crystals at 1.9Å resolution, was used for an independent refinement of the structure. The final Fourier map was considerably improved and gave a picture of the hinge peptide which was full in agreement with the previous interpretation [13]. No significant residual density was found in this region.

3. Potential energy calculations

X-ray analysis does not give an exhaustive description of the structural features of the seminal enzyme. Indeed, in order to explain the behaviour in solution of the enzyme when the interchain disulfides are selectively reduced and alkylated, D'Alessio and coworkers [15] proposed that the native enzyme is a mixture of two isomers. According to their results, while a large fraction of the native enzyme can be represented by a dimer having the structure found in the solid state, there must be a small fraction of dimers formed by chains that do not have their tails interchanged. This is not in contrast with the crystallographic results, as it may well occur that in crystals only the most abundant conformational isomer is selected. On the other hand, their suggestion is supported by the fact that the monomeric derivative of BS-RNase is fully active [16]. This property necessarily requires that the N-terminus segment in this species be folded on its own body in order to reconstitute the active site. In addition, a model of the non-interchanged dimer can be easily built, assuming for the hinge peptide the conformation found in pancreatic RNase (Fig.7). This alternative conformation of the peptide appears to be sterically compatible with the remaining part of the dimer. It must be pointed out, however, that the model in Fig.7 does not take into account the differences in the sequence between the pancreatic and seminal enzyme in this region (Table 1).

Table 1.   Sequence of the hinge peptide (16-22)

RNase A	Ser16-Thr17-Ser18-Ala19-Ala20-Ser21-Ser22
BS-RNase	Gly16-Asn17-Ser18-Pro19-Ser20-Ser21-Ser22

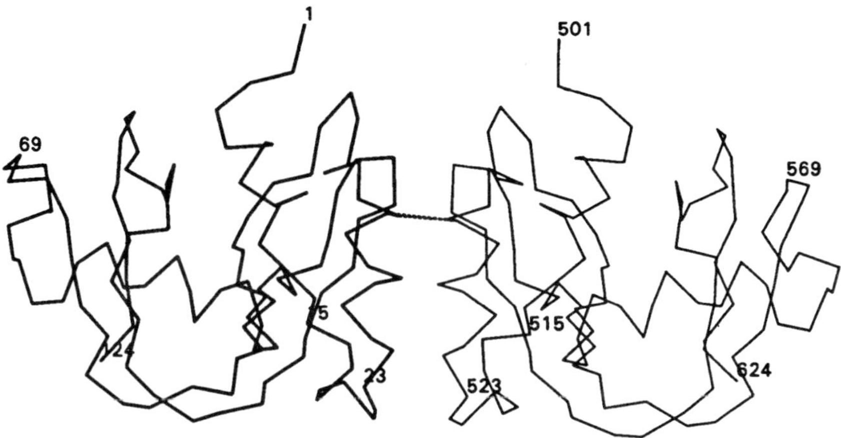

Figure 7. C$^\alpha$ drawing of a model for the non-interchanged dimer, based on the conformation of the hinge peptide found in RNase A.

We are trying to grow crystals of the non-interchanged dimer in order to determine the conformation of this peptide by X-ray diffraction analysis. In the meantime, we have also addressed the problem by means of potential energy calculations, with the aim of predicting the structure of the hinge peptide in the non-interchanged dimer. This means generating the pancreatic-like structure using the sequence of the seminal enzyme. In particular, it would be interesting to investigate the role of Pro19 on the relative stability of the two alternative conformations of the hinge peptide. In the following discussion, we outline the method and give some preliminary results. A detailed account of the results will be published further on.

On the basis of the structural features of RNase A and BS-RNase, it is reasonable to assume that differences between the two dimers are strictly limited to the heptapeptides16- 22 of the two chains. Therefore, the position of Ser15 and Ser23 can be considered to be accurately known in both dimers and the energy calculations can be restricted to all possible conformations of the peptide which properly connect these two residues (Fig.8). For this purpose, we have used the program described by Palmer and Scheraga [17,18], based on the algorithm of Go and Scheraga [19] for closing loops in the framework of ECEPP/3 standard geometry and potentials [20].

In a fixed geometry approach, the generation of a protein model, which has the chosen standard geometry and is sufficiently close to the X-ray model, is not a trivial problem. Indeed, when torsion angles derived from the X-ray coordinates are used together with standard bond lengths and bond angles, the resulting model can be, on average, even more than 5 Å distant from the experimental model. To cope with this problem we have used the program PROLSQ [21], which refines the crystallographic residual applying appropriately weighted constraints on bond lengths, bond angles and other stereochemical parameters of the molecule. The weights are usually chosen to give a refined model which presents a r.m.s. deviation from the mean of 0.2-0.3 Å for bond lengths and 2-3° for valence angles. However, by gradually increasing the weight of the stereochemical residual, it is possible to direct the refinement toward a highly idealized model, under the condition of a minimal increase of the crystallographic disagreement R factor.

From the coordinates obtained with this procedure, the main chain dihedral angles ϕ, ψ and ω were calculated and used to build up an ECEPP model for each chain, which had a r.m.s. deviation from the X-ray model of ~0.4 Å, for all non-hydrogen atoms. This model was used as a reference structure for the prediction of the conformation of the hinge peptide. In order to save computer time, at the first stage, only the fragment 10-27 was considered and the interactions with the remaining part of the molecule, were analyzed successively. Conformational search and energy minimization of the allowed conformations were performed for the peptide 17-21, whereas the conformation of the other residues was kept fixed during all the calculations, except for residues 16 and 22 which were allowed to vary during the energy minimization. To search loop closure between Gly16 and Ser22, three different starting points were considered: i) segments 10-16 and 22-27 of subunit 1, ii) the same segments of subunit 2, iii) segments 10-16 and 22-27 assembled as in selection i) and ii), respectively. In the first two cases, the allowed conformations of the pentapeptide 17-21 represent possible conformations of the hinge peptide for the interchanged dimer and can be compared to the experimental structure found in the crystal of BS-RNase for subunit 1 and subunit 2, respectively. In the last case, they represent a model for the non-interchanged dimer.
The minimum-energy conformation, found for case i) superimposed on the experimental structure, is shown in Fig.9. The r.m.s. deviation between the predicted and observed X-ray structure is 0.4Å. Similar results were obtained for case ii), indicating that the method can predict the correct conformations of the hinge peptide for the interchanged dimer.

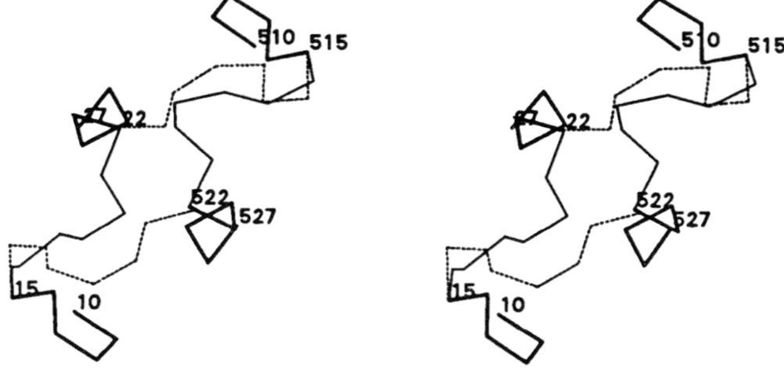

Figure 8. Stereodrawing of the C$^\alpha$ carbon atoms of the fragments 10-27, viewed approximately down the molecular twofold axis. The pancreatic conformation of the hinge peptides (see legend to Fig.7) for the non-interchanged dimer (broken lines) is superimposed on the conformation observed for the interchanged dimer. Residue numbers, less and greater than 500, refer to residues of subunit 1 and 2, respectively, for the latter dimer. Thick lines represent residues which were assumed to be structurally invariant in the two dimers.

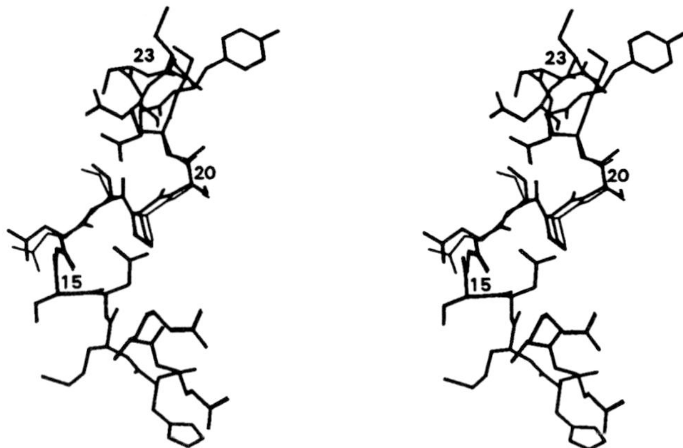

Figure 9. Stereoview of minimum-energy conformation predicted for the hinge peptide of one subunit of BS-RNase (thin lines) superimposed on the crystal stucture (thick lines).

Calculations for case iii) gave a less clearcut result. For the *trans* conformation of Pro19, all the low energy conformations of the peptide 16-22 calculated in the context of the fragment 10-27, give rise to strongly repulsive interactions with the remaining part of the chain or with the other chain. This also occurs for the lowest energy conformation obtained for the *cis* conformation of the proline. However, in this case, few conformations, close in energy to the lowest one, are consistent with the structure of the full dimer.

4. Conclusion

Among oligomeric proteins, ribonucleases possess a unique property: the polipeptide chain can adopt two distinct foldings which both produce active molecules. In particular, seminal ribonuclease can exist as a fully active monomer and two covalent active dimers which differ for the interchange of the N-terminal segments between the two subunits. Native enzyme is a mixture of the two dimers with a prevalent concentration of the species which has the N-termini interchanged. Although the three species have similar active sites, the interchanged dimer has composite sites formed by residues from different chains. This is the species responsible for the cooperative effects and the high antitumoral activity of the native enzyme[4]. Therefore, these functional features can be ascribed to the interchange of the tails rather than to the covalent bonds between the chains, which are present also in the non-interchanged dimer. The interchain disulfides may be important in order to induce a quaternary structure favourable to such interchange.

On the other hand, the primary structure of the hinge peptide is certainly important in determining which conformation of this peptide is favoured. In this region several mutations occur between the pancreatic and seminal enzymes. On the whole, the most important one seems to be the replacement in position 19 of alanine by proline in the last molecule. The preliminary results of energy calculations suggest that while the conformation found in the interchanged dimer can be easily built using the *trans* conformation for this residue, the pancreatic-like structure of the non-interchanged dimer could require the proline to be in the *cis* conformation. These results could have been determined by the fact that only a relatively short fragment had been allowed to vary in the loop search procedure, and for these reasons we are extending the calculations also to include residues 16 and 22 in the variable fragment. However, they support the results of an earlier experiment, suggesting the presence of a *cis-trans* conformational isomerism as a slow step in the reacquisition of the quaternary structure of the seminal enzyme [22].

Acknowledgements

We are indebted to Dr. K.A.Palmer for making available the computer program described in Ref. [17]. This work was supported by Italian CNR and MURST.

References

1. P. Blackburn, & S. Moore,."Pancreatic Ribonuclease" in <u>The enzymes</u>, Vol. 15, pp.317-433, edited by P.D.Boyer, 3rd ed.,Academic Press, New York, (1982).
2. S. Benner, R.K. Allemann, "The Return of Pancreatic Ribonucleases", *Trends Biochem. Sci.* **14**, 396-397, (1989).
3. G. D'Alessio, E. Leone, "The Action of Seminal Enzymes on Ribonucleic Acid" *Biochem. J.* **89**, 7P, (1963).
4. G. D'Alessio, A. Di Donato, A. Parente, & R. Piccoli, "Seminal RNase: a unique member of the ribonuclease superfamily". *Trends Biochem. Sci.* **16**, 104-106, (1991).
5. R.G. Fruchter, A.M. Crestfield, "On the Structure of Ribonuclease Dimer". *J. Biol. Chem.* **240**, 3875-3882, (1965).
6. G. D'Alessio, M.C. Malorni, A. Parente, "Dissociation of Bovine Seminal Ribonuclease into Catalytically Active Monomers by Selective Reduction and Alkylation of the Intersunit Disulfide Bridges". *Biochemistry* **14**, 1116-1122, (1975).
7. R. Piccoli, A. Di Donato, & G. D'Alessio, "Co-operativity in Seminal Ribonuclease Function". *Biochem. J.* **253**, 329-336, (1988).
8. H.W. Wyckoff, D. Tsernoglou, A.W. Hanson, J.R. Knox, B. Lee, F.M. Richards, "The 3-Dimensional Structure of Ribonuclease -S. Interpretation of an Electron-Density Map at a Nominal Resolution of 2Å Resolution" *J. Biol. Chem.* **245**, 305-328, (1970).

9. A. Wlodawer, "Structure of Bovine Pancreatic Ribonuclease by X-ray and Neutron Diffraction" in <u>Biological Macromolecules and Assemblies</u>, Vol. II, Nucleic Acids and Interactive Proteins (Jurnak & McPherson, Eds.) pp. 393-439,   Wiley, New York, (1985) .

10. A. Wlodawer, L.A. Svensson, L. Sjölin, & G.L. Gilliland, "Structure of Phosphate-Free Ribonuclease A Refined at 1.26 Å" . *Biochemistry*, **27**, 2705-2717, (1988).

11. R.L. Campbell, & G.A. Petsko, "Ribonuclease Structure and Catalysis: Crystal Structure of Sulfate-Free Native Ribonuclease A at 1.5 Å Resolution" *Biochemistry* **26**, 8579-8584, (1987).

12. L. Mazzarella, C.A. Mattia, S. Capasso, & G. Di Lorenzo. "Composite Active Sites in Bovine Seminal Ribonuclease". *Gazz. Chim. Ital.*, **117**, 91-97, (1987).

13. L. Mazzarella, S. Capasso, D. Demasi, G. Di Lorenzo, C.A. Mattia, A. Zagari, "Bovine Seminal Ribonuclease at 1.9 Å Resolution", *Acta Cryst. D* , in press (1993).

14. A. Wlodawer, & L. Sjölin, "Structure of Ribonuclease A: Results of Joint Neutron and X-ray Refinement at 2.0 ". *Biochemistry* **22**, 2720-2728, (1983).

15. R. Piccoli, M. Tamburrini, G. Piccialli, A. Di Donato,A. Parente, & G. D'Alessio, "The Dual-Mode Quaternary Structure of Seminal Ribonuclease". *Proc. Natl. Acad. Sci. USA*. **89**,1870-1874, (1992).

16. R. Piccoli, G. D'Alessio, " Relationships between Nonhyperbolic Kinetics and Dimeric Structure in Ribonucleases" *J. Biol. Chem.* **259**, 693-695, (1984).

17. K.A. Palmer, H.A. Scheraga, "Standard-Geometry Chains Fitted to X-Ray Derived Structures: Validation of the Rigid-Geometry Approximation. I.Chain Closure through a Limited Search of "Loop" Conformations" *J. Comput. Chem.* **12**, 505-526, (1991).

18. K.A. Palmer, H.A. Scheraga, "Standard-Geometry Chains Fitted to X-Ray Derived Structures: Validation of the Rigid-Geometry Approximation. II Systematic Searches for Short Loops in Proteins: Applications to Bovine Pancreatic Ribonuclease A and Human Lysozyme " *J. Comput. Chem.* **13**, 329-350, (1992).

19. N. Go, H.A. Scheraga, "Ring Closure and Local Conformational Deformations of Chain Molecules", *Macromolecules* **3**, 178-187, (1970).

20. G. Némethy, K.D. Gibson, K.A. Palmer, C.N. Yoon, G. Paterlini, A. Zagari, S. Rumsey, H.A. Scheraga, "Energy Parameters in Polypeptides. 10. Improved Geometrical Parameters and Nonbonded Interactions for Use in the ECEPP/3 Algorithm, with Application to Proline-Containing Peptides." *J. Phys. Chem.* **96**, 6472-6484, (1992).

21.W.A. Hendrickson, J.H. Konnert, "Stereochemically Restrained Crystallographic Least-Squares Refinement of Macromolecule Structures" in <u>Biomolecular Structure, Conformation, Function and Evolution,</u> Vol. I, pp. 43-57, Srinivasav, R. Ed., Pergamon Press, Oxford, (1981).

22. A. Parente, G. D'Alessio, "Reacquisition of quaternary structure by fully reduced and denatured seminal ribonuclease" *Eur. J. Biochem.* **149**, 381-387, (1985).

Conformational Studies in Solution on Cyclolinopeptide A Analogs. A Two-D NMR Study of cyclo{Pro^1-Pro-phe-Phe-Ac$_6$c-Ile-ala-Val^8}

M. MAZZEO, C. PEDONE, L. PAOLILLO, V. PAVONE, C. ISERNIA, F. ROSSI,
M. SAVIANO and E. BENEDETTI
*Department of Chemistry, University of Naples, Via Mezzocannone 4, 80134 Naples,
Italy*

1. Introduction

Cyclolinopeptide A and antamanide belong to a class of cyclic peptides well known
for their competitive inhibitory activity toward the cholate uptake in hepatocytes [1].
It has recently been shown that they can assume several conformational states in
various organic solvents [2,3,4]. CD studies have also shown their ability to bind
mono and bivalent metal ions [5,6]. Binding properties have been explained in terms
of conformational flexibility, as revealed by NMR studies. The sequence
Pro-Pro-Phe-Phe, common to this class of peptides, has been hypothesized as
essential for biological activity.

In order to obtain a structurally rigid peptide, and to further ascertain the role of the
common sequence, we have designed and synthesized a cyclic octapeptide containing
a (D)-Phe residue in position 3 and a bulky aminoacid in position 5:

$$\text{cyclo}\{\text{Pro}^1\text{-Pro-phe-Phe-Ac}_6\text{c-Ile-ala-Val}^8\}$$

A D-Ala residue was also introduced in position 7, since diffraction studies on native
cyclolinopeptide A have shown that the aminoacid in this position has structural
features typical of D residues [2].

The conformation of this novel cyclooctapeptide has been examined by two-D NMR
and molecular dynamics computer simulation.

2. Experimental

2.1. SYNTHESIS

The synthesis of the linear octapeptide H-ala-Val-Pro-Pro-phe-Phe-Ac$_6$c-Ile-OH was
achieved by solid phase methods with t-Boc protected aminoacids using a 430A
Applied Biosystem automatic synthesizer.

313

N. Russo et al. (eds.), Properties and Chemistry of Biomolecular Systems, 313–319.

© *1994 Kluwer Academic Publishers. Printed in the Netherlands.*

t-Boc--Ac$_6$c-OH was prepared according to procedure [7]; dicyclohexylcarbodiimide (DCC) and 1-hydroxybenzotriazole (HOBt) were used as coupling reagents using the standard procedure. t-Boc-Ac$_6$c-OH was dissolved in CH_2Cl_2 and coupled as an asimmetric anhydride twice in double excess. After release of the linear peptide from the resin by HF [8], the crude peptide (yield 90%) was then purified by preparative HPLC using a Delta Prep Millipore-Waters system with a linear gradient H_2O 0.1% TFA: CH_3CN 0.1% TFA and characterized by FAB mass spectroscopy (MH^+=915). Peptide cyclization was obtained in a diluted solution of DMF and CH_2Cl_2 in the presence of isobutylchloroformate and N-methylmorpholine at pH=8.2. The reaction mixture left overnight at room temperature was then reduced in volume, diluted with water, extracted in CH_3Cl, dried over Na_2SO_4 and evaporated in vacuo. The purification by preparative HPLC was carried by a linear gradient H_2O 0.1% TFA:CH_3CN 0.1% TFA, and gave the pure cyclopeptide (yield 73%). FAB mass for MH^+ was 897 as expected.

2.2. NMR

NMR measurements were carried out on a VARIAN UNITY 400MHz spectrometer equipped with a SUN computer SPARC 330. All ^{1}H spectra were recorded at 298K and referenced to internal TMS. NMR samples were prepared by dissolving 16 mg and 2.2 mg of peptide in 0.75 ml of CD_3CN (99.96% ^{2}H atom) and in 0.75 ml of $CDCl_3$ (99.96% ^{2}H atom), respectively. Two-D homonuclear NOESY, ROESY, TOCSY and DQCOSY and two-D heteronuclear HMQC and HMBC in reverse detection were acquired with the States-Haberkorn method. Mixing time values were of 70 ms for TOCSY, from 50 to 400 ms for NOESY and from 30 to 200 ms for ROESY spectra. Two-dimensional experiments were typically acquired with 256 increments and 2048 data points in t_2 and zero filled to 1K in F_1.

2.3. RESTRAINED MOLECULAR DYNAMICS (RMD)

RMD simulations were carried out on a Silicon Graphics (Personal Iris 4D25 GT Turbo) workstation, employing the INSIGHT/DISCOVER program. CVFF force field was utilized for all the simulations. All clearly distinguishable NOE effects were used for the MD calculations. In each RMD simulation, performed with a time step of 0.5 fs, the molecule was equilibrated for 40 ps. After this first step, an additional 60 ps of simulation without rescaling was carried out.

3. Results and Discussion

The cyclo{Pro^1-Pro-phe-Phe-Ac$_6$c-Ile-ala-Val^8} was studied in chloroform and acetonitrile solutions. Sequential assignments of all spin systems were achieved by the conventional procedure using TOCSY and NOESY spectra. Selective 1D-TOCSY experiments were acquired to confirm the assignment and to measure vicinal coupling constants.In fig.1 the spectra obtained for the spin system of Phe^3 and Val^8 in CD3CN are shown. Coupling constants and temperature coefficients of amide protons have been measured in both solvents.

The wide dispersion of the chemical shift of ^{13}C makes heteronuclear correlation experiments valuable for assignment and, in some cases, for conformational purposes. The ^1H-^{13}C correlation experiment (HMQC) was used to unambiguously assign the Val^8-Pro^1 and Pro^1-Pro^2 peptide bond isomerism. In both solvents a trans and a cis peptide linkage for the Val^8-Pro^1 and Pro^1-Pro^2 bonds were found, respectively. These results were further confirmed by NOE contacts between $\alpha CHVal^8$-$\delta CHPro^1$ and between $\alpha CHPro^1$-$\alpha CHPro^2$. HMQC spectrum is shown in fig.2.

In chloroform solution the sequential assignment of the two prolines was also obtained acquiring an HMBC spectrum in reverse mode. In fig.3 two regions of the HMBC experiment, showing the CO-NH and CO-αCH correlations, are presented.

Conformational parameters, coupling constants, temperature coefficients, relaxation times T_1 and NOE effects point to a very rigid structure with intramolecular hydrogen bonds. The NOESYs and ROESYs spectra, at different mixing times, were used to evalutate the NOE buildup rates and to calculate interproton distances. These distances were used as constraints in molecular dynamics calculations.

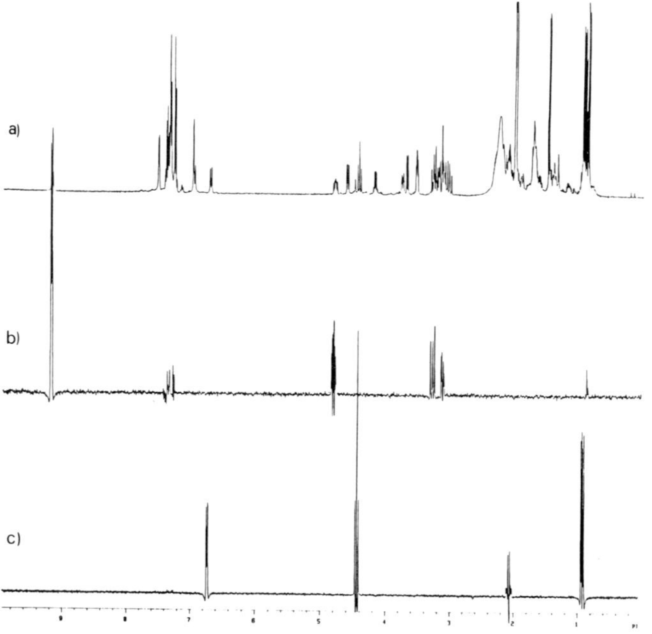

Figure 1. Monodimensional (a) and 1D-TOCSY spectra for the spin system of phe^3 (b) and Val^8 (c) in CD3CN at 298 K. TOPHAT shapes pulses were used for selective excitation.

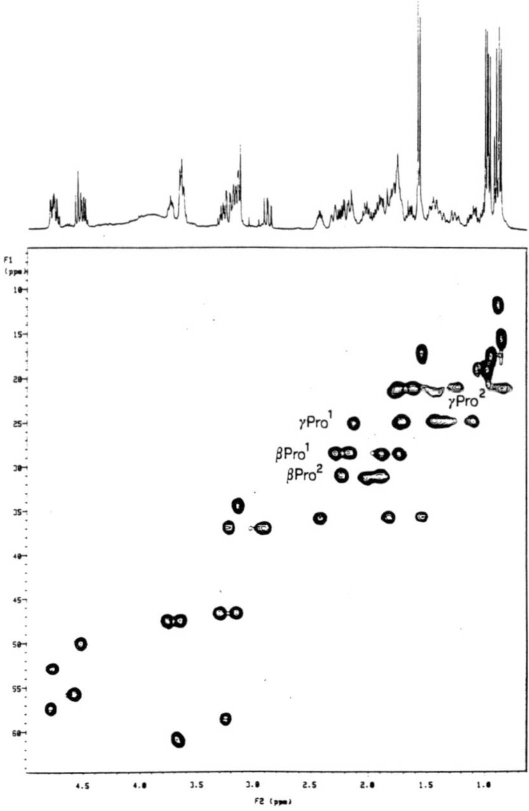

Figure 2. High field region of the HMQCR spectrum in CDCL3 at 298 K.

The values of interproton distances calculated from NOESY and ROESY spectra, and measured on the averaged model obtained for the molecule in CD3CN, are reported in table I. On the basis of the restrained molecular dynamics simulation, we have been able to determine that only one model is consistent with NMR data. This structure is stabilized by three intramolecular H-bonds involving the NH groups of Phe[4], ala[7], and Val[8] with the CO groups of Val[8], Ac6c, and Phe[4], respectively. A comparison of this structure with that of cyclolinopeptide A is reported in fig.4.

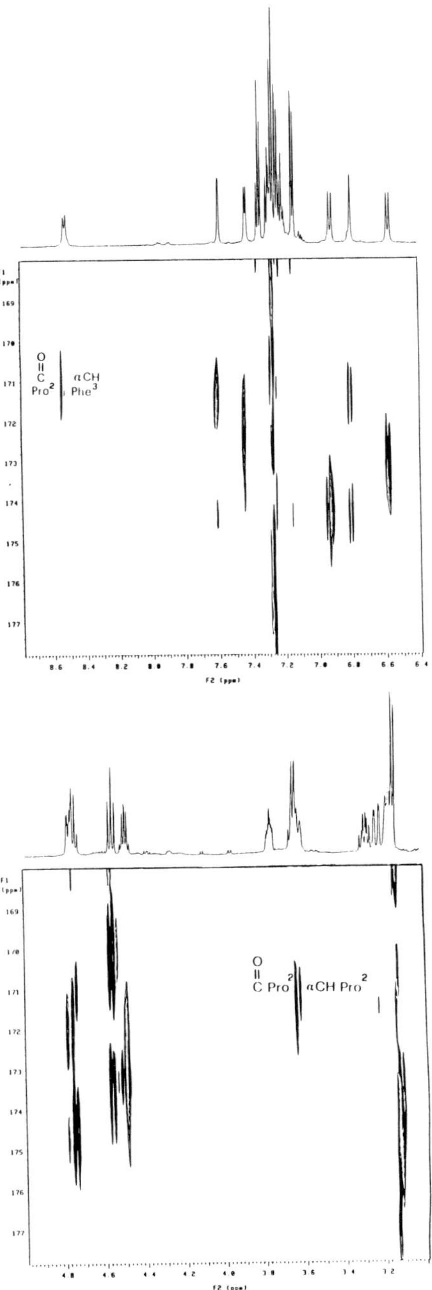

Figure 3. Regions of the HMBCR spectrum in CDCL3 showing CO-NH and CO-αCH correlations (long-range coupling constant=7 Hz).

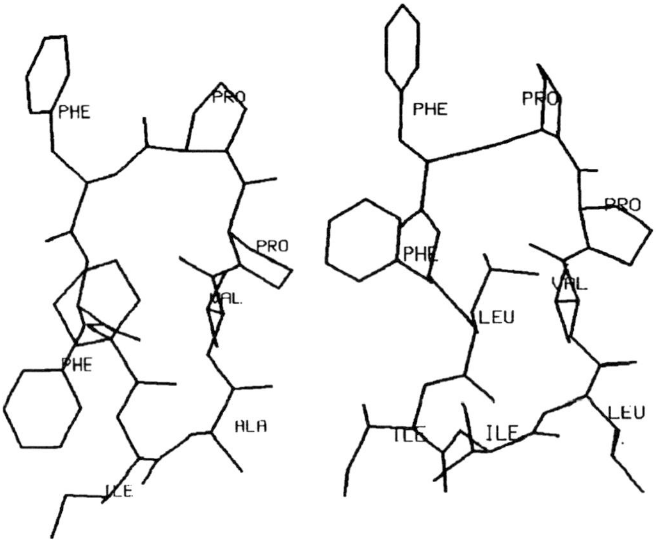

Figure 4. Comparison of the model obtained in CD3CN (left) with the solid state structure of cyclolinopeptide A (right)

Table 1. Interproton distances calculated from σ^n, σ^r, and MD in CD3CN for cyclo{Pro[1]-Pro-phe-Phe-Ac6c-Ile-ala-Val [8]}

Cross peak	r_n (Å)	r_r (Å)	r_{MD} (Å)
NHPhe[4]-αCHphe[3]	2.2	2.3	2.2
NHphe[3]-αCHPro[1]	2.3	2.3	2.1
NHPhe[4]-αCHPhe[4]	2.4	2.6	2.9
NHPhe[4]-βCHPhe[4]	2.7	2.9	2.9
NHphe[3]-βCHphe[3]	2.6	2.7	2.7
αCHVal[8]-δCHPro[1]	2.1	2.1	2.2
αCHPro[1]-αCHPro[2]	2.0	2.1	2.2
αCHIle[6]-βCHIle[6]	2.4	2.5	2.6
αCHPro[2]-βCHPro[2]	2.6	2.7	2.6
αCHPro[2]-β'CHPro[2]	2.1	2.4	2.3
NHphe[3]-αCHphe[3]		3.0	3.0
NHala[7]-αCHIle[6]	2.9		2.6
NHPhe[4]-β'CHPhe[4]	2.4		2.5
αCHPro[1]-βCHPro[1]	2.2		2.2
αCHVal[8]-βCHVal[8]	2.8		3.1

standard cross-peak

γCHPro[2]-γ'CHPro[2]   r=1.78 Å

4. Conclusions

The present study confirms our expectation that, in organic solvents at room temperature, the reduction of the main chain length (from deca or nona to octa peptide), coupled with the insertion of sterically hindered aminoacids and appropriate D residues, gives rise to a highly rigid molecule.
As far as the overall topology of the Pro-Pro-Phe-Phe segment is concerned, the model of the octapeptide derived from our data shows considerable similarity with that found in solution and in the solid state for cyclolinopeptide A.
Furthermore we can confirm that the complexation capability of cyclic peptides relates well with conformational flexibility. In fact preliminary data indicate that the rigid compound presently investigated is unable to bind monovalent and bivalent metal ions in organic solvents. Further studies are presently in progress to evaluate the biological activity of this molecule.

References

1. H. Kessler, M. Klein , A. Muller, K. Wagner, J. Bats , K. Ziegler and M. Frimmer, *Angew. Chem. Int. Ed. Engl.,* **25**, 997, (1986).
2. B. Di Blasio, F. Rossi , E. Benedetti, V. Pavone, C. Pedone, P.A. Temussi, G. Zanotti, T. Tancredi, *J. Am. Chem. Soc.* **111**, 9089, (1989).
3. T. Tancredi, G. Zanotti, F. Rossi, E. Benedetti, C. Pedone, P.A. Temussi, *Biopolymers* **28**, 513, (1989).
4. D.J. Patel , *Biochemistry* **12**, 661, (1973).
5. T. Tancredi, E. Benedetti, M. Grimaldi, C. Pedone, F. Rossi, M. Saviano, P.A. Temussi, G. Zanotti, *Biopolymers* **31**, 761, (1991).
6. V.T. Ivanov, *Ann. New York Acad. Sci.,* **264**, 221, (1975).
7. M. Bodanszky, A. Bodanszky, <u>In The Practice of Peptide Synthesis.</u> Springer-Verlag, Berlin , . vol. **21**, p.12, (1984).
8. J.P. Tam, W.P. Heath, R.B. Merrifield, *J. Am. Chem. Soc.* **105**, 6442, (1983).

Homogeneous Catalytic Hydrogen Formation Using Dinuclear Multiply Bonded Complexes of Molybdenum (III) and Tungsten (III) and Low-Valency Metal Ions, M=Cr(II), V(II), in Aqueous Acidic Solutions

C. MERTIS*, M. KRAVARITOY, M. CHORIANOPOULOU, S. KOINIS, and N. PSAROUDAKIS
*University of Athens, Inorganic Chemistry Laboratory, Panepistimiopolis, Kouponia, 15701 Athens, Greece

1. Introduction

In the last twenty years intensive research has been done in the important field of multiply bonded dinuclear transition metal complexes. The chemistry of the $(M\text{-}M)^{n+}$ containing compounds is extraordinarily rich and complex [1], and the factors governing the reactivity of metal-metal multiple bonds are now better understood. Additionally, some of the commonly observed reactions of these dimers are not found in mononuclear chemistry and point out to the importance of the $(M\text{-}M)^{n+}$ unit as an inorganic functional group. In fact, the whole field of multiple bonds between metal atoms provides a distinct departure from classical coordination chemistry [2]. These species posses several attractive properties for their application as multielectron reagents or photoreagents. Firstly, their electronic structure is well defined, and often, the lowest energy excited states of many $(M\text{-}M)^{n+}$ compounds are sufficiently long lived to permit their subsequent chemical reaction. Secondly, the binuclear metal core is an electron source or sink in oxidation-reduction processes. Addition of electrons to, or removal of, electrons from the metal-metal bond is accompanied by facile interconversion between $(M\text{-}M)^{n+}$ dimers of different bond orders. Finally, substartes readily add to the coordinatively unsaturated metal-metal core. The ability to coordinate substrate molecules to redox-active binuclear metal centers has important implications in the ultimate application of $(M\text{-}M)^{n+}$ dimers as multi-electron catalysts or photocatalysts. Proton oxidative addition to mononuclear or cluster transition metal complexes leading to hydride formation [3] is a key step in many catalytic processes including chemical [4], photochemical [5], or biochemical [6], reactions in which dihydrogen is evolved. Molybdenum and tungsten compounds are receiving special attention in this context because of the involvement of the first metal ion in such reactions and the attempts to exploit the high reactivity of the $(W^4\text{-}W)^{4+}$ bond [2] in stoichiometric or catalytic reactions of particular importance. The $[M_2X_8]^{4-}$, $[M_2HX_8]^{3-}$ and $[M_2X_9]^{3-}$ halides of Molybdenum and Tungsten. Dimolybdenum acetate reacts with hydrogen halides (X=Cl, Br) in the absence of oxygen to give the quadruply bonded octahaloderivative 1 (X=Cl, 1a; X=Br, 1b), eq. (1):

321

N. Russo et al. (eds.), Properties and Chemistry of Biomolecular Systems, 321–329.

© 1994 Kluwer Academic Publishers. Printed in the Netherlands.

$$Mo_2(O_2CMe)_4 + 8HX \longrightarrow [Mo_2X_8]^{4-} + 4CH_3COOH + 4H^+ \tag{1}$$
$$1$$

Proton oxidative addition to this ion yields the formally triply bonded hydride [7] $[Mo_2(\mu\text{-}H)(\mu\text{-}X)_2X_6]^{3-}$ (X=Cl, 2a; X=Br, 2b), eq. (2):

$$[Mo_2X_8]^{4-} \xrightarrow{HX} [Mo_2HX_8]^{3-} \tag{2}$$
$$1 \qquad\qquad\qquad 2$$

Further reaction of 2 with HX and loss of hydrogen to give the triply bonded nonahaloderivative $[Mo_2(\mu\text{-}X)_3X_6]^{3-}$ (X=Cl, 3a; X=Br 3b), eq. 3, requires more drastic conditions e. q heating at 90 °C for several hours in 12M HX in strict absence of oxigen [8]:

$$[Mo_2HX_8]^{3-} \xrightarrow[\;-H_2\;]{HX} [Mo_2X_9]^{3-} \tag{3}$$
$$2 \qquad\qquad\qquad\qquad 3$$

Complex 3 is in equilibrium [9] with the monomer eq. (4):

$$[Mo_2X_9]^{3-} + 3Cl^- \rightleftarrows 2[MoX_6]^{3-} \tag{4}$$

In solution, the hydride 2 is very reactive towards oxygen [8,10] producing the mononuclear oxo-compound $[MoOX_4]^-$. At lower acid concentrations, complex 1 undergoes hydrolysis [2,11] eq. (5):

$$[Mo_2X_8]^{4-} \longrightarrow [Mo_2(X)_{8-x}(H_2O)_x]^{4-x} \tag{5}$$

The structures both of the reactants and products of these reactions are known [2]. Whereas the $(Mo\text{---}4\text{---}Mo)^{4+}$ dimers and the $(Mo\text{---}3\text{---}Mo)^{8+}$ hydrides are stable, the isostructural and isoelectronic ditungsten analogues $[W_2X_8]^{4-}$, 4, and $[W_2(\mu\text{-}X)(\mu\text{-}X)_2X_3]^{3-}$, 5, remained elusive, despite the efforts of nearly twenty years. The isolation and structural characterization of the $[W_2Me_{8-x}Cl_x]^{4-}$ ion [12] and the interception and identification of the $W_2Cl(PBu_3^n)(OOCPh)_2(\mu\text{-}H)(\mu\text{-}Cl)$, resulting from the facile oxidative addition of the generated HCl to the initially formed $W_2Cl_2(PBu_3^n)_2(OOCPh)_2$, derived from the reaction of $W_2Cl_4(PBu_3^n)_4$ with benzoic acid [13], renewed the optimism for the isolation of 4, which eventually was prepared under non-oxidizing conditions [14] eq. (6):

$$W_2Cl_6(THF)_4 + 2Na/Hg \xrightarrow[\;-30°C\;]{THF} 1/2\ Na_4(THF)x\ W_2Cl_8 + "WCl_2" \tag{6}$$

The product is thermally unstable at room temperature, in any solvent it dissolves, eq. (7):

$$Na_4(THF)_x\ W_2Cl_8 \xrightarrow[\theta > 0°C]{THF} 4NaCl + 2"WCl_2" \qquad (7)$$

Complex 4 gives [15] reactions (2) and (3) even at -78°C producing the stable triply bonded $[W_2(\mu\text{-}Cl)_3Cl_6]^{3-}$, 6, presumably *via* the hydride $[W_2(\mu\text{-}H)(\mu\text{-}Cl)_2Cl_6]^{3-}$, 5. Reaction (2) and (3) are easily reversible for molybdenum [16] but not for tungsten [2,4,17]. This enhanced susceptibility of $(W\text{---}4\text{---}W)^{4+}$ bonds to protons, or other oxidants, and irreversibility of these reactions has been the reason for the slow development in the chemistry of the W_2^{4+} unit, and the belief, that these allegedly important catalytic species, are in fact, kinetic and thermodynamic sinks and therefore relatively uninteresting in terms of their catalytic properties. In our efforts to understand better, to control and further exploit this M_2^{4+}/M_2^{6+} interconversion, we have attempted a kinetic and mechanistic study of reaction (2) (M=Mo, X= Cl, Br) and the

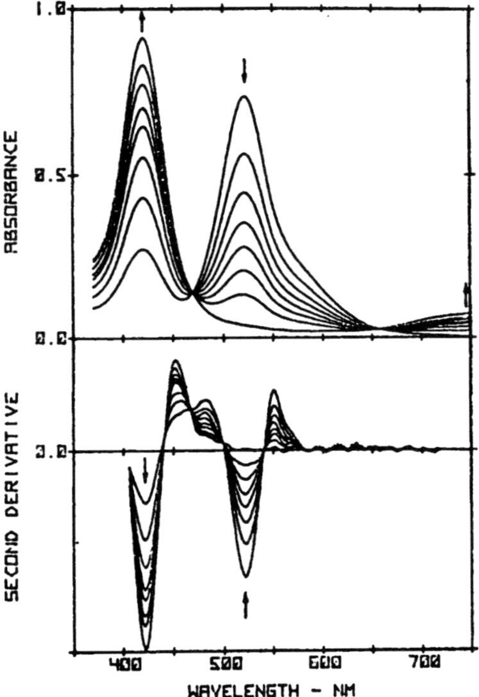

Figure 1. Spectra of a solution $K_4Mo_2Cl_8$ (0.5 x 10^{-3} M) recorded over a period of 4hrs (scan speed 0.5 nm/s). Second derivative spectra were calculated by using numerical methods [29].

homogeneous reduction of the diamagnetic dimers 2a, 2b and 6 by the low-valency paramagnetic Cr(II), 7 and V(II), 8, ions in aqueous acidic solutions.

2. Results and discussion

Kinetics and mechanism of reaction (2), X=Cl. When the potassium salt of 1a (peak at 520 nm) is dissolved in HCl (6-12M) two new peaks start to grow, one at 420nm and a weaker one at 750 nm, both attributed to the hydride, 2a. The two isosbestic points are maintained throughout the reaction indicating that 2a is the only product, Fig. 1. The second derivative, which depends inversely on the square of the half-width (for a Gauss peak), is sensitive to narrow peaks, and it is seen (Fig. 1) that no such peaks appear. Kinetically the formation of the hydride was followed by monitoring the decrease in absorbance at 520 nm and the increase at 420 and 750nm. The first order kinetics of this reaction, and the dependence of the pseudo-first order rate constants on acid concentration, have been previously reported [18]. The energy of activation at 6M HCl is 80.5 KJ·mol^{-1}. Kinetic evidence that there are no side products in appreciable amount is provided by the fact that the plots of ln(A_t-$A\infty$) vs time, at the three wavelenghts where the reaction was monitored, are parallel, and from the internal linearity plot, A_t at 750 nm vs A_t at 520 nm, which is a straight line. Thus, under the conditions of our experiments, there is no evidence for the formation of 3a (peaks at 752, 658, 52, and 420 nm) or of [MoCl$_6$]$^{3-}$ (peaks at 673, 513 and 410 nm). Reaction (2), X=Br. The reaction is considerably more facile with the bromo-derivative, and when we attempted to isolate 1b [7], eq.(1), from a concentrated HBr solution (7.2 M), we obtained exclusively the hydride. Thus, the reactions of the bromo-derivative were studied by dissolving Mo$_2$(O$_2$CMe)$_4$ in thermostated solutions of HBr (4.7 - 6 M) at   25 °C. At concentrations higher than 6M the reactions are too fast to be followed by simpl mixing techniques. The development of the absorbance is similar to that of the chloroderivative, Fig. 1, except that here the maxima are slightly shifted (530, 430, and 780 nm). Typical pseudo-first order kinetics and rate constants on acid concentration are observed. The energy of activation in 6M HBr was found equal to 33.0 KJmol^{-1}. The plot of ln(A_t-$A\infty$) vs time at the two wavelenghts, where the reaction was monitored, are parallel and also the internal linearity plot, A_t, at 530 and 430 nm is a straight line, indicating that here too, as for the chloride, there are no side products in appreciable amounts. It has been argued [18] that the activation required for the addition of the proton to the quadruple bond of 1a originates in the extensive reorganization involved, and that it is also related to the nucleophilicity of this bond. Accordingly, the increased reactivity of the bromo-derivative, compared to the corresponding chloro-derivative is attributed to a larger accumulation of effective negative charge on the multiple bond and an increased nucleophilicity in this region. Also, more repulsion is expected between the bulkier bromine atoms, which will facilitate the disruption of the weakened 8-component of the quadruple bond. The situation here is similar to complex 4 where the destabilization of the 8-component is caused by the higher core-core repulsion at these extremely short internuclear distances of the markedly denser W atoms. Indeed, the W-W distance in 4 is 12 pm longer than in the corresponding 1a ion. Theoretical SCF-X$_a$-SW calculations and photoelectron spectroscopy show that while the electronic structures of Mo$_2$$^{4+}$ and W$_2$$^{4+}$are very similar, there is a difference in the stability of the 8-bonding electrons

[19]. The formation of the hydride can be simply viewed as the protonatin of the quadruple bond and subsequent fast rearrangement or as a step-wise addition of HCl across the metal-metal bond followed by loss of chloride ion and chloride bridge formation. The dependence of the rate constants upon the hydrogen and halide ion concentrations are in favour of the former pathway [18,20].

Reduction of the Mo_2^{6+} dimers (M=Mo, W) with Cr(II) and V(II). The electrochemistry of the Mo_2^{4+} and Mo_2^{6+} dimers 1a, 2a, and 3a in aqueous HCl solutions is complicated by the media, but they have been well studied in ambient temperature malten salts [21]. Reduction of 2a and 3a heteogeneously (amalgamized zinc) gives 1a but attempts to reduce the analogous tungsten complex 6 to 4 (heterogeneously or homogeneously) have not been successful. Since the reduction of carbon-carbon double and triple bonds by the Cr(II) and V(II) ions in acidic solutions to give single and double bonds respectively is well documented [22,23], we have studied the reaction of these low-valency ions with the triply bonded molybdenum and tungsten dimers. When thoroughly deoxygenated freshly prepared HCl 6M solutions of an excess of chromous chloride and a mixture of $K_3Mo_2HCl_8$/KCl are mixed at room temperature, the colour changes from light yellow-green to dark-green-brown and gas evolution (H_2) is observed. The reaction can be followed spectrophotometrically, Fig. (2), and shows the formation of 1a and a mixture [24] of $[Cr(H_2O)_5Cl]^{2+}$ and $[Cr(H_2O)_4Cl_2]^+$. A similar mixture of chromium species is obtained from reaction (8), which occurs very slowly even at concentrations as high as 12M eq. (8):

$$Cr^{2+} + H^+ \longrightarrow Cr^{3+} + 1/2H_2 \tag{8}$$

At the end of the reaction, by suitable work-up procedures complex 1a and Cr(III) in the form of the dichloro-complex can be isolated. The reaction taking place is:

$$Mo_2HCl_8^{3-} + 2Cr^{2+} + H^+ \longrightarrow Mo_2Cl_8^{4-} + 2Cr^{3+} + H_2 \tag{9}$$

$$\uparrow\underline{\hspace{2cm}H^+\hspace{3cm}}\rfloor$$

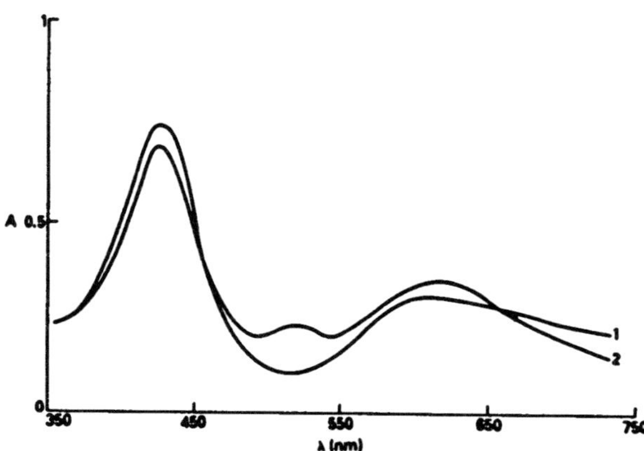

Figure 2 Spectra of a mixture of $K_3Mo_2HCl_8$, 8.5x 10^{-3} M and $CrCl_2$, 3.4x 10^{-4} M HCl at 25°C; (a) after 15 min from the start of the reaction, (b) after 1h.

but the measured stoichiometry is little more than two Cr^{2+} ions for each $Mo_2HCl_8^{-3}$ because of the parallel reaction (8). In HCl the product 1a is converted back to 2a, hence, the dimolybdenum couple participates in the cycle shown in eq. (9). The Mo_2^{6+} core effectively oxidized Cr(II) to Cr(III) comparing to the behaviour of the olefinic and acetylenic compounds, except that here the bond order increases rather than decreases, from three to four. Analogous results are obtained in aqueous HBr 6M with $Cs_3Mo_2HBr_8$/CsBr and $CrBr_2$ or $VBr_2 \cdot 4H_2O$, but due to the rapidity of reaction (2) the hydride was recovered unchanged instead of 1b. At 5M HBr reaction (1) is slower and the 2b $\Leftrightarrow$ 1b conversion can be followed spectrophotometrically. Similarly, when carefully deoxygenated, aqueous solutions, of one equivalent of $K_3W_2Cl_9$/KCl and an excess of $CrCl_2$, 7, were mixed at room temperature or at -5°C, the colour quickly changed and a precipitate was formed. The mixture was filtered, washed with water and tetrahydrofuran leaving behind a grey-black powder insoluble in most common organic solvents. The light green filtrate contains a mixture of $Cr_{(aq)}^{2+}$ (broad peak at 650-700 nm) and $Cr_{(aq)}^{3+}$ (peak at 412 and 590 nm) [24]. Similar results are obtained when an aq. 0.125 M HCl solution of VCl_2, 8, was employed. The composition of the precipitate dependes on the reaction conditions (acid concentration) and approximately analyses as "$WCl_{2-x}(OH)_x$ ", x=0...1. The insolubility of various salts of 6 [K^+, $(Ph_4As)^+$, $(Bu_4^n-N)_4$] in tetrahydrofuran precluded the reaction at low temperature in this medium where the intensely blue-coloured 4 is more stable [14] (598 nm, 0°C) and attempts to detect or to isolate it at its TMEDA adduct [25]. In mixture (s) of tetrahydrofuran-water, in which all the reactants are soluble, a transient formation of a bluish colour at -20°C is observed but it is followed by fast decomposition as described above. However, the same reaction when carried out in aq. HCl solutions (from 6 to 12 M) produced only small amounts of the insoluble precipitate and hydrogen evolution was observed (identified by GC). The reaction can be followed spectroscopically (Fig. 3). The intensity of the peak at 452 nm attributed to 6 increases whereas a peak at 615 nm due to $Cr_{(aq)}^{3+}$ probably as a mixture of $[Cr(H_2O)_5Cl_2]^{2+}$ (peaks at 430, 605 nm) and $[Cr(H_2O)_4Cl_2]^{2+}$ (peaks at 450, 635 nm), appears. Also, the peak due to $V_{aq)}^{2+}$ (860, 560 nm) at 860 nm disappears and a shoulder at 400 nm characteristic of the $V_{(aq)}^{2+}$ ion (peaks at 588, 400, 263) appears next to the peak of 6 at 452 nm [28]. After removing the solvent *in vacuo*, the residues were extracted with tetrahydrofuran leaving behind a deep green solid identified as 6; the filtrates contain Cr(III) in the form of the dichloro complex $[Cr(THF)_4Cl_2]^+$ (peaks at 445 and 640 nm) or V(III) (peaks at 502, 751 nm) [27]. If a large excess of the reducing metal ion was used (from 4 to 200 mol) under identical conditions, the reaction proceeds exothermically and cooling is necessary. Again hydrogen is evolved, 6 was recovered unchanged, and 7, or 8 were oxidized quantitatively to their trivalent states. In view of the instability of 4, expected to be even greater in water than in organic solvents, we propose that in aqueous media a two-electron reduction occurs to give 4 [eq. (10)] followed by its fast decomposition and / or hydrolysis:

$$[W_2Cl_9]^{3-} + 2Cr^{2+} \longrightarrow [W_2Cl_8]^{4-} + 2Cr^{3+} + Cl^- \qquad (10)$$

In acid decomposition it is prevented because reactions (2) and (3) intercept reproducing 6, thus the W_2^{6+}/W_2^{4+} couple participates in the catalytic cycle [28] (Fig. 4). The hydrogen evolved comes from reactions (2) and (3).

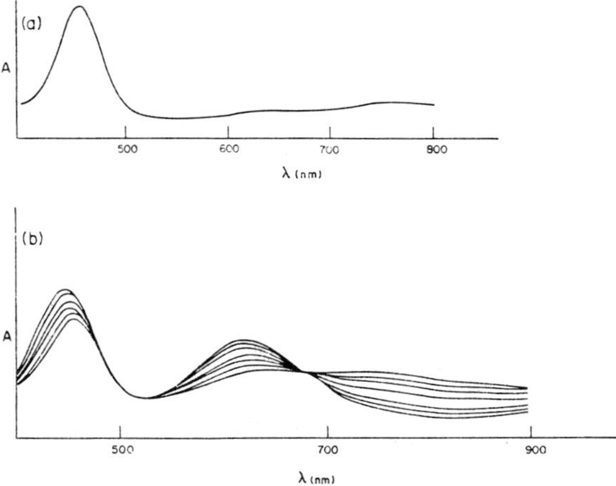

Figure 3 (a) Spectrum   of $K_3W_2Cl_9$ (6.5×10^{-4} M) in 6 M HCl at 25°C. (b) Spectral changes of a mixture of $K_3W_2Cl_9$ (6.9×10^{-4} M) and $CrCl_2$(1.6×10^{-2} M) in 6 M HCl at 25°C recorded at 15-min intervals.

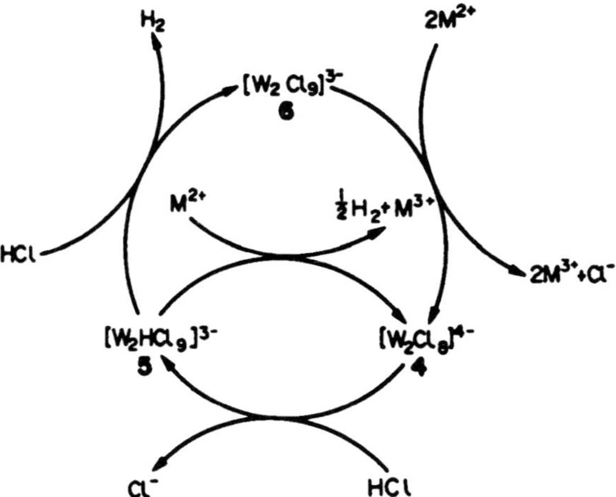

Figure 4. The binuclear tungsten complex **6** catalyses the evolution of H_2 in HCl solutions in the presence of excess $M_{(aq)}^{2+}$, M=Cr, V.

The electrons required are provided by the low-valency metal ions which are consumed (oxidized), respectively. Parallel formation of hydrogen from the reduction of the tungsten hydride (**5**), similarly as for molybdenum [eq. (9)] is possible, and may occur to a small extent at low acid concentrations in which reaction (3) is slower. The mechanistic aspects of these homogeneous electron transfer reactions and the implications to chemical or biochemical catalysis are further examined.

References

1. M.H. Chisholm, *Angew. Chem., Int. Ed. Engl.* **25**, 21, (1986).
2. F.A. Cotton, R.A. Walton, "Multiple bonds between metal atoms", J. Wiley and Sons: New York, (1982).
3. Inorganic Reactions and Methods, Ed. J.J. Zucherman, VCH, Vol. **2** and references therein.
4. P. Escaffre, A. Thorez and P. Kolck, *J. Chem. Soc., Chem. Commun.***146**, and references therein, (1987).
5. (a) I.S. Segal, K.R. Mann, H.B. Gray, *J. Am. Chem. Soc.* **102**, 7252, (1980); W.C. Trogler, D.K. Erwin, G.L. Geoffron, H.B. Gray, *ibid.* **100**, 1160, (1978).
6. (a) E.I. Stiefel, W.E. Newton, G.D. Natt, K.L. Hadfield and W.A. Bullen, Bio-Inorganic II, Ed. K.N. Raymond, American Chemical Society, Washington D.C., 353, (1977). (b) J. Chatt, J.R. Dilworth and R. Richarsds, *Chem.Rev.* **78**, 589, (1977). (c) R.A. Henderson, G. Leigh and J.C. Pickett, *Adv. Inorg. Chem. , Radiochem.* **27**, 198, (1983).
7. (a) A. Bino and F.A. Cotton, *Angew Chem. Int. Ed. Engl.* **18**, 332, (1979). (b) F.A. Cotton and B.J. Kalbacher, *Inorg. Chem.* **15**, 521, (1976); (c) A. Bino *Inorg. Chim. Acta* **106**, 17, (1985); (d) W.H. Delphin and R.A. Wentworth, *Inorg. Chem.* **13**, 2037, (1974).
8. C. Mertis et al, unpublished results.
9. J. Lewis, R.S. Nyholm and P.W. Smith, *J. Chem. Soc. (A)*, **57**, (1969).
10. J.V. Brencie and F.A. Cotton, *Inorg. Syn.* **13**, 170, (1972).
11. A.R. Bowen and H. Taube, *Inorg. Chem.* **13**, 2245, (1974).
12. F.A. Cotton, S. Koch, K. Mertis, M. Millard and G. Wilkinson, *J. Am. Chem. Soc.* **99**, 4989, (1977).
13. F.A. Cotton and G.N. Mott, *J. Am. Chem. Soc.* **104**, 5978, (1982).
14. R.R. Schrock, L.G. Sturgeoff and P.R. Sharp, *Inorg. Chem.* **22**, 2801, (1983).
15. A.P. Sattelberger, K.W. McLanghlin and J.C. Huffman, *J. Am. Chem. Soc.* **103**, 2280, (1981).
16. A. Bino and D. Gibson, *J. Am. Chem. Soc.* **102**, 4277, (1980).
17. F.A. Cotton and G.N. Mott, *J. Am. Chem. Soc.* **104**, 5978, (1982).
18. C. Mertis, M. Kravaritou, A. Shehadeh and D. Katakis, *Polyhedron* **6**, 1975, (1987).
19. F.A. Cotton, J.L. Hubbard, D.L. Lichtenberger and I. Shim, *J. Am. Chem. Soc.* **104**, 679, (1982).
20. C. Mertis et al, work in progress.
21. R.T. Carlin and R.A. Osteryoung, *Inorg. Chem.* **27**, 1482, (1988).
22. (a) C.E. Castro and R.D. Stephens, *J. Am. Chem. Soc.* **86**, 4358, (1964); (b) R.S. Bottei and W.A. Joern, *ibid.* **90**, 297, (1968).
23. D. Katakis, J. Konstantatos and E. Vrachnou-Astra, *J. Organoment. Chem.* **279**, 131, (1985).
24. P.J. Elving and B. Zemzl, *J. Am. Chem. Soc.* **97**, 1281, (1957).

25. F.A. Cotton, G.N. Mott, R.R. Schrock and L. Sturgeoff, *J. Am. Chem. Soc.* **104**, 6781, (1982).
26. J.J. Lingane and L.A. Small, *J. Am. Chem. Soc.* **71**, 973, (1949).
27. R.J.H. Clark, <u>The Chemistry of Titanium and Vanadium,</u> p. 166-172, Elsevier, Amsterdam, (1968).
28. C. Mertis and N. Psaroudakis, *Polyhedron* **8**, 469, (1989).
29. S.P. Koinis, Ph.D. Thesis, University of Athens, (1987).

A Theoretical Study on the Protonation of Nucleic Acid Pyrimidine and Purine Bases

V. MILANO [a], N. RUSSO [a] and M. TOSCANO [b]
a) Dipartimento di Chimica, Universita' della Calabria, I-87030 Arcavacata di Rende (CS), Italy; b) Facolta' di Farmacia, Universita' della Calabria, I-87030 Arcavacata di Rende (CS), Italy.

1. Introduction

The understanding of the behaviour of tautomeric equilibria and of protonation processes of nucleic acid bases is of great importance. In fact the protonation has a relevant effect on the hydrogen bonding ability of the bases, and hence on the variety of pairing schemes in which they participate [1, 2]. For these reasons protomeric tautomerism and the intrinsic basicity of nucleic acid bases have attracted considerable attention from both experimental [3-9] and theoretical [10-14] points of view. The protonation of DNA pyrimidine and purine bases has been studied theoretically at ab-initio level and employing minibal basis set considering only the most stable tautomers. Recently the gas-phase proton affinity of nucleo-bases and deoxyribonucleosides have been determined by fast atom bombardment tandem mass spectrometry (FABS) [5] extending the high-pressure mass spectrometric (MS) measurements limited to the study of thymine, cytosine and adenine [6]. Notwithstanding these investigations, some significant problems, such as the protonation sites, the influence of other low energy isomers, the influence of full geometry optimization, are not clearly established.
In this contribution we report the result of a theoretical study of protonation behaviour and the proton affinities determined by using the advanced AM1 semiempirical method [15] and considering different neutral tautomers and all possible protonation sites , for guanine, adenine, cytosine, uracil and thymine nucleo-bases.

2. Computational details

We have employed the AM1 hamiltonian, together with full geometry optimization procedures using analytical gradient technique, as implemented in the MOPAC code [16]. For each nucleo-bases several low energy (in the range of about 20 Kcal/mol above the absolute minimum) tautomers other than the most stable one, have been considered. In particular for guanine, adenine, cytosine and thymine we have considered on the whole 9, 6, 3 and 3 tautomeric forms respectively.Three uracil tautomers have been investigated. The proton affinities have been calculated considering the process $B + H^+ \rightarrow BH^+ + \Delta Hr$ as follows:

N. Russo et al. (eds.), Properties and Chemistry of Biomolecular Systems, 331–351.
© 1994 *Kluwer Academic Publishers. Printed in the Netherlands.*

$$PA = -\Delta Hr = -[(\Delta H_{BH^+}) - (\Delta H_{H^+} + \Delta H_B)]$$

For the heat of formation of H^+ (ΔH_{H^+}) we have used the experimental value of 367.2 Kcal/mol [17]. ΔH_{BH^+} and ΔH_B are the calculated heat of formation of protonated and neutral nucleic acid bases respectively.

3. Results and Discussion

In this section we briefly report the predicted tautomeric equilibria of nucleic acid bases obtained by AM1 method. After this, we report and discuss our results on the protonation process for the individual nucleo bases. For further informations about the reliability of AM1 in the reproduction of protonation processes see refs. 18 and 19.

3.1. RELATIVE TAUTOMERIC STABILITIES

The tautomeric equilibria for nucleic acid bases have been extensively studied with a variety of quantum-mechanical methods [20-25]. Previous AM1 computations [24] show that for each nucleic acid base, the tautomer found by experiment [26] is also the most stable one. Results are also consistent with the available ab-initio investigations [21, 22]. Because our interest is focussed on the protonation process of both nucleo bases and deoxiribonucleosides we have investigated a number of selected tautomers (see Figure 1 for adenine and Figure 2 for cytosine, thymine, uracil and guanine).

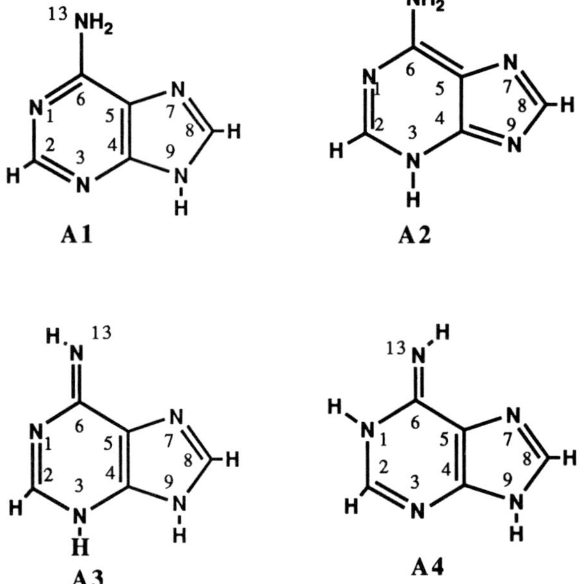

Figure 1. The four studied adenine tautomers.

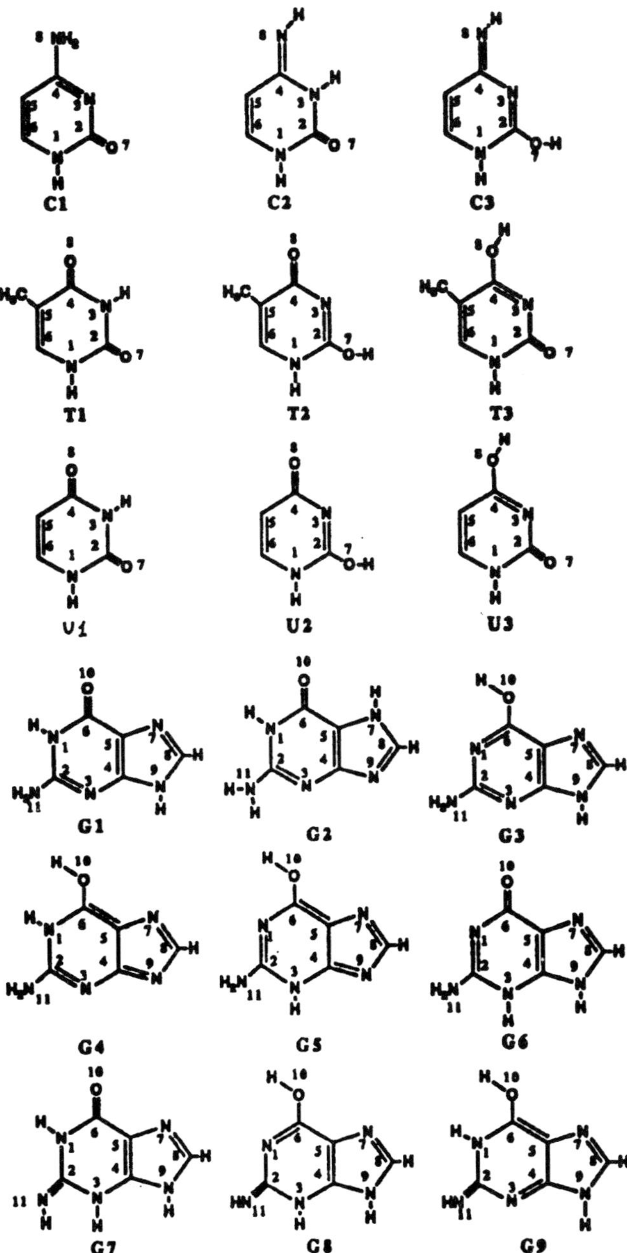

Figure 2. The studied cytosine, thymine, uracil and guanine tautomers

The relative energies of the considered systems, obtained by different theoretical methods are collected in table 1. From the table it is clear that the diketo tautomers of uracil (U1) and thymine (T1) are the most stable at AM1 and 3-21-G [21,24] level in agreement with the experiment [26]. All recent computations agree in the prediction of the 1-H-oxo-amino (C1) as the most stable form of cytosine. Other tautomers (i.e. C2) that lie at very low energies above the absolute minimum, can be found in a substantial amount in the gas phase. For the adenine, the AM1 method indicates the 9-H-amino form A1 as the most stable one. The AM1 and PM3 [24, 25] methods propose the amino-oxo form G1 as the most stable tautomer for guanine. The same form is also observed experimentally [26]. Other tautomers (i.e. G2 and G3) lie in a range of few Kcal/mol above G1 minimum.

Table 1. AM1 calculated relative energies (Kcal/mol) of selected low energy nucleobases tautomers. For the labelling of the tautomers see Figures 1 and 2. STO-3G and 3-21G data are taken from ref. 21. For MNDO value see ref. 23.

Tautomer	MNDO	AM1	STO-3G	3-21G
		Cytosine		
C1	0.0	0.0	0.0	0.0
C2	4.8	1.5	11.1	0.4
C3	10.3	16.3	23.7	22.7
		Thymine		
T1	0.0	0.0	0.0	0.0
T2	-8.9	20.2	-15.1	25.3
T3	7.3	12.7	15.4	32.7
		Uracil		
U1	0.0	0.0	0.0	0.0
U2	3.0	20.0	6.6	19.6
U3	-	12.6		
		Adenine		
A1	0.0	0.0	-	-
A2	12.0	11.6	-	-
A3	22.4	22.5	-	-
A4	11.3	13.6	-	-
		Guanine		
G1	0.0	0.0		-
G2	2.6	1.5		-
G3	-7.2	3.3		-
G4	12.9	21.3		-
G5	8.0	17.3		-
G6	14.1	14.2		-
G7	13.4	11.1		-
G8	11.1	18.7		-
G9	12.5	19.6		-

3.2. PROTONATION PROCESS OF THE NUCLEIC ACID BASES

Although the protonation process in the plane of the base is favoured [27] , no restrictions have been imposed in the optimization procedure. Results show that the attach of proton in the plane is fovoured in all cases.

In the following sections, protonation of the individual bases will be discussed. Some observation about the trends in geometric and energetic changes due to the protonation will proposed. Comparison between our results with previous theoretical and experimental works will be also made.

3.2.1. *Thymine*

For thymine two distinct protonation sites (the two carbonil groups for the T1 form and other two sites for the T2 and T3 tautomers) have been considered (see figure 3). The calculated proton affinities are reported in Table 2.

The data in this table show that in the case of T1 the O8 protonation site is preferred to the O7 one (205.0 versus 198.7 Kcal/mol). The more favourable geometry is that in which the added proton lies on the C5 side of the C4=O group. In this position the repulsion energy between the extra proton and the in-plane hydrogen of the methyl group is smaller than the interaction between O8-H proton and the more acid imide hydrogen (N4-H).

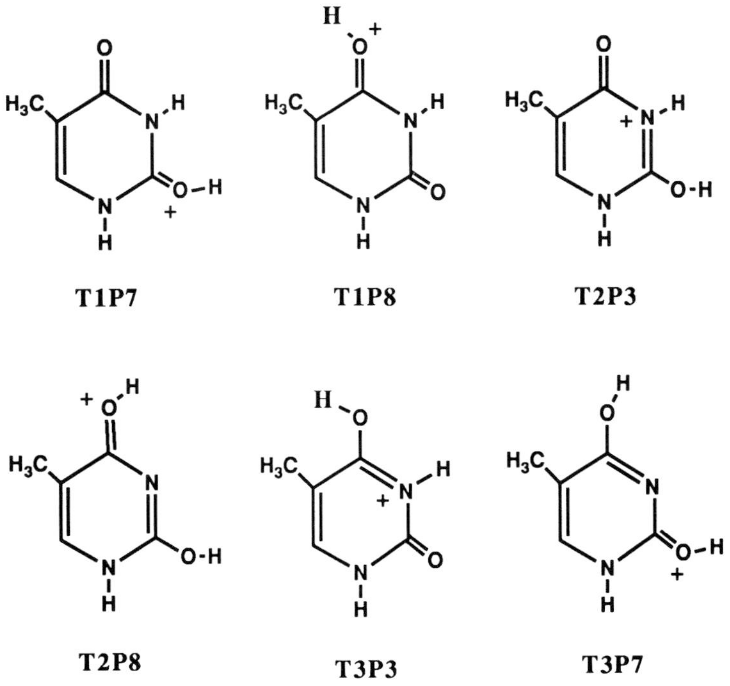

Figure 3. Calculated protonotated thymine tautomers.

Table 2. AM1 heat of formation (ΔH) and proton affinities (PA) of selected numbers of thymine tautomers. All values are in Kcal/mol.

Molecule	ΔH	PA
T1	-61.1	-
T1P7	107.3	198.7
T1P8	101.1	205.0
T2	-40.9	-
T2P3	107.3	219.0
T2P8	104.0	222.3
T3	-48.4	-
T3P3	101.1	217.1
T3P7	104.0	214.8

In the case of T2 tautomer we found that the protonation on the O8 is more favoured than on nitrogen N3 (222.3 versus 219.0 Kcal/mol). On the contrary, in the case of T3 form, the protonation on nitrogen (N3) is preferred with respect to that on oxygen (O7) site (217.1 versus 214.8 Kcal/mol). On the basis of this result we can conclude that the protonation process is energetically favoured in the case of O8 site of T2 form. The neutral T2 tautomer is found at 20.2 Kcal/mol above T1. The calculated equilibrium geometries are reported in Table 3.

Table 3. Optimized geometries for neutral and protonated thymine tautomers. The x-ray data are taken from ref. 3.

Distance	X-ray	T1	T1P7	T1P8	T2	T2P8	T3
1-2	1.355	1.413	1.365	1.422	1.397	1.387	1.436
2-3	1.361	1.402	1.367	1.431	1.324	1.364	1.410
3-4	1.391	1.407	1.446	1.362	1.418	1.363	1.329
4-5	1.447	1.476	1.479	1.432	1.489	1.451	1.455
5-6	1.349	1.364	1.361	1.392	1.364	1.384	1.378
1-6	-	1.380	1.401	1.359	1.389	1.381	1.368
2-7	1.234	1.249	1.362	1.233	1.381	1.357	1.245
4-8	1.231	1.242	1.227	1.348	1.241	1.345	1.367
5-9	1.503	1.475	1.476	1.477	1.474	1.474	1.472
Angle							
1-2-3	118.0	117.9	121.9	116.9	126.9	124.5	119.3
1-2-7	122.0	120.2	113.8	122.5	111.4	113.2	116.8
2-3-4	126.0	123.0	121.2	120.8	117.6	116.4	117.7
1-6-5	-	122.2	121.9	121.9	121.2	121.3	121.5
3-4-8	121.0	118.6	116.4	113.8	120.1	120.7	120.0
3-2-7	-	121.9	124.4	120.5	121.7	122.3	123.9
9-5-6	-	123.3	123.6	122.0	122.6	122.8	123.3

The inspection of this table reveals a change in the geometry upon protonation. In particular the main changes concern the bond lengths and valence angles around the protonation sites. The other geometrical parameters remain essentially the same.

3.2.2. Cytosine

For the three considered cytosine tautomers two possible protonation sites have been considered (see figure 4 and Table 4) .

Table 4. AM1 heat of formation (ΔH) and proton affinities (PA) for a selected numbers of cytosine tautomers. All values are in Kcal/mol.

Molecule	ΔH	PA
C1	2.8	-
C1P3	142.7	227.2
C1P7	146.4	223.5
C2	4.3	-
C2P8	142.7	228.8
C2P7	169.5	202.0
C3	19.0	-
C3P8	146.4	239.8
C3P3	169.5	216.7

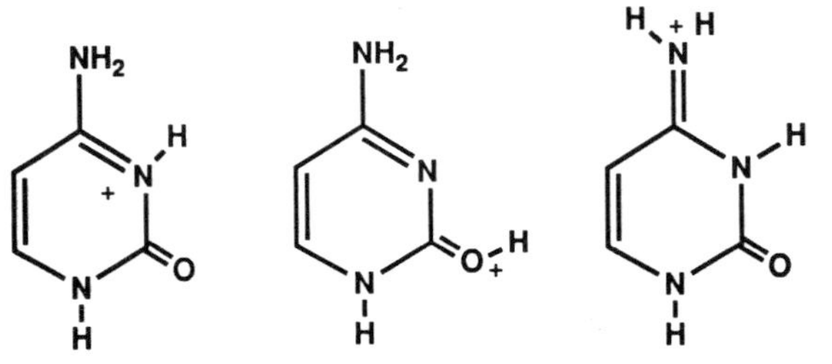

C1P3   C1P7   C2P8

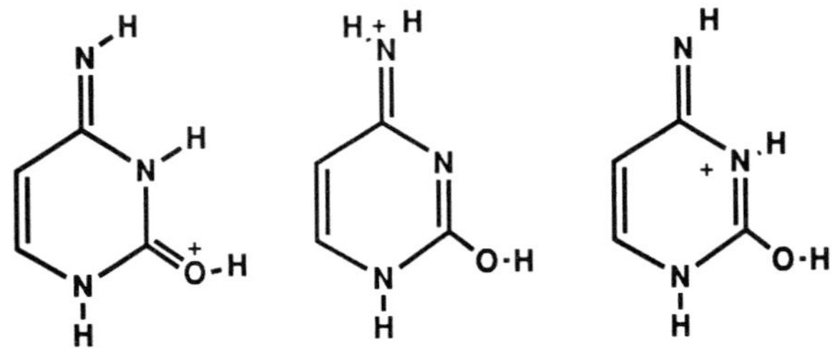

C2P7   C3P8   C3P3

Figure 4. Protonated cytosine tautomers.

Table 5. AM1 optimized geometries for cytosine neutral and protonated tautomers. X-ray values are taken from ref. 3.

Distance	X-ray	C1	C1P3	C1P7	C2	C3	C3P3
1-2	1.381	1.442	1.416	1.393	1.407	1.399	1.366
2-3	1.364	1.398	1.427	1.352	1.401	1.319	1.363
3-4	-	1.349	1.384	1.390	1.424	1.436	1.466
4-5	1.410	1.465	1.452	1.461	1.487	1.488	1.485
5-6	1.340	1.373	1.377	1.373	1.355	1.356	1.351
1-6	1.353	1.364	1.364	1.381	1.383	1.392	1.406
2-7	1.241	1.247	1.236	1.360	1.251	1.384	1.364
4-8	1.342	1.374	1.353	1.347	1.297	1.297	1.278
Angle							
1-2-3	118.2	119.9	117.9	125.5	117.9	127.1	123.1
1-2-7	-	115.9	122.5	112.6	120.7	112.3	113.2
2-3-4	-	118.4	121.8	117.3	123.6	117.9	121.2
1-6-5	120.8	120.7	121.7	121.4	122.1	121.4	121.5
3-4-8	-	119.9	121.4	119.4	125.5	124.1	125.1
3-2-7	122.2	124.2	119.7	121.9	121.4	121.6	123.7

For the C1 form results show that the most stable protonation process occurs on the nitrogen N3 atom (227.2 Kcal/mol). The proton affinity obtained by protonation of oxygen atom (O7) is 223.5 Kcal/mol. For both C2 and C3 tautomers the protonation is fovoured when the NH group is considered. The proton affinities are, for this site, 228.8 and 239.8 Kcal/mol. As it is evident, the more stable protonation site is the NH group of C3 tautomer (see table 4). In the neutral form this tautomer lies at 16.3 Kcal/mol above the C1 one.

As in the case of thymine, the change on geometries upon protonation are concentrated in the neighbourhood of the atoms directly involved in this process (see table 5).

3.2.3. Uracil

As is shown in Figure 5 two protonation sites for the considered uracil tautomers have been investigated. Table 6 show that the most stable cation is the U2P3 in which the N3 atom of neutral uracil is protonated. At only 3.2 Kcal/mol above this form we found the U2P8 isomers in which the protonation process involve the O8 atom. Considering the most stable neutral uracil tautomer we note that the favoured protonation site is that in which the O8 atom is involved.

The protonation in the O7 atom requires a more higher energy. Finally, considering the U3 tautomer we found that, as in the U2 form, the more stable protonation site involved the N3 atom rather than the O8 one.

The variation of geometrical parameters going from neutral to protonated forms are shown in table 7. Also in this case the most significant variations on the bond and valence angles occurs near the protonation site.

Table 6. AM1 heat of formation (ΔH) and proton affinities (PA) for selected numbers of uracil tautomers. All values are in Kcal/mol.

Molecule	ΔH	PA
U1	-53.9	-
U1P7	117.4	195.9
U1P8	111.3	202.0
U2	-33.3	-
U2P3	117.4	216.5
U2P8	120.6	213.3
U3	-41.3	-
U3P3	120.6	205.3
U3P7	11.3	214.6

U1P7 U1P8 U2P3

U2P8 U3P3 U3P7

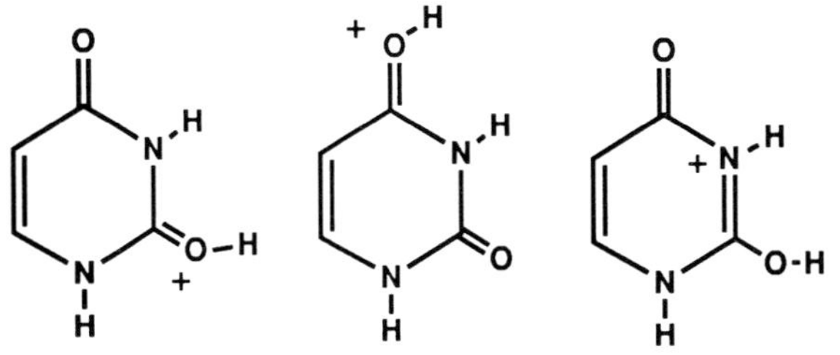

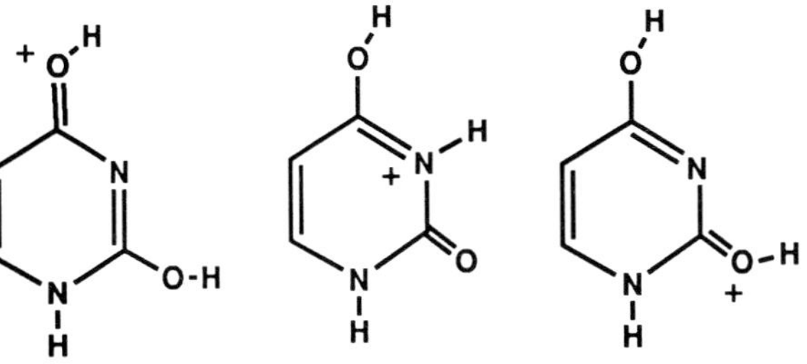

Figure 5. Protonated uracil tautomers.

Table 7. Optimized geometries for neutral and protonated uracil tautomers. The x-ray data are taken from ref. 6 .

Distance	X-ray	U1	U1P7	U1P8	U2	U2P8	U3
1-2	1.371	1.415	1.377	1.423	1.401	1.397	1.438
2-3	1.376	1.402	1.358	1.434	1.323	1.362	1.410
3-4	1.370	1.409	1.451	1.368	1.421	1.367	1.330
4-5	1.429	1.469	1.469	1.429	1.481	1.444	1.450
5-6	1.340	1.359	1.354	1.386	1.359	1.377	1.374
1-6	1.358	1.380	1.404	1.360	1.389	1.383	1.368
2-7	1.215	1.249	1.360	1.233	1.380	1.357	1.245
4-8	1.245	1.241	1.225	1.346	1.241	1.342	1.367
Angle							
1-2-3	114.0	118.2	122.5	117.2	126.9	124.1	119.4
1-2-7	123.7	119.9	123.2	122.6	121.4	121.4	116.7
2-3-4	126.7	122.6	121.1	120.6	117.5	116.3	117.5
1-6-5	122.3	121.7	121.3	121.6	121.0	120.9	121.1
3-4-8	119.2	118.8	116.9	122.9	120.3	120.7	120.4
3-2-7	122.3	121.9	114.3	120.2	121.7	114.5	123.9

3.2.4. *Adenine*

As mentioned before, four adenine tautomers have been considered. Three nitrogen protonation sites are available for the protonation reaction (N1, N3, N7) (see Figure 6) on the A1 tautomer. The data of table 8 show that protonation at either of the nitrogen in the

Table 8. AM1 heat of formation and proton affinity for selected adenine tautomers. All values are in Kcal/mol.

Molecule	ΔH	PA
A1	86.9	-
A1P7	239.7	214.4
A1P3	231.2	222.9
A1P1	232.0	222.1
A2	98.4	-
A2P9	231.2	234.4
A3	110.7	-
A3P7	263.3	214.6
A3P1	256.6	221.3
A3P13	231.2	246.7
A4	100.4	-
A4P7	251.8	215.8
A4P3	256.6	211.0
A4P13	232.0	235.6

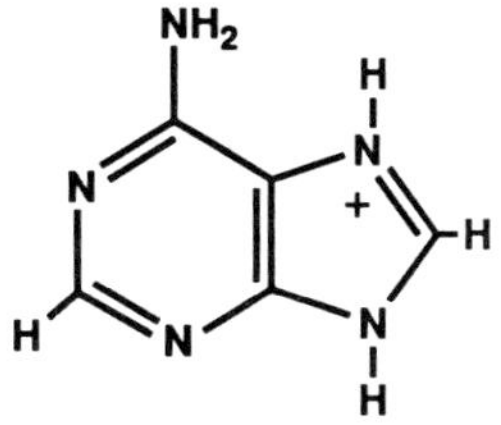

Figure 6. Protonated adenine tautomers.

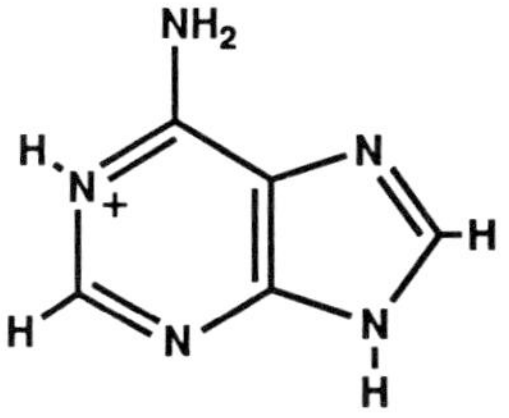

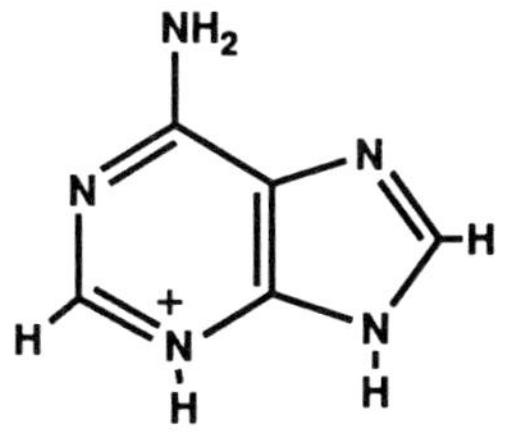

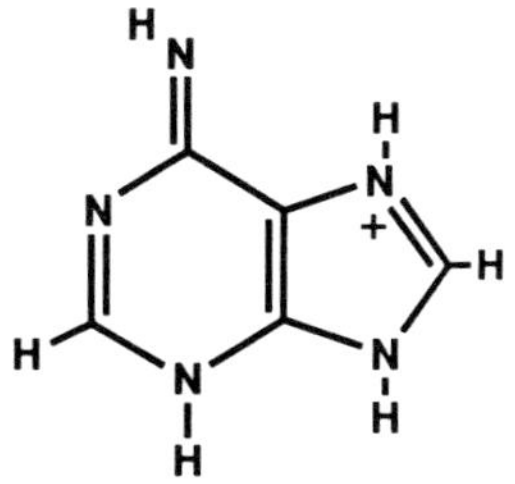

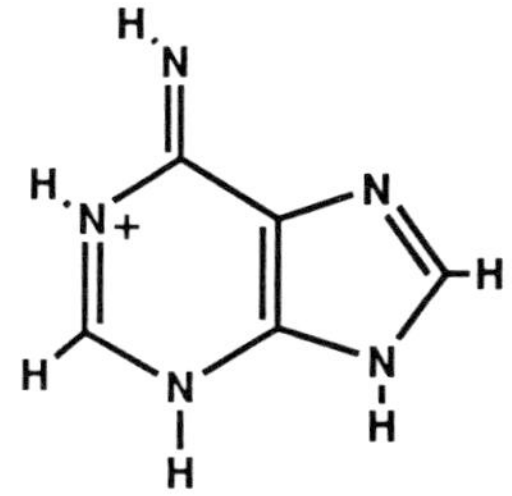

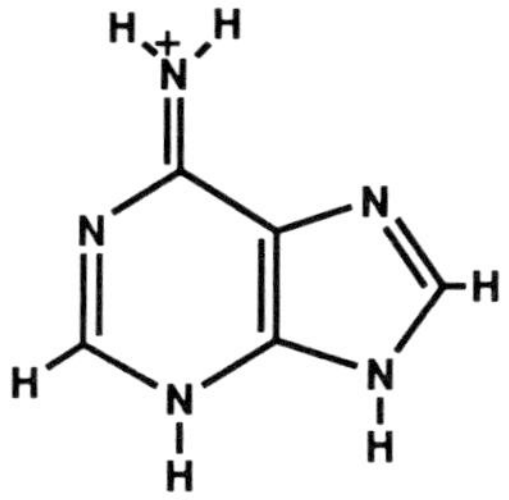

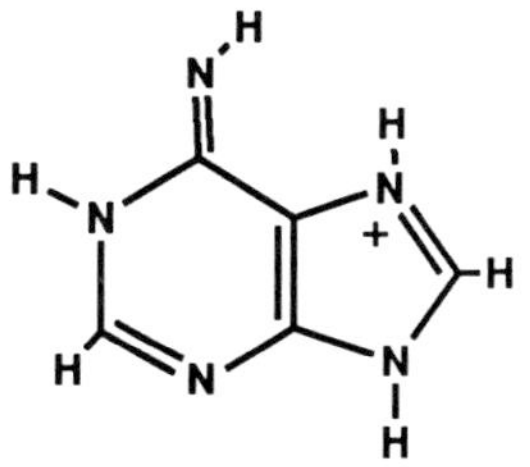

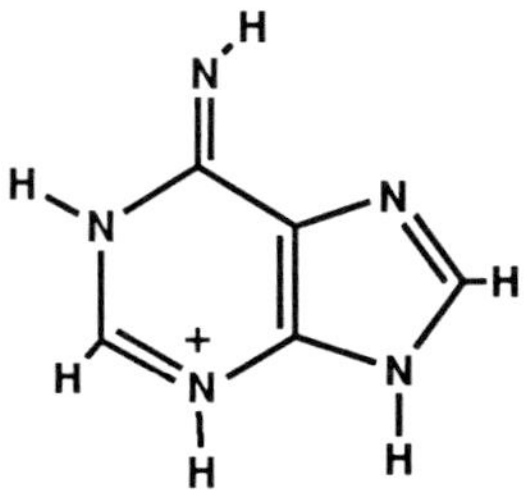

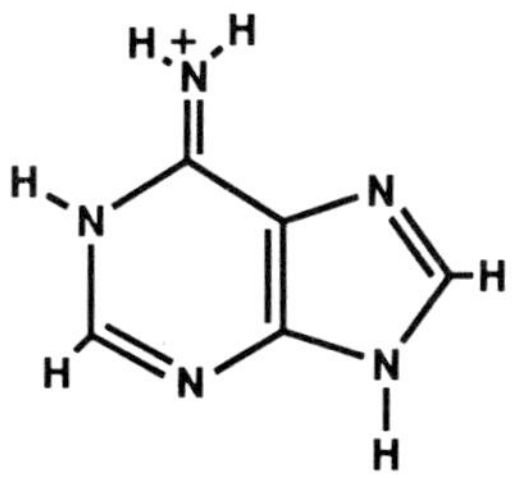

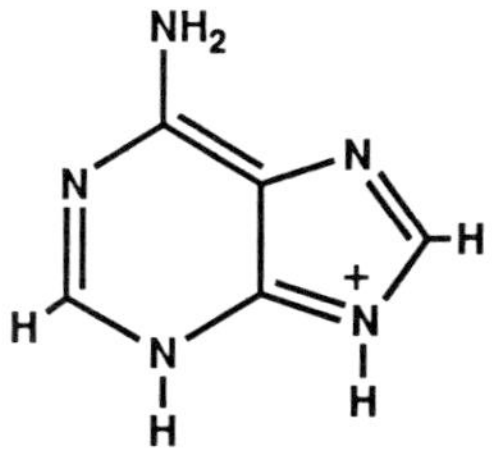

six-membered (N1 and N3) ring is preferred over protonation at nitrogen on the five-membered ring (N7). The proton affinities are 222.9, 222.1 and 214.4 Kcal/mol for N3, N1 and N7 respectively.

Table 9. AM1 optimized structures for different neutral and protonated adenine tautomers. The X-ray results are taken from ref.3.

Distance	X-ray	A1	A1P7	A1P1	A1P3	A2	A3	A4	A4P7	A4P3
1-2	1.340	1.361	1.350	1.403	1.322	1.327	1.297	1.383	1.373	1.344
2-3	1.330	1.354	1.368	1.320	1.390	1.376	1.414	1.322	1.334	1.359
3-4	1.349	1.368	1.352	1.387	1.381	1.388	1.392	1.395	1.383	1.400
4-5	1.381	1.460	1.457	1.458	1.452	1.491	1.433	1.437	1.433	1.432
5-6	1.415	1.437	1.440	1.432	1.442	1.415	1.480	1.469	1.476	1.468
1-6	-	1.376	1.382	1.396	1.406	1.403	1.441	1.431	1.435	1.458
5-7	1.385	1.401	1.405	1.400	1.400	1.392	1.393	1.392	1.397	1.387
7-8	1.308	1.342	1.367	1.343	1.341	1.359	1.348	1.351	1.372	1.355
8-9	1.362	1.413	1.376	1.418	1.421	1.408	1.412	1.410	1.380	1.407
4-9	-	1.399	1.419	1.385	1.389	1.366	1.397	1.395	1.408	1.396
6-13	1.332	1.369	1.359	1.354	1.342	1.364	1.294	1.297	1.290	1.283
Angle										
1-6-5	-	118.8	117.3	116.4	117.5	118.9	114.3	112.2	111.3	111.9
2-1-6	-	117.5	118.9	120.2	119.9	118.9	120.7	121.8	122.5	123.6
3-4-9	-	130.9	130.0	131.0	132.7	132.0	131.2	130.1	129.4	131.9
5-7-8	103.5	105.1	108.3	105.1	105.5	103.9	105.6	105.2	108.1	105.3
6-5-7	-	133.0	135.2	131.9	131.8	133.6	130.1	129.5	131.6	128.7
7-8-9	114.3	113.5	110.8	113.2	113.1	116.9	112.8	112.7	110.0	112.8
1-6-13	-	121.0	120.8	122.7	121.0	119.8	120.1	119.8	120.0	118.1
5-6-13	124.5	120.2	121.9	120.9	121.6	121.3	125.7	128.0	128.8	130.0

The inspection of HOMO composition reveals a larger contribution coming from atomic orbitals of N1 and N3 atoms. This fact can explain the preference for these protonation sites. Previous ab-initio study gives the N1 and N3 as preferred protonation sites in adenine, with N1 slightly favoured [12]. Experimentally, the preferred protonation sites are N1 and N3 [12]. As is shown in figure 6 also for the A3 and A4 tautomers three protonation sites have been considered (N1, N7 and NH for A3, and N3, N7 and NH for A4). In the case of A3 tautomer, table 8 indicates that NH group is the preferred protonation site with a proton affinity of 246.7 Kcal/mol.

Compared to protonation at NH, protonation at N1and N7 is less favourable by 25.4 and by 32.1 Kcal/mol respectively. Finally for the A4 tautomer we found that the fovoured protonation site occurs at NH group (235.6 Kcal/mol) followed by N7 (215.8 Kcal/mol) and N3 (211.0 Kcal/mol) ones.

The absolute minimum for the protonation process is located at NH position of A3 tautomer that lies at 22.5 Kcal/mol above the absolute A1 neutral minimum. For the A2 tautomer only the N4 site has been considered. The other positions give protonation systems that we found also in the protonation process of the other considered tautomers. The proton affinity in this case is found to be 234.4 Kcal/mol.

Data in table 5 show also the geometric variations upon the protonation process. Again, the protonation process induces changes essentially around the attached sites.

3.2.5. *Guanine*

Nine guanine tautomers have been investigated (see Figure 7a and 7b). The considered protonation sites are : N3, N7 and O10 for G1; N9 for G2, G4 and G5; N1, N3, N7 for G3; N1, N7 and O10 for G6; N7, NH and O10 for G7; N1, N7 and NH for G8 and N3, N7 and NH for G9.

The PA are reported in table 10. In the case of the most stable G1 neutral form the preferred protonation site is the P7 one (223.5 Kcal/mol). The proton affinity at N3 is 212.7 Kcal/mol. The protonation of oxygen is less favoured by 12.7 Kcal/mol. Also for the protonation of G3 tautomer the most favourable site is the N7, for which the proton affinity is 222.4 Kcal/mol.

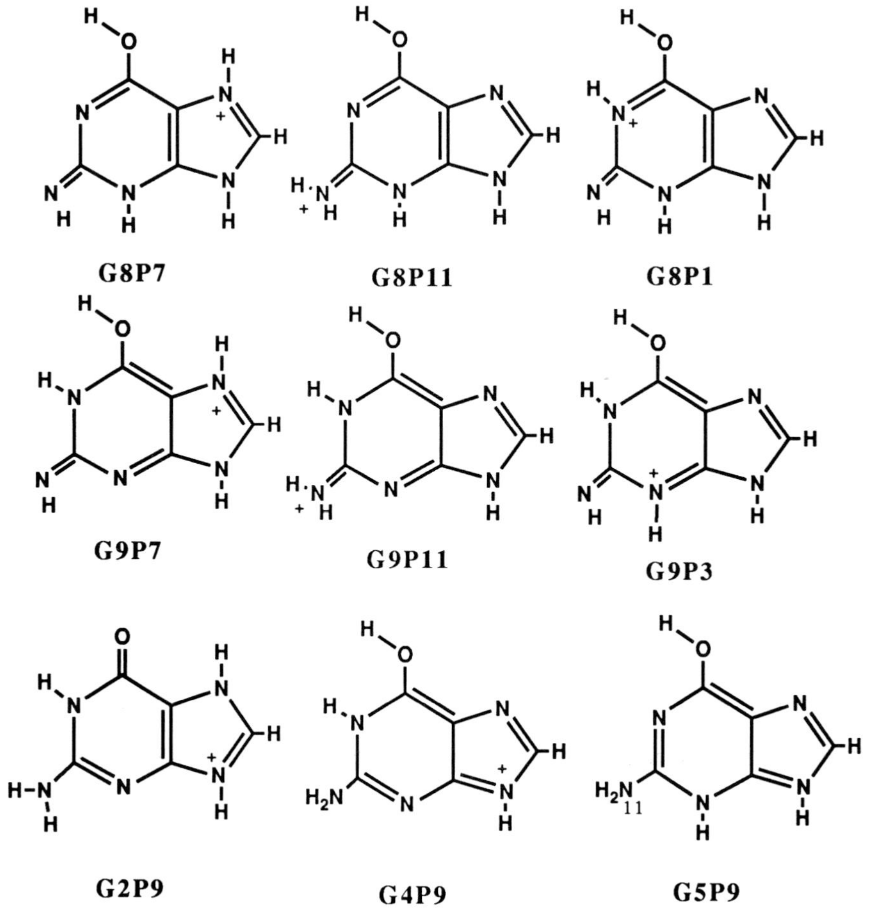

Figure 7a. G8 and G9 protonated guanine tautomers.

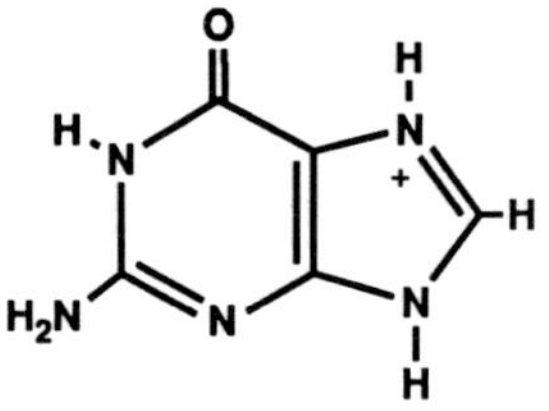

G1P7 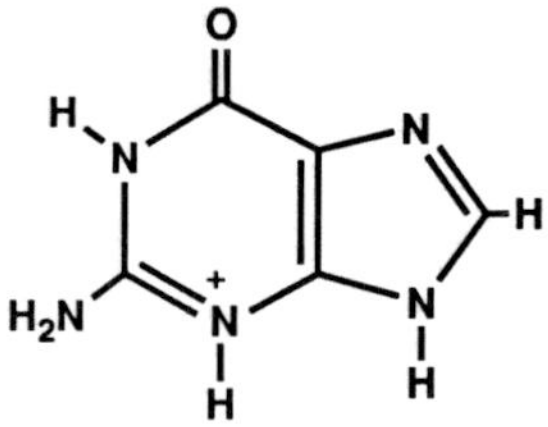 G1P3 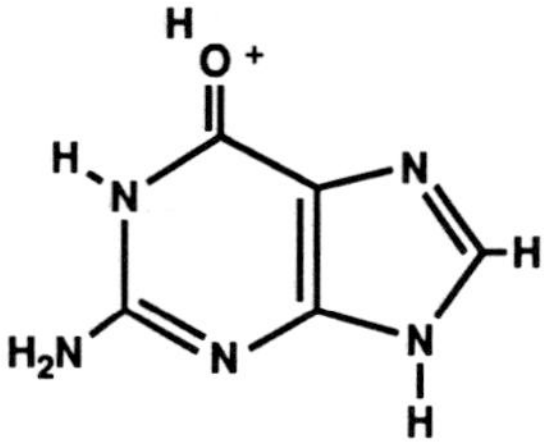 G1P10

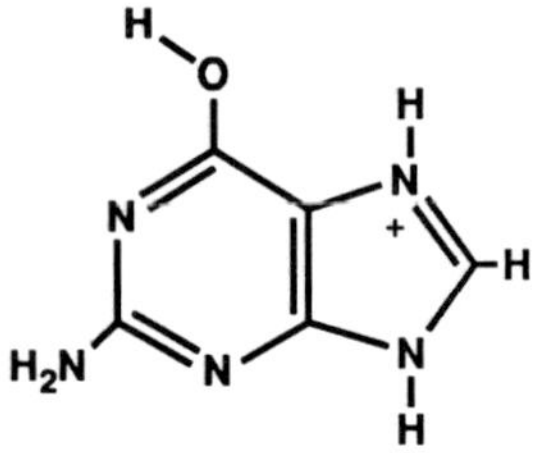

G3P7 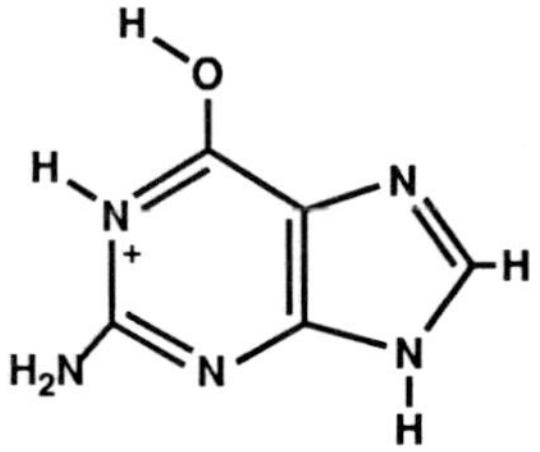 G3P1 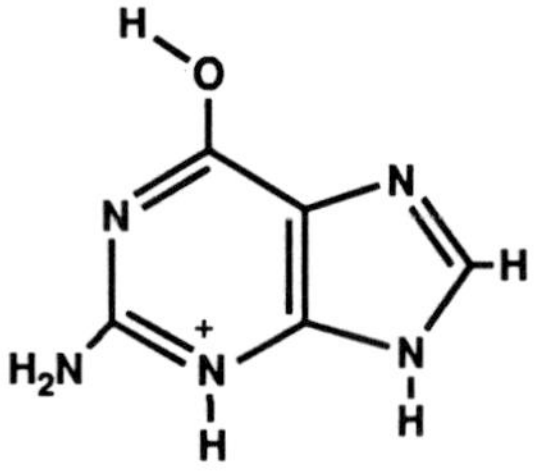 G3P3

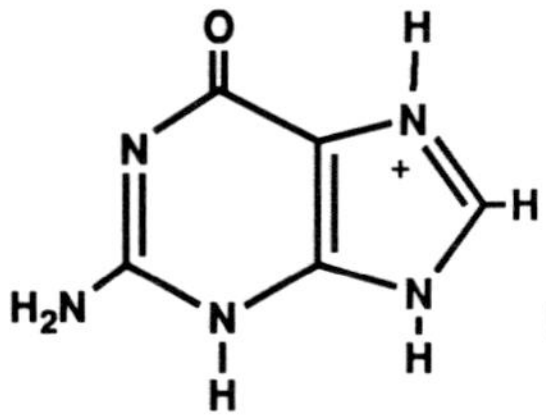

G6P7 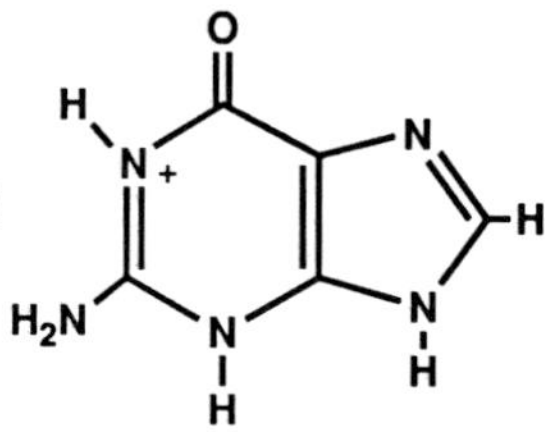 G6P1 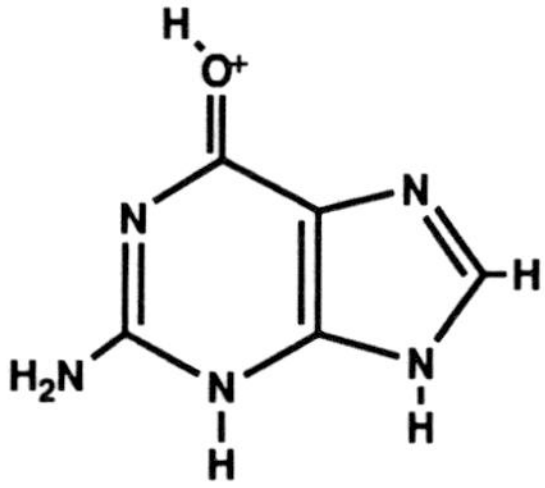 G6P10

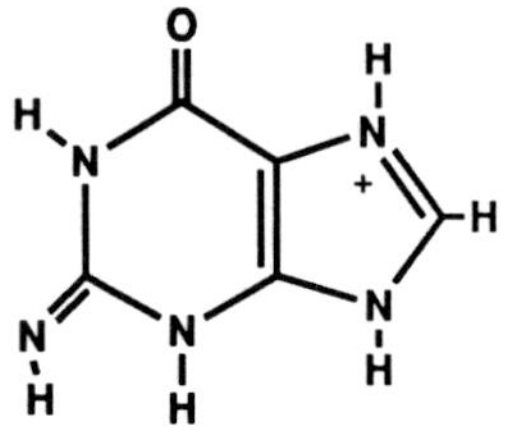

G7P7 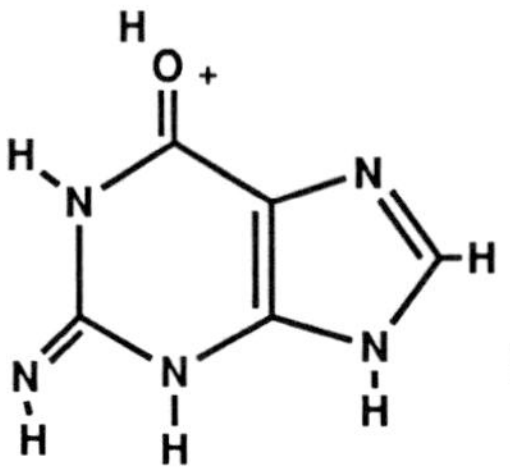 G7P10 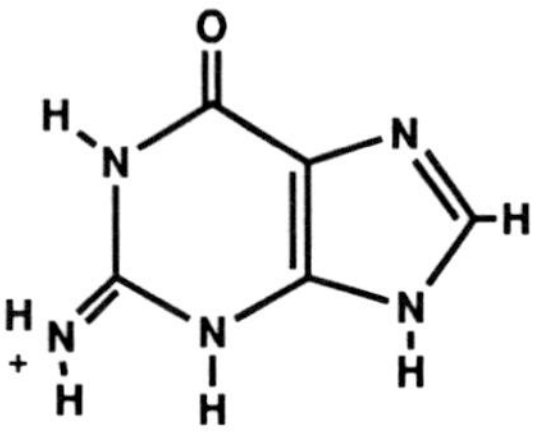 G7P11

Figure 7b. G1, G3, G6 and G7 protonated guanine tautomers.

This value is 4.8 and 7.2 Kcal/mol greater than the PA for N3 and N1 sites. In the case of G6 form the proton attach on the O10 position gives the more stable system with a PA of 228.4 Kcal/mol. The PA for N1 and N7 are 226.9 and 219.9 Kcal/mol.

Table 10. AM1 heat of formation and proton affinities for selected guanine tautomers. all values are in Kcal/mol.

Molecule	ΔH	PA
G1	48.8	-
G1P7	192.5	223.5
G1P3	203.3	212.7
G1P10	204.2	211.8
G2	50.3	-
G2P9	192.5	225.0
G3	52.2	-
G3P7	197.0	222.4
G3P1	204.2	215.2
G3P3	201.8	217.6
G4	70.4	-
G4P9	204.2	233.2
G5	67.9	-
G5P9	201.8	233.3
G6	63.0	-
G6P7	210.3	219.9
G6P1	203.3	226.9
G6P10	201.8	228.4
G7	59.9	-
G7P7	211.2	215.9
G7P11	203.3	223.8
G7P10	212.1	215.0
G8	67.5	-
G8P7	218.8	215.9
G8P11	201.8	232.9
G8P1	212.1	222.6
G9	68.4	-
G9P7	224.2	211.4
G9P11	204.2	231.4
G9P3	212.1	223.5

Concerning the protonation of G7, G8 and G9 tautomers we found that the absolute protonation mimimum is on NH group (223.8, 232.9 and 231.4 Kcal/mol for G7, G8 and G9 respectively). For the G7 the PA at N7 and O10 atoms are quite similar (215.9 and 215.0 respectively), while for G8 the secondary minimum is that protonated at N1 position (222.6 Kcal/mol). The protonation on N7 gives a PA of 215.9 Kcal/mol. The second minimum for G9 is found adding the proton on N3 (223.5 Kcal/mol) and the third on N7 (211.4 Kcal/mol). For G2, G4 and G5 the obtained PA are 225.0, 233.2 and 233.3 Kcal/mol respectively.

The higher values of PA are found in the case of the protonation of G4 and G5 tautomers. In the neutral guanine these two tautomers lie at 18.7 and 19.6 Kcal/mol above the G1 absolute minimum.

Comparison of geometrical parameters for all the systems involved in the protonation study shows (see table 11a and 11b) that also in this case the main changes are observed for the bond lengths and valence angles nearest to the protonation sites.

3.2.6. *Trends in geometries, electronic structures and proton affinities*

Significant structural changes accompany the protonation of nucleic acid nucleobases. In particular, the protonation of carbonyl oxygen results in a significant lengthening of the C-O bond (an average of 0.13 Å), which indicates, as expected, a loss of double bond character. Simultaneously, the O-C-C angle increases by about 9°. Protonation on nitrogen on both five- and six-membered rings leads to significant changes of C-N bonds (a lengthening of about 0.01-0.03 Å) and C-N-C angles (an increase of about 5°). Inspection of the C-O distances (about 1 Å) and H-O-C angles (about 109°), in the case of oxygen protonation, shows that the protonation occurs at the oxygen lone pair electrons. The value of the N-H distance and H-N-C angles shows also that, in the case of nitrogen protonation sites, the H^+ adds at the nitrogen lone pair of electrons. From these results it is clear that reliable computed proton affinities cannot be obtained without a full geometry optimization of the bases upon proton addition. Similar trends have been observed experimentally by X-ray measurements [3].

The Mulliken population analysis reveals that the protonation process is acconpanied by a larger redistribution of electron density. The main effect is a charge transfer of about $0.7e^-$ to the proton. This charge transfer occurs through the σ electron system of the nucleic acid bases and has a stabilizing effect on the cations. For oxygen protonation an increase in the π electron population is observed. The same increase of π density is found for the nitrogen protonation.

Considering the most stable tautomers of neutral nucleic acid bases the following proton affinity trend is observed at AM1 level:

cytosine (227.2 Kcal/mol) > guanine (223.5 Kcal/mol) > adenine (222.9 Kcal/mol) >> thymine (205.0 Kcal/mol > uracil (202.0)

The ab-initio 4-31G study [9] gives the same trend but with different absolute values

cytosine (249.2 Kcal/mol) > guanine (244.7 Kcal/mol) > adenine (240.0 Kcal/mol) >> thymine (211.8 Kcal/mol)

Recent gas phase mass spectrometry experiment [5] states that

guanine (227.4 Kcal/mol) > cytosine (225.9 Kcal/mol) > adenine (224.2 Kcal/mol) >> thymine (209.6 Kcal/mol)

Table 11a. AM1 optimized structures for different neutral and protonated guanine tautomers. The X-ray results are taken from ref. 3.

Distance	G1	G1P7	G1P3	G2	G3	G3P7	G3P1	G4	G5
1-2	1.371	1.408	1.410	1.382	1.418	1.399	1.410	1.429	1.423	1.355
2-3	1.315	1.355	1.388	1.398	1.348	1.385	1.408	1.374	1.337	1.393
3-4	1.364	1.383	1.360	1.397	1.386	1.359	1.343	1.361	1.383	1.381
4-5	1.392	1.443	1.438	1.431	1.443	1.466	1.456	1.479	1.510	1.504
5-6	1.405	1.448	1.451	1.450	1.445	1.418	1.426	1.398	1.386	1.397
1-6	-	1.423	1.417	1.450	1.409	1.349	1.341	1.388	1.374	1.376
5-7	1.405	1.396	1.397	1.393	1.391	1.402	1.401	1.411	1.394	1.390
7-8	1.319	1.346	1.364	1.349	1.395	1.339	1.364	1.335	1.358	1.360
8-9	1.369	1.414	1.387	1.412	1.357	1.416	1.382	1.428	1.409	1.409
9-4	-	1.395	1.413	1.393	1.408	1.398	1.412	1.384	1.363	1.365
2-11	1.333	1.413	1.370	1.373	1.421	1.397	1.364	1.371	1.442	1.444
6-10	1.239	1.239	1.235	1.223	1.245	1.366	1.359	1.352	1.369	1.363
Angle										
1-6-5	-	113.8	113.4	113.0	113.4	122.1	121.2	118.4	119.9	121.8
2-1-6	-	122.9	123.1	125.4	122.1	117.8	118.1	120.9	119.6	119.6
3-4-9	-	129.0	128.9	130.2	128.8	129.9	129.7	130.4	131.4	131.3
5-7-8	104.2	104.9	108.0	105.2	105.8	105.0	107.8	104.7	103.2	103.8
6-5-7	-	130.6	131.7	129.8	131.5	135.0	135.8	132.7	134.1	135.9
7-8-9	113.0	113.1	110.2	112.9	113.6	113.8	111.0	114.0	117.2	117.1
1-2-11	127.7	117.2	118.6	120.7	116.2	116.4	117.0	118.0	113.3	120.9
1-6-10	-	117.6	120.4	116.3	119.7	120.0	122.3	122.3	114.1	119.1
5-6-10	115.3	130.8	126.5	130.8	126.9	117.9	116.5	119.8	125.9	119.1

Table 11b. AM1 optimized structures for different neutral and protonated guanine tautomers. The X-ray results are taken from ref. 3.

Distance	X-Ray	G6	G6P7	G6P10	G7	G7P7	G8	G8P7	G8P1	G9	G9P7
1-2	1.371	1.324	1.338	1.386	1.425	1.432	1.433	1.435	1.456	1.471	1.464
2-3	1.315	1.438	1.450	1.419	1.439	1.451	1.451	1.470	1.441	1.429	1.444
3-4	1.364	1.390	1.373	1.381	1.385	1.359	1.367	1.354	1.363	1.324	1.306
4-5	1.392	1.435	1.427	1.427	1.432	1.429	1.442	1.439	1.459	1.489	1.484
5-6	1.405	1.470	1.491	1.431	1.456	1.467	1.446	1.454	1.407	1.383	1.398
1-6	-	1.421	1.398	1.355	1.409	1.394	1.319	1.311	1.362	1.369	1.357
5-7	1.405	1.393	1.393	1.400	1.395	1.395	1.397	1.394	1.409	1.416	1.412
7-8	1.319	1.346	1.368	1.342	1.345	1.364	1.341	1.364	1.333	1.330	1.360
8-9	1.369	1.414	1.387	1.422	1.417	1.390	1.414	1.390	1.433	1.421	1.374
9-4	-	1.398	1.411	1.390	1.397	1.413	1.397	1.414	1.386	1.407	1.435
2-11	1.333	1.424	1.383	1.366	1.309	1.297	1.306	1.295	1.292	1.304	1.294
6-10	1.239	1.236	1.233	1.347	1.239	1.237	1.370	1.363	1.348	1.369	1.362
Angle											
1-6-5	-	115.7	114.8	121.3	114.4	113.7	122.8	121.8	119.6	119.9	119.5
2-1-6	-	122.2	123.0	119.3	125.9	126.0	120.2	121.1	123.2	121.9	122.1
3-4-9	-	130.1	131.0	132.0	129.8	130.6	131.2	131.3	131.7	129.7	129.4
5-7-8	104.2	105.6	108.3	105.1	105.2	108.2	105.3	108.1	105.1	104.9	108.0
6-5-7	-	131.5	132.1	132.8	130.3	131.1	133.7	134.8	132.0	133.5	135.9
7-8-9	113.0	112.9	109.9	113.2	112.9	110.1	113.0	110.3	113.4	114.5	111.8
1-2-11	127.7	120.1	119.9	117.8	125.9	127.1	124.5	125.9	125.5	121.4	115.4
1-6-10	-	120.1	124.5	115.2	118.5	122.6	116.3	124.0	114.9	114.9	114.0
5-6-10	115.3	124.2	120.8	123.6	127.0	123.7	120.9	114.3	125.5	125.3	126.4

If we consider the higher proton affinities found, considering some other tautomers other than the most stable one, we obtain the order

adenine (246.7 Kcal/mol) > cytosine (239.8 Kcal/mol) > guanine (233.3 Kcal/mol) >> thymine (222.3 Kcal/mol) > uracil (216.5)

This trend is not reliable because many of the neutral tautomers that give these values of PA, lie at high energy above the most stable form and cannot be populated in the gas phase. A possible alternative and more reliable order can be obtained considering the PA values of neutral tautomers that slightly differ from the absolute minima (i.e. in a range of 3 Kcal/mol). Inspection of table 1 shows that only the C2 and G2 tautomers of cytosine and guanine satisfy this condition. Considering the value of PA obtained by the protonation of C2 and G2 tautomers the previous AM1 trend does not change, but the agreement with the experimental absolute values increases.

4. Concluding remarks

The protonation of nucleic acid nucleo bases (cytosine, thymine, uracil, adenine and guanine) have been studied employing the AM1 semiempirical hamiltonian and fully optimizing all the geometries.
From the results the following conclusions can be drawn.
i. The protonation process occurs in the molecular plane of nucleic acid bases.
ii. The preferred protonation sites for the most stable neutral tautomers are O8 on the C4 side of the C4=O group in thymine, N3 in cytosine, O8 in uracil, N3 in adenine and N7 in guanine.
iii. The protonation process induces significant structural changes in the bases. In particular the bond lengths and valence angles nearest to the protonation sites are more perturbed. Hence, reliable proton affinities can be obtained only if full geometry optimization procedures are used in the computations.
iv. Significant charge transfer to the proton is induced by the protonation.

Acknowledgements

This work was supported by Ministero dell'Universita' e della Ricerca Scientifica e Tecnologica (MURST).

References

1. T.L.V. Ulbricht, " Purines, <u>Pyrimidines and nucleotides and the Chemistry of Nucleic Acids</u>, Pergamon Press, New York, 1964.
2. T.J. Kistenmacher, M. Rossi, C.C. Chiang, J.P. Caradonna and L.G. Marzilli, *Adv. Mol. Relax. Interect. Proc.* **17,** 113-131,(1980).

3. R. Taylor and O. Kennard, "The Molecular Structures of Nucleosides and Nucleotides", *J. Mol. Struct.* **78**, 1-18, (1982).

4. S.G. Lias, J.S. Liebmann and R.D. Levine, *J. Phys. Chem. Ref. Data* **13**, 695 (1984).

5. A. Liguori, F. Greco, G. Sindona and N. Uccella, " Gas Phase Proton Affinity of Deoxiribonucleosides and related Nucleo-bases vy Fast Atom Bombardment Tandem Mass Spectrometry", *J. Am. Chem. Soc.* **112**, 9092-9096, (1990).

6. M. Meot-Ner, "Ion Thermochemistry of Low-Volatily Compounds in the Gas Phase. 2. Intrinsic Basicities and Hydrogen- Bonded Dimers of Nitrogen Heterocycles and Nucleic Bases" *J. Am. Chem. Soc.* **101**, 2396- ,(1979).

7. M.S. Wilson and J.A. McCloskey, "Chemical Ionization Mass Spectrometry of Nucleosides Mechanisms of Ion Formation and Estimations of Proton Affinity" *J. Am. Chem. Soc.* **97**, 3436- , (1975).

8. L.F. Cavalieri and B.H. Rosenberg, *J. Am. Chem. Soc.* **79**, 5352- , (1957).

9. J.J. Christensen, J.H. Rytting and R.M. Izatt, *Biochemistry* **9**, 4907- , (1970) .

10. A. Pullman and A.M. Harmbruster, "Non- Empirical SCF MO Studies on the Protonation of Biopolymer Constituents" *Theor. Chim. Acta* **45**, 249- , (1977).

11. T.G. Mezey, J.J. Ladik and M. Barry, " Non empirical SCF MO studies on the protonation of Biopolymer Constituent", *Theor. Chim. Acta* **54**, 251-258, (1980).

12. J. E. Del Bene, " Molecular Orbital Study of the Protonation of DNA bases", *J. Phys. Chem.* **87**, 367-371, (1983).

13. N. Russo and M. Toscano, "Some applications of the Quantum Mechanical Semiempirical Methods to the Gas Phase Chemistry of Bio-organic Ions", in " Mass Spectrometry in the Biological Sciences" ed. by M.L. Gross, Kluwer, Dordrecht, (1992).

14. H. Umeyana and K. Morokuma, "Origin of Alkyl Substituent Effect in the Proton Affinity of Amines, Alchols and Ethers" *J. Am. Chem. Soc.* **98**, 4400- , (1976).

15. M.J.S. Dewar, E.G. Zaebisch, E.F. Healy and J.P. Steward, "Ground- State Analogues of Transition States for Attack at Sulfonyl, Sulpfynil, and Sulfonyl Sulfur: A Sulfuranide Dioxide (10-S-5) Salt, a sulfuranide Oxide (10-S-4) Salt, and a Sulfuranide (10-S-3) Salt" *J. Am. Chem. Soc.* **107**, 3902- , (1985).

16. J.P. Steward, QCPE program 455, Bloomington, Indiana, (1987).

17. D.R. Stull and H. Prophet (eds.), JANAF Thermochemical Tables, NSRDS-NBS 37, US Govt. Print. Off., Washington D.C., (1971).

18. R. Voets, J.P. Francais, J.M.L. Martin, J. Mullens, J. Yperman and L.C. Von Pauche, "Theoretical study of Proton Affinities of 2-, 3-, and 4-monosubstituted Phenolate Ions in Gas-phase by Means of MINDO/3, MNDO and AM1", *J. Comput. Chem.* **11**, 269-290, (1990).

19. W.J. Welsh, "AM1 Molecular Orbital Studies of the Structures, Conformations, Protonation Energies, and Electronic Properties of Triazine dihydrofolate Reductase Inhibitors", *J. Comput. Chem.* **11**, 644-653, (1990).

20. J.S. Kwiatkowski, T.J. Zielinski and R. Rein, *Adv. Quant. Chem.* **18**, 85- , (1986) and references therein.

21. M. Scanlan and I. H. Hillier, "An ab-initio Study of Tautomerism of Uracil, Thymine, 5-Fluorouracil and Cytosine", *J. Am. Chem. Soc.* **106**, 3737-3745, (1984).

22. L. Adamowicz and R.J. Bartlett, "Relative Stability of Cytosine Tautomers with the Coupled Cluster Method and First- Order Correlation Orbitals", *J. Phys. Chem.* **93**, 4001- (1989).
23. A. Buda and A. Sygula, "MNDO Study of the Tautomers of Nucleic bases", *J. Mol. Struct.(Theochem)* **92**, 255-265, (1983).
24. U. Norinder, "A Theoretical Reinvestigation of the Nucleic Bases Adenine, Guanine, Cytosine, Thymine and Uracil Using AM1", *J. Mol. Struct. (Theochem)* **151**, 259-269, (1987).
25. J. P. Steward *J. Comput. Chem.* **10**, 209- , (1989).
26. J. Elguero, C. Martzin, A.R. Katrintxky and P. Linda, <u>The Tautomerism of Heterocycles</u>, Adv. Heterocy. Chem. Supple. 1, Academic Press, London (1976) and references therein.
27. R. Lavery, A. Pullman and B. Pullman, "On the Relative Acidity and Basicity of the Amino Groups of the Nucleic Acid Bases", *Theor. Chim. Acta* **50**, 67-73, (1978).

Phosphorylation of an Overexpressed Yeast Ras2 Protein During the G1 Phase of the Cell Cycle

M.G. MIRISOLA, G. SEIDITA, C. KAVOUNIS* and O. FASANO
Dipartimento di Biologia Cellulare e dello Sviluppo, Università di Palermo, via Archirafi 22, I-90123 Palermo, Italy, and (*) Biological Structures Programme, European Molecular Biology Laboratory, Meyerhofstrasse 1, D-6900 Heidelberg, Germany.

1. Introduction

RAS proteins regulate growth and differentiation in evolutionarily distant systems such as vertebrates and yeast (for reviews, see Tamanoi, 1988; Gibbs and Marshall, 1989; Broach and Deschenes, 1990). At the moleular level, a key function of the yeast RAS1 and RAS2 proteins (collectively referred to as RAS) is to positively regulate the production of cyclic AMP at the onset of the G1 phase of the cell cycle (Toda et al., 1985; De Vendittis et al., 1986). At this stage, RAS proteins are transiently activated by the noncovalent binding of a GTP molecule. Reversal of the effect occurs by the hydrolytic splitting of the γ-phosphate of GTP, that leaves a functionally inactive RAS-GDP complex, thus terminating cyclic AMP synthesis. While the mechanism and functional role of the binding of either GTP or GDP to RAS has been investigated in detail, less is known about the physiological role of covalent modifications of RAS involving phosphorylation of serine residues. A major obstacle to the analysis of the functional relevance of this covalent modification is represented by the very low concentration of RAS proteins into the cell. To bypass this difficulty, we have taken advantage of a mutated form of the RAS2 gene encoding a protein that can be overexpressed at ligh levels. This gene, which is called *ras2-ts1*, was previously isolated as an attenuated form of the wild-type RAS2 gene (Fasano et al., 1988). The corresponding ras2-ts1 protein, that is functional at 30 °C while becoming nonfunctional at 37 °C, was easily labelled metabolically with radioactive 32P under the form of orthophosphate, and was immunoprecipitated with anti-RAS specific antibodies. We have used this protein to explore RAS2 phosphorylation in cells at different stages of the cell cycle.

2. Materials and Methods

2.1. STRAINS

The yeast strains JR26-19D (*a ade2-1 can1-100 his3 leu2-3,112 lys1-1 ura3-52 ras1::URA3 RAS2*) and TS1-6 (*a ade2-1 can1-100 his3 leu2-3, 112 lys1-1 ura3-52*

353

N. Russo et al. (eds.), Properties and Chemistry of Biomolecular Systems, 353–362.
© 1994 *Kluwer Academic Publishers. Printed in the Netherlands.*

ras1::URA3 ras2-ts1) have been described previously (Fasano et al., 1988).

The strain ABE2A (relevant genotype *ras1-Δ ras2-Δ cyr1::HIS3 bcy1-11 leu2-3,112*), with the *CYR1* gene disrupted by the marker *HIS3* (Feger et al., 1991) and with deletions within the coding region of the chromosomal *RAS1* and *RAS2* genes was used as a host for the expression of plasmid-encoded wild-type *RAS2* proteins. The construnction of this strain will be described further on. Genetic manipulation of yeast cells was carried out as described by Mortimer and Schild (1981).

2.2. MEDIA AND LABELLING CONDITIONS

Standard media with carbon sources, other than glucose, were obtained by using 2% bacto-peptone and 1% yeast extract, supplemented with 2% galactose, or 3% glycerol. Complete synthetic medium contained 0.67% yeast nitrogen base without amino acids (Difco manual), and amino acids and nucleic acid bases as indicated by Sherman et al., (1986). Selective synthetic media contained all the components that were present in the cmplete synthetic medium, with the exception of the amino acid used for the selection. For *in vivo* labelling with radioactive orthophosphate, we started with a selective synthetic medium containing a 200-fold lower concentration of unlabelled potassium phosphate as compared to the standard synthetic medium. Other ingredients were as recommended by the Difco manual. A typical experiment was performed by adding 500 μCi of ^{32}P-labelled phosphate to a 10 ml culture when the A595 reached the value of 0.4. After further incubation for 30 min at 25 °C, the cells were harvested by centrifugation and frozen at -80 °C.

2.3. IDENTIFICATION OF YEAST GENES THAT COULD SUPPRESS A TEMPERATURE-SENSITIVE RAS BY OVEREXPRESSION

We constructed a S. *cerevisiae* genomic library using the multicopy plasmid vector YEp13 (Broach et al., 1979). The library was generated by cloning yeast chromosomal DNA fragments obtained by *Sau*3A partial digestion of DNA from the strain TS1-6 (Fasano et al., 1988), into the unique *Bam*HI site of YEp13. The vector YEP13 also carries the selectable marker *LEU2*, and allows the selection of Leu$^+$ transformants on synthetic medium without leucine at 30 °C. The library was used to transform competent TS1-6 cells. Leu$^+$ transformants were subsequently tested for growth at 37 °C on nonfermentable carbon sources by replica plating. Most of the transformants, like the original TS1-6 strain, could not grow at 37 °C. Only twenty colonies showed a variable degree of growth at this temperature. In most of them, loss of plasmid sequences upon growth on rich medium was associated with reversion to temperature-sensitive growth. Moreover, the plasmids isolated from the transformants were able to transfer with high frequency the temperature-tolerant phenotype to competent TS6-1 yeast recipient cells. Most of the plasmid clones that we isolated contained inserts encoding carboxy-terminal portions of the adenylyl cyclase gene, as determined by sequence analysis. However, using the same tecnique we found that the insert of the plasmid pR2H harbored the chromosomal *ras2-ts1* gene. This plasmid was used in the course of this work for the overexpression of the ras2-ts1 protein.

2.4. PLASMID VECTORS FOR THE EXPRESSION OF WILD-TYPE AND MUTATED RAS2 PROTEINS IN YEAST

The plasmid pR2H, that we used for the overexpression of the ras2-ts1 protein, was constructed as described in the preceding paragraph. For the overexpression of wild-type and mutated forms of the RAS2 protein in yeast we also used the high copy number vector YEp51, harboring the selectable marker *LEU2*. The vector carried the 1.2 kbp HpaI-HindIII fragment of the *RAS2* gene including the complete coding region (Powers et al., 1984) cloned as described by Verrotti et al., (1992). The vector directing the expression of a *RAS2* protein in which Gly19 was replaced by Val was denominated YEp51-RAS2V19.

2.5. PREPARATION OF CELL EXTRACTS

Unless indicated otherwise, cells growing logarithmically were harvested when the A595 reached the value of 0.4-0.5. Cell pellets from 10ml cultures were resuspended into 300 µl of Tris. HCl 50 mM pH 7.4, KCl mM, $MgCl_2$ 3 mM, dithiothreitol 1 mM, GDP 20 µM, 2% Triton X-100, SDS 0.1% (buffer E) at 0 °C. Following the addition of an equal volume of acid-washed glass beads (250-300 µm), the mixture was shaken on a mixer for 5 min in a cold room. The liquid phase was centrifuged for 60 min at 13000 rpm to remove insoluble material, and the supernatant was subjected to immunoprecipitation by standard techniques. Usually, 2 µl of immune serum were used for each extract.

2.6. OTHER TECHNIQUES

Total yeast DNA was prepared as described by Nasmyth and Reed (1980). Plasmid DNAs from *E. coli* cells were prepared by centrifugation in ethidium bromide-cesium chloride gradients. Preparation of yeast competent cells and yeast transformations were as described by Ito et al., (1983). SDS-PAGE, immunoblotting of proteins and immunostaining with RAS-specific antibodies were as described by De Vendittis et al., (1986).

3. Results

3.1. A MUTATED FORM OF THE RAS2 PROTEIN CAN BE OVEREXPRESSED *IN VIVO*

In the yeast *Saccharomyces cerevisiae*, the RAS2 protein is expressed at very low levels. We attempted to increase its level of expression by transforming competent yeast cells with high copy number plasmids carrying the cloned *RAS2* gene. The latter was derived from the yeast vector YEp13 (Broach et al., 1979), harboring the selectable marker *LEU2*. The presence of the marker allowed the selection of Leu[+] transformants on synthetic medium without leucine at 30 °C. However, probably because of a powerful feedback mechanism that controls RAS activity (Nikawa et al., 1987), we were unable to obtain a significant increase of membrane-bound RAS2 protein (results not shown).

We subsequently attempted the overexpression of a mutated form of the *RAS2* gene encoding an attenuated form of the RAS2 protein (*ras2-ts1*, Fasano et al., 1988).

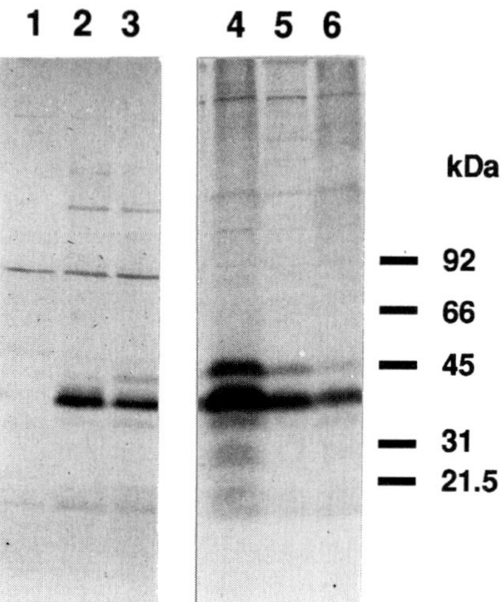

Figure 1. Expression of the wild-type RAS2 and of the mutated ras2-ts1 proteins in yeast membranes. Membrane extracts (50 µg) from the following strains were fractionated by SDS-PAGE on 12.5% gels and immunoblotted with anti-RAS specific polyclonal serum. Lane 1: ABE2A (*ras1-Δ ras2-Δ*); lane 2: JR26-19D (*ras1-Δ RAS2*); lanes 3 and 6: TS1-6 (*ras1-Δ ras2-ts1*); lanes 4 and 5: TS1-6 carrying the plasmid pR2H. Only 5 µg of membranes were loaded in lane 5.

Previous work of these authors had shown that the ras2-ts1 protein carried a double amino acid substitution at 82 and 84 positions. As a consequence, cells expressing the ras2-ts1 protein and lacking other RAS proteins (strain TS1-6, Fasano et al., 1988) could not grow at 37 °C on media containing nonfermentable carbon sources. However, the efficient growth of the same strain at 30 °C on glucose-based media showed that at this temperature the function of the protein was partially retained. Therefore, we tested the possibility that the functional attenuation of the ras2-ts1 protein could facilitate its overexpression. As a vector, we used a YEP13-derived plasmid carrying the *ras2-ts1* gene as a Sau3A insert into the BamHI site of the plasmid (pR2H, see Methods). Subsequently, we introduced pR2H into TS1-6 competent cells, and we determined the levels of membrane-bound ras2-ts1 protein in membrane extracts. The results, shown in Figure 1, indicated that a protein band with an apparent molecular weight of 42 kDa was absent in a strain in which the RAS genes had been disrupted (lane 1). A 42 kDa band, corresponding to either the RAS2 protein or to the ras2-ts1 protein was detectable in lanes 2 and 3, respectively. The level of the ras2-ts1 protein was strongly increased in cells transformed with pR2H, as compared to cells harboring only one chromosomal copy of the *ras2-ts1* gene (compare lane 4-5 with 3 and 6). Interestingly, an additional immunoreactive band of 45 kDa was observed on expression of the ras2-ts1 protein. Since no normal RAS proteins are expressed in TS1-6, both the 42 and the 45 kDa bands represent the product of the ras2-ts1 gene.

3.2. THE OVEREXPRESSION OF THE RAS2-TS1 PROTEIN FACILITATES LABELLING WITH RADIOACTIVE ORTHOPHOSPHATE

Metabolic labelling of the yeast RAS2 protein with ^{32}P has been reported by Sreenath et al., (1988) and Cobitz et al., (1989). However, the amount of radioactivity incorporated into this weakly expressed protein was relatively low. Therefore, we tested whether the increased expression that we observed with the ras2-ts1 protein (see preceding paragraph) could facilitate labelling.
We used TS1-6 cells trasformed with pR2H, as well as cells transformed with a plasmid encoding the RAS2V19 protein, for *in vivo* labelling with radioactive orthophosphate (see Methods). Cells were subsequently collected, washed, and broken with glass beads. The extracts were immunoprecipitated with RAS-specific polyclonal sera, fractionated by SDS-PAGE on 12.5% gels, and transferred onto a nitrocellulose filter. After immunostaining the filter with RAS-specific antibodies (Figure 2, left part)

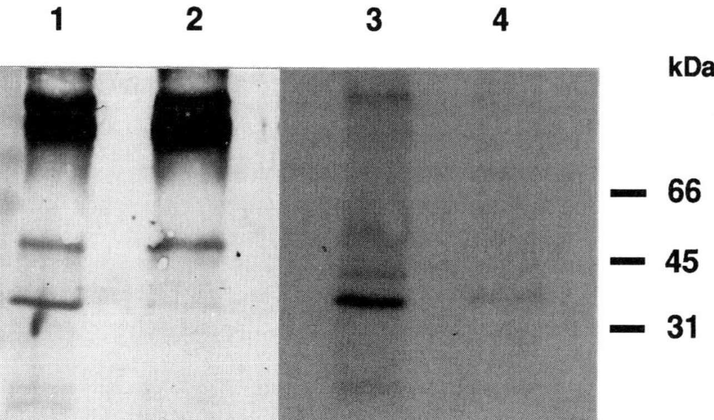

Figure 2. Levels of expression and phosphorylation of the ras2-ts1 and RAS2V19 protein in yeast extracts. Total cell extracts from a 10 ml culture were immunoprecipitated and fractionated by SDS-PAGE (see Methods). After blotting onto a nitrocellulose filter, this was either immunoblotted with anti-RAS specific polyclonal serum (lanes 1-2) or subjected to autoradiography (lanes 3-4). Extracts from the following strains were used: TS1-6 carrying the plasmid pR2H (lanes 1 and 3); ABE2A carrying the plasmid YEp51-RAS2V19 (lanes 2 and 4).

we observed, as predicted, a much higher level of expression of the ras2-ts1 protein compared to the RAS2V19 protein (compare line 1 with line 2). After of two days exposure of the filter to an X-ray sensitive film, the intensity of the radioactive bands corresponding to the ras2-ts1 and RAS2V19 proteins (Figure 2, lanes 3 and 4, respectively) indicated that the former protein contained a much higher amount of radioactivity. However, the relative increase of radioactive labelling of the ras2-ts1 protein *versus* the wild-type paralleled the increased expression of ras2-ts1 (compare lines 1-2 with 3-4, respectively). This suggested that increased incorporation of radioactive phosphate was the result of increased expression, rather than the consequence of qualitative changes of the ras2-ts1 protein induced by the mutations.

3.3. ^{32}P-LABELLED RAS2-TS1 PROTEIN CAN BE IMMUNOPRECIPITATED FROM CELLS IN THE G1 PHASE OF THE CELL CYCLE

Yeast cells at various stages of the cell cycle can be obtained by elutriation of logarithmically growing cells. To investigate the relationship between RAS2 phosphorylation and cell cycle, we elutriated a colture of TS1-6 cells trasformed with pR2H (see previous paragraphs). Eight distinct fractions were obtained. The analysis of the DNA content revealed that the first fractions contained almost exclusively cells in the G1 phase of the cell cycle (Figure 3 and Table I).

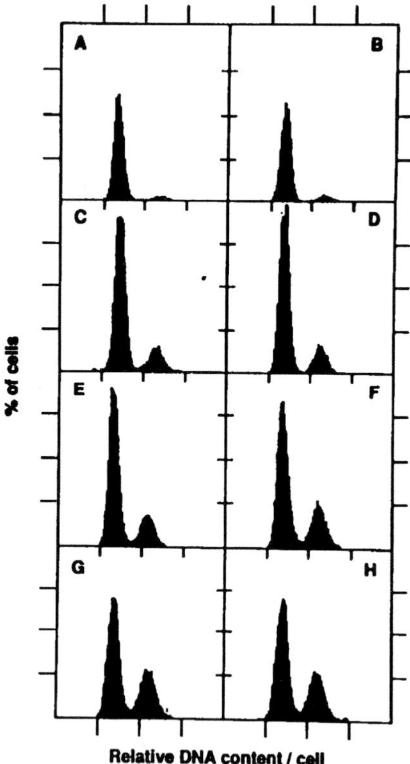

Figure 3. Analysis of the DNA content of elutriated yeast cells (TS1-6 transformed with pR2H) using a fluorescence-activated cell sorter. Fractions were sequencially labelled A-G.

Table I. Cell cycle analysis of sequential fractions of elutriated cells.

Fractions	% of cells in G1	% of cells in G2
A	94	6
B	89	11
C	84	16
D	83	17
E	80	20
F	72	28
G	66	34
H	63	37

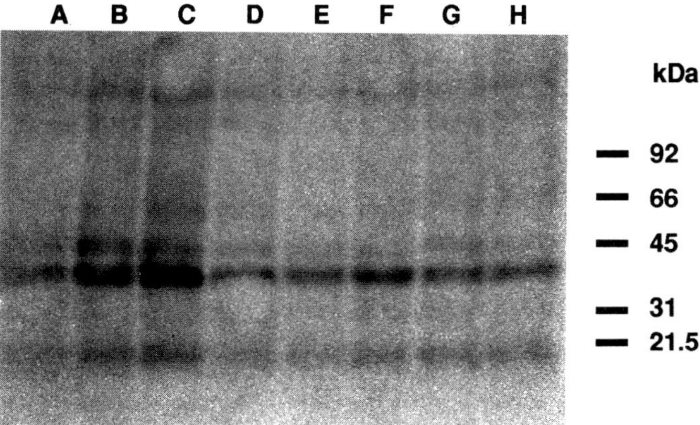

Figure 4. Phosphorylation of the ras2-ts1 protein in sequential fractions of elutriated cells. For each fraction, the relative proportion of cells in different stages of the cell cycle is reported in Table I.

Immediately after the collection of the cells, radioactive orthophosphate was added to each fraction and labelling was performed during a 10 minute incubation at 30 °C. Since this period of time is short compared to the time required for progression through the cycle, we assume that no significant desynchronization of the cells occurred during this period of time. After labelling, the cells were rapidly collected, frozen in liquid nitrogen, and broken with glass beads. The relative amount of ^{32}P-labelled ras2-ts1 protein in the different fractions was determined by immunoprecipitation with specific antibodies (see Methods). The results, shown in Figure 4, indicated that comparable amounts of the ras2-ts1 protein were present in the different fractions.

4. Discussion

We have found that yeast ras2-ts1 protein can be phosphorylated *in vivo*. This confirms earlier results obtained with the wild-type RAS2 protein (Sreenath et al., 1988; Cobitz et al., 1989). However, the high level of expression of the mutated protein used in this work greatly facilitated, not only the detection of RAS phosphorylation, but also further analysis.

We have taken advantage of the enhanced sensitivity of the phosphorylation assay using the mutated protein to explore the relationship between phosphorylation and progression through the cell cycle. We have observed a detectable phosphorylation of the ras2-ts1 protein in elutriated fractions containing a small number of cells (Figure 4). The first fractions contained almost exclusively cells in the G1 phase of the cell cycle, as determined by the DNA content/cell (Figure 3 and Table I). Since the phosphorylation of the ras2-ts1 protein was comparable in the different fractions, we can conclude that cells in G1 express phosphorylated ras2-ts1 protein.

The heterogeneity of the other fractions did not allow us to reach a similar conclusion for cells in other stages of the cell cycle. We do not know the fraction of RAS molecules that were phosphorylated in G1, and therefore we cannot make any inference about the functional role of this covalent modification. In the future, the engineering of yeast strains expressing mutated RAS proteins that cannot be phosphorylated will help solving this point . A prerequisite for the engineering of such strains is the determination of all the RAS2 protein phosphorylation sites. In this respect, the great efficiency of labelling of the ras2-ts1 protein with radioactive phosphate should greatly facilitate the identification of such phosphorylation sites using biochemical procedures.

During this work, we could confirm that the expression of the ras2-ts1 protein *in vivo* results in the production of two polypeptides with apparent molecular weights of 45 and 42 kDa, as shown by Fasano et al., (1988). Only the 42 kDa species could be detected upon expression of the wild-type RAS2 protein. However, we suspect that the low level of expression of the latter would, in any case, prevent the observation of a less abundant 45kDa protein. Elucidating the molecular mechanism, whereby the amino acid substitutions present into ras2-ts1 protein lead to the production of an additional 45 kDa protein, is important before asking if a similar pattern can be observed with wild-type RAS2. A novel finding of this work was that the 45 kDa ras2-ts1 protein, like the canonical 42 kDa form, is phosphorylated. It was previously established that the 3 kDa apparent mobility shift can only be explained by additional posttranslational modifications (Fasano et al., 1988). Therefore, it is possible that a novel site of phosphorylation/modification is generated by the two amino acid substitutions present in the ras2-ts1 protein (G82S and G84R, Fasano et al., 1988). Alternatively, a minor physiological phosphorylation/modification site could be unmasked by the mutations. The phosphopeptide analysis of the labelled ras2-ts1 protein will in help answering these questions.

Acknowledgements

We wish to thank Dr. B. Ducommun for discussions and for help in elutriation of yeast cells, and Mr. Graham Smith for DNA content analysis. Work done in Italy was supported by a grant from the Aassociazione Italiana per la Ricerca sul Cancro.

References

1. J.R. Broach, J.N. Strathern, and J.B. Hicks "Transformation in yeast: development of a hybrid cloning vector and isolation of the *CAN1* gene", *Gene* **8**, 121-133, (1979).
2. J.R. Broach and R.J. Deschenes "The function of RAS genes in *Saccharomyces cerevisiae*", *Adv. Cancer Res.* **54**, 79-139, (1990).
3. A.R. Cobitz, E.H. Yim, W.R. Brown, C.M. Perou, and F. Tamanoi Phosphorylation of RAS1 and RAS2 proteins in *Saccharomyces cerevisiae*", *Proc. Natl. Acad. Sci.* **86**, USA, 858-862, (1989).
4. E. De Vendittis, A. Vitelli, R. Zahn and O. Fasano "Suppression of defective *RAS1* and *RAS2* functions in yeast by an adenylate cyclase activated by a single amino acid change", *EMBO J.* **5**, 3657-3663, (1986).
5. O. Fasano, J.B. Crechet, E. De Vendittis, R. Zahn, G. Feger, A. Vitelli and A. Parmeggiani "Yeast mutants temperature-sensitive for growth after random mutagenesis of the chromosomal *RAS2* gene and deletion of the *RAS1* gene" *EMBO J.* **7**, 3375-3383, (1988).
6. G. Feger, E. De Vendittis, A. Vitelli, P. Masturzo, R. Zahn, A.C. Verrotti, C. Kavounis, G.P. Pal, and O. Fasano "Identification of regulatory residues of the yeast adenylyl cyclase" *EMBO J.* **10**, 349-359, (1991).
7. J.B. Gibbs and M.S. Marshall "The *ras* oncogene - an important regulatory element in lower eucaryotic organism" *Microbiol. Rev.* **53**, 171-185, (1989).
8. H. Ito, Y. Fukuda, K. Murata, and A. Kimura "Transformation of intact yeast cells treated wih alkali cations" *J. Bacteriol.* **153**, 163-168, (1983).
9. R. Mortimer, and D. Schild "<u>Genetic mapping in *Saccharomyces cerevisiae*. In The molecular biology of the yeast *Saccharomyces*. Life cycle and inheritance</u> (J.N. Strathern, E.W. Jones, and J.R. Broach, eds.) pp. 11-26. Cold Spring Harbor Laboratory, Cold Spring Harbor, N.Y, (1981).
10. K.A. Nasmyth, and I. Reed "Isolation of genes by complementation in yeast: molecular cloning of a cell cycle gene" *Proc. Natl. Acad. Sci.* **77**, USA, 2119-2123, (1980).
11. J. Nikawa, S. Cameron , T. Toda, K.M. Ferguson and M. Wigler "Rigorous feedback control of cAMP levels in *Saccharomyces cerevisiae*" *Genes Dev.* **1**, 931-937, (1987).
12. F. Sherman, G.R. Fink, and J.B. Hicks, "<u>In Laboratory course manual for methods in yeast genetics</u>" (ed. F. Sherman, G.R. Fink, and J.B. Hicks) Cold Spring Harbor Laboratory Press, Cold Spring Harbor, New York, (1986).
13. T.L.V. Sreenath, D. Breviario, N. Ahmed, and R. Dhar "Two different protein kinase activities phosphorylate RAS2 protein in *Saccharomyces cerevisiae*" *Biochem. Biophys. Res. Comm.* **157**, 1182-1189, (1988).
14. F. Tamanoi "Yeast RAS genes" *Biochim Biophys. Acta* **948**, 1-15, (1988).
15. T. Toda, I. Uno, T. Ishikawa, S. Powers, T. Kataoka, D. Broek, S. Cameron, J. Broach, K. Matsumoto and M. Wigler "In yeast, RAS proteins are controlling elements of adenylate cyclase" *Cell* **40**, 27-36, (1985).
16. A.C. Verrotti, J.B. Crechet, F. Di Blasi, G. Seidita, M. Mirisola, C. Kavounis, V. Nastopoulos, E. Burderi, E. De Vendittis, A. Parmeggiani, and O. Fasano "Ras Residues that are Distant from the GDP Binding Site Play a Critical Role in Dissociation Factor-Stimulated Release of GDP" *EMBO J.* **11**, 2855-2862, (1992).

Effect of Isomerism at Non-leaving Ligands in Platinum Anticancer Drugs

G. NATILE
Dipartimento Farmaco-Chimico, Università di Bari, via E. Orabona, 4, I-70125 Bari, Italy.
and M. COLUCCIA
Dipartimento di Scienze Biomediche e Oncologia Umana, Università di Bari, Piazza Giulio Cesare, 11, I-70125 Bari, Italy

1. Introduction

It is generally understood that the platinum drug *cis*-$[PtCl_2(NH_3)_2]$ (*cis*-DDP), on its way to the target, loses the two chloride ions which are replaced by other nucleophiles, and keeps the amminic ligands [1]. While the anionic ligands are likely to play an important role in determining the transport of the complex throughout the living organism, the amminic ligands are likely to be more responsible for the drug-receptor interaction. Therefore it is of great interest to see how a different conformation or configuration of the non-leaving ligands can influence the DNA binding properties and the biological activity of platinum complexes.

The search for platinum complexes, with isomeric non-leaving ligands and having different biological properties, was the first aim of this investigation the second, was the structural characterization of their adducts with DNA.

2. Platinum complexes with chiral monoamines

It is well known that enantiomeric molecules which are similar to the extent that one is the mirror image of the other, can have a very different biological activity. Platinum complexes with enantiomeric primary amines were examined and no significant difference in their biological activity was found [2]. One compound of this class, that is, the platinum complex with phenethylamine (phetam), is shown in Chart I.

A possible explanation for this result is that the free rotations of the amine about the platinum-nitrogen bond and of the chiral radical about the nitrogen-carbon bond average the steric effect of the ligands and offset any stereospecificity in the interaction with a biological substrate.

3. Platinum complexes with chiral N-substituted ethylenediamines

The degree of rotational freedom, in a complex of the type described above, can be reduced by bridging together the two nitrogens of the cis amines. A ligand which fulfils

N. Russo et al. (eds.), Properties and Chemistry of Biomolecular Systems, 363–368.
© 1994 *Kluwer Academic Publishers. Printed in the Netherlands.*

$[PtCl_2(phetam)_2]$

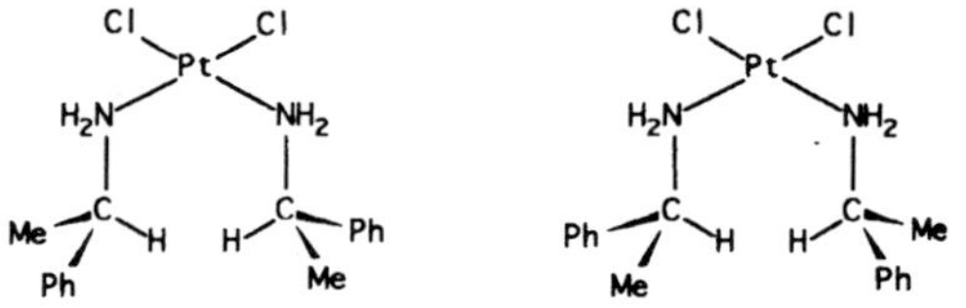

Configuration at carbons   *R*   *S*

Chart I

these requisites is the ethambutol. This molecule was already used in medicine as an anti-TBC and, very interestingly, only the *S,S* isomer was found very active while the *R,R* enantiomer was completely inactive [3,4]. The coordination of this diamine to platinum led to the formation of different isomers even if the complexing ligand was isomerically pure. The reason for this is that, upon coordination to platinum, the nitrogens also become stable chiral centres which can have either *R* or *S* configuration. The complete list of isomers is given in Chart II (a and b, e and i, f and h, and g and l are couples of enantiomers).

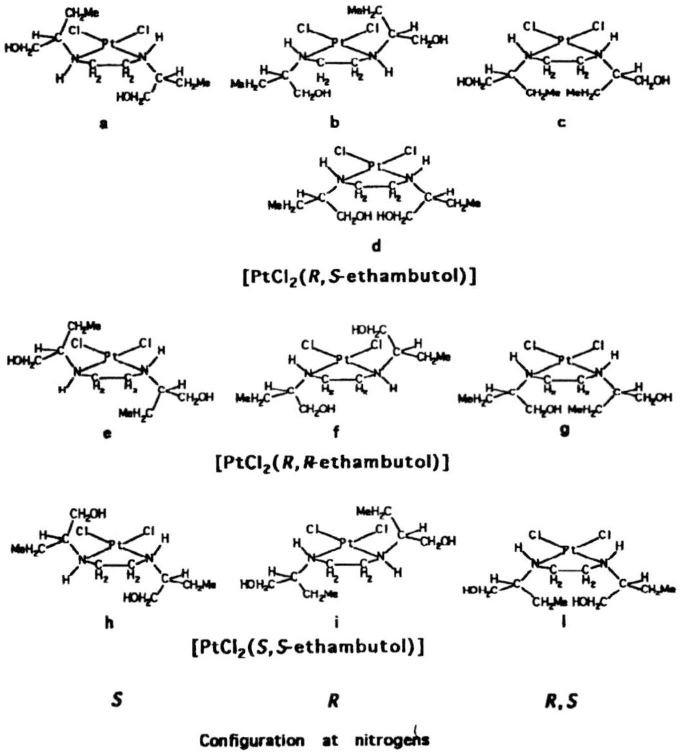

a   b   c

d

$[PtCl_2(R,S\text{-ethambutol})]$

e   f   g

$[PtCl_2(R,R\text{-ethambutol})]$

h   i   l

$[PtCl_2(S,S\text{-ethambutol})]$

S   *R*   *R,S*

Configuration at nitrogens

Chart II

It is interesting to note that the bridging of the two nitrogens with the ethylene chain not only blocks the rotation about the Pt-N bond but also hinders, to some extent, the rotation of the asymmetric 1-butanol-2-yl radical with respect to the C-N bond.[5] The methylene protons of the -CH$_2$Me and -CH$_2$OH groups are adjacent to an asymmetric carbon and are diastereotopic. Therefore in the NMR they can give two signals, the separation of which is as greater as slower is the rotation of the methylene group with respect to the asymmetric centre. In the isomer c the diastereotopic splitting was greater only for the methylene protons of the ethyl groups and very small for the methylene protons of the hydroxymethyl groups.

This indicates that the average orientation of the 1-butanol-2-yl radicals is such that only the ethyl residues are hindered in their rotation. Similarly, in isomer g the diastereotopic splitting was greater for one, ethyl and one hydroxymethyl group indicating that the average orientation of the two 1-butanol-2-yl radicals is such that the ethyl in one and the hydroxymethyl in the other, are hindered in their rotation.

The steric rigidity of these complexes could lead to a different biological activity of the different isomers. Indeed, the isomer l was less mutagenic and less toxic than the g enantiomer, but, in contrast, exhibited a good antitumor activity towards P388 sarcoma and Lewis lung carcinoma [6]. Apparently compound l can couple a small mutagenic activity to a good antitumor activity, and this is a rather noteworthy result.

Although the configuration at the nitrogen atoms was stable at neutral pH for days, at room temperature, it could undergo isomerization if the pH was kept either basic or acidic. This prevented further studies on complexes of this class.

4. Platinum complexes with chiral C-substituted ethylenediamines

The complication of isomerization at the nitrogen atoms could be avoided by using primary diamines (in this case the nitrogen is no longer a chiral centre) and inserting the chiral carbon(s) in the organic chain bridging the two nitrogens. In this way the steric rigidity of the non-leaving ligands is further increased since the chiral groups are no longer free to rotate about the C-N bond.

Kidani and coworkers reported that platinum complexes with 1,2-diamino-cyclohexane had biological activities depending upon the chirality of the diamine ligand. The R,R isomer apparently was endowed with greater antitumor activity, and was less mutagenic than the S,S isomer [7].

A comparative study of three platinum complexes with chiral diamines [PtCl$_2$(N-N)] (N-N = 1,2-diaminopropane, 1,2-DAP; 2,3-diaminobutane, 2,3-DAB; and 1,2-diaminocyclohexane, 1,2-DAC) was carried on by us (Chart III) [8].

The biological tests, *in vitro*, revealed a marked difference among isomers. For instance, the mutagenic activity, which is strictly related to the interaction of the drug with DNA, could be even ten times greater in one isomer with respect to the corresponding enantiomer. In all cases examined the S,S isomer was by far the most mutagenic indicating that the different isomers give adducts with DNA which can be discriminated by the enzymatic systems involved in mutagenesis.

The indications gained from mutagenic data on the relevance of the configuration of non-leaving ligands in platinated DNA were confirmed by experiments of inhibition of restriction enzyme activity. The extent of inhibition of enzymes cutting at G-rich sites

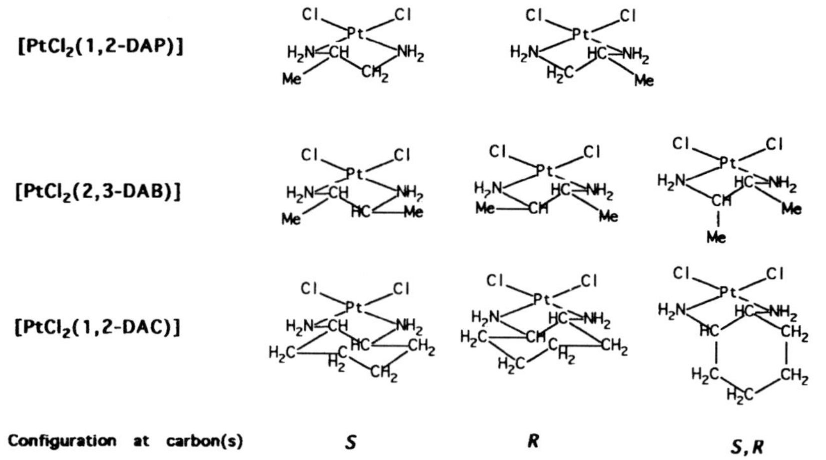

[PtCl₂(1,2-DAP)]

[PtCl₂(2,3-DAB)]

[PtCl₂(1,2-DAC)]

Configuration at carbon(s)   *S*   *R*   *S,R*

Chart III

was significantly different for the different isomers, the *R,R* form being more active than the others.

A different mode of interaction with DNA of the different isomers was also proved by NMR spectroscopy. The ^{31}P NMR of natural DNA shows a single broad band. The spectrum of DNA platinated with *cis*-DDP exhibits a phosphodiester resonance shifted downfield by 1 ppm or more. This downfield-shifted resonance arises from the central phosphate ester in 1,2-G,G-intrastrand cross-link which is deshielded due to the closure of the O-P-O angle [9]. Performing the reaction with platinum complexes of *S,S* and *R,R*-DAB, only the *S,S* enantiomer gave a deshielded band having the same chemical shift and the same intensity of that given by DNA platinated with *cis*-DDP.

Non-leaving ligands of the type described above are also capable to exert a steric control on the coordination to platinum of nucleotides. A ^{1}H NMR investigation of the complex [Pt(*S,R,R,S*-L)(5'GMP)₂] (L = N,N'-dimethyl-2,3-diaminobutane; *S,R,R,* and *S* are the configurations at the four asymmetric centres N, C, C, and N, respectively) has shown that: [10]

a) The chirality of the diamine deeply influences the equilibrium among the three possible atropisomers (head-to-tail, HT1 and HT2, and head-to-head, HH) the order of stability being HT1 >> HH > HT2.

b) The greater stability of HT1 isomer stems from 06···NH hydrogen bonding predominating over PO···NH hydrogen bonding.

c) In the HH atropisomer one of the two nucleotides has the H8 signal *ca.* 1 ppm downfield with respect to the other and this guanine would be in the 3' position if the two nucleotides were allowed to couple together by sugar phosphate condensation. Also, in platinum complexes with GpG the 3'G H8 resonance was shifted at lower field [11]. However, in the present case, the two nucleotides are not tethered and therefore the interligand interactions within the coordination sphere of platinum and not the constraints of the sugar-phosphate backbone, are responsible for the guanine orientation leading to the observed shift differences.

5. Platinum complexes with *E,Z* isomerism at the non-leaving ligands

A different type of isomerism at the non-leaving ligands is shown by platinum complexes with iminoethers. Iminoethers, like amines, are potential N-donor ligands and have a N-bound hydrogen suitable for hydrogen bond formation, moreover, they can have geometrical isomerism (*E* or *Z*) about the C=N double bond. The *E* or *Z* configuration of the iminoether allows the formation of three different isomers for *cis*-$[PtCl_2(iminoether)_2]$. A scheme of different isomers is given in Chart IV [12].

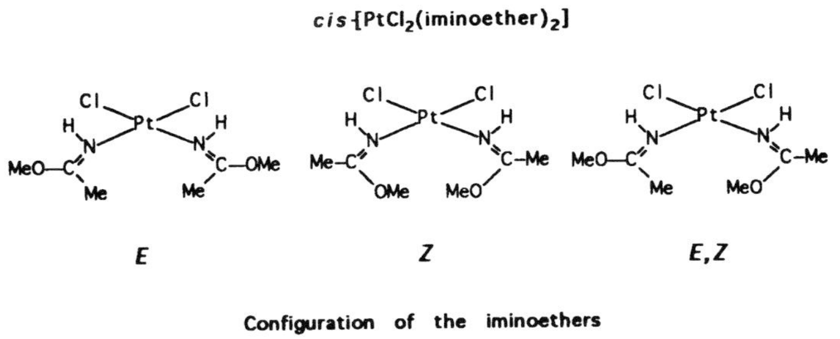

cis-$[PtCl_2(iminoether)_2]$

Chart IV

The *in vitro* inhibitory effect of increasing doses of platinum-iminoether complexes on the growth of P388 cells was investigated. The ID_{50} values of *cis-EE* (7.5) and *cis-ZZ* (18) are greater than that of *cis*-DDP (2) by a factor of 4 and 9, respectively. As far as the *in vivo* effects are concerned, the results obtained with P388 leukemia-bearing mice indicate that *cis-EE* is endowed with significant antitumor activity (% T/C = 145), while *cis-ZZ* is inactive. Primer extension footprinting assays have shown that *cis-EE* has the greatest binding specificity for d(pGG) sites while *cis-ZZ* is able to react with the same efficacy with d(pGG) and d(pAG) sites.

6. Conclusions

The complexes of platinum (or other metals) with isomeric ligands pose an interesting theme for structure/activity investigations. Determining the modes of interaction of different isomers with DNA and which of them inhibits DNA replication and transcription can be useful for a deeper insight into the mechanism of action of *cis*-DDP, and can help in the designing new platinium-anticancer drugs.

References

1. C. A. Lepre, S.J. Lippard, "Interaction of platinum antitumor compounds with DNA", *Nucleic Acids and Molecular Biology* **4**, F. Eckstein, D.M.J. Lilley eds, Springer-Verlag Berlin, 9-38, (1990).

2. M. Coluccia, M. Correale, D. Giordano, M.A. Mariggiò, S. Moscelli, F.P. Fanizzi, G. Natile, L. Maresca, "Mutagenic activity of some platinum complexes with monodentate and bidentate amines" *Inorg. Chim. Acta,* **123**, 225-229, (1986).

3. R.G. Wilkinson, R.G. Shepherd, J.P. Thomas, C. Baughn,"Stereospecificity in a new type of synthetic antituberculous agent" *J. Am. Chem. Soc.* **83**, 2212-2213, (1961).

4. A.M. Kritsyn, A.M. Likhoshertov, T.V. Protopopova, A.P. Skoldinov, "Ethambutol and related compounds. Synthesis and stereochemical relationships" *Dokl. Akad. Nauk. S.S.S.R.* **145**, 332-335, (1962).

5. G. Giannini, G. Natile, "Steric constraints inside the metal-coordination sphere as revealed by diastereotopic splitting of methylene protons" *Inorg. Chem.* **30**, 2853-2855, (1991).

6. M. Coluccia, F.P. Fanizzi, G. Giannini, D. Giordano, F.P. Intini, G. Lacidogna, F. Loseto, M.A. Mariggiò, A. Nassi, G. Natile, "Synthesis, mutagenicity, binding to pBR322 DNA and antitumor activity of platinum(II) complexes with ethambutol" *Anticancer Res.* **11**, 281-288, (1991).

7. Y. Kidani, K. Inagaki, R. Saito, S. Tsukagoshi, "Synthesis and antitumor activities of platinum(II) complexes of 1,2-diaminocyclohexane isomers and their related derivatives" *J. Clin. Hematol. Oncol.* **7**, 197-208, (1977).

8. F. P. Fanizzi, F. P. Intini, L. Maresca, G. Natile, R. Quaranta, M. Coluccia, L. Dibari, D. Giordano, M.A. Mariggiò, "Biological activity of platinum complexes containing chiral centres on the nitrogen or carbon atoms of a chelate diamine ring" *Inorg. Chim. Acta,* **137**, 45-51, (1987).

9. S.E. Sherman, S.J. Lippard, "Structural aspects of platinum anticancer drug interaction with DNA" *Chem. Rev.* **87**, 1153-1181, (1987).

10. Y. Xu, G. Natile, F.P. Intini, L.G. Marzilli, "Stereochemically controlled influence atropisomerization of Pt(II) nucleotide complexes. Evidence for head-to-tail and stable L head-to-tail atropisomers" *J. Am. Chem. Soc.* **112**, 8177-8179, (1990).

11. J.H.J. den Hartog, C. Altona, J. -C. Chottard, J. -P. Girault, J. -Y. Lallemand, F. A.A.M. de Leeuw, A.T.M. Marcelis, J. Reedijk, "Conformational analysis of the adduct *cis*-[Pt(NH$_3$)$_2$d(GpG)] in aqueous solution. A high field (500-300 MHz) nuclear magnetic resonance investigation" *Nucleic Acids Res.* **10**, 4715-4730, (1982).

12. F.P. Fanizzi, F.P. Intini, G. Natile, "Nucleophilic attack of methanol on bis(benzonitrile)dichloroplatinum(II): formation of mono- and bis-imido ester derivatives" *J. Chem. Soc. Dalton Trans.* 947-951, (1989).

A Photochemical Approach to Study the Antimitotic-Drugs Tubulin Interaction.

G. PALUMBO
Dipartimento di Biologia e Patologia Cellulare e Molecolare "L.Califano" Università Federico II Centro di Endocrinologia ed Oncologia Sperimentale del CNR Napoli

1. Introduction

At least three chemically distinct types of drugs, **colchicine, vinblastine, podophyllotoxin,** and their derivatives, are known to inhibit mitosis in dividing cells (1) **Colchicine,** the active principle of *Cochicum autumnale* is found in various plants and has been known since ancient times, as a poison. Its use in medicine is probably very old, since it was known to Greek physicians as early as the fifth century A.D. under the name of "hermodactyl" (finger of Hermes). Its use was introduced to north-western medicine in 1763 by a german physician, to relief the symptoms of "gutta". The colchicine was purified in 1883, while the first experimental work on its pharmacological properties was done by a sicilian scientist in 1889. At the beginning of this century antimitotic properties of colchicine were discovered: since then, this molecule has been the center of thousands of scientific papers dealing with both animal and plant biology and medicine.

Vincristine and Vinblastine. Popular medicine attributed to the extracts of the *Vinca rosea* antidiabetic properties. This has been proved to be wrong, but a severe leukopenia (as mitotic arrest of blood-forming cells) was observed in treated animals. The active principles were two indole alkaloids Vincristine and Vinblastine. Up to date, these alkaloids and some of their derivatives, are actively used as anticancer drugs.

Podophyllotoxin is contained in an extract from the dried leaves of the *Juniperus sabina.* Podophyllotoxin is the active principle of an old liniment which was used as early as 1860 to cure skin tumors.

The only known common mode of action of these molecules is their interaction with the major proteic component of "microtubules" . Microtubules are found in eucaryotic cells, where they participate in a wide variety of functions such as mitosis, cell shaping, axonal growth, secretion, motility, transport, etc. (2). Indeed these structures are polymers of a protein called tubulin, which is itself a dimer composed of two similar but not identical subunits named a and b, respectively. The interaction between colchicine, vinblastine and podophyllotoxin and tubulin has been widely studied and many details, particularly on the binding parameters and stoichiometry, are known. Tubulin

N. Russo et al. (eds.), Properties and Chemistry of Biomolecular Systems, 369–379.
© 1994 *Kluwer Academic Publishers. Printed in the Netherlands.*

vinblastine and podophyllotoxin and tubulin has been widely studied and many details, particularly on the binding parameters and stoichiometry, are known. Tubulin has not less than three "binding sites": one for colchicine and podophyllotoxin, and at least one (and probably two) for vinblastine [3].

A more detailed understanding of tubulin-drug interaction at a molecular level will require knowledge of the chemistry of the individual drugs and their derivatives, information on the stoichiometry and thermodynamic parameters that characterize the binding reactions between tubulin and each drug, and a thorough understanding of the chemistry of those regions of tubulin polypeptides that serve as drug-binding sites. A great deal of information is available on the first two issues, particularly in the case of colchicine and its derivatives. On the contrary, very little is known about the chemical nature of the colchicine (and vincristine, podophyllotoxin) binding site.

2. Tubulin properties

Tubulin is a heterodimer formed by two subunits indicated as α and β subunits. The two polypeptide chains are extremely similar (aminoacid composition, sequence and a molecular weight of about 55,000) but are not identical. Notwithstanding, for reasons not completely understood, the two subunits can be easily resolved electrophoretically in Na-dodecyl sulphate, containing trace amounts of undecyl and decyl moieties.

Tubulin is able to polymerize *in vitro*, providing that some requirements are fulfilled. For example, the presence of GTP and the absence of Calcium ions is quite mandatory; temperature has to be kept at 37°C. On the contrary, dilution of polymerized microtubules, addition of Calcium ions, and low temperatures (0°C), all induce a rapid tubulin depolymerization [2].

3. Colchicine-tubulin interaction

Tubulin heterodimer binds colchicine very slowly at least 30 minutes of incubation is needed for full interaction [4]. It is generally accepted that the binding is constituted by two subsequent processes: the first consists in a fast reversible association, the second in a slow ligand-induced conformational change. The overall binding equilibrium constant (37°C) is about 0.1 μM [5]. Podophyllotoxin, which shares with colchicine a trimethoxy benzene ring, competes with this compound, binds tubulin with an affinity constant of the same order of magnitude, but does not induce any protein conformational change [4].

To date, it is widely accepted that colchicine is a bifunctional ligand that binds principally through the trimetoxybenzene and tropolone rings. The second moiety (tropolone ring), which is absent in podophyllotoxin, is responsible of the conformational change of tubulin [6]. Vincaleucoblastine (Vinblastine) binds tubulin interacting with a different site; this fact is suggested by the observation that it does not compete with colchicine. Very scarce information, however, is available about the binding properties of these alkaloids with tubulin.

4. Material and methods

Tubulin was prepared from rat brain by polymerization cycling followed by a phosphocellulose chromatography step and was >98% pure [7]. Ring-C-methoxy (^{3}H) Colchicine was obtained from Amersham (4.2 Ci/mmol) diluted in ethanol. The stock was divided into siliconized Eppendorf tubes and dried under vacuum. Tubulin was added to these tubes and the mixture was incubated for 30-60 min at 37°C in the dark. The complex was irradiated directly in the same Eppendorf tube at 0-4°C with a high pressure mercury lamp at about 90 W. Samples were irradiated under 2 cm (pathlength) of a 20% CuSO$_4$ solution. This provided <1% transmission below 305, about 50% at 322 and more than 97 at 353 nm, i.e. the absorption maximum of colchicine. In addition CuSO$_4$ solution was an excellent infrared filter above 570 nm, thus reducing the evaporation of samples. In the geometry of the system the samples were receiving about 35 mW/cm^2. The best radioactive yields were obtained at high tubulin-to-colchicine ratios, presumably because the improved rate of covalent binding with respect to lumicolchicine formation and the protective effect of tubulin with respect of its photodecay (data not shown). Following the irradiation the samples were boiled for two minutes with an SDS/mercaptoethanol loading solution and electrophoresed (two series of identical samples) on a 8% polyacrylamide gel. After the electrophoresis half of the gel was stained, destained, photographed and impregnated with salicylate and exposed for autoradiography; the other, duplicate half, was extensively washed with water, and 7 mm bands were then cut from each line, dissolved in 300 μl of 30% H$_2$O$_2$ at 55°C for 3-5 hr, mixed with scintillation liquid, allow to stand to reduce chemiluminescence and counted for ^{3}H. In these conditions the yield in the monomer band was consistently 5% or more. Direct photolabelling with tritiated podophyllotoxin was realized in identical conditions except that, as the excitation maximum was near 290 nm, a 2 cm quartz cuvette containing 0.3% BSA was used as a filter. Instead for vinblastine, a filter constituted by a solution of potassium hydrogen phtalate was used.

5. Results and Discussion

The effects of irradiation time on photolabelling of ^{3}H-tubulin were studied. Fig. 1 depicts the effect of irradiation time on the ^{3}H-colchicine incorporation in α e β subunits. In this case colchicine and tubulin were incubated for 45 min at 37° to allow full non-covalent interaction before irradiation. The need for this incubation before irradiation (fixed time) to enhance the binding of colchicine to tubulin, is illustrated in Fig. 2. It appears that the efficiency of photoaffinity labelling increases with the time of preincubation, reaching a plateau within 30-50 min. See figure legends for other details.

A second set of experiments were tailored to study how the protein "conformational status" might influence the distribution of the colchicine between the two subunits. Fig. 3a shows the effect of increasing concentrations (up to 4M) of glycerol, a known tubulin conformation stabilizer. It is evident that, while the overall incorporation is remarkably enhanced, the labelling of β-subunit progressively increases with respect to that of α subunit which remains substantially unchanged.

                                                G. PALUMBO

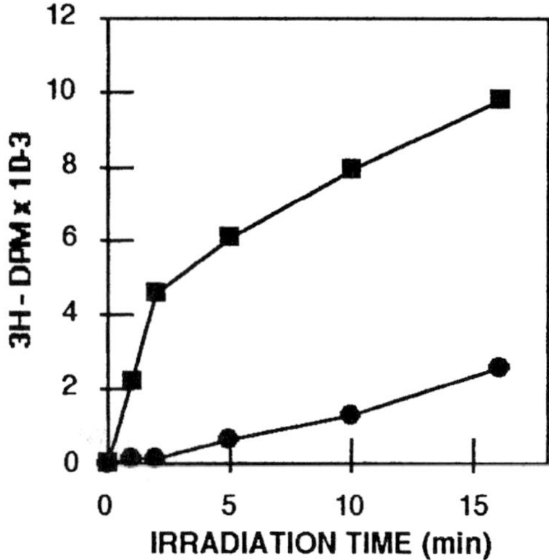

Figure 1. 32 μM tubulin was incubated with 2.4 μM colchicine for 45 min at 37°C and irradiated at 0-4°C at 87 W for various times (-■- β, -●- α).

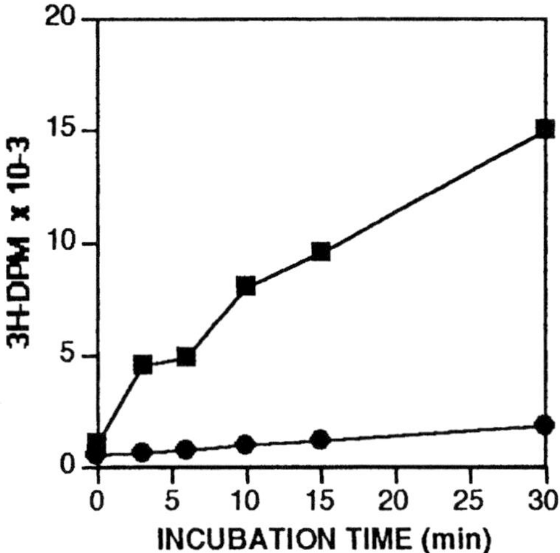

Figure 2. 32 μM tubulin was incubated with 2.4 μM colchicine for various times at 37°C and then irradiated at 0-4°C at 87 W for five min (-■- β, -●- α).

Fig. 3b shows the change in the α/β ratio as a function of glycerol concentration. When the labelling was performed in the presence of a destabilizing agent, i.e. at increasing urea

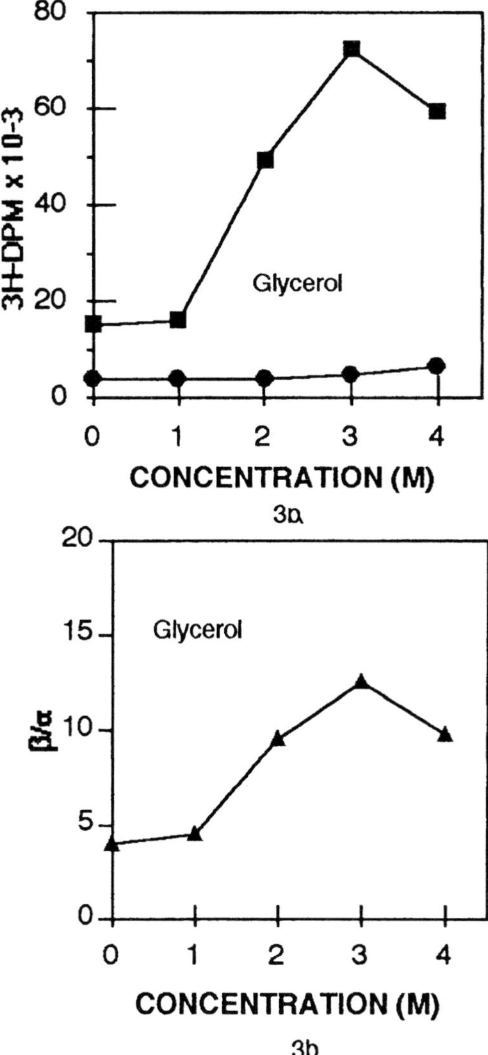

Figure 3. a) Distribution of labelled colchicine between α e β, subunits as a function of glycerol concentration. 49 μM tubulin was incubated with 5.3 μM ³H-colchicine for 60 min at 37°C in presence of indicated concentrations. Samples were irradiated for 5 min at 4°. (-■- β, -●- α); b) β/α ratio as a function of glycerol concentration.

concentrations, a different effect is seen. Up to 2.5 M urea, colchicine incorporation diverges in the two subunits.: it increases in the α-subunit and decreases in the β.

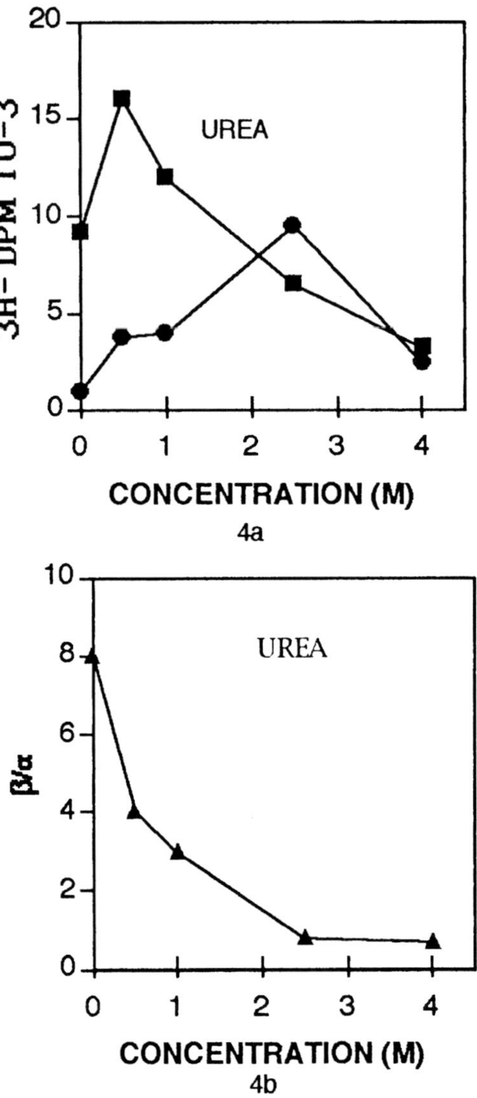

Figure 4. a) Distribution of labelled colchicine between α e β, subunits as a function of urea concentration. 49 μM tubulin was incubated with 5.3 μM ^{3}H-colchicine for 60 min at 37°C in presence of indicated concentrations.Samples were irradiated for 5 min at 4°. (-■- β, -●- α); b) β/α ratio as a function of urea concentration.

At higher urea concentrations, i.e. at concentrations which induce a massive protein unfolding, the colchicine incorporation decreases in both subunits. (Fig. 4a). The ratio β/α decreases also (Fig. 4b).

A different picture is obtained if the labelling is performed in presence of increasing concentrations of a neutral salt (NaCl). In this case the amount of tritiated colchicine rises in both subunits, but the rate of labelling of the α-subunit is significantly higher (Fig. 5a).

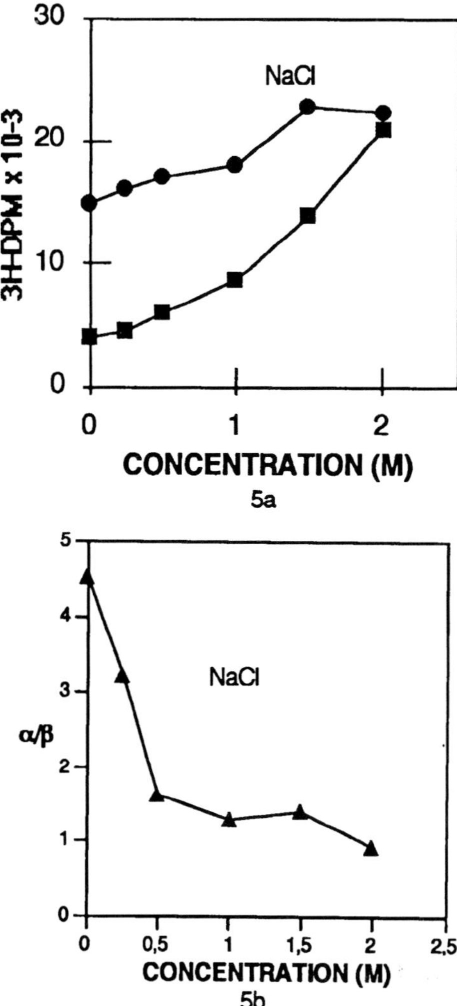

Figure 5. a) Distribution of labelled colchicine between α e β, subunits as a function of NaCl concentration. 49 μM tubulin was incubated with 5.3 μM ^{3}H-colchicine for 60 min at 37°C in presence of indicated concentrations.Samples were irradiated for 5 min at 4°.(-■- β, -●- α); b) β/α ratio as a function of NaCl concentration.

The net effect is a decrease in the β/α ratio as shown in Fig. 5b. It is known that tubulin ability to polymerize is progressively lost if the protein is left for a few hours at room temperature [8]. Also tubulin kept at 4°C undergoes a similar decay within one or two days. This is generically indicated as tubulin ageing. It has been observed that labelling of tubulin aged from 0 up to 6 hs results in a decrease in the β/α (Fig. 6). Aging must induce structural modifications leading to the exposure of low affinity binding site (α-subunit). All these observations are in accord with the conclusion that any conformational status of tubulin determine the distribution of the covalently bound colchicine between its two subunits.

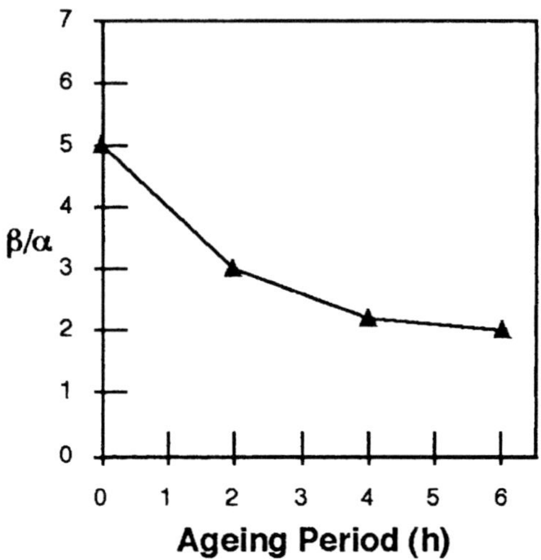

Figure 6. Effect of ageing on tubulin photolabeling. Frozen tubulin samples were thowed at various times (0-6 h) and incubated with appropriate amounts of tritiated colchicine for 60 min. Irradiation time was 5 min in standard conditions. The ratio β/α is plotted versus the ageing period.

Previous observations with photosensitive colchicine derivatives have indicated that the binding site of the molecule on tubulin is on its α-subunit. Nevertheless this might have resulted from the use of large photosensitive groups (i.e. change in the protein conformation) [9,10]. Subsequent use of shorter spacers indeed yielded β-tubulin labelling as well [10]; this was, however, ascribed to a second, low affinity colchicine binding site. In contrast to these findings, the present data suggest that β-tubulin labelling is strongly favored, while a low affinity binding site is localized on the α-subunit. In accordance with such hypothesis it has been observed that, when the irradiation was performed on a sample, in which the molar ratio of colchicine to tubulin was increased from 0.1 to 30, the β/α ratio decreased five folds (Fig. 7). This result implies the presence on the β-subunit of a high affinity binding site and the presence of a low affinity binding site on the α-subunit. It is known that tubulin may be dissociated by dilution and that this dissociation is regulated on the basis of the simple equilibrium laws [11]. Labelling of dilute solutions of tubulin (not shown) indicated that the overall incorporation of colchicine into the protein is reduced to minimum. This result strongly suggests that both subunits are needed for an efficient binding and that the binding site may span the two subunits. Various experiments have been made in attempting to saturate the labelling with colchicine. Indeed the efficiency of incorporation of a fixed amount of ^{3}H-colchicine was markedly reduced in presence of increasing concentrations of unlabelled colchicine (Fig. 7). The rapid photodecay of unbound colchicine and possible inner filter effects, however, complicate and obscure the interpretation of these experiments. To circumvent this, tubulin-^{3}H-colchicine mixtures containing increasing amounts of podophyllotoxin, have been irradiated. Indeed, podophyllotoxin competes with colchicine for the same site, but it does not absorb at the wavelenght used (above 305 nm). As it can be observed in Fig. 8, inhibition of binding with podophyllotoxin decreases the yield of covalently bound colchicine. Furthermore, the labelling of the protein results in a significant reduction of incorporation in the β-subunit, while the incorporation in the α remains substantially unchanged. This is in accordance with the known finding that the podophyllotoxin binding is competitive [4]. Finally, direct photolabelling of tubulin with podophillotoxin and vinblastine (the relative photolabelling conditions have be detailed in the "methods" section) has been carried out. The data (not shown) indicate that, while podophyllotoxin binds as expected, the β-subunit vinblastine binds, with high specificity the α-subunit.

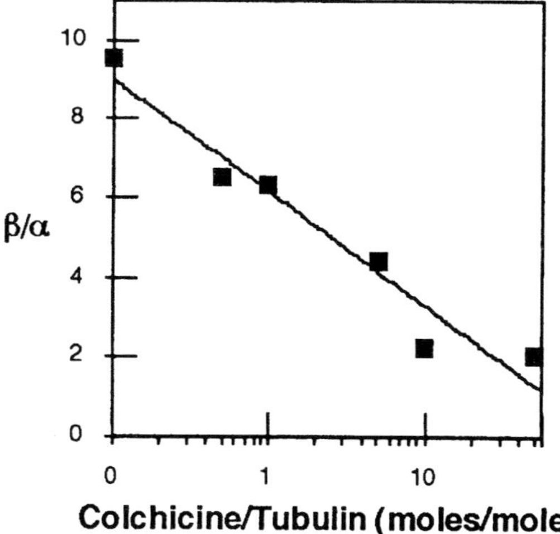

Figure 7. Competition of ^{3}H-colchicine binding. 49 μM tubulin was incubated with 4.9 μM ^{3}H-colchicine for 60 min at 37°C in presence increasing amounts of unlabeled colchicine. The molat ratio colchicineItubulin varies from 0.1 to 30. Samples were irradiated for 5 min at 4°.

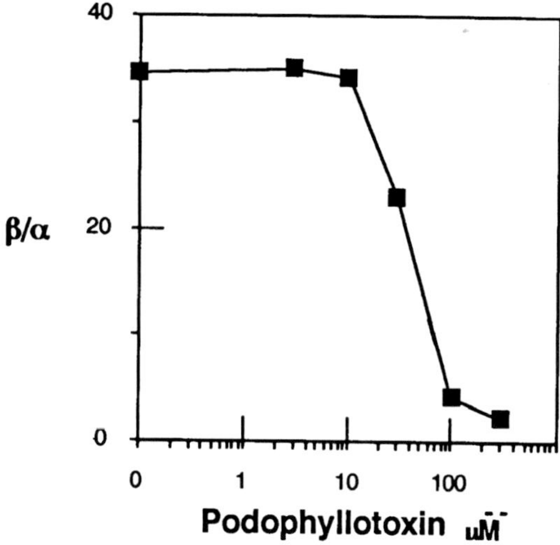

Figure 8. Photolabeling of tubulin by colchicine in presence of increasing amounts of podophyllotoxin. The two compounds compete for the same binding site. The distribution is indicated by the change in the ratio β/α versus podophyllotoxin concentration.

6. Conclusions

The body of data presented leads to some simple conclusions which are now summarized.

i. UV irradiation of the colchicine-tubulin complex leads to direct photolabelling, but a preincubation step is necessary. The yield is a function of incubation and irradiation times.

ii. β-tubulin labels first and constitutes 70 to >90% of the tubulin conformation stabilizers (glycerol).

iii. α-tubulin labels with delay or after manipulations that damage the protein as aging, urea or ionic strength.

iv. Podophyllotoxin blocks photolabelling by colchicine, especially in β-tubulin. ^{3}H-podophyllotoxin labels β-tubulin primarily.

v. Vinblastine reacts preferentially with α-tubulin (α/β ratios of about 3 to 1).

vi. It is proposed that the colchicine binding site spans the α-β interface, with preferential binding to β-tubulin that can be displaced toward α-tubulin by large spacers or structural changes that loosen the α-β interface.

Acknowledgements

This work has been partly supported by CNR grant "Progetto finalizzato Biotecnologia e Biostrumentazione".

References

1. L.J. Wilson & J. Brian, <u>Biochemical and pharmacological properties of microtubules</u>. In "Advances in Cell and Molecular Biology." DuPraw E.J. (Ed.) Vol. III Academic Press Inc. New York N.Y., (1974).
2. J.B. Olmsted & G.G. Borisy, *Annu. Rev. Biochem.* **42**, 507-540, (1973).
3. P. Dustin In "<u>Microtubules</u>" 2nd (Ed.), Spring-Verlag Inc. New York, N.Y., (1984).
4. F. Cortese , B. Bhattacharyya & J. Wolff, *J. Biol. Chem.* **252**, 1134-1140, (1974).
5. D.L. Garland, *Biochemistry* **17**, 4266-4272, (1978).
6. F.J. Medrano, J.M. Andreu, M.J. Gorbunoff & SN. Timasheff, *Biochemistry* **28**, 5589-5599, (1989).
7. R.D. Sloboda, & J.L. Rosenbaum, *Methods in Enzymol.* **85**, 409-416, (1982).
8. B. Rousset , F. Bernier-Valentin, J. Wolff & B. Roux, *Eur. J. Biochem.* **131**, 31-39, (1983).
9. R.F.Williams, C.L. Mumford, G.A. Williams, L.J. Floyd, M.J. Aivaliotis, R.A. Martinez, A.K. Robinson & L.D. Barnes, *J.Biol.Chem.* **260**, 13794-13802, (1985).
10. L.J. Floyd, L.D. Barnes & R.F. Williams, *Biochemistry* **28**, 8515-8525, (1989).
11. D.L. Sackett, J.R. Knutson & J. Wolff, *J. Biol. Chem.* **265**, 14899-14906, (1990).

A QSAR Model of the Isoelectric Points and of the Atomic Charges of Amino Acids

L. POGLIANI
Dipartimento di Chimica, Università della Calabria, 87030 Rende (CS), Italy

1. Introduction

Molecular connectivity indexes have been employed in many structure-activity relationship studies from a wide range of research areas in chemistry [1-5]. As certain types of structural informations are directly related to the presence of functional groups in a molecule, we will describe here a method to correlate the connectivity indexes to the pH at which the amino acid has no net charge, called the isoelectric point, denoted by pI, a point which is highly dependent on the presence of ionizable groups in the side-chain [6, 7]. Tested connectivity index model is based solely on the functional groups of the amino acids. As atomic charges are also strongly dependent on the presence of functional groups in the molecule, we will also present in this paper a molecular connectivity model for an approximate evaluation of the atomic charges in amino acids with the aim, also, to extend the understanding of the valence delta indexes [8].

2. Connectivity Model for pI Calculations

The simple molecular connectivity index of the first-order

$$^1X = \Sigma(\delta_i\delta_j)^{-1/2} \tag{1}$$

and the corresponding one-bond valence molecular connectivity index $^1X^v$ where the valence delta δ^v replaces δ, are used in this study [1-5]; i and j denote two adjacent atoms, other than hydrogen. Following sum-delta index is also used

$$D = \Sigma\,\delta_i \tag{2}$$

together with the corresponding valence sum-delta index, D^v, where δ^v replaces δ [7]. Summation runs over the nonhydrogen atoms of the molecule.

Index δ is a count of nonhydrogen σ-bond electrons contributed by atom i, while the valence delta index, δ^v, is a count of all nonhydrogen valence electrons contributed by atom i, $\delta^v_i = \delta_i + p_i + n_i$, where σ, p and n are sigma, p and lone-pair electrons respectively.

381

N. Russo et al. (eds.), Properties and Chemistry of Biomolecular Systems, 381–387.
© 1994 *Kluwer Academic Publishers. Printed in the Netherlands.*

Two types of main fragments are considered: MF1 (fig.1a), valid for most of the amino acids and MF2 (fig. 1b), valid for cyclic amino acids, such as Pro and ProOH.

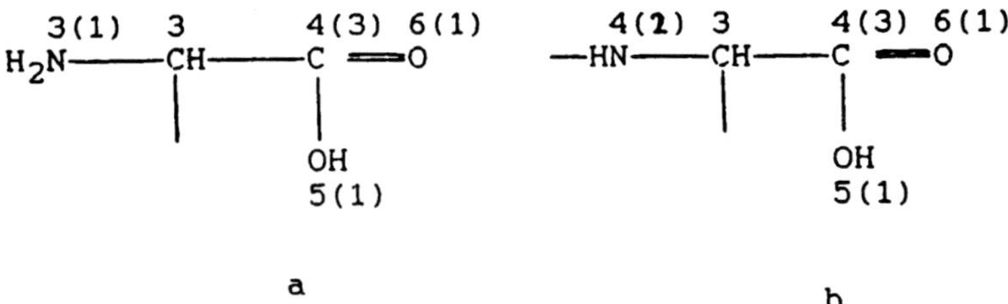

Figure 1. Assignment of δ^V and (δ) values to the atoms of (A) MF1 and (B) MF2 where MF2 is valid for Pro and ProOH (when only one index is given, then $\delta^V=\delta$).

Calculated indexes for MFi (summation for D values runs over the atoms of the ionizable group only) are:

MF1) $^1X = 2.065$, $^1X^V = 1.050$, $D = 6$ and $D^V = 18$
MF2)   " $= 1.896$,   " $= 1.005$, " $= 7$ and   " $= 19$

To these two basic sets of indexes we add or subtract the connectivity and D values of the side-chain ionizable fragments (SFi). Five different types of side-chain fragments were considered, following Lehninger's classi fication [6] for R groups in amino acids:
SF1) Non polar fragment: present in Ala, Val, Leu, Ile, Pro, Phe and Trp. This kind of fragment gives no contribution to the calculated connectivity and D values of MF1,2 (MF2 for Pro).
SF2) Negatively charged polar fragment at pH=6: present in Asp and Glu (fig.2). The calculated connectivity and D values of this fragment, $^1X = 1.563$, $^1X^V = 0.781$, $D = 5$ and $D^V = 15$, yield   a positive contribution to the MF1 indexes.

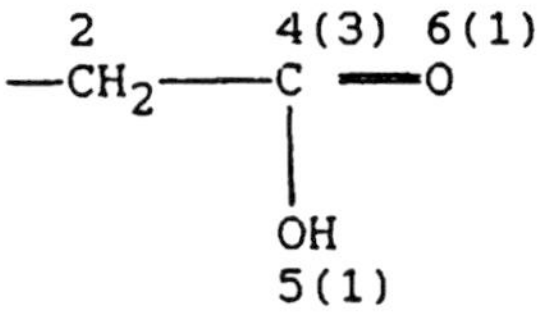

Figure 2. Assignment of δ and δ to SF2, which is valid for Asp and Glu..

SF3) Uncharged polar fragments at pH=6: present in Gly, Ser, Thr, ProOH, Tyr, Asn and Gln (fig. 3a,b,c, and d). The calculated indexes of these fragments that give also a positive contribution to the MF1,2 (MF2 for ProOH) connectivity and D values are:

SF3a) $^1X = 0.707$, $^1X^V = 0.316$, $D = 1$ and $D^V = 5$
SF3b) " $= 0.577$, " $= 0.258$, " $= 1$ "   " $= 5$
SF3c) " $= 0.577$, " $= 0.224$, " $= 1$ "   " $= 5$
SF3d) " $= 0.408$, " $= 0.269$, " $= 3$ "   " $= 7$

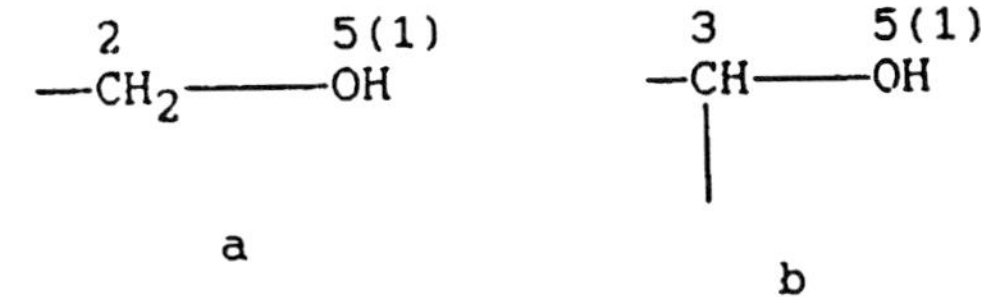

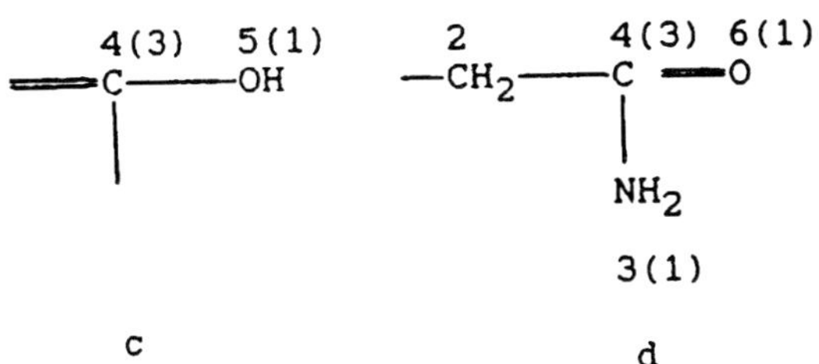

Figure 3. Assignment of δ values to SF3a, wich is valid for Ser; SF3b wich is valid for Thr and PrOH; SF3c wich is valid for Tyr; and SF3d, wich is valid for Asn and Glu..

Connectivity and D values of this last fragment result from the difference between the connectivity and D values of the two subfragments: $-CH_2-C=O$ and $-C-NH_2$. Gly, with no side-chain fragment, adds no contribution to the MF1 connectivity values.

SF4) Positively charged polar fragments at pH=6: present in Lys, Arg and His (fig. 4a,b,c).

Figure 4. Assignment of δ values to SF4a wich is valid for Lys; SF4b, wich is valid for Arg; and SF4c, wich is valid for His. The Gnd 11(5) results from the sum of the δ indexes of the guanidinium group —— C 4(3) —— NH2 3(1)

$$\begin{matrix} \| \\ NH\ 4(1) \end{matrix}$$

The following indexes of these fragments give rise to a negative contribution to the corresponding MF1 values:

SF4a) $^1X = 0.707$, $^1X^v = 0.408$, $D = 1$ and $D^v = 3$
SF4b) " $= 0.816$, " $= 0.504$, " $= 2$ " " $= 4$
SF4c) " $= 0.500$, " $= 0.258$, " $= 2$ " " $= 5$

In table 1 are collected the corresponding ionization index values of 19 amino acids, together with the corresponding experimental and theoretical pI values.

The best fit between pIexp and the connectivity indexes is given by the following equation:

$$pI = 10.998 - 10.85\ ^1X^V + 0.55\ (D^V - D) \tag{3}$$

r=0.98 s=0.36 n=19, F = 194

With the aid of this equation were obtained the calculated pI values plotted vs. the corresponding experimental ones in fig. 5.

Multiple linear regression analysis underlines the importance of σ, π and lone-pair electrons in the ionization process of the functional groups of the amino acids. In fact, whereas $^1X^V$ encodes information on all nonhydrogen valence electrons (all σ, π and lone-pair electrons) through the corresponding δ^V dependence, the difference D^V-D counts all of the nonhydrogen π and lone-pair electrons of the heteroatom N and O [1, 4].

Table 1: Amino Acid pI and $^1X^V$, D^V, ^1X , D ionization values.

AA	D^V-D	$^1X^V$	D^V	^1X	D	pIexp	pIcalc
Asp	22	1.831	33	3.628	11	3.0	3.2
Glu	22	1.831	33	3.628	11	3.1	3.2
Asn	16	1.319	25	2.473	9	5.4	5.5
Thr	16	1.308	23	2.642	7	5.6	5.6
Tyr	16	1.274	23	2.642	7	5.6	5.9
Gln	16	1.319	25	2.473	9	5.7	5.5
Ser	16	1.366	23	2.772	7	5.7	4.9
ProOH	14	1.263	22	2.473	8	5.8	5.0
Trp	12	1.050	18	2.065	6	5.9	6.2
Phe	"	"	"	"	"	"	"
Val	"	"	"	"	"	6.0	"
Leu	"	"	"	"	"	"	"
Ile	"	"	"	"	"	"	"
Ala	"	"	"	"	"	6.1	"
Gly	"	"	"	"	"	"	"
Pro	"	1.005	19	1.896	7	6.3	6.7
His	9	0.792	13	1.565	4	7.6	7.3
Lys	11	0.642	15	1.358	5	9.5	9.5
Arg	10	0.546	14	1.249	4	10.8	10.6

3. Connectivity Model for Atomic Charge Evaluation

Years ago Del Re et al. [9] presented a systematic study on the electronic structure of the a-amino acids. In spite of the many crude approximations of the theoretical treatment, based on a parametric method to evaluate the σ-charge distribution of the neutral,

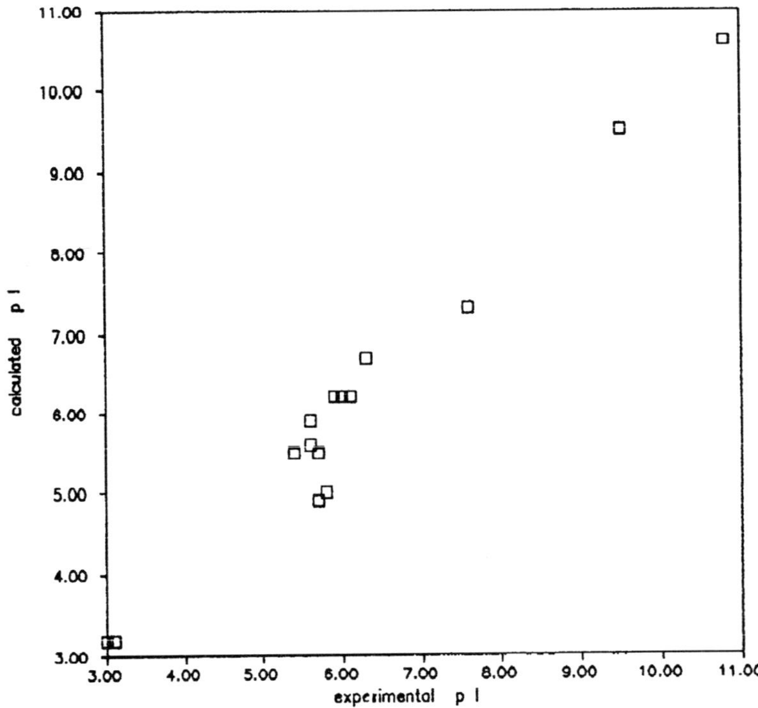

Figure 5. Plot of calculated vs. experimental pI for 19 amino acids.

zwitterionic, cationic and anionic form of the amino acids, that study led to charge distributions, Q_I, which permitted a fairly qualitative and quantitative interpretation of the nmr chemical shifts in amino acids. If charge distribution of the second row atoms (I = C, N, O) of the four forms of the amino acids is considered to be 20% accurate, mean Q_I values of the only neutral form can be considered. Table 2 shows the second row atoms of a-amino acids on a Q_I, δ^V matrix.

In this table some crude correlations, ie, are emphasized by straght lines. The slopes of all correlations have the same sign and many correlations also have nearly the same slope. The most significant ones are those linking 1) NH_2, NH and N with slope s $\approx$ 0.2, 2) C_z (terminal carbon) and the different C_is (internal carbons) with s $\approx$ 0.05, 3) the different Cs bonded to a NH function with s $\approx$ 0.05, 4) the different Cs bonded to a OH function with an average slope s $\approx$ 0.05.

Table 2: Q_I and δ^V Matrix for second row Atoms of α-Amino Acids.

Q_I / δ^V	1	2	3	4	5	6
-.500			NH_2		OH	
-.310				NH		
-.120	C_z				N	O
-.071		C_i				
-.025		C(NH)	C_i			
-.012		$C(NH_2)$				
.003		$C\alpha$				
.016				C_i		
.037			C(N,NH)			
.051		C(OH)	$C\alpha$			
.084			C(OH)/ C(N)	C(NH)		
.140				C(OH)		
.240				$C^\circ/C^{\circ\prime}$		

The two smallest correlations are those concerning the Ca atoms with $s \approx 0.05$ also and the OH and O groups with $s \approx 0.4$. If a value of 7 for $(\delta^V + \delta)$ is assigned to a C° (and $C^{\circ\prime}$) atom (C° is the carboxyl carbon and $C^{\circ\prime}$ is the carboxylamide carbon) then the correlation, with $s \approx 0.05$, that links Ca carbons can be extended to include C° and $C^{\circ\prime}$ carbons (solid and dotted lines in table 2).

Atomic charge distribution of the second row atoms of a-amino acids show a crude correlation with δ^V (for sp^3 carbons $\delta^V = \delta$); therefore, charge distribution in amino acids can be better encoded by an index that counts all valence electrons, including sigma electrons ($\delta^V = \sigma + p + n - h$) [1, 4]. The only exception is for the C° carbon. This terminal carbon bonded to two highly electronegative atoms is a special case, and its highly positive charge can be fairly well correlated with the Ca charges assuming a Γ

index value of 7; $\Gamma = \delta^v + \delta$. Because a similar sum succeeded in encoding the Bondi atomic volumes (4), C° may be consider as a special atom with a cardinal number of Γ. Despite the low n values of our correlations, the similar s value of the correlations of the different classes of carbon atoms in amino acids is striking, while the different slopes for the correlations of the hetero-atoms (0.4, 0.2, and 0.05 for O, N and C functions respectively) can be qualitatively rationalized in terms of the decreasing electronegativities of O, N and C.

4. Conclusions

The two cardinal numbers, the delta and the valence delta, elaborated in the frame of molecular connectivity theory, can be used to encode molecular properties, such as pI in amino acids, that are directly related to the presence of functional groups in a molecule as, in fact, these numbers match in some way the single atomic charges in polifunctional molecules. It should also be stressed that, as most of the receptor maps rely strongly on localized charge patterns, that mirror the supposed distribution in the molecule to be bound, an effort to obtain, in an easy and direct way (like here done for amino acids) charges on particular atoms of the substrate molecule can be highly rewarding.

References

1. L.B. Kier and L. H. Hall, <u>Molecular Connectivity in Structure-Activity Analysis</u>, Wiley, New York, (1986).
2. L.B. Kier, L.H. Hall, W.J. Murray and M. Randic, "Molecular Connectivity I: Relationships to Nonspecific Local Anesthyesia", *J. Pharm. Sci.* **64**, 1971-1974, (1975).
3. L. B. Kier, L. H. Hall and W. J. Murray, "Molecular Connectivity II: Water Solubility and Boiling Point", *J.Pharm.Sci.* **64**, 1974-1977, (1975).
4. L.B. Kier and L.H. Hall, "Molecular Connectivity VII: Specific Treatment of Heteroatoms", *J.Pharm.Sci.* **70**, 583-589, (1981).
5. M. Randic and C. Wilkins, "Graph Theoretical Approach to Recognition of Structural Similarity in Molecules", *J. Chem. Inf. Comp. Sci.* **19**, 31-37, (1979).
6. A. Lehninger, <u>Biochemistry</u>, Worth, New York, 1977.
7. L. Pogliani, "Molecular Connectivity Model for Determination of Isoelectric Point of Amino Acids", *J. Pharm. Sci.* **81**, 334-336, (1992).
8. L. Pogliani, "Molecular Connectivity: Treatment of the Electronic Structure of Amino Acids", *J. Pharm. Sci.* **81**, 967-969, (1992).
9. G. Del Re, B. Pullman and T. Yonezawa, "Electronic Structure of the _-Amino Acids of Proteins. I. Charge Distributions and Proton Chemical Shifts", *Biochim. Biophys. Acta* **75**, 153-182, (1963).

On the Molecular and Supramolecular Structure of Elastin

A.M. TAMBURRO°, D. DAGA GORDINI⁺, V. GUANTIERI* and A. DE STRADIS∞
° *Department of Chemistry, Università della Basilicata, Potenza, Italy.*
⁺ *Institute of Histology, Università di Padova, Italy*
* *Department of Inorganic Chemistry, Università di Padova, Italy*
∞ *Centre for Microscopy, Università della Basilicata, Potenza, Italy*

1. Introduction

Elastin, the protein responsible for elasticity in most tissues of vertebrates, displays peculiar aminoacid composition and primary sequence. Actually, about 90% of aminoacid residues are apolar, and quite unusual cross-links (desmosine and isodesmosine) [1] and many repetitive sequences are present [2-6]. the mature, insoluble protein originates from a soluble precursor called tropoelastin by post-translational cross-linking. In recent years gene analysis has revealed the primary sequence of tropoelastins from different sources, such as the human [2], bovine [3],, chick [4], and murine [5] species. While these findings have considerably deepened our understanding of the elastin system, there are still several unanswered questions: is elastin a "classical" elastomer according to Flory theory [7] or not ?, is the supramolecular, fibrous structure of the protein determined by some preferred molecular conformation? Other questions concern the possible biological role played by specific sequences of elastin. In particular, tropoelastin itself and some peptides such as VGVAPG, AGVPGFGVG, GFGVGAGVP and GFGVG have been shown to possess chemotactic activity towards fibroblasts [8-10]. Interestingly, one of these peptides is a fragmentation product of elastase attack. Therefore, one wonders whether they could play a physiological and/or pathological role. in addition, recent studies have evidenced several immunogenic regions of the proteins including the chemotactic sequence VGVAPG [8,9]. Interestingly, this sequence is also chemotactic for certain tumor cell lines such as the M 27 line of the Lewis lung carcinoma [11]. One current approach toward elastin structure-function relationships has been based on investigations concerning synthetic models and fragments of the protein. As mentioned, the primary sequence of elastins is characterized by many repetitive sequences. The repetition, when cosidered in a statistical sense, can be found at different scales: at the lowest scale XGG and XPG (X=A, V, L, I) sequences are very frequently found, at the highest scale the entire protein has been interpreted as the repetition of few particular domains [4]. Given that, one can reasonably approach the problem of elastin structure through the synthesis of appropriate "monomers" and repeating polypeptides (polimers).

389

N. Russo et al. (eds.), Properties and Chemistry of Biomolecular Systems, 389–403.
© 1994 *Kluwer Academic Publishers. Printed in the Netherlands.*

Accordingly, several studies utilizing this approach have been carried out during the last twenty years [12-15, and references therein], mostly in Birmingham (USA) and in Padova and Potenza (Italy).

It is important to add that those investigations can give, and indeed gave, useful information, not only on the molecular structure, but also on the supramolecular structures of elastin. The most relevant and recent results will be reviewed in the following sections.

2. Molecular structure

The classical theory of elasticity [7] implies an entropic origin for the retractive force according to the equilibrium:

$$\text{Relaxed State} \underset{+ \Delta s}{\overset{- \Delta s}{\longleftrightarrow}} \text{Stretched State}$$

As a corollary, the native, relaxed structure should be a high entropy one. The fundamental question to be answered is: could an elastomeric protein, say elastin, adopt a high entropy native structure characterized by a network of random chains with a random distribution of end-to-end chain lengths? *Prima facie*, the classical model, essentially devised for synthetic polymers, seem hardly acceptable in the context of traditional views on protein structure. In fact, dynamic, kinetically free polypeptide chains are antithetic to the usual, thermodynamically controlled, (relative) minimum energy structures currently assumed for protein systems. In other words, the problem is whether a protein might assume a "native structure" poorly describable in thermodynamical terms in that it is fluctuating among several conformational states at approximately the same energy.

2.1. ELASTIN-LIKE SYNTHETIC POLYPEPTIDES

Many peptides and polypeptides mimicking the physico-chemical properties of elastin have been synthesized. A non exhaustive list includes poly (VPG), poly (VPGG), poly (VPGVG), poly (IPGVG), poly (VAPGVG) [12,13,16-23] for the proline-rich regions of elastin, and poly (AGG), poly (VGG), poly (LGG), poly (IGG), poly (VGGLG) and Boc-(GVGGL)$_n$-OMe (n=1,2,3) [15, 24-27] for the glycine-rich regions. The proline-containing polymers were mainly studied by Urry and co-workers. The obtained results led to the proposal of a new structure, the so called β-spiral, and to a new mechanism of elasticity, the so called librational entropy mechanism. Briefly, it is proposed that the entropic elasticity of elastin derives from regular structures, with fixed end-to-end chain lengths, in which internal librational motion provides the entropic elastomeric force. According to studies carried out on synthetic poly (VPGVG), a repetitive pentapeptide sequence present in elastin, the β-spirals are characterized by type II β-turns acting as spacers between the turn of the spiral. Between the β-turns, are suspended dipeptide segments within which large-amplitude, low-frequency librations can occur. The decrease in amplitude of the librations on extension constitutes a large decrease in the entropy of the segment, providing the driving force for return to the relaxed state [28]. Proline-devoid, glycine-rich peptides and polypeptides were synthesized and studied by Tamburro and co-workers. The results can be summarized as follows:

i) for pentapeptide "monomers" such as Boc-GVGGL-OMe or Boc-GLGGV-OMe, the structures present in these regions are the type II β-turn having X-Gly at the corners and the two flanking glycyl residues connected by the $4 \rightarrow 1$ hydrogen bond, and C7 cycles such as those shown in figure 1 and a possible Gly^3 C=0 ... HN Leu^5.

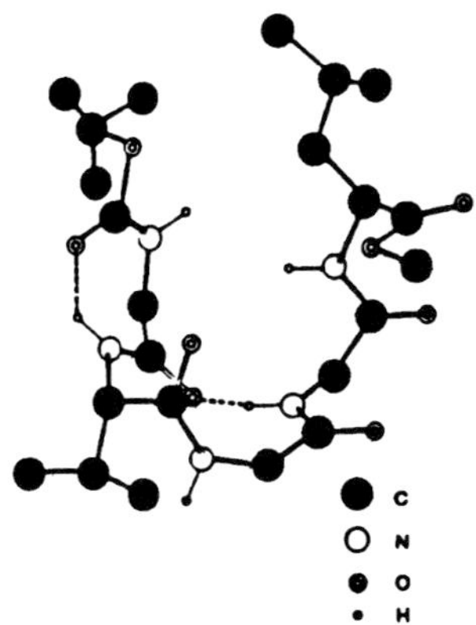

Figure 1. Proposed structure for Boc-GVGGL-OMe showing the C_{10} hydrogen-bonded cycle (type II β-turn) and the C7 hydrogen-bonded cycle.

ii) the stability of the turns was found to be largely dependent on several factors including the nature of the X residue, the concentration, the polarity of the solvent and the temperature. In particular, the folded conformations are destabilized by increasing the water activity.

iii) the stability of the turns dramatically decreases on increasing the chain length. This result would indicate that the related regions in elastin contain only few, isolated β-turns.

iv) very interestingly, NMR data also suggests the possibility of "sliding" β-turns according to the equilibrium

Gly^1 C=0 ... HN Gly^4 $\leftrightarrow$ Val^2 C=0 ... HN Leu^5 and/or one additional C7 cycle Gly^3 C=0 ... HN Leu^5.

Accordingly, the emerging picture from experimental studies would be that of a very labile structure containing isolated and dynamic (possibly sliding along the chain) β-turns. To get further, detailed insights into the low-energy conformational possibilities of the pentapeptide and, what is most important, into the dynamical processes of interconversion between them, theoretical studies using both molecular mechanics and molecular dynamics were carried out [29,30].

In figure 2 the minimum-energy conformation found by molecular mechanics is reported. Five intramolecular hydrogen bonds are present, three of them, including the

Gly[1] C=0 ... HN Gly[4] typical of the type II β-turn, being those evidenced by the experimental results.

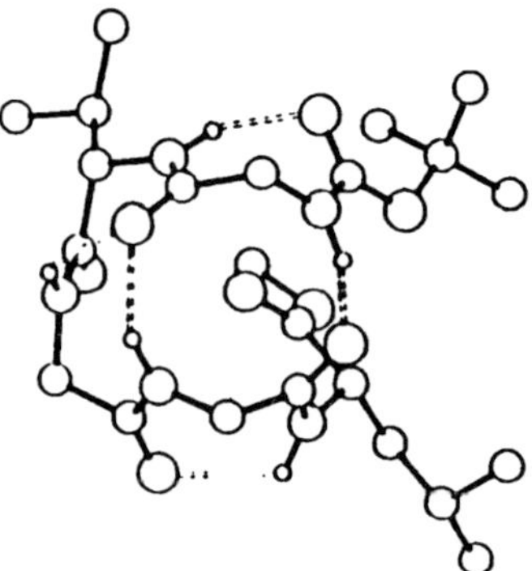

Figure 2. The minimum energy structure derived from molecular mechanics calculations for Boc-GVGGL-OMe.

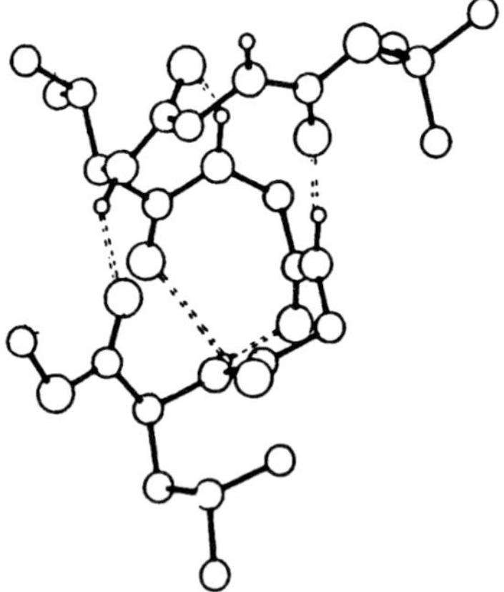

Figure 3. Another low energy structure for Boc-GVGGL-OMe.

Quite interestingly (figure 3) another conformation, only ~ 0.7 Kcal/mole higher than the minimum, is characterized by the Val[2] C=0 ... HN Leu[5] hydrogen bond suggested by the experimental results. It is to be noted that the two ten-membered type II β-turns are mutually exclusive and this could point towards a conformational equilibrium as hypothesized in the experimental structure. The results of molecular dynamics simulations were of particular interest: f.e. the β-turn due to the H-bond Val[2] C=0 ... HN Leu[5], absent in the starting conformation, appeared afterwards. Following the simulation, it was possible to observe that the reduction of the strength of this H-bond is partially compensated by formation of the C_7 ring Gly[3]C=0 ... HN Leu[5], the modification of the distances which characterize this H-bond pattern being strongly correlated. Furthermore, the correlated motion relative to the two H-bond dicstances has a clear counterpart in the two dihedral angles ϕ_4 and ψ_4. Finally, dielectric relaxation measurements [31] gave additional information concerning the influence of the hydration on the polypeptide chain dynamics. As a matter of fact, thermostimulated current spectroscopy showed that a low-temperature relaxation mode (related to localized motions of the chain) was shifted to higher temperatures by deydration of the polytripeptides. This indicates that the dipoles need more energy to relax when dehydrated. Moreover, water was found to preferentially bind to the amorphous regions, acting as a plasticizer, i.e., facilitating the localized (low frequency) motions of the polypeptide chain. Quite interestingly, preliminary studies carried out on elastin evidenced the striking similarity between the localized motions of the model polytripeptides and elastin itself.

2.2. A CLASSICAL MODEL FOR ELASTIN MOLECULAR STRUCTURE

The above results allowed an approach to the elasticity mechanism of elastin based on the following structural scheme [32]: the deydrated protein could be described according to different structural domains including:

i) the α-helical poly-alanine sequences of the cross-links regions;

ii) the repetitive sequences containing a regular array of PG segments and then probably arranged in β-spiral structures because of the stabilization of the constituent type II β-turns;

iii) the regions containing the flexible GG bond where only isolated β-turns and some cross β-structures are possibly present together with amorphous regions. Water binds to the amorphous regions and also acts as destructuring towards the cross β-structures, by destabilizing the incorporated β-turns and disrupting the intrachain hydrogen bonds. This, in its turn, would facilitate the localized, low-frequency motions, thus greatly increasing the entropy of the system and allowing the onset of the entropic elastomeric force. In fact, elastin is elastic only when swelled in water (or very polar solvents) [33]. On the other hand, if the β-spirals are indeed stable in an aqueous environment, as suggested by molecular dynamics simulations [34], then they could contribute to the whole entropy of the system through the librational mechanism proposed by Urry [28]. The emerging structural picture of the hydrated protein, apart from finer details, points to a collection of very labile structrures (quickly) interchanging among themselves, with polypeptide chains characterized by variable end-to-end distances (i.e. they are kinetically free), internal chain dynamics and high entropy.

2.3. A FRACTAL DESCRIPTION OF ELASTIN MOLECULAR STRUCTURE

Obviously, the above model has to be seen as an arbitrary "freezing" of a series of possible "trajectories". Here the language typical of dynamical systems (motion trajectories) is intentionally used in order to emphasize this analogy and to introduce a description of elastin in terms of fractal geometry. As it is well known [35], fractals are mathematical, but also real, objects whose non topological dimension is a real number not necessarily an integer. In simpler words, if one examines these objects at different size scales, one repeatedly encounters the same fundamental elements. The repetitive pattern defines the fractional, or fractal, dimension of the structure. Fractal examples are non differentiable curves like typical brownian trails. Quite interestingly, also the so called random walk, which typically identifies the random network of an elastomeric chain in the classical theory, belongs to these fractal curves with a fractal dimension D_F

$= \frac{E+2}{3}$ in an E-dimensional space (~1.66 for E=3). Recent correlations [36] established for proteins the range of values D_F = 1.2 - 1.8 (D_F = 1.0 for a linear chain, D_F = 2.0 for unrestricted random walk in 3-D space), the fractal dimension increasing with an increase in compacteness, i.e. the amount of β-turns in our case, of the molecular conformation. Accordingly, a fractal dimension of 1.6 - 1.7 for (tropo) elastin monodispersed in solution seems quite reasonable. This value more properly refers to the chain fractal dimension, that is the scaling exponent of the number of monomers (aminoacid residues) N of length r, or contour length, with respect to end-to-end length R [37]:

$$N = (R/r)^{D_F} \tag{1}$$

3. Supramolecular Structures

3.1. ELECTRON MICROSCOPY STUDIES

As it is well known, soluble elastin derivatives (tropoelastin α-elastin) undergo an entropy (hydrophobic) driven, reversible aggregation, the so called coacervation, on increasing the temperature [38]. At the equilibrium, the resulting supramolecular structure is constituted by fibrils and fibres quite similar to fibrous elastin [39]. Moreover, in some cases, under conditions of decreased water activity, a staggered lateral alignment of filaments yielding banded fibrils has been observed [40-41]. Of paramount importance, given the abundance of repeating sequences in the protein, even simple sequential polypeptides, mimicking those sequences, have been shown to possess self-organizing properties similar to those of elastin and its derivatives. Among these are poly (VAPGVG), poly (VPGVG), poly (VGGLG), poly (VPGG), and simple polytripeptides such as poly (VPG) [27, 42-46]. Recent results of our group extended previous studies to simple pentapeptides considered as "monomers" of elastin polypeptides chain. Here, some results obtained by scanning electron microscopy (SEM) and transmission electron microscopy (TEM) on VGGVG, Boc-VGGLG-OEt and poly (VGGLG) will be discussed. Figs 4 - 9 clearly show that all samples are able to self organize in filamentous, fibrillar and fibrous supramolecular structures. This is quite similar to what was found for elastin. A previously unstressed feature is the apparent hierarchy of the structures. As a matter of act, the interaction of filaments to give fibrils and fibril bundles seems to occur according to a repeating pattern involving a twisted-rope arrangement. In addition poly (VGGLG) exhibits characteristic clusters of aggregates (fig. 10). Interestingly, mature elastin has never shown a cluster organization under physiological conditions. However, under pathological conditions such as elastoderma, a disease of elastin accumulation within the skin, "grape-like globular structures" were found [47]. It is to be noted that both tropoelastin [48] and our

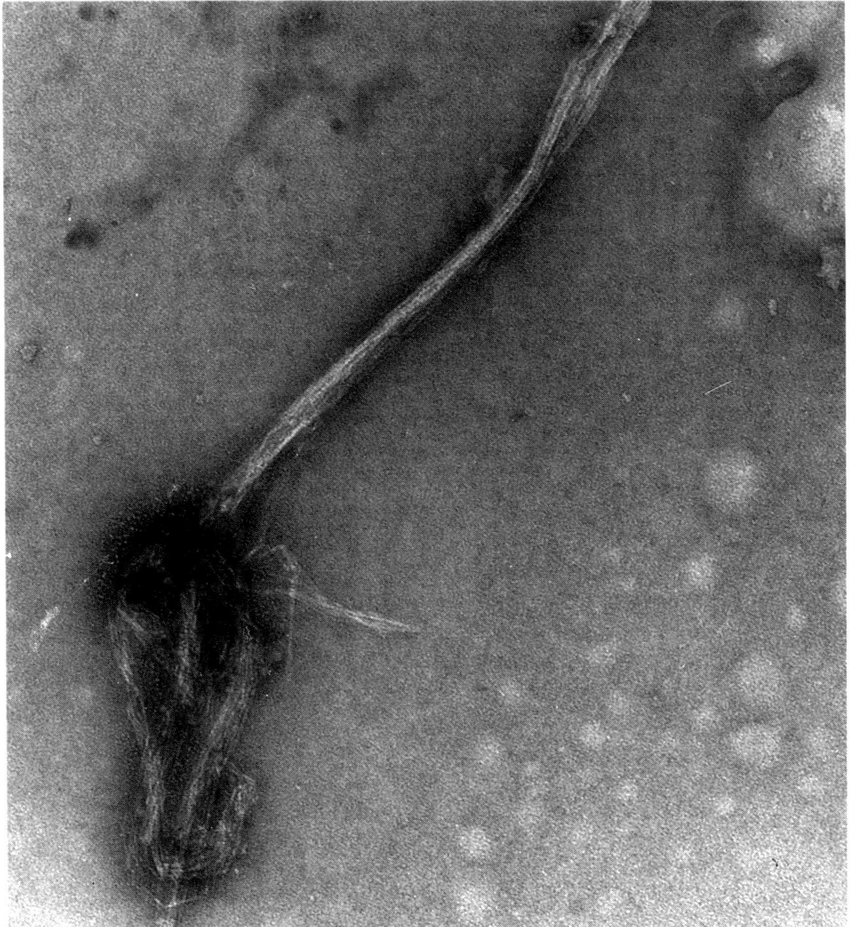

Figure 4. TEM micrograph of VGGVG in 50% ethyl alcohol at 50 °C. Negative staining with uranyl acetate, pH 3.5 . X 117,500.

elastin-related polypeptides initially organize themselves in "beads" (quasi-globular subunits) that subsequently form either filaments or globular aggregates. The beads revealed intimate substructures whose minimal dimensions appeared to correlate adequately with conformationally averaged molecular dimensions.

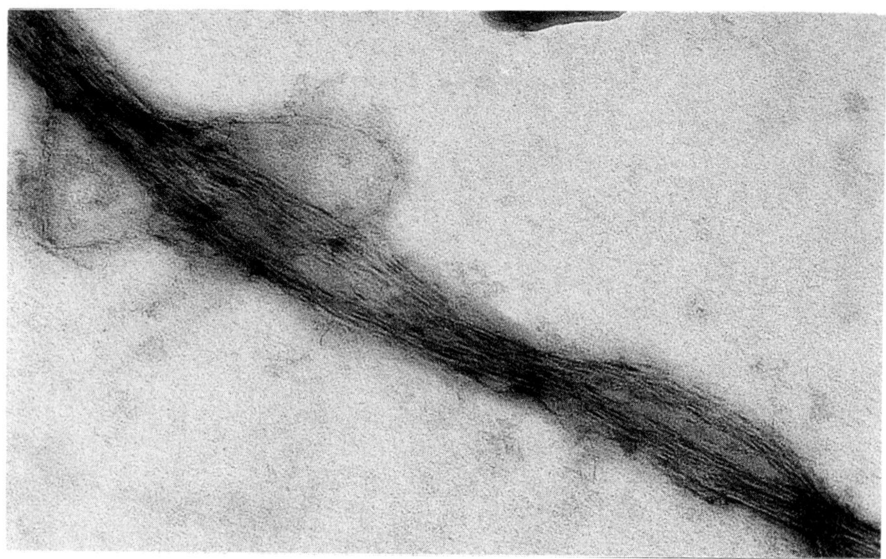

Figure 5. TEM micrograph of Boc-VGGLG-OEt in ethyl alcohol at room temperature. Negative staining with uranyl acetate, pH 3.5 . X 154,000.

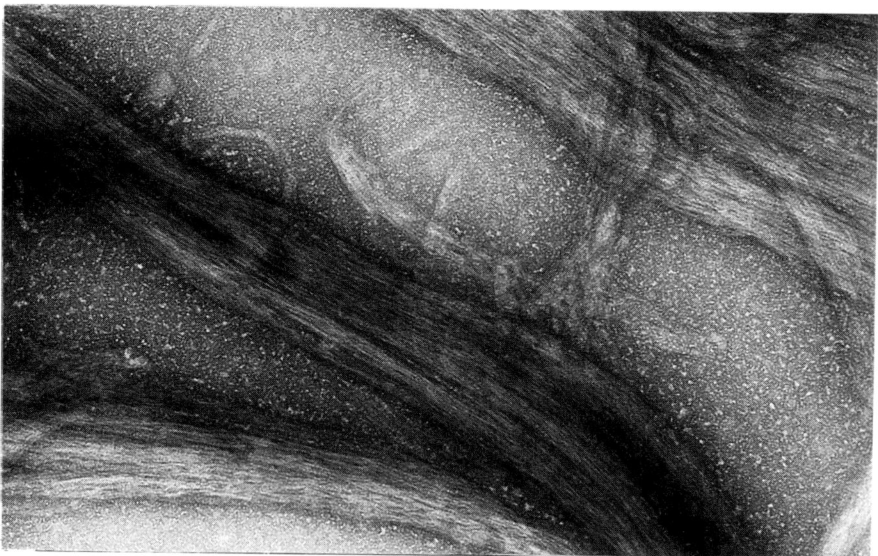

Figure 6. TEM micrograph of poly (VGGLG) at 50 °C. Negative staining with uranyl acetate, pH 3.5 . X 101,200.

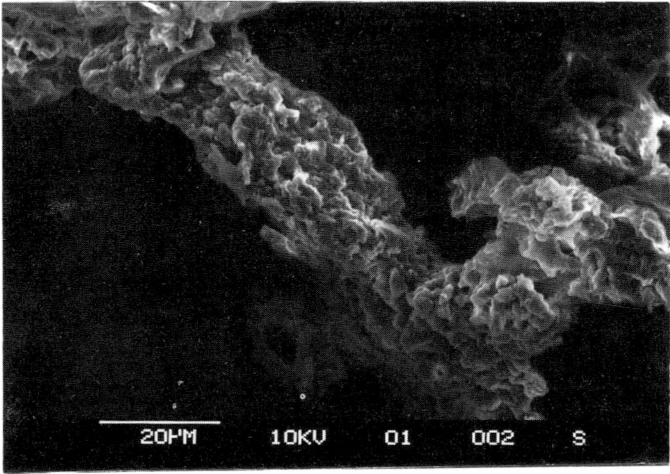

Figure 7. SEM micrograph of VGGVG in 50% ethyl alcohol at 50 °C. The sample was coated by a carbon layer of 15 nm thickness.

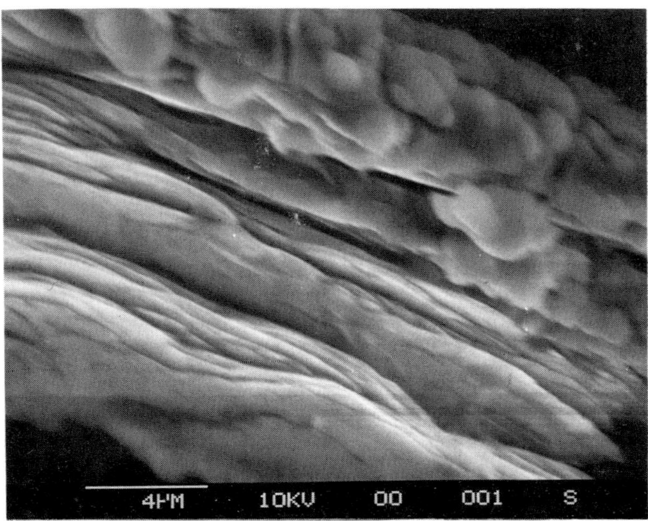

Figure 8. SEM micrograph of Boc-VGGLG-OEt in 50% ethyl alcohol at room temperature. The sample was coated by a carbon layer of 15 nm thickness.

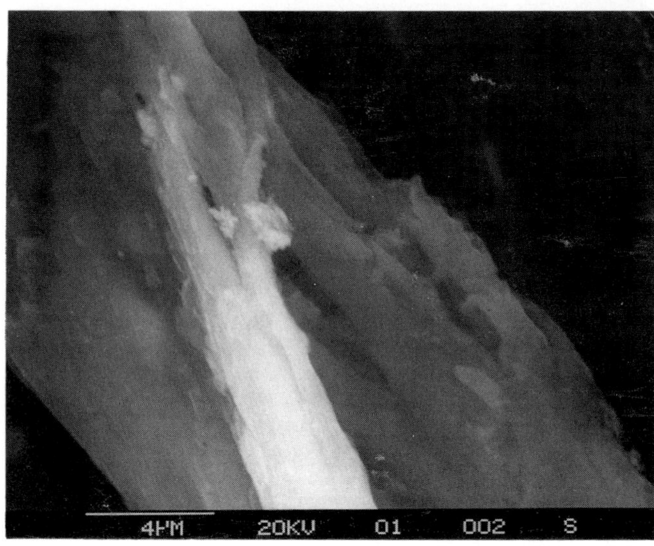

Figure 9. SEM micrograph of poly (VGGLG) in 50% ethyl alcohol at room temperature. The sample was coated by a carbon layer of 15 nm thickness.

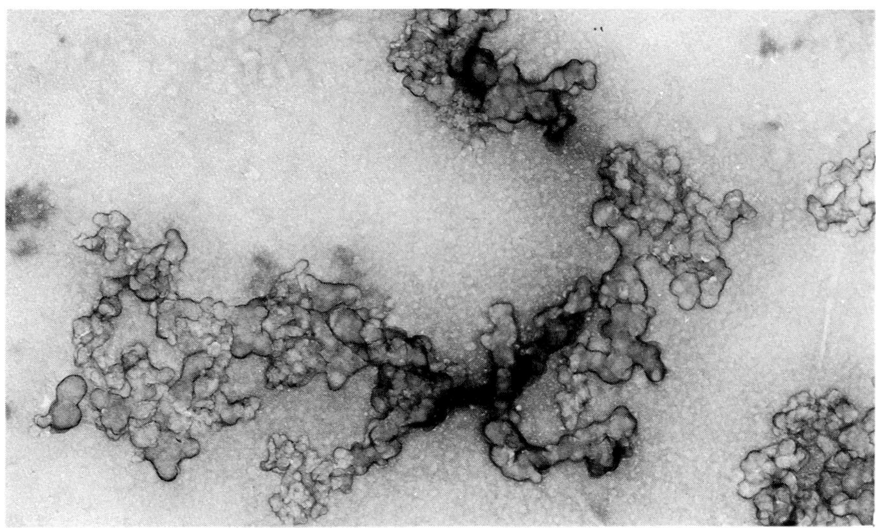

Figure 10. TEM micrograph of poly (VGGLG) at 0 °C. Negative staining with uranyl acetate, pH 3.5 . X 78,000.

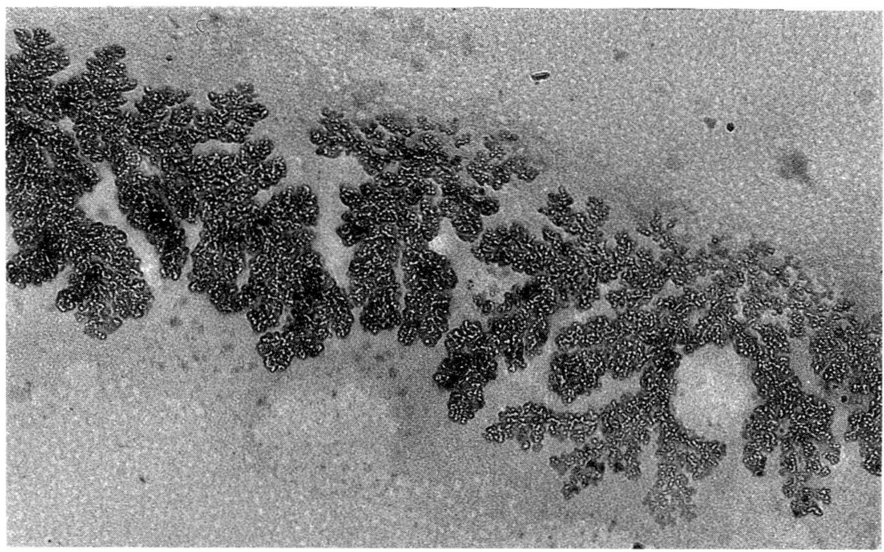

Figure 11. TEM micrograph of α-elastin coacervate at 38 °C. Negative staining with uranyl acetate, pH 3.5 . X 49,500.

3.2. A FRACTAL DESCRIPTION OF ELASTIN SUPRAMOLECULAR STRUCTURES

An initial approach [32] was to experimentally determine the fractal dimension of α-elastin coacervate by using turbidity (τ) measuremants according to Teixera [49] and Horne [50]:

$$D_F = 4 - \gamma + \frac{d\log\tau}{d\log\tau} \tag{2}$$

For mesurements extending over the wavelength (λ) range 400 - 800 nm, a value of -0.2 for the γ term was used [50]. A value of $D_F = 2.24$ was obtained for α-elastin coacervate. This figure is related to the scaling exponent of the mass within a sphere with respect to the radius of the sphere. Treating the fractal as a set of points, we should expect a fractal dimension equalling the Euclidean dimension E of the embedding space for a uniform distribution of points. For $D_F \neq E$ the density ρ of the fractal must vary exponentially with distance r according to

$$\rho(r) \propto r^{D_F-E} \tag{3}$$

For $D_F < E$ one could imagine the formation of clusters. Actually, the value found for α-elastin coacervate (2.24) compares well with those found for casein (2.21) [50] and for clusters of gold colloids (2.2) [51]. Moreover, it is not far from the theoretically expected (2.4 - 2.5) value for diffusion-limited particle-cluster aggregation (DLA) [52] in a 3-D space. Deeper insights have been recently obtained by analysis of the EM results on synthetic elastin-related polypeptides. As a matter of fact, both twisted-rope structures and globular clusters described in section 3.1. are characterized by a self-similarity (dilation symmetry) typical of fractal objects. In the case of clusters of aggregates, the obtained forms (Fig. 10) show an impressive analogy to those simulated in three dimensions by computer according to DLA mechanism [53]. Even more impressive are the results obtained very recently for α-elastin coacervate (Fig. 11). The characteristic leaf-like or fern-like aggregates are practically identical to those obtained by simulation of a DLA in a 2-D space [52, 53]. This mechanism of self-organization implies that the aggregates are indeed statistical, non deterministic, fractal objects, whose formation is only limited by the diffusion rate. In particular, the flow rate of particles approaching the growing aggregate is only dependent on the concentration gradient, according to Fick's law. Interestingly, the process is quite aleatory (particles describe random walks) but irreversible, that is, the aggregate cannot undergo any rearrangement. On the contrary, the formation of filaments, and their alignment, either staggered or not, should require and entirely different mechanism. Assuming a reversible, stepwise process for the aggregation of the polypeptide chains, one can suggest an aggregation along a somewhat preferre direction, as in the case of mesomorphic polymers [54]. As a matter of fact, some "polymeric" species could well have a relatively high lenght-to-diameter ratio, and they should therefore be able to self organize in an ordered phase originated by the same forces responsible for the primary linear aggregation. In vitro, and with model polypeptides, both processes, the reversible and the irreversible one, compete with each other. In vivo, under physiological conditions where tropoelastin concentration is low and presumably in a steady state, the fibre formation dominates. When "pathological" elastin accumulates, and then the system should be far from equilibrium, the irreversible formation of globule clusters and/or leaf-like aggregates may became active.

4. Conclusion

An attempt has been made to evaluate the structure-function relationships of elastin from both classical (molecular) and fractal points of view. Basically, the two approaches are mutually exclusive, nevertheless they do also reveal complementary aspects. In particular, the dynamics of the protein chains is explained either in terms of internal chain dynamics, such as sliding β-turns, or by a collection of Brownian trails ("freezed" trajectories) that display a fractal behaviour. Both approaches seem to agree with classical theory of rubber elasticity. It is to be appreciated the key role that synthetic polypeptide models and fragments have played in this context as well as in previous classical studies [12-27]. Furthermore, the fractal approach has discovered non trivial aspects of the form and growth of the elastomeric aggregate. Clearly, current descriptions are confined to a rather qualitative level, nevertheless new perspectives are now possible when looking at the details of supramolecular structure of elastin. Most importantly, one could glimpse a reasonable experimental approach to the understanding of the (supra) molecular mechanisms underlying some pathologies involving elastin.

Acknowledgements

This work was partially supported by the CNR Progetto Finalizzato Chimica Fine e Secondaria and by MURST grants.

References

1. J.M. Thomas, D.F. Elsden and S.M. Partridge, *Nature* **200**, 651-652, (1963).
2. Z. Indik, H. Yeh, N. Ornstein-Goldstein, P. Sheppard, N. Anderson, J.C. Rosenbloom, L. Peltonen and J. Rosenbloom, *Proc. Natl. Acad. Sci.* USA **84**, 5680-5684, (1987).
3. K. Raju and R.A. Anwar, *J. Biol. Chem.* **262**, 5755-5762, (1987).
4. G.M. Bressan, P. Argos and K.K. Stanley, *Biochemistry* **26**, 1497-1503 (1987).
5. R.A. Pierce, S.B. Deak, C.A. Stolle and C.D. Boyd, *Biochemistry* **29**, 9677-9683, (1990).
6. L.B. Sandberg, J.G. Leslie, C.T. Leach, V.L. Alvarez, A.R. Torres, and D.V. Smith, *Pathol. Biol.* **33**, 266-274, (1985).
7. P.J. Flory, <u>Statistical Mechanics of Chain Molecules</u>, Wiley-Interscience, N.Y., (1969).
8. D.S. Wrenn, G.L. Griffin, R.M. Senior and R.P. Mecham, *Biochemistry* **25**, 5172-5176, (1986).
9. R. Mecham, A. Hinek, R. Entwistle, D.S. Wrenn, L.Griffin and R.M. Senior, *Biochemistry* **28**, 3716-3722, (1989).
10. M.M. Long, V.J. King, K.U. Prasad and D.W. Urry, *Biochim. Biophys. Acta* **968**, 300-311, (1988).
11. C.H. Blood, J. Sasse, P. Brodt, and B.R. Zetter, *J. Cell. Biol.* **107**, 1987-1993, (1988).
12. D.W. Urry, *J. Protein Chem.* **3**, 403-436, (1984).
13. D.W. Urry, *J. Protein Chem.* **7**, 1-34, (1988).
14. A.M. Tamburro, in <u>Connective Tissue Research: Chemistry, Biology, and Physiology</u>, eds. Z. Deyl and M. Adam, A.R. Liss., Inc., N. Y., 45-62, (1981).
15. A.M. Tamburro, in <u>Elastin: Chemical and Biological Aspects</u>, eds. 125-145, (1990).
16. A.M. Tamburro and V. Guantieri, *Int. J. Biol. Macrom.* **8**, 62-63, (1986).
17. V. Guantieri, A.M. Tamburro, D. Cabrol, H. Broch and D. Vasilescu, *Int. J. Peptide Protein Res.* **29**, 216-230, (1987).
18. M.A. Khaled, V. Renugopalakrishnan and D.W. Urry, *J. Am. Chem. Soc.* **98**, 7547-7553, (1986).
19. M.M. Long, R.S. Rapaka, D. Volpin, I. Pasquali-Ronchetti and D.W. Urry, *Arch. Biochem. Biophys.* **201**, 445-452, (1980).
20. D.W. Urry, *Methods Enzymol.* **82**, 673-716, (1982).

21. D.W. Urry, M.M. Long, R.D. Harris and K.V. Prasad, *Biopolymers* **25**, 1939-1953, (1986).

22. D.W. Urry, *Perspect. Biol. Med.* **21**, 265-295, (1978).

23. D.W. Urry, T. Ohnishi, M.M. Long and L.W. Mitchell, *Int. J. Peptide Protein Res.* **7**, 367-378, (1975).

24. A.M. Tamburro, V. Guantieri, L. Pandolfo and A. Scopa, *Biopolymers* **29**, 855-870, (1990).

25. M.A. Castiglione-Morelli, A. Scopa, V. Guantieri and A.M. Tamburro, *Int. J. Biol. Macromol.* **12**, 363-368, (1990).

26. A.M. Tamburro, V. Guantieri, A. Scopa and J.M. Drabble, *Chirality* **3**, 318-323, (1991).

27. A.M. Tamburro, V. Guantieri and D. Daga Gordini, *J. Bimol. Struct. Dyn.* **10**, 441-454, (1992).

28. D.W. Urry, C.M. Venkatachalam, M.M. Long and K.U. Prasad, in <u>Conformation in Biology</u>, eds. R. Srinivasan and R.H. Sarma, G.N. Ramachaudran Festschrift Volume, Adenine Press, USA, 11-27, (1982).

29. F. Lelj, A.M. Tamburro, V. Villani, P. Grimaldi and V. Guantieri, *Biopolymers* **32**, 159-170, (1992).

30. V. Villani and A. M. Tamburro, *J. Chem. Soc. Perkins Trans.* **2**, in the press (1992).

31. C. Mégret, A. Lamure, M.T. Pieraggi, V. Guantieri, A.M. Tamburro and C. Lacabanne, submitted to *Int. J. Biol. Macromol.*

32. A.M. Tamburro and V. Guantieri, in <u>Ecological Physical Chemistry</u>, eds. C. Rossi and E. Tiezzi, Elsevier, Amsterdam, 391-400, (1991).

33. L. Gotte, M. Mammi and G. Pezzin, in <u>Symposium on Fibrous Proteins</u>, ed. W.G. Crewther, Butterworths, Australia, 236-245, (1968).

34. Z.R. Wassermann and F.R. Salemme, *Biopolymers* **29**, 1613-1631, (1990).

35. B. Mandelbrot, <u>The Fractal Geometry of Nature</u>, W.H. Freeman N.Y., (1983).

36. G.C. Wagner, J.T. Colvin, J.P. Allen and H.J. Stapleton, *J. Am. Chem. Soc.* **107**, 5589-5594, (1985).

37. J.T. Colvin and H.J. Stapleton, *J. Chem. Phys.* **32**, 4699-4706.

38. S.M. Partridge and H.F. Davis, *Biochem. J.* **61**, 21-30, (1955).

39. D.W. Urry, M.M. Long, B.A. Cox, T. Ohnishi, L.W. Mitchell and M. Jacobs, *Biochim. Biophys. Acta* **371**, 597-602, (1974).

40. G.M. Bressan, I. Castellani, M.G. Giro, D. Volpin, C. Fornieri and I.Pasquali-Ronchetti, *J. Ultrastruct. Res.* **82**, 335-340, (1983).

41. V. Velebny, A. Kadar, M. Ledvina and V. Lankesova, *Conn. Tissue. Res.* **12**, 197-201, (1984).

42. D. Volpin, D.W. Urry, I. Pasquali-Ronchetti and L. Gotte, *Micron* **7**, 193-198, (1976).

43. D.W. Urry, and M.M. Long, in <u>Elastin and Elastic Tissue</u>, eds. L.B. Sandberg, W.R. Gray and C. Franzblau, Plenum press, N.Y. 685-714, (1977).

44. M.M. Long, R.S. Rapaka, D. Volpin, I. Pasquali-Ronchetti and D.W. Urry, *Arch. Biochem. Biophys.* **201**, 445-452, (1980).

45. D. Daga-Gordini, V. Guantieri and A.M. Tamburro, *Conn. Tissue. Res.* **19**, 27-34, (1989).

46. V. Guantieri, D. Daga-Gordini and A.M. Tamburro, in <u>Elastin: Chemical and Biological Aspects</u>, eds. A.M. Tamburro and J.M. Davidson, Congedo, Galatina (Italy), 170-178, (1990).
47. R.L. Korneberg, S. Hendler, A. I. Oikarinen, L.Y. Mitsuoka and J. Uitto, *New Engl. J. Med.* **312**, 771-774, (1985).
48. G.M. Bressan, I. Pasquali-Ronchetti, C. Fornieri, F. Mattioli, I. Castellani and D. Volpin, *J. Ultrasctruct. Mol. Struct. Res.* **94**, 209-216, (1986).
49. J. Teixeira, in <u>On Growth and Form</u>, eds. H.E. Stanley and N. Ostrowsky, M. Nijhoff, Dordrecht, 145-162, (1986).
50. D.S. Horne, *Faraday Discuss. Chem. Soc.* **83**, 259-270, (1987).
51. P. Dimon, S.K. Sinha, D.A. Weitz, C.R. Safinya, G.S. Smith, W.A. Varaday and H.M. Lindsay, *Phys. Rev. Lett.* **57**, 595-597, (1986).
52. R.C. Ball, in <u>On Growth and Form</u>, eds. H.E. Stanley and N. Ostrowsky, M. Nijhoff, Dordrecht, 69-78, (1986).
53. E. Guyon and H.E. Stanley, <u>Fractal Forms</u>, Elsevier, Amsterdam, (1991).
54. P.J. Flory, in <u>Montedison Nobel Lectures</u>, ed; bassetti, Imago Milan (Italy), Vol. III, 25-51, (1985).

Conformational Analysis of Dolavaline, Dolaisoleuine, Dolaproine and Dolaphenine Unusual Amino Acids

M. TOSCANO
*Dipartimento di Chimica, Facolta' di Farmacia, Universita' della Calabria, I-87030
Arcavacata di Rende (CS), Italy.*

1. Introduction

As the conformational analysis plays a significant role in molecular biology, different
theoretical and experimental methods have been devoted to this subject in the last 30
years. The conformational properties of the most common amino acids have been
extensively investigated employing different theoretical methods and experimental
techniques. From a theoretical point of view, the empirical methods (i.e molecular
mechanics, non-bonded) have demonstrated to give results close to those obtained from
more sophisticated quantum mechanical ones with very low computational efforts. So,
the theoretical investigation of structure, conformational flexibility, solvent effects and
dynamical properties of amino acid residues, peptides, polipeptides, proteins and
nucleic acid segments is now possible by using empirical potentials and parameters
largely tested in literature.The knowledge of the conformational properties of amino
acid residues becomes very useful for the study of more complex peptide and protein
structures In this paper we report the conformational analysis of four unusual amino
acid residues: (dolavaline, dolaisoleuine, dolaproine and dolaphenine performed
employing the non bonded method. These unusual amino acids have been recently
found in the dolastatin 10 pentapeptide extracted from sea hare *Dolabella Auricularia* [1-
4]. Since dolastatin 10 is one of the most powerful (i.e., lowest in vivo dose)
antineoplastic substance known to date [2], the study of its constituent can contribute to
the investigation of its structure and to the possible explanation of its biological activity.

2. Computational details

Due to the absence of experimental information on the geometries of the amino acid
residues we have optimized the structures using the MODEL package [5] and the MMX
force field [6]. Optimized structures have been then employed to obtain the potential
energy surfaces at non-bonded level [7]. The conformational analysis has been performed
as a function of the three torsional angles $\Phi 1$, $\Phi 2$, $\Phi 3$ and by employing the NB/80 [8]

405

N. Russo et al. (eds.), Properties and Chemistry of Biomolecular Systems, 405–413.
© 1994 *Kluwer Academic Publishers. Printed in the Netherlands.*

package which evaluates the conformational energy as a sum of nonbonded, electrostatic, hydrogen-bond and intrinsic torsional contributions.

$$E_{total} = E_{stretch} + E_{bend} + E_{torsion} + E_{VDW} + E_{elec}$$

The net atomic charges as well as the non bonded parameters have been taken from ref. 6. After an initial search into the whole conformational space (employing a grid mesh of $10°$) minimum-energy conformations has been obtained by ICER package using algorithms with analytical evaluation of both gradient and Hessian [9] and optimizing all the torsional angles including also the methyl groups torsion. Conformational energies (in Kcal/mole) are expressed as $\Delta E = E - E_0$, where E_0 is the energy of the most stable conformation.

3. Results and Discussion

With the aim to further test the reliability of the employed method, as a first step of the work, we have redone the conformational analysis of L-alanine for which different theoretical studies, performed at different levels of theory, exist in the literature [10-16]. Our $\Phi_{C\alpha-C}$, $\Psi_{C\alpha-N}$ conformational energy map is very similar to those previously found by Liquori [10], Ramachandran [11], Zimmerman [12], Wiener et al. [13], Sheraga et al. [14], Opfinger et al. [15], and Ooi et al. [16] using different geometries and potentials as well as with that obtained using ab-initio methods [17,18]. In particular we can note that the influence of the used L-alanine geometry is less significant than that obtained employing empirical parameters.

3.1. DOLAVALINE

Dolavaline (DOV) is the first amino acid of the dolastatin 10 peptide chain. This unusual amino acid is characterized by the presence of two CH_3 groups attached to the nitrogen atom. The $-CH(CH_3)_2$ group has the S configuration.

Figure 1 shows the ψ_1, φ_1 potential energy map of dolavaline together with the definitions of the torsional angles. Four low energy minima are present in a range of 10Kcal/mol. The first three occur for a value of $\chi_1 = 180°$ and the fourth for $\chi_1 = 80°$. With respect to L-valine we note that two minima of this amino acid are missing. This is essentially due to the presence, in the DOV, of an extra methyl group that reduces the low energy conformational space.

The refined structure and the energy difference of the four minima with respect to the absolute one are reported in table 1.

From this table it is evident that the four minima fall now in a range of energy that is less than 2 Kcal/mol. This means that the residue has high conformational flexibility. The analysis of energy reveals that the most significant contributions arise from Van der Waals and electrostatics interactions for all four minima.

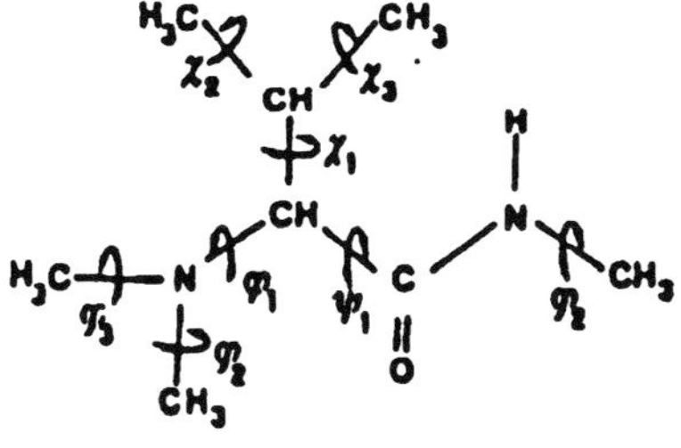

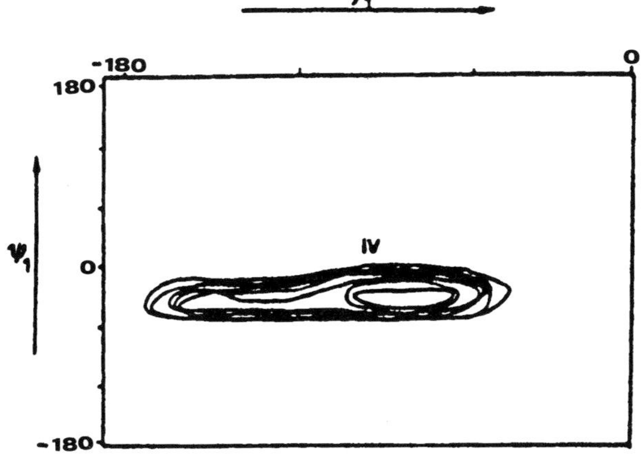

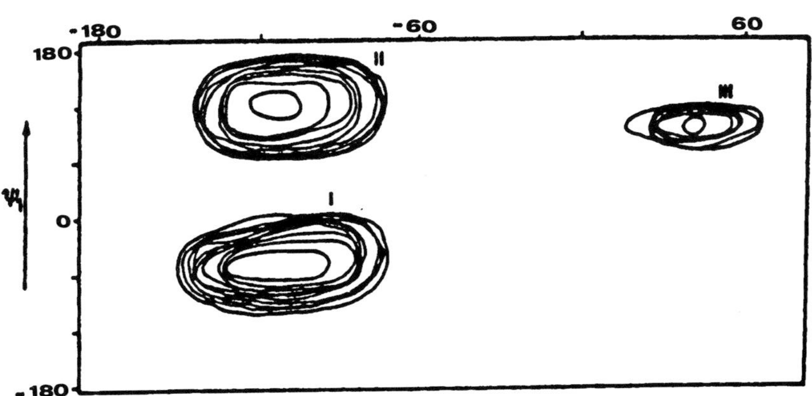

Figure 1. Minimum regions of dolavaline conformational map.

Table 1. Torsional angles (degree) and ΔE value (Kcal/mol) for dolavaline residue after optimization.

Mini-mum	φ_1	ψ_1	χ_1	φ_2	φ_3	φ_4	χ_2	χ_3	ΔE
I	-89.3	-43.3	175.3	157.3	180.0	188.8	171.5	167.0	0.0
II	-107.2	118.7	178.9	170.9	180.0	183.0	181.9	175.5	1.5
III	36.5	86.2	187.8	162.4	180.1	186.8	196.8	193.6	1.7
IV	-74.0	-30.4	73.0	137.1	180.0	206.6	165.3	196.0	1.2

3.2. DOLAISOLEUINE

With respect to isoleucine (ILE), the dolaisoleuine (DIL) has the following significative modifications: a methylated nitrogen atom; the presence of $-CHCH_3CH_2-CH_3$ mojety instead of the $CH(CH_3)_2$ group, in S configuration; the presence of an $O-CH_3$ group on the Cb with R configuration; a further CH_2 group in the main chain. In figure 2 the schematic structure of DIL residue and the low energy minimum regions of the conformational energy map are reported.

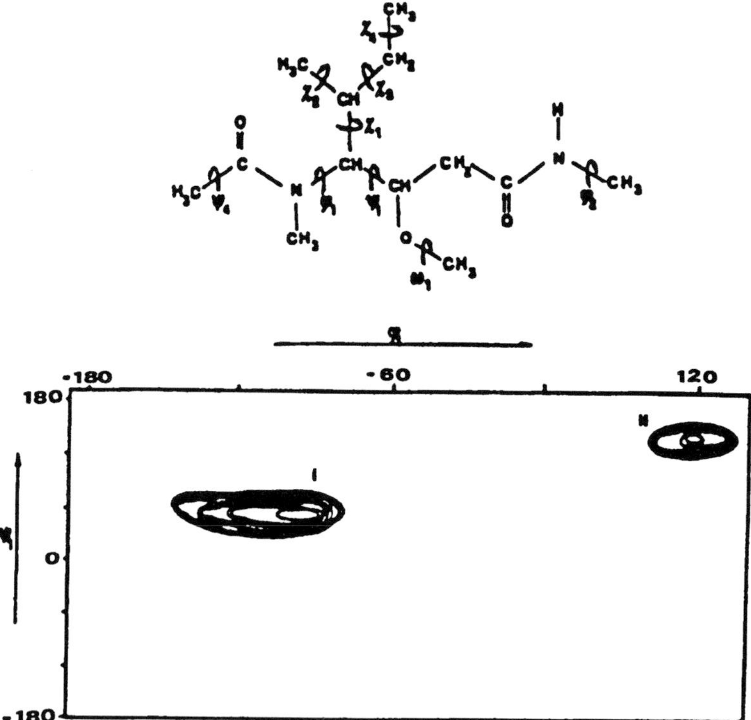

Figure 2. Minimum regions of dolaisoleuine conformational map.

The conformational analysis reveals the presence of only two minima in a range of 10 Kcal/mol .With respect to the corresponding isoleucine map, four minima disappear. This drastic reduction of low energy conformational space is probably due to the presence, in the DIL skeleton, of groups that increase the steric hindrance and consequently the conformational mobility. Due to the significant structural difference of DIL with respect to ILE, the full torsional angle optimization becomes significant. Results are reported in table 2.

The optimization gives two minima practically isoenergetic (DE= 1.07 Kcal/mol). Also in this case the dominant energetic contributions are those coming from electrostatic and Van der Waals effects.

Table 2. Torsional angles (degree) and DE value (Kcal/mol) for dolaisoleuine residue after the optimization.

Minimum	φ_1	ψ_1	χ_1	ψ_4	φ_2	χ_2	χ_3	χ_4	ω_1	ΔE
I	-100.5	124.2	181.4	180.5	181.0	159.6	145.9	157.2	131.0	0.0
II	64.0	134.9	185.0	180.5	181.0	166.4	142.2	175.5	131.0	1.1

3.3. DOLAPROINE

The unusual amino acid dolaproine (DAP) (see Figure 3) is very different from parent proline (PRO) compound. The main resemblance resides in the presence of a five member ring in the skeleton of both systems.

The conformational characteristic of proline is governed by its mobility around the CH-C(O) torsional angle that gives, as it is well known, two conformational minima. The largest chain of DAP, increases the available conformational space with respect to proline residue. In Figure 3 we report the minimum regions of the ψ_1, ψ_2 conformational map of DAP. We found four conformational minima in an energy range of 15 Kcal/mol . The optimization of these minima indicates that the structure II is the absolute one in the potential energy surface taking into account all torsional degrees of freedom (see Table 3). The minima I and III lie at 6.6 and 4.3 Kcal/mol while the structure IV has a higher DE (12.4 Kcal/mol).

Table 3. Torsional angles (degree) and DE value (Kcal/mol) for dolaproine residue after optimization.

Minimum	ψ_1	ψ_2	ψ_3	ψ_4	ψ_5	ω_1	ψ_6	ΔE
I	-25.1	-86.7	-75.9	184.5	180.9	207.3	160.2	6.6
II	174.0	-80.9	-74.4	183.8	180.8	141.6	159.3	0.0
III	174.3	-81.42	112.9	182.2	180.5	141.9	160.0	4.3
IV	-26.7	-83.7	131.7	182.4	171.7	204.5	160.3	12.4

The inspection of energy contributions reveals that, other than the electrostatic and Van der Waals interactions, the torsional effects are significant also.
A comparison of the conformational behaviour between PRO and DAP underlines the higher flexibility of the latter.

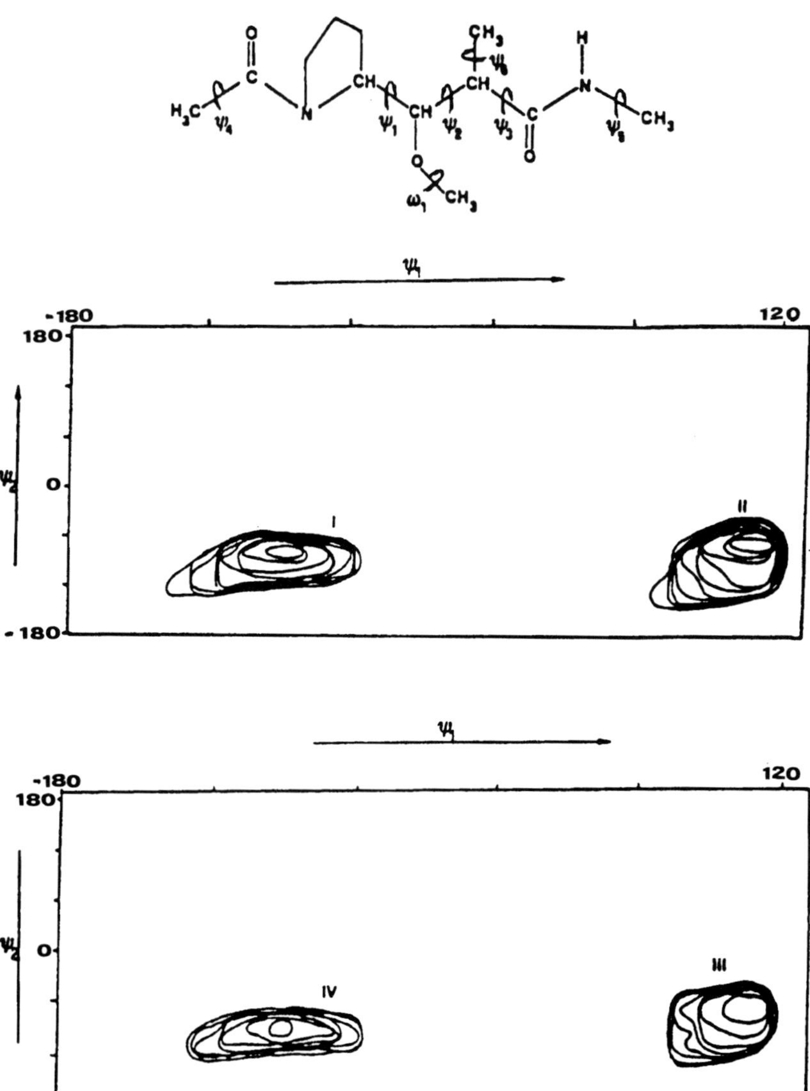

Figure 3. Minimum regions of dolaproine conformational map.

3.4. DOLAPHENINE

Dolaphenine (DOE) is the last residue of dolastatin 10 pentapeptide.

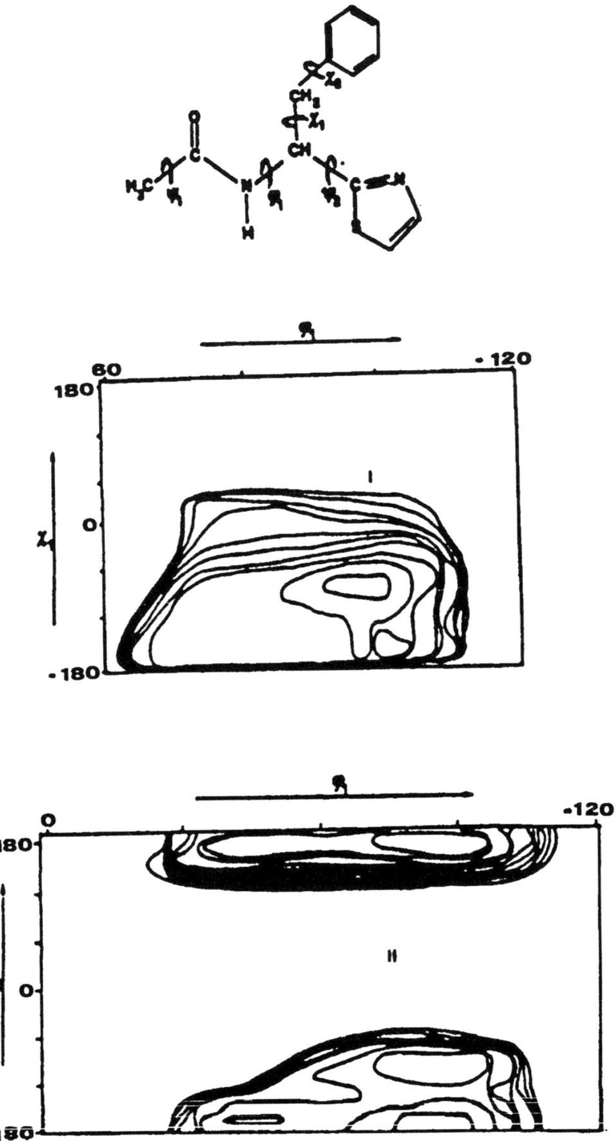

Figure 4. Minimum regions of dolaphenine conformational map.

As it is shown in Figure 4 this unusual aminoacid is very different from the fenilalanina (PHE) parent compound because of the presence, at the end of its chain, of a thiazole functional group. The potential energy surface with respect to the φ_1 , χ_1 torsional angles is also shown in Figure 4. Two minima are present for χ_2 =60° and for χ_2 =100°. The relative minimum II lies at about 1.1 Kcal/mole with respect to the absolute one. In both cases, when other torsional angles (see Figure 4) are considered in the optimization procedure, the energy difference between the two minima is only of 0.87 Kcal/mol (see Table 4). Comparison between the conformational properties between PHE and DOE reveals that the former has four minima and the latter only two practically isoenergetic. As in the other cases considered, the main energetic contributions to the conformational energy is due to the electrostatic and Van der Waals interactions, with a small contribution coming from the torsional effects.

Table 4. Torsional angles (degree) and DE value (Kcal/mol) for dolapheine residue after optimization.

Minimum	φ_1	χ_1	χ_2	ψ_1	ψ_2	ΔE
I	-188.1	-95.5	69.0	-47.3	179.7	0.0
II	-187.9	178.0	67.5	-47.1	179.8	0.87

4. Conclusions

On the basis of the conformational analysis performed for the four unusual amino acids (DOV, DIL, DAP, DOE) of antineoplastic dolastatin 10 natural pentapeptide, the following conclusions can be drawn:
i. the presence of additional groups, in the chain of all four studied unusual amino acids, increases the steric hindrance and, consequently reduces the low energy confomational space with respect to the parent usual residues.
ii. the optimization of all torsional degrees of freedom is important to obtain more realiable energies for the minima.
iii. the most significant contributions to the total energy arises from the electrostatic and Van der Waal interactions.

Acknowledgements

The financial support was provided by the Ministero dell' Universita' e della Ricerca Scientifica e Tecnologica (MURST). I thank Dr. N. Russo for the helpful discussion.

References

1. G. R. Pettit , Y. Kamano, C.L. Herald , A.A. Tuinman, F.E. Boettner, H. Kizu, J. M. Schmidt , L. Baczynskyj, K.B. Tomer and R.J. Bontems, " The Isolationan and Structure of a Remarkable Marine Animal Antineoplastic Constituent: Dolastatin 10", *J. Am. Chem. Soc.* **109**, 6883-6885, (1987).

2. R.P. Pettit, S.B. Singh, F. Hogan, P. Lloyd-Williams, D. L. Herald, D.D. Burkett and P.L. Clewlow " The Absolute Configuration and Synthesis of Natural (-)-Dolastatin 10" *J. Am. Chem. Soc.* **111**, 5463-5465, (1989).

3. M. Toscano, E. Zasa and N. Russo, " Conformational Analysis of Antineoplastic Pentapeptide Dolastatin 10 and its Analogue Without Nonstandard Amino Acids" in <u>Recent Advance in the Chemistry and Molecular Biology of Cancer Research</u> D. Qianhuan and M-A. Armour (eds), Springer Verlag and Science Press, Berlin, 1993.

4. P. Fantucci, E. Mattioli, T. Marino and N. Russo "The Conformational Properties of (-)-Dolastatin 10, a Powerful Antineoplastic Agent", see this book.

5. Program Model, Version 9.2 of Steliou K, Universite de Montreal (Canada).

6. S.J. Weiner, P.A. Kollman, D.T. Nguyen and D.A. Case, *J. Comput. Chem* **7**, 230-, (1986) and reference therein.

7. V. Barone, F. Fraternali and P.L. Cristinziano "Sensitivity of Peptide Conformation to Methods and Geometrical Parameters. A Comparative ab Initio and Molecular Mechanics Study of Oligomers of α-Aminoisobutyric Acid" *Macromolecules* **23**, 2038- , (1990) andreferences therein.

8. Barone V, Cristinziano PL, Lelj F, Programs NB/80 and ICER, unpublished.

9. H.B. Schlegel "Optimization of Equilibrium Geometries and Transition Structures" *Adv. Chem. Phys.* **67**, 249-266 , (1987).

10. A.M. Liguori " The Stereochemical Code and the Logic of a Protein Molecule", *Quart. Rev. Biophys.* **2**, 65-92, (1969).

11. G.N. Ramachandran <u>Aspects of Protein Structures</u>, Academic Press, London, 1963.

12. S.S. Zimmerman, M.S. Pottle, G. Nemethy and H.A. Sheraga " Conformational Analysis of the 20 Naturally Occurring Amino Acid Residues Using ECEPP", *Macromolecules*, **10**, 1-9, (1977).

13. S.J. Wiener, P.A. Kollman, D.A. Case, U. Chandra Singh, C. Ghio, G. Alagona, S. Profeta and P. Wiener " A New Force Field for Molecular Mechanical Simulation of Nucleic Acids and Proteins", *J. Am. Chem. Soc.*, **106**, 765-784, (1984).

14. I.K. Roterman, M.H. Lambert, K.D. Gibson and H.A. Scherage " A Comparison of the CHARMM, AMBER and ECEPP Potentials for Peptides", *J. Biol. Struct. Dynam.***7**, 421-453, (1989).

15. R.J. Opfinger <u>Conformational Analysis</u>, Academic Press, New York, 1973.

16. T. Ooi, R.A. Scott, G. Vanderkooi and H.A. Sheraga, " ", *J. Chem. Phys.*, **46**, 4410-4420 , (1967).

17. B. Pullman, B. Maigret and D. Perahia "Molecular Orbital Calculations on the Conformation of Polypeptides and Proteins", *Theor. Chim. Acta*, **18**, 44-56, (1970).

18. T. Head-Gordon, M. Head-Gordon, M.J. Frisch, C. Brooks III and J. Pople " A Theoretical Study of Alanina Dipeptide and Analogs", *Int. J. Quantum Chem. Quant. Biol. Symp.*, **16**, 311-322, (1989).

INDEX

TOPICS IN
MOLECULAR ORGANIZATION AND ENGINEERING

Honorary Chief Editor: W. N. Lipscomb, Harvard, U.S.A.
Executive Editor: Jean Maruani, Paris, France

1. J. Maruani (ed.): *Molecules in Physics, Chemistry, and Biology.*
    Vol. 1: General Introduction to Molecular Sciences. 1988
    ISBN 90-277-2596-9

2. J. Maruani (ed.): *Molecules in Physics, Chemistry, and Biology.*
    Vol. 2: Physical Aspects of Molecular Systems. 1988
    ISBN 90-277-2597-1

3. J. Maruani (ed.): *Molecules in Physics, Chemistry, and Biology.*
    Vol. 3: Electronic Structure and Chemical Reactivity. 1989
    ISBN 90-277-2598-5

4. J. Maruani (ed.): *Molecules in Physics, Chemistry, and Biology.*
    Vol. 4: Molecular Phenomena in Biological Sciences. 1989
    ISBN 90-277-2599-3

5. E. Schoffeniels and D. Margineanu: *Molecular Basis and Thermodynamics of Bioelectrogenesis.* 1990   ISBN 0-7923-0975-8

6. A. Lund and M. Shiotani (eds.): *Radical Ionic Systems.* Properties in Condensed Phases. 1991   ISBN 0-7923-0988-X

7. P.I. Lazarev (ed.): *Molecular Electronics.* Materials and Methods. 1991
    ISBN 0-7923-1196-5

8. E. Rizzarelli and T. Theophanides (eds.): *Chemistry and Properties of Biomolecular Systems.* 1991   ISBN 0-7923-1393-3

9. L.A. Montero and Y.G. Smeyers (eds.): *Trends in Applied Theoretical Chemistry.* 1992   ISBN 0-7923-1745-9

10. M.T. Pope and A. Müller (eds.): *Polyoxometalates: From Platonic Solids to Anti-Retroviral Activity.* 1994   ISBN 0-7923-2421-8

11. N. Russo, J. Anastassopoulou and G. Barone (eds.): *Properties and Chemistry of Biomolecular Systems.* 1994   ISBN 0-7923-2666-0

KLUWER ACADEMIC PUBLISHERS – DORDRECHT / BOSTON / LONDON